Principles of Foundation Engineering

www.brookscole.com

brookscole.com is the World Wide Web site for Brooks/Cole and is your direct source to dozens of online resources.

At *brookscole.com* you can find out about supplements, demonstration software, and student resources. You can also send email to many of our authors and preview new publications and exciting new technologies.

brookscole.com

Changing the way the world learns®

Principles of Foundation Engineering

Fifth Edition

Braja M. Das
California State University, Sacramento

THOMSON

BROOKS/COLE

Australia • Canada • Mexico • Singapore • Spain
United Kingdom • United States

THOMSON

BROOKS/COLE

Engineering Editor: *Bill Stenquist*
Editorial Assistant: *Valerie Boyajian*
Marketing Manager: *Tom Ziolkowski*
Marketing Assistant: *Mona Weltmer*
Advertising Project Manager: *Margaret Parks*
Project Manager, Editorial Production: *Tom Novack*
Print/Media Buyer: *Jessica Reed*
Permissions Editor: *Sue Ewing*
Production Service: *WestWords, Inc.*

Text Designer: *Carmela Pereira*
Photo Researcher: *Kathleen Olson*
Copy Editor: *Brian Baker*
Illustrator: *Lotus Art; WestWords, Inc.*
Cover Designer: *Denise Davidson*
Cover Image: © *Case Foundation Company*
Cover Printer: *Phoenix Color Corp.*
Compositor: *WestWords, Inc.*
Printer: *Phoenix Color Corp.*

For more information about our products, contact us at:
Thomson Learning Academic Resource Center
1-800-423-0563

For permission to use material from this text, contact us by: **Phone:** 1-800-730-2214
Fax: 1-800-730-2215
Web: http://www.thomsonrights.com

Library of Congress Cataloging-in-Publication Data

Das, Braja M.
    Principles of foundation engineering / Braja M. Das.—5th ed.
        p.   cm.
    Includes bibliographical references and index.
    ISBN 0-534-40752-8
        1. Foundations.   I. Title.

TA775 .D37 2004
624.1'5—dc21

2002038362

Brooks/Cole–Thomson Learning
511 Forest Lodge Road
Pacific Grove, CA 93950
USA

Asia
Thomson Learning
5 Shenton Way #01-01
UIC Building
Singapore 068808

Australia
Nelson Thomson Learning
102 Dodds Street
Southbank, Victoria
Australia 3006

Canada
Nelson
1120 Birchmount Road
Toronto, Ontario M1K 5G4
Canada

Europe/Middle East/Africa
Thomson Learning
High Holborn House
50/51 Bedford Row
London WC1R 4LR
United Kingdom

In the memory of my father, and to Janice and Valerie

Contents

3   Shallow Foundations: Ultimate Bearing Capacity   **123**

4   Ultimate Bearing Capacity of Shallow Foundations: Special Cases   **164**

5 Shallow Foundations: Allowable Bearing Capacity and Settlement 189

6 Mat Foundations 255

7   Lateral Earth Pressure   293

8   Retaining Walls   331

9 Sheet Pile Walls 387

10 Braced Cuts 439

11   Pile Foundations   471

12   Drilled-Shaft Foundations   578

Preface

The first edition of *Principles of Foundation Engineering,* published in 1984, was intended for use as a text by undergraduate civil engineering students. The text was well received by students and practicing geotechnical engineers. Their encouragement and comments helped to develop the second edition in 1990, the third edition in 1995, the fourth edition in 1999, and, finally, this fifth edition of the text.

The core of the original text has not changed greatly. There has been some rearrangement of the contents in this edition compared with the fourth edition. The book now contains 14 chapters. A brief overview of the changes follows:

- A conscious attempt was made to reduce the length of the text without compromising the content.
- SI units are used as the primary units, with English units in parentheses.
- Most graphs and equations have been made nondimensional (unitless) by using atmospheric pressure or unit weight of water (whichever is applicable) if at all possible. This eliminates the need for giving equations or graphs in dual units.
- Some new photographs have been added, including pictures of a field plate load test; a pile load test; pile driving in the field; a Chicago subway cut; a Washington, DC, metro cut; the construction of stone columns; a PVC installation, and auger drilling for drilled shafts.
- Chapter 3 of the fourth edition has been split into two chapters—Chapters 3 and 4. Chapter 4 in the current edition contains some new topics, such as the bearing capacity of a shallow foundation with a rigid base at a shallow depth. The stress characteristic solution for foundations on a slope has been added. Topics on foundations on geogrid-reinforced soil have been deleted.
- Chapter 5, on the settlement of shallow foundations (Chapter 4 in the fourth edition), has been greatly modified. Elastic settlement calculations using Steinbrenner (1934) shape factors and Fox (1948) depth factors have been introduced. Also, elastic settlement calculations using Mayne and Poulos' (1999) method has been added. The three-dimensional effect on primary consolidation for overconsolidated clay has been introduced with examples. Secondary consolidation settlement has been discussed in greater detail.

- Chapter 10, on braced cuts, was removed from the chapter on sheet pile walls in the fourth edition and is presented in a separate chapter in the current edition.
- Chapter 11, on piles, has gone through a major revision. Correlations for the calculation of point-bearing capacity with SPT and CPT results have been added, and the LCPC method and Dutch method using CPT results have been expanded. Correlations on skin friction with the results of cone penetration tests have been added as well. Methods for calculating skin friction in saturated clay have been modified. Results from several recent studies to determine pile capacity with time (after driving) are discussed. Formulas for pile capacity for vibration-driven piles have been added. Pullout resistance of piles has been deleted.
- Chapter 12, on drilled shafts, has gone through a major revision. The procedure for the construction of drilled shafts using the dry method, casing method, and slurry method is discussed. Compressibility factors for point-bearing capacity are introduced. The theory of Berezantzev et al. (1961) for point-bearing capacity has been added, as has the calculation of lateral load-bearing capacity by the characteristic load method. Topics on caissons have been deleted.

Because the text introduces civil engineering students to the fundamental concepts and application of foundation analysis and design, the mathematical derivations of some equations are not presented; instead, only the final forms are given. However, a list of references for further information and study is included in every chapter.

Multiple theories and empirical correlations have been presented where applicable. The reason for presenting different theories and correlations is to acquaint the reader with the chronological development of the subject and the answers obtained from theories based on different assumptions. This approach also convinces the reader that the soil parameters obtained from different empirical correlations will not always be the same, with some being better than others for a given soil condition. The discrepancy arises primarily because soil profiles are seldom homogeneous, elastic, and isotropic. The judgment needed to properly apply the theories, equations, and graphs to the evaluation of soils and foundation design cannot be overemphasized or completely taught by any textbook. Field experience must supplement classroom work.

Acknowledgments

Thanks are due to my wife, Janice, who provided the primary motivation for the completion of this edition. She also typed the revisions and completed the original graphs and figures.

The previous four editions were reviewed by 19 individuals; their helpful comments and suggested have improved the quality of the book. For their reviews and helpful suggestions in the development of this fifth edition, I would like to thank the following people:

Paul W. Mayne, Georgia Institute of Technology, Georgia
Thomas F. Zimmee, Rensselaer Polytechnic Institute, New York
Khaled Sobhan, New Mexico State University, New Mexico
Nagaratnam Sivakugan, James Cook University, Queensland, Australia

I am truly grateful to Dr. Sivakugan, who was kind enough to suggest the inclusion of pictures of some instruments that are used for geotechnical studies in the field; these are presented in Appendix A. He most generously supplied the photographs and guidelines for descriptions of the instruments.

Finally, I would like to thank Publisher Bill Stenquist and the production staff at Brooks/Cole for their cooperation and assistance in the final development and production of the book.

Braja M. Das
Sacramento, California

1

Geotechnical Properties of Soil

1.1 Introduction

The design of foundations of structures such as buildings, bridges, and dams generally requires a knowledge of such factors as (a) the load that will be transmitted by the superstructure to the foundation system, (b) the requirements of the local building code, (c) the behavior and stress-related deformability of soils that will support the foundation system, and (d) the geological conditions of the soil under consideration. To a foundation engineer, the last two factors are extremely important because they concern soil mechanics.

The geotechnical properties of a soil—such as its grain-size distribution, plasticity, compressibility, and shear strength—can be assessed by proper laboratory testing. In addition, recently emphasis has been placed on the *in situ* determination of strength and deformation properties of soil, because this process avoids disturbing samples during field exploration. However, under certain circumstances, not all of the needed parameters can be or are determined, because of economic or other reasons. In such cases, the engineer must make certain assumptions regarding the properties of the soil. To assess the accuracy of soil parameters—whether they were determined in the laboratory and the field or whether they were assumed—the engineer must have a good grasp of the basic principles of soil mechanics. At the same time, he or she must realize that the natural soil deposits on which foundations are constructed are not homogeneous in most cases. Thus, the engineer must have a thorough understanding of the geology of the area—that is, the origin and nature of soil stratification and also the groundwater conditions. Foundation engineering is a clever combination of soil mechanics, engineering geology, and proper judgment derived from past experience. To a certain extent, it may be called an art.

When determining which foundation is the most economical, the engineer must consider the superstructure load, the subsoil conditions, and the desired tolerable settlement. In general, foundations of buildings and bridges may be divided into two major categories: (1) *shallow foundations* and (2) *deep foundations. Spread footings, wall footings,* and *mat foundations* are all shallow foundations. In most shallow foundations, *the depth of embedment can be equal to or less than three to four times the width of the foundation. Pile* and *drilled shaft* foundations are deep foundations. They are used when top layers have poor load-bearing capacity and when the use of shallow foundations will cause considerable structural damage or instability. The problems relating to shallow foundations and mat foundations are considered in Chapters 3, 4, 5, and 6. Chapter 11 discusses pile foundations, and Chapter 12 examines drilled shafts.

This chapter serves primarily as a review of the basic geotechnical properties of soils. It includes topics such as grain-size distribution, plasticity, soil classification, effective stress, consolidation, and shear strength parameters. It is based on the assumption that you have already been exposed to these concepts in a basic soil mechanics course.

1.2   *Grain-Size Distribution*

In any soil mass, the sizes of the grains vary greatly. To classify a soil properly, you must know its *grain-size distribution*. The grain-size distribution of *coarse-grained* soil is generally determined by means of *sieve analysis*. For a *fine-grained* soil, the grain-size distribution can be obtained by means of *hydrometer analysis*. The fundamental features of these analyses are presented in this section. For detailed descriptions, see any soil mechanics laboratory manual (e.g., Das, 2002).

Sieve Analysis

A sieve analysis is conducted by taking a measured amount of dry, well-pulverized soil and passing it through a stack of progressively finer sieves with a pan at the bottom. The amount of soil retained on each sieve is measured, and the cumulative percentage of soil passing through each is determined. This percentage is generally referred to as *percent finer*. Table 1.1 contains a list of U.S. sieve numbers and the corresponding size of their openings. These sieves are commonly used for the analysis of soil for classification purposes.

Table 1.1   U.S. Standard Sieve Sizes

Sieve no.	Opening (mm)
4	4.750
6	3.350
8	2.360
10	2.000
16	1.180
20	0.850
30	0.600
40	0.425
50	0.300
60	0.250
80	0.180
100	0.150
140	0.106
170	0.088
200	0.075
270	0.053

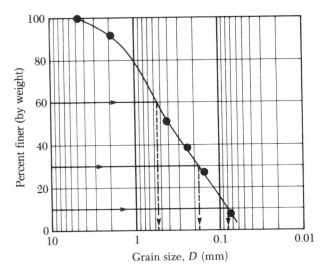

Figure 1.1 Grain-size distribution curve of a coarse-grained soil obtained from sieve analysis

The percent finer for each sieve, determined by a sieve analysis, is plotted on *semilogarithmic graph paper*, as shown in Figure 1.1. Note that the grain diameter, D, is plotted on the *logarithmic scale* and the percent finer is plotted on the *arithmetic scale*.

Two parameters can be determined from the grain-size distribution curves of coarse-grained soils: (1) the *uniformity coefficient* (C_u) and (2) the *coefficient of gradation*, or *coefficient of curvature* (C_c). These coefficients are

$$C_u = \frac{D_{60}}{D_{10}} \tag{1.1}$$

and

$$C_c = \frac{D_{30}^2}{(D_{60})(D_{10})} \tag{1.2}$$

where D_{10}, D_{30}, and D_{60} are the diameters corresponding to percents finer than 10, 30, and 60%, respectively

For the grain-size distribution curve shown in Figure 1.1, $D_{10} = 0.08$ mm, $D_{30} = 0.17$ mm, and $D_{60} = 0.57$ mm. Thus, the values of C_u and C_c are

$$C_u = \frac{0.57}{0.08} = 7.13$$

and

$$C_c = \frac{0.17^2}{(0.57)(0.08)} = 0.63$$

Parameters C_u and C_c are used in the *Unified Soil Classification System,* which is described later in the chapter.

Hydrometer Analysis

Hydrometer analysis is based on the principle of sedimentation of soil particles in water. This test involves the use of 50 grams of dry, pulverized soil. A *deflocculating agent* is always added to the soil. The most common deflocculating agent used for hydrometer analysis is 125 cc of 4% solution of sodium hexametaphosphate. The soil is allowed to soak for at least 16 hours in the deflocculating agent. After the soaking period, distilled water is added, and the soil–deflocculating agent mixture is thoroughly agitated. The sample is then transferred to a 1000-ml glass cylinder. More distilled water is added to the cylinder to fill it to the 1000-ml mark, and then the mixture is again thoroughly agitated. A hydrometer is placed in the cylinder to measure the specific gravity of the soil–water suspension in the vicinity of the instrument's bulb (Figure 1.2), usually over a 24-hour period. Hydrometers are calibrated to show the amount of soil that is still in suspension at any given time t. The largest diameter of the soil particles still in suspension at time t can be determined by Stokes' law,

$$D = \sqrt{\frac{18\eta}{(G_s - 1)\gamma_w}}\sqrt{\frac{L}{t}} \tag{1.3}$$

where D = diameter of the soil particle
   G_s = specific gravity of soil solids
   η = viscosity of water

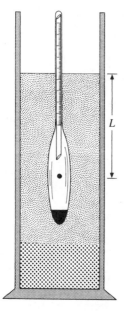

Figure 1.2   Hydrometer analysis

γ_w = unit weight of water
L = effective length (i.e., length measured from the water surface in the cylinder to the center of gravity of the hydrometer; see Figure 1.2)
t = time

Soil particles having diameters larger than those calculated by Eq. (1.3) would have settled beyond the zone of measurement. In this manner, with hydrometer readings taken at various times, the soil *percent finer* than a given diameter D can be calculated and a grain-size distribution plot prepared. The sieve and hydrometer techniques may be combined for a soil having both coarse-grained and fine-grained soil constituents.

1.3 Size Limits for Soils

Several organizations have attempted to develop the size limits for *gravel, sand, silt,* and *clay* on the basis of the grain sizes present in soils. Table 1.2 presents the size limits recommended by the American Association of State Highway and Transportation Officials (AASHTO) and the Unified Soil Classification systems (Corps of Engineers, Department of the Army, and Bureau of Reclamation). The table shows that soil particles smaller than 0.002 mm have been classified as *clay*. However, clays by nature are cohesive and can be rolled into a thread when moist. This property is caused by the presence of *clay minerals* such as *kaolinite, illite,* and *montmorillonite.* In contrast, some minerals, such as *quartz* and *feldspar,* may be present in a soil in particle sizes as small as clay minerals, but these particles will not have the cohesive property of clay minerals. Hence, they are called *clay-size particles,* not *clay particles.*

1.4 Weight–Volume Relationships

In nature, soils are three-phase systems consisting of solid soil particles, water, and air (or gas). To develop the *weight–volume relationships* for a soil, the three phases can be separated as shown in Figure 1.3a. Based on this separation, the volume relationships can then be defined.

The *void ratio, e,* is the ratio of the volume of voids to the volume of soil solids in a given soil mass, or

$$e = \frac{V_v}{V_s} \tag{1.4}$$

Table 1.2 Soil-Separate Size Limits

Classification system	Grain size (mm)
Unified	Gravel: 75 mm to 4.75 mm Sand: 4.75 mm to 0.075 mm Silt and clay (fines): <0.075 mm
AASHTO	Gravel: 75 mm to 2 mm Sand: 2 mm to 0.05 mm Silt: 0.05 mm to 0.002 mm Clay: <0.002 mm

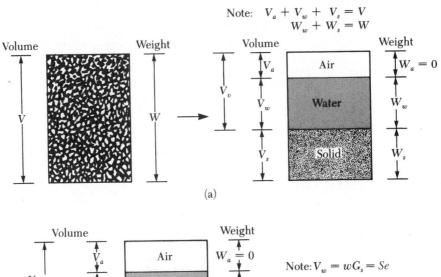

Note: $V_a + V_w + V_s = V$
$W_w + W_s = W$

(a)

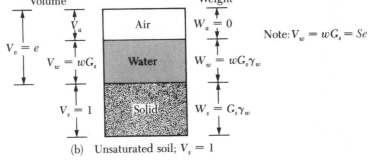

Note: $V_w = wG_s = Se$

(b) Unsaturated soil; $V_s = 1$

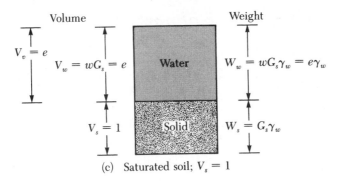

(c) Saturated soil; $V_s = 1$

Figure 1.3 Weight–volume relationships

where V_v = volume of voids
V_s = volume of soil solids

The *porosity, n,* is the ratio of the volume of voids to the volume of the soil specimen, or

$$n = \frac{V_v}{V} \tag{1.5}$$

where V = total volume of soil

Moreover,

$$n = \frac{V_v}{V} = \frac{V_v}{V_s + V_v} = \frac{\dfrac{V_v}{V_s}}{\dfrac{V_s}{V_s} + \dfrac{V_v}{V_s}} = \frac{e}{1 + e} \tag{1.6}$$

The *degree of saturation, S,* is the ratio of the volume of water in the void spaces to the volume of voids, generally expressed as a percentage, or

$$S(\%) = \frac{V_w}{V_v} \times 100 \tag{1.7}$$

where V_w = volume of water

Note that, for saturated soils, the degree of saturation is 100%.

The weight relationships are *moisture content, moist unit weight, dry unit weight,* and *saturated unit weight,* often defined as follows:

$$\text{Moisture content} = w(\%) = \frac{W_w}{W_s} \times 100 \tag{1.8}$$

where W_s = weight of the soil solids
W_w = weight of water

$$\text{Moist unit weight} = \gamma = \frac{W}{V} \tag{1.9}$$

where W = total weight of the soil specimen = $W_s + W_w$

The weight of air, W_a, in the soil mass is assumed to be negligible.

$$\text{Dry unit weight} = \gamma_d = \frac{W_s}{V} \tag{1.10}$$

When a soil mass is completely saturated (i.e., all the void volume is occupied by water), the moist unit weight of a soil [Eq. (1.9)] becomes equal to the *saturated unit weight* (γ_{sat}). So $\gamma = \gamma_{\text{sat}}$ if $V_v = V_w$.

More useful relations can now be developed by considering a representative soil specimen in which the volume of soil solids is equal to *unity,* as shown in Figure 1.3b. Note that if $V_s = 1$, then, from Eq. (1.4), $V_v = e$, and the weight of the soil solids is

$$W_w = G_s \gamma_w$$

where G_s = specific gravity of soil solids
γ_w = unit weight of water (9.81 kN/m^3, or 62.4 lb/ft^3)

Also, from Eq. (1.8), the weight of water $W_w = wW_s$. Thus, for the soil specimen under consideration, $W_w = wW_s = wG_s\gamma_w$. Now, for the general relation for moist unit weight given in Eq. (1.9),

$$\gamma = \frac{W}{V} = \frac{W_s + W_w}{V_s + V_v} = \frac{G_s \gamma_w (1 + w)}{1 + e} \tag{1.11}$$

Similarly, the dry unit weight [Eq. (1.10)] is

$$\gamma_d = \frac{W_s}{V} = \frac{W_s}{V_s + V_v} = \frac{G_s \gamma_w}{1 + e} \tag{1.12}$$

From Eqs. (1.11) and (1.12), note that

$$\gamma_d = \frac{\gamma}{1 + w} \tag{1.13}$$

If a soil specimen is completely saturated, as shown in Figure 1.3c, then

$$V_v = e$$

Also, for this case,

$$V_v = \frac{W_w}{\gamma_w} = \frac{w G_s \gamma_w}{\gamma_w} = w G_s$$

Thus,

$$e = w G_s \qquad \text{(for saturated soil } only\text{)} \tag{1.14}$$

The saturated unit weight of soil then becomes

$$\gamma_{\text{sat}} = \frac{W_s + W_w}{V_s + V_v} = \frac{G_s \gamma_w + e \gamma_w}{1 + e} \tag{1.15}$$

Relationships similar to Eqs. (1.11), (1.12), and (1.15) in terms of porosity can also be obtained by considering a representative soil specimen with a unit volume. These relationships are

$$\gamma = G_s \gamma_w (1 - n)(1 + w) \tag{1.16}$$
$$\gamma_d = (1 - n) G_s \gamma_w \tag{1.17}$$

and

$$\gamma_{\text{sat}} = [(1 - n)G_s + n]\gamma_w \tag{1.18}$$

Except for peat and highly organic soils, the general range of the values of specific gravity of soil solids (G_s) found in nature is rather small. Table 1.3 gives some representative values. For practical purposes, a reasonable value can be assumed in lieu of running a test.

Table 1.3   Specific Gravities of Some Soils

Type of Soil	G_s
Quartz sand	2.64–2.66
Silt	2.67–2.73
Clay	2.70–2.9
Chalk	2.60–2.75
Loess	2.65–2.73
Peat	1.30–1.9

Table 1.4   Typical Void Ratio, Moisture Content, and Dry Unit Weight for Some Soils

Type of soil	Void ratio e	Natural moisture content in saturated condition (%)	Dry unit weight, γ_d (kN/m³)	(lb/ft³)
Loose uniform sand	0.8	30	14.5	92
Dense uniform sand	0.45	16	18	115
Loose angular-grained silty sand	0.65	25	16	102
Dense angular-grained silty sand	0.4	15	19	120
Stiff clay	0.6	21	17	108
Soft clay	0.9–1.4	30–50	11.5–14.5	73–92
Loess	0.9	25	13.5	86
Soft organic clay	2.5–3.2	90–120	6–8	38–51
Glacial till	0.3	10	21	134

Table 1.4 presents some representative values for the void ratio, dry unit weight, and moisture content (in a saturated state) of some naturally occurring soils. Note that in most cohesionless soils the void ratio varies from about 0.4 to 0.8. The dry unit weights in these soils generally fall within a range of about 14–19 kN/m³ (90–120 lb/ft³).

1.5 Relative Density

In *granular soils*, the degree of compaction in the field can be measured according to the *relative density*, defined as

$$D_r(\%) = \frac{e_{max} - e}{e_{max} - e_{min}} \times 100 \qquad (1.19)$$

where e_{max} = void ratio of the soil in the loosest state
e_{min} = void ratio in the densest state
e = *in situ* void ratio

The values of e_{max} are determined in the laboratory in accordance with the test procedures outlined in the American Society for Testing and Materials, *ASTM Standards* (2000, Test Designation D-4254).

Table 1.5 Denseness of a Granular Soil

Relative density, D_r (%)	Description
0–20	Very loose
20–40	Loose
40–60	Medium
60–80	Dense
80–100	Very dense

The relative density can also be expressed in terms of dry unit weight, or

$$D_r(\%) = \left\{ \frac{\gamma_d - \gamma_{d(\min)}}{\gamma_{d(\max)} - \gamma_{d(\min)}} \right\} \frac{\gamma_{d(\max)}}{\gamma_d} \times 100 \qquad (1.20)$$

where γ_d = *in situ* dry unit weight

$\gamma_{d(\max)}$ = dry unit weight in the *densest* state; that is, when the void ratio is $e_{\min}$

$\gamma_{d(\min)}$ = dry unit weight in the *loosest* state, that is, when the void ratio is $e_{\max}$

The denseness of a granular soil is sometimes related to the soil's relative density. Table 1.5 gives a general correlation of the denseness and D_r. For naturally occurring sands, the magnitudes of $e_{\max}$ and $e_{\min}$ [Eq. (1.19)] may vary widely. The main reasons for such wide variations are the uniformity coefficient, C_u, and the roundness of the particles.

Example 1.1

A representative soil specimen collected from the field weighs 1.8 kN and has a volume of 0.1 m³. The moisture content, as determined in the laboratory, is 12.6%. Given that G_s = 2.71, determine the following:

a. Moist unit weight
b. Dry unit weight
c. Void ratio
d. Porosity
e. Degree of saturation

Solution
Part a: Moist Unit Weight
From Eq. (1.9),

$$\gamma = \frac{W}{V} = \frac{1.8 \text{ kN}}{0.1 \text{ m}^3} = \textbf{18 kN/m}^3$$

Part b: Dry Unit Weight
From Eq. (1.13),

$$\gamma_d = \frac{\gamma}{1 + w} = \frac{18}{1 + \dfrac{12.6}{100}} = \textbf{15.99 kN/m}^3$$

Part c: Void Ratio
From Eq. (1.12),

$$\gamma_d = \frac{G_s \gamma_w}{1 + e}$$

or

$$e = \frac{G_s \gamma_w}{\gamma_d} - 1 = \frac{(2.71)\,(9.81)}{15.99} - 1 = \mathbf{0.66}$$

Part d: Porosity
From Eq. (1.6),

$$n = \frac{e}{1 + e} = \frac{0.66}{1 + 0.66} = \mathbf{0.398}$$

Part e: Degree of Saturation
Referring to Figure 1.3b, we obtain

$$S = \frac{V_w}{V_v} = \frac{wG_s}{e} = \frac{(0.126)\,(2.71)}{0.66} \times 100 = \mathbf{51.7\%}$$ ■

Example 1.2

A granular soil (sand) was tested in the laboratory and found to have maximum and minimum void ratios of 0.84 and 0.38, respectively. The value of G_s was determined to be 2.65. A natural soil deposit of the same sand has 9% moisture, and its moist unit weight is 18.64 kN/m³. Determine the relative density of the soil in the field.

Solution
From Eq. (1.13),

$$\gamma_d = \frac{\gamma}{1 + w} = \frac{18.64}{1 + \dfrac{9}{100}} = 17.1 \text{ kN/m}^3$$

Also,

$$\gamma_d = \frac{G_s \gamma_w}{1 + e}$$

or

$$e = \frac{G_s \gamma_w}{\gamma_d} - 1 = \frac{(2.65)\,(9.81)}{17.1} - 1 = 0.52$$

From Eq. (1.19),

$$D_r = \frac{e_{max} - e}{e_{max} - e_{min}} = \frac{0.84 - 0.52}{0.84 - 0.38} = 0.696 = \mathbf{69.6\%}$$ ■

1.6 *Atterberg Limits*

When a clayey soil is mixed with an excessive amount of water, it may flow like a *semi-liquid*. If the soil is gradually dried, it will behave like a *plastic, semisolid,* or *solid* material, depending on its moisture content. The moisture content, in percent, at which the soil changes from a liquid to a plastic state is defined as the *liquid limit* (LL). Similarly, the moisture content, in percent, at which the soil changes from a plastic to a semisolid state and from a semisolid to a solid state are defined as the *plastic limit* (PL) and the *shrinkage limit* (SL), respectively. These limits are referred to as *Atterberg limits* (Figure 1.4):

- The *liquid limit* of a soil is determined by Casagrande's liquid device (ASTM Test Designation D-4318) and is defined as the moisture content at which a groove closure of 12.7 mm (1/2 in.) occurs at 25 blows.
- The *plastic limit* is defined as the moisture content at which the soil crumbles when rolled into a thread of 3.18 mm (1/8 in.) in diameter (ASTM Test Designation D-4318).
- The *shrinkage limit* is defined as the moisture content at which the soil does not undergo any further change in volume with loss of moisture (ASTM Test Designation D-427).

The difference between the liquid limit and the plastic limit of a soil is defined as the *plasticity index* (PI), or

$$PI = LL - PL \tag{1.21}$$

Table 1.6 gives some representative values of the liquid limit and the plastic limit for several clay minerals and soils. However, Atterberg limits for different soils will vary considerably, depending on the origin of the soil and the nature and amount of clay minerals in it.

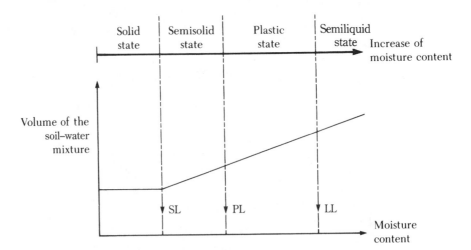

Figure 1.4 Definition of Atterberg limits

Table 1.6 Typical Liquid and Plastic Limits for Some Clay Minerals and Soils

Description	Liquid limit	Plastic limit
Kaolinite	35–100	25–35
Illite	50–100	30–60
Montmorillonite	100–800	50–100
Boston Blue clay	40	20
Chicago clay	60	20
Louisiana clay	75	25
London clay	66	27
Cambridge clay	39	21
Montana clay	52	18
Mississippi gumbo	95	32
Loessial soils in north and northwest China	25–35	15–20

1.7 *Soil Classification Systems*

Soil classification systems divide soils into groups and subgroups based on common engineering properties such as the *grain-size distribution, liquid limit,* and *plastic limit.* The two major classification systems presently in use are (1) the *American Association of State Highway and Transportation Officials (AASHTO) System* and (2) the *Unified Soil Classification System* (also *ASTM*). The AASHTO system is used mainly for the classification of highway subgrades. It is not used in foundation construction.

AASHTO System

The AASHTO Soil Classification System was originally proposed by the Highway Research Board's Committee on Classification of Materials for Subgrades and Granular Type Roads (1945). According to the present form of this system, soils can be classified according to eight major groups, A-1 through A-8, based on their grain-size distribution, liquid limit, and plasticity indices. Soils listed in groups A-1, A-2, and A-3 are coarse-grained materials, and those in groups A-4, A-5, A-6, and A-7 are fine-grained materials. Peat, muck, and other highly organic soils are classified under A-8. They are identified by visual inspection.

The AASHTO classification system (for soils A-1 through A-7) is presented in Table 1.7. Note that group A-7 includes two types of soil. For the A-7-5 type, the plasticity index of the soil is less than or equal to the liquid limit minus 30. For the A-7-6 type, the plasticity index is greater than the liquid limit minus 30.

For qualitative evaluation of the desirability of a soil as a highway subgrade material, a number referred to as the *group index* has also been developed. The higher the value of the group index for a given soil, the weaker will be the soil's performance as a subgrade. A group index of 20 or more indicates a very poor subgrade material. The formula for the group index is

$$\text{GI} = (F_{200} - 35)[0.2 + 0.005(\text{LL} - 40)] + 0.01(F_{200} - 15)(\text{PI} - 10) \qquad (1.22)$$

Table 1.7 AASHTO Soil Classification System

General classification	Granular materials (35% or less of total sample passing no. 200 sieve)						
	A-1		**A-3**	**A-2**			
Group classification	A-1-a	A-1-b		A-2-4	A-2-5	A-2-6	A-2-7
Sieve analysis (% passing)							
No. 10 sieve	50 max						
No. 40 sieve	30 max	50 max	51 min				
No. 200 sieve	15 max	25 max	10 max	35 max	35 max	35 max	35 max
For fraction passing No. 40 sieve							
Liquid limit (LL)				40 max	41 min	40 max	41 min
Plasticity index (PI)	6 max		Nonplastic	10 max	10 max	11 min	11 min
Usual type of material	Stone fragments, gravel, and sand		Fine sand	Silty or clayey gravel and sand			
Subgrade rating				Excellent to good			

General classification	Silt–clay materials (More than 35% of total sample passing no. 200 sieve)			
Group classification	**A-4**	**A-5**	**A-6**	**A-7** A-7-5[a] A-7-6[b]
Sieve analysis (% passing)				
No. 10 sieve				
No. 40 sieve				
No. 200 sieve	36 min	36 min	36 min	36 min
For fraction passing No. 40 sieve				
Liquid limit (LL)	40 max	41 min	40 max	41 min
Plasticity index (PI)	10 max	10 max	11 min	11 min
Usual types of material	Mostly silty soils		Mostly clayey soils	
Subgrade rating		Fair to poor		

[a] If PI $\leq$ LL $-$ 30, the classification is A-7-5.
[b] If PI $>$ LL $-$ 30, the classification is A-7-6.

where F_{200} = percent passing no. 200 sieve, expressed as a whole number
LL = liquid limit
PI = plasticity index

When calculating the group index for a soil belonging to group A-2-6 or A-2-7, use only the partial group-index equation relating to the plasticity index:

$$GI = 0.01(F_{200} - 15)(PI - 10) \tag{1.23}$$

The group index is rounded to the nearest whole number and written next to the soil group in parentheses; for example, we have

$$\underbrace{A-4}_{\text{Soil group}} \qquad \underbrace{(5)}_{\text{Group index}}$$

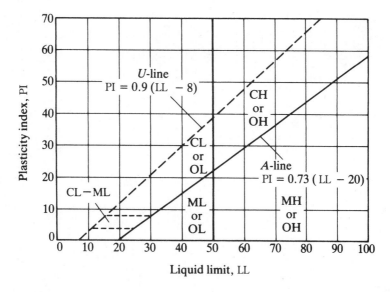

Figure 1.5 Plasticity chart

Unified System

The Unified Soil Classification System was originally proposed by A. Casagrande in 1942 and was later revised and adopted by the United States Bureau of Reclamation and the U. S. Army Corps of Engineers. The system is currently used in practically all geotechnical work.

In the Unified System, the following symbols are used for identification:

Symbol	G	S	M	C	O	Pt	H	L	W	P
Description	Gravel	Sand	Silt	Clay	Organic silts and clay	Peat and highly organic soils	High plasticity	Low plasticity	Well graded	Poorly graded

The plasticity chart (Figure 1.5) and Table 1.8 show the procedure for determining the group symbols for various types of soil. When classifying a soil be sure to provide the group name that generally describes the soil, along with the group symbol. Figures 1.6, 1.7, and 1.8 give flowcharts for obtaining the group names for coarse-grained soil, inorganic fine-grained soil, and organic fine-grained soil, respectively.

Example 1.3

Classify the following soil by the AASHTO classification system:

Percent passing no. 4 sieve = 82
Percent passing no. 10 sieve = 71
Percent passing no. 40 sieve = 64
Percent passing no. 200 sieve = 41
Liquid limit = 31
Plasticity index = 12

Table 1.8 Unified Soil Classification Chart (after ASTM, 2000)

Criteria for Assigning Group Symbols and Group Names Using Laboratory Tests[a]				Group Symbol	Group Name[b]
Coarse-grained soils More than 50% retained on No. 200 sieve	Gravels More than 50% of coarse fraction retained on No. 4 sieve	Clean Gravels Less than 5% fines[c]	$C_u \geq 4$ and $1 \leq C_c \leq 3$[e]	GW	Well-graded gravel[f]
			$C_u < 4$ and/or $1 > C_c > 3$[e]	GP	Poorly graded gravel[f]
		Gravels with Fines More than 12% fines[c]	Fines classify as ML or MH	GM	Silty gravel[f,g,h]
			Fines classify as CL or CH	GC	Clayey gravel[f,g,h]
	Sands 50% or more of coarse fraction passes No. 4 sieve	Clean Sands Less than 5% fines[d]	$C_u \geq 6$ and $1 \leq C_c \leq 3$[e]	SW	Well-graded sand[i]
			$C_u < 6$ and/or $1 > C_c > 3$[e]	SP	Poorly graded sand[i]
		Sand with Fines More than 12% fines[d]	Fines classify as ML or MH	SM	Silty sand[g,h,i]
			Fines classify as CL or CH	SC	Clayey sand[g,h,i]
Fine-grained soils 50% or more passes the No. 200 sieve	Silts and Clays Liquid limit less than 50	inorganic	PI > 7 and plots on or above "A" line[j]	CL	Lean clay[k,l,m]
			PI < 4 or plots below "A" line[j]	ML	Silt[k,l,m]
		organic	$\dfrac{\text{Liquid limit} - \text{oven dried}}{\text{Liquid limit} - \text{not dried}} < 0.75$	OL	Organic clay[k,l,m,n] / Organic silt[k,l,m,o]
	Silts and Clays Liquid limit 50 or more	inorganic	PI plots on or above "A" line	CH	Fat clay[k,l,m]
			PI plots below "A" line	MH	Elastic silt[k,l,m]
		organic	$\dfrac{\text{Liquid limit} - \text{oven dried}}{\text{Liquid limit} - \text{not dried}} < 0.75$	OH	Organic clay[k,l,m,p] / Organic silt[k,l,m,q]
Highly organic soils			Primarily organic matter, dark in color, and organic odor	PT	Peat

[a] Based on the material passing the 75-mm. (3-in) sieve.

[b] If field sample contained cobbles or boulders, or both, add "with cobbles or boulders, or both" to group name.

[c] Gravels with 5 to 12% fines require dual symbols: GW-GM well-graded gravel with silt; GW-GC well-graded gravel with clay; GP-GM poorly graded gravel with silt; GP-GC poorly graded gravel with clay

[d] Sands with 5 to 12% fines require dual symbols: SW-SM well-graded sand with silt; SW-SC well-graded sand with clay; SP-SM poorly graded sand with silt; SP-SC poorly graded sand with clay

[e] $C_u = D_{60}/D_{10} \quad C_c = \dfrac{(D_{30})^2}{D_{10} \times D_{60}}$

[f] If soil contains ≥15% sand, add "with sand" to group name.

[g] If fines classify as CL-ML, use dual symbol GC-GM or SC-SM.

[h] If fines are organic, add "with organic fines" to group name.

[i] If soil contains ≥15% gravel, add "with gravel" to group name.

[j] If Atterberg limits plot in hatched area, soil is a CL-ML, silty clay.

[k] If soil contains 15 to 29% plus No. 200, add "with sand" or "with gravel," whichever is predominant.

[l] If soil contains ≥30% plus No. 200, predominantly sand, add "sandy" to group name.

[m] If soil contains ≥30% plus No. 200, predominantly gravel, add "gravelly" to group name.

[n] PI ≥ 4 and plots on or above "A" line.

[o] PI < 4 or plots below "A" line.

[p] PI plots on or above "A" line.

[q] PI plots below "A" line.

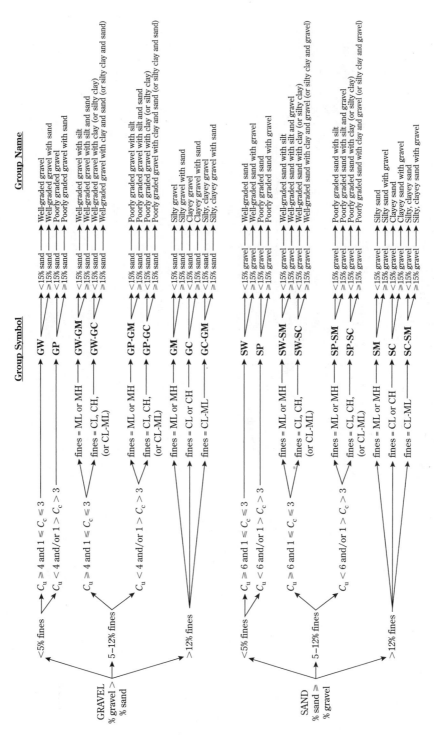

Figure 1.6 Flowchart for Classifying Coarse-Grained Soils (More Than 50% Retained on No. 200 Sieve) (after ASTM, 2000)

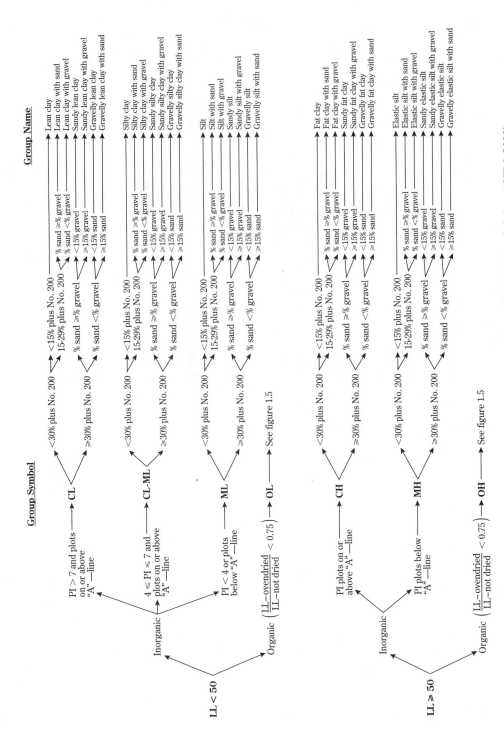

Figure 1.7 Flowchart for Classifying Fine-Grained Soil (50% or More Passes No. 200 Sieve) (after ASTM, 2000)

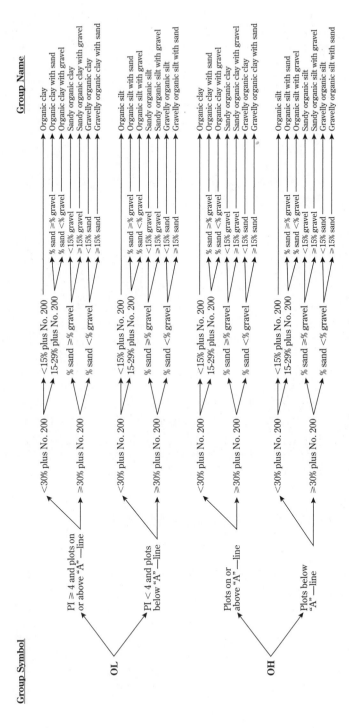

Figure 1.8 Flowchart for Classifying Organic Fine-Grained Soil (50% or More Passes No. 200 Sieve) (after ASTM, 2000)

Solution

Referring to Table 1.7, we see that since more than 35% of our test soil passes through a No. 200 sieve, it is a silt–clay material. It could be A-4, A-5, A-6, or A-7. Because, for this soil, LL = 31 (i.e., LL is less than 40) and PI = 12 (i.e., PI is greater than 11), the soil falls into group A-6. From Eq. (1.22),

$$GI = (F_{200} - 35)[0.2 + 0.005(LL - 40)] + 0.01(F_{200} - 15)(PI - 10)$$

so

$$GI = (41 - 35)[0.2 + 0.005(31 - 40)] + 0.01(41 - 15)(12 - 10)$$
$$= 1.45$$

Hence, the soil is **A-6(1)**. ∎

Example 1.4

Classify the soil described in Example 1.3 according to the Unified Soil Classification System.

Solution

We are given that $F_{200} = 41$, LL = 31, and PI = 12. Since 59% of the sample is retained on a No. 200 sieve, the soil is a coarse-grained material. The percentage passing a No. 4 sieve is 82, so 18% is retained on No. 4 sieve (gravel fraction). The coarse fraction passing a No. 4 sieve (sand fraction) is 59 − 18 = 41% (which is more than 50% of the total coarse fraction). Hence, the specimen is a sandy soil.

Now, using Table 1.8 and Figure 1.5, we identify the group symbol of the soil as **SC**.

Again from Figure 1.6, since the gravel fraction is greater than 15%, the group name is **clayey sand with gravel.** ∎

1.8   *Hydraulic Conductivity of Soil*

The void spaces, or pores, between soil grains allow water to flow through them. In soil mechanics and foundation engineering, you must know how much water is flowing through a soil per unit time. This knowledge is required to design earth dams, determine the quantity of seepage under hydraulic structures, and dewater foundations before and during their construction. Darcy (1856) proposed the following equation (Figure 1.9) for calculating the velocity of flow of water through a soil:

$$v = ki \tag{1.24}$$

In this equation, v = Darcy velocity (unit: cm/sec)
k = hydraulic conductivity of soil (unit: cm/sec)
i = hydraulic gradient

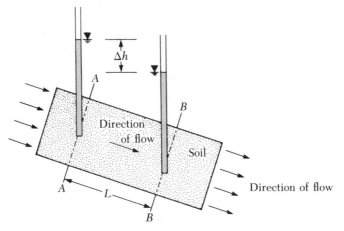

Figure 1.9 Definition of Darcy's law

Table 1.9 Range of the Hydraulic Conductivity for Various Soils

Type of soil	Hydraulic conductivity, k (cm/sec)
Medium to coarse gravel	Greater than 10^{-1}
Coarse to fine sand	10^{-1} to 10^{-3}
Fine sand, silty sand	10^{-3} to 10^{-5}
Silt, clayey silt, silty clay	10^{-4} to 10^{-6}
Clays	10^{-7} or less

The hydraulic gradient is defined as

$$i = \frac{\Delta h}{L}$$

(1.25)

where Δh = piezometric head difference between the sections at AA and BB
   L = distance between the sections at AA and BB

(*Note:* Sections AA and BB are perpendicular to the direction of flow.)

Darcy's law [Eq. (1.24)] is valid for a wide range of soils. However, with materials like clean gravel and open-graded rockfills, the law breaks down because of the turbulent nature of flow through them.

The value of the hydraulic conductivity of soils varies greatly. In the laboratory, it can be determined by means of *constant-head* or *falling-head* permeability tests. The constant-head test is more suitable for granular soils. Table 1.9 provides the general range for the values of k for various soils. In granular soils, the value depends primarily on the void ratio. In the past, several equations have been proposed to relate the value of k to the void ratio in granular soil:

$$\frac{k_1}{k_2} = \frac{e_1^2}{e_2^2}$$

(1.26)

$$\frac{k_1}{k_2} = \frac{\left(\dfrac{e_1^2}{1 + e_1}\right)}{\left(\dfrac{e_2^2}{1 + e_2}\right)} \tag{1.27}$$

$$\frac{k_1}{k_2} = \frac{\left(\dfrac{e_1^3}{1 + e_1}\right)}{\left(\dfrac{e_2^3}{1 + e_2}\right)} \tag{1.28}$$

In these equations, k_1 and k_2 are the hydraulic conductivities of a given soil at void ratios e_1 and e_2, respectively.

According to their experimental observations, Samarasinghe, Huang, and Drnevich (1982) suggested that the hydraulic conductivity of normally consolidated clays could be given by the equation

$$k = C\frac{e^n}{1 + e} \tag{1.29}$$

where C and n are constants to be determined experimentally.

For clayey soils in the field, a practical relationship for estimating the hydraulic conductivity (Tavenas et al., 1983) is

$$\log k = \log k_0 - \frac{e_0 - e}{C_k} \tag{1.30}$$

where k = hydraulic conductivity at void ratio e
$\quad\quad\quad k_0$ = *in situ* hydraulic conductivity at void ratio e_0
$\quad\quad\quad C_k$ = conductivity change index $\approx 0.5e_0$

For clayey soils, the hydraulic conductivity for flow in the vertical and horizontal directions may vary substantially. The hydraulic conductivity for flow in the vertical direction (k_v) for *in situ* soils can be estimated from Figure 1.10. For marine and other massive clay deposits,

$$\frac{k_h}{k_v} < 1.5 \tag{1.31}$$

where k_h = hydraulic conductivity for flow in the horizontal direction

For varved clays, the ratio of k_h/k_v may exceed 10.

1.9   *Steady-State Seepage*

For most cases of seepage under hydraulic structures, the flow path changes direction and is not uniform over the entire area. In such cases, one of the ways of determining the rate of seepage is by a graphical construction referred to as the *flow net,*

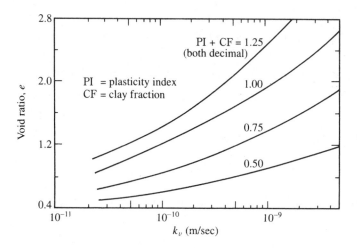

Figure 1.10 Variation of *in situ* k_v for clay soils (after Tavenas et al., 1983)

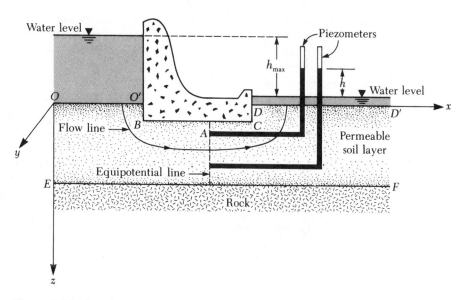

Figure 1.11 Steady-state seepage

a concept based on Laplace's theory of continuity. According to this theory, for a steady flow condition, the flow at any point A (Figure 1.11) can be represented by the equation

$$k_x \frac{\partial^2 h}{\partial x^2} + k_y \frac{\partial^2 h}{\partial y^2} + k_z \frac{\partial^2 h}{\partial z^2} = 0 \qquad (1.32)$$

where k_x, k_y, k_z = hydraulic conductivity of the soil in the x, y, and z directions, respectively

h = hydraulic head at point A (i.e., the head of water that a piezometer placed at A would show with the *downstream water level* as *datum,* as shown in Figure 1.11)

For a two-dimensional flow condition, as shown in Figure 1.11,

$$\frac{\partial^2 h}{\partial^2 y} = 0$$

so Eq. (1.32) takes the form

$$k_x \frac{\partial^2 h}{\partial x^2} + k_z \frac{\partial^2 h}{\partial z^2} = 0 \qquad (1.33)$$

If the soil is isotropic with respect to hydraulic conductivity, $k_x = k_z = k$, and

$$\frac{\partial^2 h}{\partial x^2} + \frac{\partial^2 h}{\partial z^2} = 0 \qquad (1.34)$$

Equation (1.34), which is referred to as Laplace's equation and is valid for confined flow, represents two orthogonal sets of curves known as *flow lines* and *equipotential lines.* A flow net is a combination of numerous equipotential lines and flow lines. A flow line is a path that a water particle would follow in traveling from the upstream side to the downstream side. An equipotential line is a line along which water, in piezometers, would rise to the same elevation. (See Figure 1.11.)

In drawing a flow net, you need to establish the *boundary conditions.* For example, in Figure 1.11, the ground surfaces on the upstream (OO') and downstream (DD') sides are equipotential lines. The base of the dam below the ground surface, $O'BCD$, is a flow line. The top of the rock surface, EF, is also a flow line. Once the boundary conditions are established, a number of flow lines and equipotential lines are drawn by trial and error so that all the flow elements in the net have the same length-to-width ratio (L/B). In most cases, L/B is held to unity, that is, the flow elements are drawn as curvilinear "squares." This method is illustrated by the flow net shown in Figure 1.12. Note that all flow lines must intersect all equipotential lines at *right angles.*

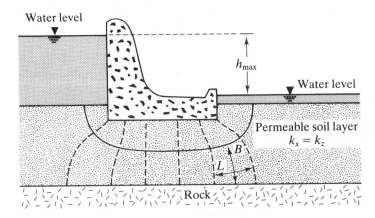

Figure 1.12   Flow net

Once the flow net is drawn, the seepage, in unit time per unit length of the structure, can be calculated as

$$q = k h_{\max} \frac{N_f}{N_d} n \qquad (1.35)$$

where N_f = number of flow channels
N_d = number of drops
n = width-to-length ratio of the flow elements in the flow net (B/L)
$h_{\max}$ = difference in water level between the upstream and downstream sides

The space between two consecutive flow lines is defined as a *flow channel,* and the space between two consecutive equipotential lines is called a *drop.* In Figure 1.12, $N_f = 2$, $N_d = 7$, and $n = 1$. When square elements are drawn in a flow net,

$$q = k h_{\max} \frac{N_f}{N_d} \qquad (1.36)$$

1.10   *Effective Stress*

Consider the vertical stress at a point A located at a depth $h_1 + h_2$ below the ground surface, as shown in Figure 1.13a. The total vertical stress at A is

$$\sigma = h_1 \gamma + h_2 \gamma_{\mathrm{sat}} \qquad (1.37)$$

where γ and γ_{sat} are unit weights of soil above and below the water table, respectively.

The total stress is carried partially by the *pore water* in the void spaces and partially by the *soil solids* at their points of contact. For example, consider a wavy plane AB drawn through point A (see Figure 1.13a) that passes through the points of contact of soil grains. The plan of this section is shown in Figure 1.13b. The small dots in the figure represent the areas in which there is solid-to-solid contact. If the sum of these areas equals A', the area filled by water equals $XY - A$. The force carried by the pore water over the area shown is then

$$F_w = (XY - A')u \qquad (1.38)$$

where u = pore water pressure = $\gamma_w h_2$ $\qquad (1.39)$

Now let $F_1, F_2, \ldots$ be the forces at the contact points of the soil solids, as shown in Figure 1.13a. The sum of the vertical components of these forces over a horizontal area XY is

$$F_s = \Sigma F_{1(v)} + F_{2(v)} + \cdots \qquad (1.40)$$

where $F_{1(v)}, F_{2(v)}, \ldots$ are the vertical components of forces $F_1, F_2, \ldots,$ respectively

According to the principles of statics,

$$(\sigma) XY = F_w + F_s$$

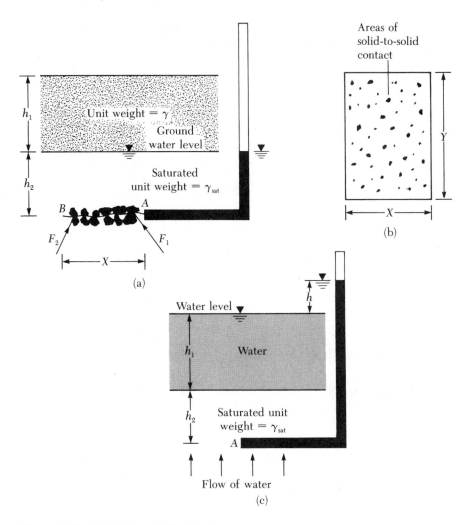

Figure 1.13 Calculation of effective stress

or

$$(\sigma)XY = (XY - A')u + F_s$$

so

$$\sigma = (1 - a)u + \sigma' \qquad (1.41)$$

where $a = A'/XY$ = fraction of the unit cross-sectional area occupied by solid-to-solid contact

$\sigma' = F_s/(XY)$ = verticle component of forces at solid-to-solid contact points over a unit cross-sectional area

The term σ' in Eq. (1.41) is generally referred to as the *vertical effective stress*. Note that the quantity a is very small; thus,

$$\sigma = u + \sigma' \tag{1.42}$$

Observe that effective stress is a *derived* quantity. Also, because the effective stress σ' is related to the contact between the soil solids, changes in effective stress will induce changes in volume. Effective stress is responsible for producing *frictional resistance* in soils and rocks. For dry soils, $u = 0$; hence, $\sigma = \sigma'$.

For the problem under consideration in Figure 1.13a, $u = h_2\gamma_w(\gamma_w = $ unit weight of water). Thus, the effective stress at point A is

$$\sigma' = \sigma - u = (h_1\gamma + h_2\gamma_{sat}) - h_2\gamma_w$$

$$= h_1\gamma + h_2(\gamma_{sat} - \gamma_w) = h_1\gamma + h_2\gamma' \tag{1.43}$$

where $\gamma' = $ effective, or submerged, unit weight of soil

$$= \gamma_{sat} - \gamma_w$$

From Eq. (1.15),

$$\gamma_{sat} = \frac{G_s\gamma_w + e\gamma_w}{1 + e}$$

so

$$\gamma' = \gamma_{sat} - \gamma_w = \frac{G_s\gamma_w + e\gamma_w}{1 + e} - \gamma_w = \frac{\gamma_w(G_s - 1)}{1 + e} \tag{1.44}$$

For the problem in Figures 1.13a and 1.13b, there was *no seepage of water* in the soil. Figure 1.13c shows a simple condition in a soil profile in which there is upward seepage. For this case, at point A,

$$\sigma = h_1\gamma_w + h_2\gamma_{sat}$$

and

$$u = (h_1 + h_2 + h)\gamma_w$$

Thus, from Eq. (1.42),

$$\sigma' = \sigma - u = (h_1\gamma_w + h_2\gamma_{sat}) - (h_1 + h_2 + h)\gamma_w$$

$$= h_2(\gamma_{sat} - \gamma_w) - h\gamma_w = h_2\gamma' - h\gamma_w$$

or

$$\sigma' = h_2\left(\gamma' - \frac{h}{h_2}\gamma_w\right) = h_2(\gamma' - i\gamma_w) \tag{1.45}$$

Note in Eq. (1.45) that h/h_2 is the hydraulic gradient i. If the hydraulic gradient is very high, so that $\gamma' - i\gamma_w$ becomes zero, *the effective stress will become zero.* In other words, there is no contact stress between the soil particles, and the soil will break up. This situation is referred to as the *quick condition,* or *failure by heave.* So, for heave,

$$i = i_{cr} = \frac{\gamma'}{\gamma_w} = \frac{G_s - 1}{1 + e} \tag{1.46}$$

where i_{cr} = critical hydraulic gradient

For most sandy soils, i_{cr} ranges from 0.9 to 1.1, with an average of about unity.

Example 1.5

Figure 1.14 shows a soil profile. Determine the total vertical stress, pore water pressure, and effective vertical stress at points *A, B*, and *C*. Draw the variation of effective stress with depth.

Solution

Determination of Unit Weights of Soil

$$\gamma_{d(\text{sand})} = \frac{G_s \gamma_w}{1 + e} = \frac{(2.65)(9.81)}{1 + 0.6} = 16.25 \text{ kN/m}^3$$

$$\gamma_{\text{sat(clay)}} = \frac{G_s \gamma_w + w G_s \gamma_w}{1 + w G_s}$$

Note: For saturated soils, $e = w G_s$ [Eq. (1.14)]; so, for this case, $e = (0.3)(2.7) = 0.81$. Hence,

$$\gamma_{\text{sat(clay)}} = \frac{(2.7)(9.81) + (0.81)(9.81)}{1 + 0.81} = 19.02 \text{ kN/m}^3$$

Total Stress Calculation

At A: $\sigma_A = \mathbf{0}$

At B: $\sigma_B = \gamma_{d(\text{sand})} \times 3 = 16.25 \times 3 = \mathbf{48.75 \text{ kN/m}^2}$

At C: $\sigma_C = \sigma_B + \gamma_{\text{sat(clay)}} \times 3 = 48.75 + (19.02)3 = \mathbf{105.81 \text{ kN/m}^2}$

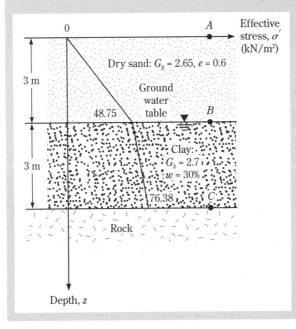

Figure 1.14 Variation of effective stress in a soil profile.

Pore Water Pressure Calculation

$$\text{At } A\!:\! u_A = \mathbf{0}$$

$$\text{At } B\!:\! u_B = \mathbf{0}$$

$$\text{At } C\!:\! u_C = 3 \times \gamma_w = 3 \times 9.81 = \mathbf{29.43 \ kN/m^2}$$

Effective Stress Calculation

$$\text{At } A\!:\! \sigma'_A = \sigma_A - u_A = \mathbf{0}$$

$$\text{At } B\!:\! \sigma'_B = 48.75 - 0 = \mathbf{48.75 \ kN/m^2}$$

$$\text{At } C\!:\! \sigma'_C = 105.81 - 29.43 = \mathbf{76.38 \ kN/m^2}$$

The variation of effective stress with depth can now be drawn on Figure 1.14 ■

1.11 *Consolidation*

In the field, when the stress on a saturated clay layer is increased—for example, by the construction of a foundation—the pore water pressure in the clay will increase. Because the hydraulic conductivity of clays is very small, some time will be required for the excess pore water pressure to dissipate and the increase in stress to be transferred to the soil skeleton. According to Figure 1.15, if $\Delta\sigma$ is a surcharge at the ground surface over a very large area, the increase in total stress at any depth of the clay layer will be equal to $\Delta\sigma$.

However, at time $t = 0$ (i.e., immediately after the stress is applied), the excess pore water pressure at any depth Δu will equal $\Delta\sigma$, or

$$\Delta u = \Delta h_i \gamma_w = \Delta\sigma \ (\text{at time } t = 0)$$

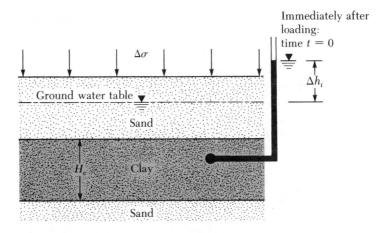

Figure 1.15 Principles of consolidation

Hence, the increase in effective stress at time $t = 0$ will be

$$\Delta\sigma' = \Delta\sigma - \Delta u = 0$$

Theoretically, at time $t = \infty$, when all the excess pore water pressure in the clay layer has dissipated as a result of drainage into the sand layers,

$$\Delta u = 0 \quad (\text{at time } t = \infty)$$

Then the increase in effective stress in the clay layer is

$$\Delta\sigma' = \Delta\sigma - \Delta u = \Delta\sigma - 0 = \Delta\sigma$$

This gradual increase in the effective stress in the clay layer will cause settlement over a period of time and is referred to as *consolidation.*

Laboratory tests on undisturbed saturated clay specimens can be conducted (ASTM Test Designation D-2435) to determine the consolidation settlement caused by various incremental loadings. The test specimens are usually 63.5 mm (2.5 in.) in diameter and 25.4 mm (1 in.) in height. Specimens are placed inside a ring, with one porous stone at the top and one at the bottom of the specimen (Figure 1.16a). A load on the specimen is then applied so that the total vertical stress is equal to σ. Settlement readings for the specimen are taken periodically for 24 hours. After that, the load on the specimen is doubled and more settlement readings are taken. At all times during the test, the specimen is kept under water. The procedure is continued until the desired limit of stress on the clay specimen is reached.

Based on the laboratory tests, a graph can be plotted showing the variation of the void ratio e at the *end* of consolidation against the corresponding vertical effective stress σ'. (On a semilogarithmic graph, e is plotted on the arithmetic scale and σ' on the log scale.) The nature of the variation of e against $\log \sigma'$ for a clay specimen is shown in Figure 1.16b. After the desired consolidation pressure has been reached, the specimen can be gradually unloaded, which will result in the swelling of the specimen. The figure also shows the variation of the void ratio during the unloading period.

From the e–$\log \sigma'$ curve shown in Figure 1.16b, three parameters necessary for calculating settlement in the field can be determined:

1. The *preconsolidation pressure,* σ'_c, is the *maximum past effective overburden pressure* to which the soil specimen has been subjected. It can be determined by using a simple graphical procedure proposed by Casagrande (1936). The procedure involves five steps (see Figure 1.16b):
   a. Determine the point O on the e–$\log \sigma'$ curve that has the sharpest curvature (i.e., the smallest radius of curvature).
   b. Draw a horizontal line OA.
   c. Draw a line OB that is tangent to the e–$\log \sigma'$ curve at O.
   d. Draw a line OC that bisects the angle AOB.
   e. Produce the straight-line portion of the e–$\log \sigma'$ curve backwards to intersect OC. This is point D. The pressure that corresponds to point D is the preconsolidation pressure σ'_c.

   Natural soil deposits can be *normally consolidated* or *overconsolidated* (or *preconsolidated*). If the present effective overburden pressure $\sigma' = \sigma'_o$ is

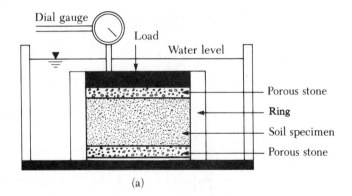

(a)

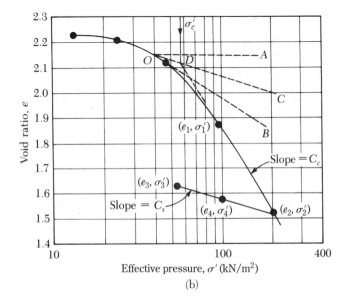

(b)

Figure 1.16
(a) Schematic diagram of consolidation test arrangement; (b) e–log σ' curve for a soft clay from East St. Louis, Illinois (*Note:* At the end of consolidation, $\sigma = \sigma'$)

equal to the preconsolidated pressure σ'_c the soil is *normally consolidated*. However, if $\sigma'_o < \sigma'_c$, the soil is *overconsolidated*.

The preconsolidation pressure (σ'_c) has been correlated with the index parameters by several investigators. The U.S. Department of the Navy (1982) also provided generalized relationships among σ'_c, LI, and the sensitivity of clayey soils (S_t). The definition of sensitivity is given in Section 1.19. Figure 1.17 shows the relationship between σ'_c and LI. Note that the latter—the liquidity index of a soil—is defined as

$$\text{LI} = \frac{w - \text{PL}}{\text{LL} - \text{PL}} \tag{1.47}$$

where $w = $ *in situ* moisture content
          LI $= $ liquid limit
          PL $= $ plastic limit

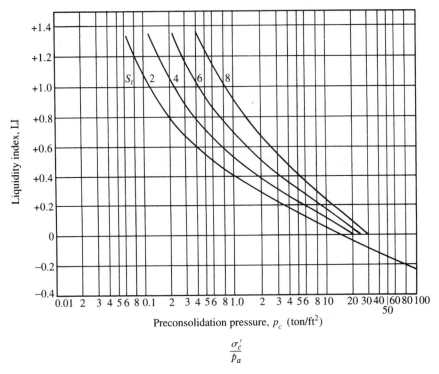

$$\frac{\sigma'_c}{p_a}$$

Figure 1.17   Variation of σ'_c with LI (after U.S. Department of the Navy, 1982)
Note: p_a = atmospheric pressure $[\approx 100 \text{ kN/m}^2 (1 \text{ U.S. ton/ft}^2)]$

2. The *compression index, C_c,* is the slope of the straight-line portion (the latter part) of the loading curve, or

$$C_c = \frac{e_1 - e_2}{\log \sigma'_2 - \log \sigma'_1} = \frac{e_1 - e_2}{\log\left(\dfrac{\sigma'_2}{\sigma'_1}\right)} \qquad (1.48)$$

where e_1 and e_2 are the void ratios at the end of consolidation under effective stresses σ'_1 and σ'_2, respectively

The *compression index,* as determined from the laboratory e–$\log \sigma'$ curve, will be somewhat different from that encountered in the field. The primary reason is that the soil remolds itself to some degree during the field exploration. The nature of variation of the e–$\log \sigma'$ curve in the field for a normally consolidated clay is shown in Figure 1.18. The curve, generally referred to as the *virgin compression curve,* approximately intersects the laboratory curve at a void ratio of $0.42e_o$ (Terzaghi and Peck, 1967). Note that e_o is the void ratio of the clay in the field. Knowing the values of e_o and σ'_c, you can easily construct the virgin curve and calculate its compression index by using Eq. (1.48).

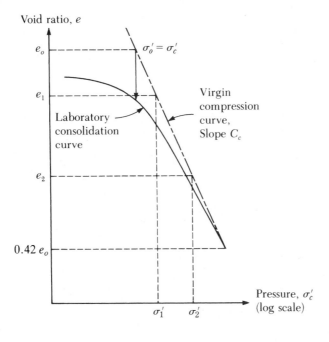

Figure 1.18 Construction of virgin compression curve for normally consolidated clay

The value of C_c can vary widely, depending on the soil. Skempton (1944) gave an empirical correlation for the compression index in which

$$C_c = 0.009(LL - 10) \tag{1.49}$$

where   LL = liquid limit

Besides Skempton, several other investigators have also proposed correlations for the compression index.

3. The *swelling index, C_s,* is the slope of the unloading portion of the e–log σ' curve. In Figure 1.16b, it is defined as

$$C_s = \frac{e_3 - e_4}{\log\left(\dfrac{\sigma_4'}{\sigma_3'}\right)} \tag{1.50}$$

In most cases, the value of the swelling index is $\frac{1}{4}$ to $\frac{1}{5}$ of the compression index. Following are some representative values of C_s/C_c for natural soil deposits:

Description of soil	C_s/C_c
Boston Blue clay	0.24–0.33
Chicago clay	0.15–0.3
New Orleans clay	0.15–0.28
St. Lawrence clay	0.05–0.1

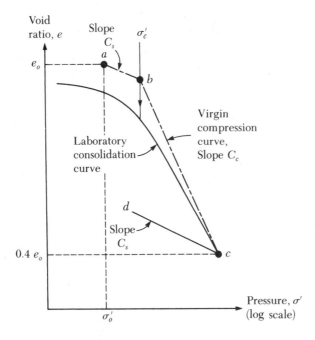

Figure 1.19 Construction of field consolidation curve for overconsolidated clay

The swelling index is also referred to as the *recompression index.*

The determination of the swelling index is important in the estimation of consolidation settlement of *overconsolidated clays.* In the field, depending on the pressure increase, an overconsolidated clay will follow an e–log σ' path *abc,* as shown in Figure 1.19. Note that point *a,* with coordinates σ'_o and e_o, corresponds to the field conditions before any increase in pressure. Point *b* corresponds to the preconsolidation pressure (σ'_c) of the clay. Line *ab* is approximately parallel to the laboratory unloading curve *cd* (Schmertmann, 1953). Hence, if you know $e_o, \sigma'_o, \sigma'_c, C_c,$ and C_s, you can easily construct the field consolidation curve.

1.12 Calculation of Primary Consolidation Settlement

The one-dimensional primary consolidation settlement (caused by an additional load) of a clay layer (Figure 1.20a) having a thickness H_c may be calculated as

$$S_c = \frac{\Delta e}{1 + e_o} H_c \tag{1.51}$$

where S_c = primary consolidation settlement
Δe = total change of void ratio caused by the additional load application
e_o = void ratio of the clay before the application of load

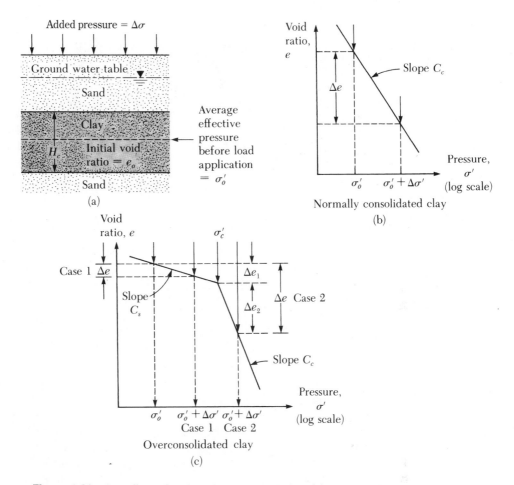

Figure 1.20 One-dimensional settlement calculation: (b) is for Eq. (1.53); (c) is for Eqs. (1.55) and (1.57)

Note that

$$\frac{\Delta e}{1 + e_o} = \varepsilon_v = \text{vertical strain}$$

For normally consolidated clay, the field e–log σ' curve will be like the one shown in Figure 1.20b. If σ'_o = initial average effective overburden pressure on the clay layer and $\Delta\sigma'$ = average effective increase in pressure on the clay layer caused by the added load, the change in the void ratio caused by the increase in load is

$$\Delta e = C_c \log\frac{\sigma'_o + \Delta\sigma'}{\sigma'_o} \qquad (1.52)$$

Now, combining Eqs. (1.51) and (1.52) yields

$$S_c = \frac{C_c H_c}{1 + e_o} \log \frac{\sigma'_o + \Delta \sigma'}{\sigma'_o} \tag{1.53}$$

For overconsolidated clay, the field e–$\log \sigma'$ curve will be like the one shown in Figure 1.20c. In this case, depending on the value of $\Delta \sigma'$, two conditions may arise. First, if $\sigma'_o + \Delta \sigma' < \sigma'_c$ then

$$\Delta e = C_s \log \frac{\sigma'_o + \Delta \sigma'}{\sigma'_o} \tag{1.54}$$

Combining Eqs. (1.51) and (1.54) gives

$$S_c = \frac{H_c C_s}{1 + e_o} \log \frac{\sigma'_o + \Delta \sigma'}{\sigma'_o} \tag{1.55}$$

Second, if $\sigma'_o < \sigma'_c < \sigma'_o + \Delta \sigma'$, then

$$\Delta e = \Delta e_1 + \Delta e_2 = C_s \log \frac{\sigma'_c}{\sigma'_o} + C_c \log \frac{\sigma'_o + \Delta \sigma'}{\sigma'_c} \tag{1.56}$$

Now, combining Eqs. (1.51) and (1.56) yields

$$S = \frac{C_s H_c}{1 + e_o} \log \frac{\sigma'_c}{\sigma'_o} + \frac{C_c H_c}{1 + e_o} \log \frac{\sigma'_o + \Delta \sigma'}{\sigma'_c} \tag{1.57}$$

1.13 *Time Rate of Consolidation*

In Section 1.11 (see Figure 1.15), we showed that consolidation is the result of the gradual dissipation of the excess pore water pressure from a clay layer. The dissipation of pore water pressure, in turn, increases the effective stress, which induces settlement. Hence, to estimate the degree of consolidation of a clay layer at some time t after the load is applied, you need to know the rate of dissipation of the excess pore water pressure.

Figure 1.21 shows a clay layer of thickness H_c that has highly permeable sand layers at its top and bottom. Here, the excess pore water pressure at any point A at any time t after the load is applied is $\Delta u = (\Delta h)\gamma_w$. For a vertical drainage condition (that is, in the direction of z only) from the clay layer, Terzaghi derived the differential equation

$$\frac{\partial(\Delta u)}{\partial t} = C_v \frac{\partial^2(\Delta u)}{\partial z^2} \tag{1.58}$$

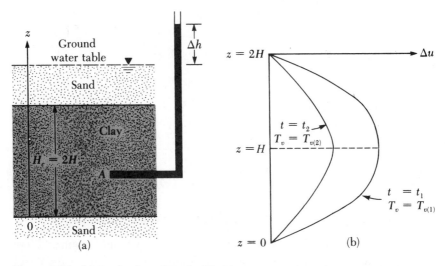

Figure 1.21 (a) Derivation of Eq. (1.60); (b) nature of variation of Δu with time

where C_v = coefficient of consolidation, defined by

$$C_v = \frac{k}{m_v \gamma_w} = \frac{k}{\dfrac{\Delta e}{\Delta \sigma' (1 + e_{av})} \gamma_w} \qquad (1.59)$$

in which k = hydraulic conductivity of the clay
Δe = total change of void ratio caused by an effective stress increase of $\Delta \sigma'$
e_{av} = average void ratio during consolidation
m_v = volume coefficient of compressibility = $\Delta e / [\Delta \sigma' (1 + e_{av})]$

Equation (1.58) can be solved to obtain Δu as a function of time t with the following boundary conditions:

1. Because highly permeable sand layers are located at $z = 0$ and $z = H_c$, the excess pore water pressure developed in the clay at those points will be immediately dissipated. Hence,

$$\Delta u = 0 \quad \text{at} \quad z = 0$$

and

$$\Delta u = 0 \quad \text{at} \quad z = H_c = 2H$$

where H = length of maximum drainage path (due to two-way drainage condition—that is, at the top and bottom of the clay)

2. At time $t = 0$,

$\Delta u = \Delta u_o$ = initial excess pore water pressure after the load is applied

With the preceding boundary conditions, Eq. (1.58) yields

$$\Delta u = \sum_{m=0}^{m=\infty} \left[\frac{2(\Delta u_o)}{M} \sin\left(\frac{Mz}{H}\right) \right] e^{-M^2 T_v} \tag{1.60}$$

where $M = [(2m + 1)\pi]/2$

m = an integer = $1, 2, \ldots$

T_v = nondimensional time factor = $(C_v t)/H^2$ (1.61)

The value of Δu for various depths (i.e., $z = 0$ to $z = 2H$) at any given time t (and thus T_v) can be calculated from Eq. (1.60). The nature of this variation of Δu is shown in Figures 1.22a and b. Figure 1.22c shows the variation of $\Delta u / \Delta u_o$ with T_v and H/H_c using Eqs.(1.60) and (1.61).

Determining the field value of C_v is difficult. Figure 1.23 provides a first-order determination of C_v, using the liquid limit (U.S. Department of the Navy, 1971).

The *average degree of consolidation* of the clay layer can be defined as

$$U = \frac{S_{c(t)}}{S_{c(max)}} \tag{1.62}$$

where $S_{c(t)}$ = settlement of a clay layer at time t after the load is applied

$S_{c(max)}$ = maximum consolidation settlement that the clay will undergo under a given loading

If the initial pore water pressure (Δu_0) distribution is constant with depth, as shown in Figure 1.22a, the average degree of consolidation can also be expressed as

$$U = \frac{S_{c(t)}}{S_{c(max)}} = \frac{\displaystyle\int_0^{2H} (\Delta u_o)dz - \int_0^{2H} (\Delta u)dz}{\displaystyle\int_0^{2H} (\Delta u_o)dz} \tag{1.63}$$

or

$$U = \frac{(\Delta u_o)2H - \displaystyle\int_0^{2H} (\Delta u)dz}{(\Delta u_o)2H} = 1 - \frac{\displaystyle\int_0^{2H} (\Delta u)dz}{2H(\Delta u_o)} \tag{1.64}$$

Now, combining Eqs. (1.60) and (1.64), we obtain

$$U = \frac{S_{c(t)}}{S_{c(max)}} = 1 - \sum_{m=0}^{m=\infty} \left(\frac{2}{M^2}\right) e^{-M^2 T_v} \tag{1.65}$$

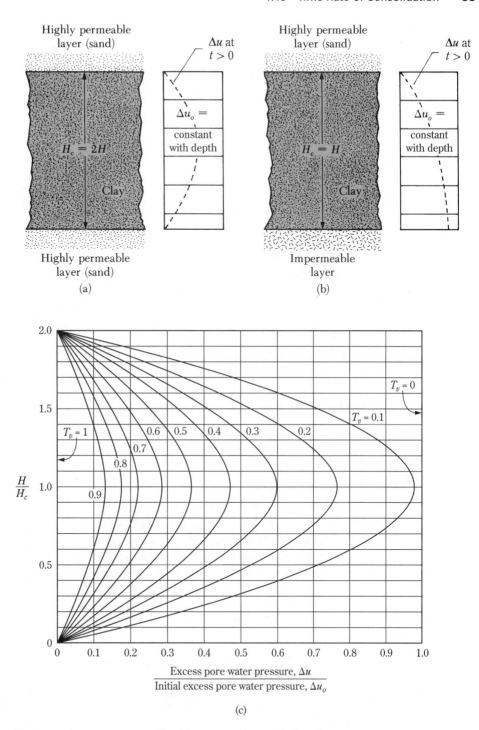

Figure 1.22 Drainage condition for consolidation: (a) two-way drainage; (b) one-way drainage; (c) plot of $\Delta u / \Delta u_o$ with T_v and H/H_c

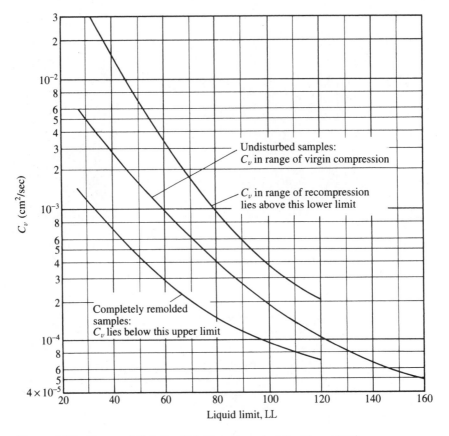

Figure 1.23   Range of C_v (after U.S. Department of the Navy, 1971)

The variation of U with T_v can be calculated from Eq. (1.65) and is plotted in Figure 1.24. Note that Eq. (1.65) and thus Figure 1.24 are also valid when an impermeable layer is located at the bottom of the clay layer (Figure 1.22). In that case, the dissipation of excess pore water pressure can take place in one direction only. The length of the *maximum drainage path* is then equal to $H = H_c$.

The variation of T_v with U shown in Figure 1.24 can also be approximated by

$$T_v = \frac{\pi}{4}\left(\frac{U\%}{100}\right)^2 \quad (\text{for } U = 0\text{--}60\%) \tag{1.66}$$

and

$$T_v = 1.781 - 0.933 \log(100 - U\%) \quad (\text{for } U > 60\%) \tag{1.67}$$

Sivaram and Swamee (1977) also developed an empirical relationship between T_v and U that is valid for U varying from 0 to 100%. The equation is of the form

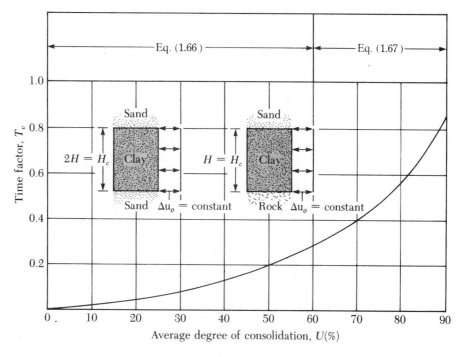

Figure 1.24 Plot of time factor against average degree of consolidation (Δu_o = constant)

$$T_v = \frac{\left(\dfrac{\pi}{4}\right)\left(\dfrac{U\%}{100}\right)^2}{\left[1 - \left(\dfrac{U\%}{100}\right)^{5.6}\right]^{0.357}}$$

$$(1.68)$$

Example 1.6

A laboratory consolidation test on a normally consolidated clay showed the following results:

Load, $\Delta\sigma'$ (kN/m²)	Void ratio at the end of consolidation, e
140	0.92
212	0.86

The specimen tested was 25.4 mm in thickness and drained on both sides. The time required for the specimen to reach 50% consolidation was 4.5 min.

A similar clay layer in the field 2.8 m thick and drained on both sides, is subjected to a similar increase in average effective pressure (i.e., $\sigma'_o = 140 \text{ kN/m}^2$ and $\sigma'_o + \Delta\sigma' = 212 \text{ kN/m}^2$). Determine

a. the expected maximum primary consolidation settlement in the field.
b. the length of time required for the total settlement in the field to reach 40 mm. (Assume a uniform initial increase in excess pore water pressure with depth.)

Solution

Part a
For normally consolidated clay [Eq. (1.48)],

$$C_c = \frac{e_1 - e_2}{\log\left(\dfrac{\sigma'_2}{\sigma'_1}\right)} = \frac{0.92 - 0.86}{\log\left(\dfrac{212}{140}\right)} = 0.333$$

From Eq. (1.53),

$$S_c = \frac{C_c H_c}{1 + e_o} \log \frac{\sigma'_o + \Delta\sigma'}{\sigma'_o} = \frac{(0.333)(2.8)}{1 + 0.92} \log \frac{212}{140} = 0.0875 \text{ m} = \textbf{87.5 mm}$$

Part b
From Eq. (1.62), the average degree of consolidation is

$$U = \frac{S_{c(t)}}{S_{c(\max)}} = \frac{40}{87.5}(100) = 45.7\%$$

The coefficient of consolidation, C_v, can be calculated from the laboratory test. From Eq. (1.61),

$$T_v = \frac{C_v t}{H^2}$$

For 50% consolidation (Figure 1.24), $T_v = 0.197$, $t = 4.5$ min, and $H = H_c/2 = 12.7$ mm, so

$$C_v = T_{50} \frac{H^2}{t} = \frac{(0.197)(12.7)^2}{4.5} = 7.061 \text{ mm}^2/\text{min}$$

Again, for field consolidation, $U = 45.7\%$. From Eq. (1.66)

$$T_v = \frac{\pi}{4}\left(\frac{U\%}{100}\right)^2 = \frac{\pi}{4}\left(\frac{45.7}{100}\right)^2 = 0.164$$

But

$$T_v = \frac{C_v t}{H^2}$$

or

$$t = \frac{T_v H^2}{C_v} = \frac{0.164\left(\dfrac{2.8 \times 1000}{2}\right)^2}{7.061} = 45{,}523 \text{ min} = \textbf{31.6 days}$$ ∎

1.14   *Degree of Consolidation Under Ramp Loading*

The relationships derived for the average degree of consolidation in Section 1.13 assume that the surcharge load per unit area ($\Delta\sigma$) is applied instantly at time $t = 0$. However, in most practical situations, $\Delta\sigma$ increases gradually with time to a maximum value and remains constant thereafter. Figure 1.25 shows $\Delta\sigma$ increasing linearly with time (t) up to a maximum at time t_c (a condition called ramp loading). For $t \geq t_c$, the magnitude of $\Delta\sigma$ remains constant. Olson (1977) considered

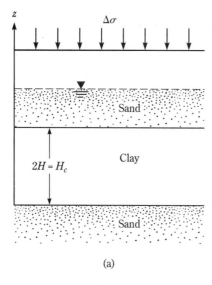

(a)

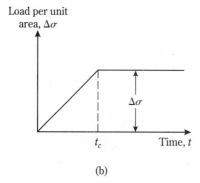

(b)

Figure 1.25 One-dimensional consolidation due to single ramp loading

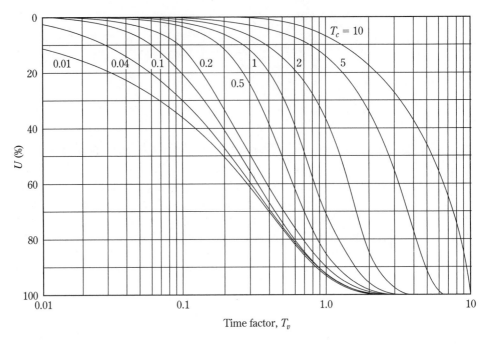

Figure 1.26 Olson's ramp-loading solution: plot of U vs. T_v (Eqs. 1.69 and 1.70)

this phenomenon and presented the average degree of consolidation, U, in the following form:

For $T_v \leq T_c$,

$$U = \frac{T_v}{T_c}\left\{1 - \frac{2}{T_v}\sum_{m=0}^{m=\infty}\frac{1}{M^4}[1 - \exp(-M^2 T_v)]\right\} \tag{1.69}$$

and for $T_v \geq T_c$,

$$U = 1 - \frac{2}{T_c}\sum_{m=0}^{m=\infty}\frac{1}{M^4}[\exp(M^2 T_c) - 1]\exp(-M^2 T_c) \tag{1.70}$$

where m, M, and T_v have the same definition as in Eq. (1.60) and where

$$T_c = \frac{C_v t_c}{H^2} \tag{1.71}$$

Figure 1.26 shows the variation of U with T_v for various values of T_c, based on the solution given by Eqs. (1.69) and (1.70).

Example 1.7

In Example 1.6, Part (b), if the increase in $\Delta\sigma$ would have been in the manner shown in Figure 1.27, calculate the settlement of the clay layer at time $t = 31.6$ days after the beginning of the surcharge.

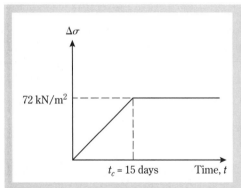

Figure 1.27 Ramp loading

Solution

From Part (b) of Example 1.6, $C_v = 7.061$ mm^2/min. From Eq. (1.71),

$$T_c = \frac{C_v t_c}{H^2} = \frac{(7.061 \text{ mm}^2/\text{min}) (15 \times 24 \times 60 \text{ min})}{\left(\dfrac{2.8}{2} \times 1000 \text{ mm}\right)^2} = 0.0778$$

Also,

$$T_v = \frac{C_v t}{H^2} = \frac{(7.061 \text{ mm}^2/\text{min}) (31.6 \times 24 \times 60 \text{ min})}{\left(\dfrac{2.8}{2} \times 1000 \text{ mm}\right)^2} = 0.164$$

From Figure 1.26, for $T_v = 0.164$ and $T_c = 0.0778$, the value of U is about 36%. Thus,

$$S_{c(t=31.6 \text{ days})} = S_{c(\max)}(0.36) = (87.5)(0.36) = \mathbf{31.5 \text{ mm}}$$ ∎

1.15 *Shear Strength*

The shear strength of a soil, defined in terms of effective stress, is

$$s = c' + \sigma' \tan \phi' \tag{1.72}$$

where σ' = effective normal stress on plane of shearing
c' = cohesion, or apparent cohesion
ϕ' = effective stress angle of friction

Equation (1.72) is referred to as the *Mohr–Coulomb failure criterion*. The value of c' for sands and normally consolidated clays is equal to zero. For overconsolidated clays, $c' > 0$.

For most day-to-day work, the shear strength parameters of a soil (i.e., c' and ϕ') are determined by two standard laboratory tests: the *direct shear test* and the *triaxial test*.

Direct Shear Test

Dry sand can be conveniently tested by direct shear tests. The sand is placed in a shear box that is split into two halves (Figure 1.28a). First a normal load is applied to the specimen. Then a shear force is applied to the top half of the shear box to cause failure in the sand. The normal and shear stresses at failure are

$$\sigma' = \frac{N}{A}$$

and

$$s = \frac{R}{A}$$

where A = area of the failure plane in soil—that is, the cross-sectional area of the shear box

Several tests of this type can be conducted by varying the normal load. The angle of friction of the sand can be determined by plotting a graph of s against σ' ($= \sigma$ for dry sand), as shown in Figure 1.28b, or

$$\phi' = \tan^{-1}\left(\frac{s}{\sigma'}\right) \tag{1.73}$$

For sands, the angle of friction usually ranges from 26° to 45°, increasing with the relative density of compaction. The approximate range of the relative density of compaction and the corresponding range of the angle of friction for various coarse-grained soils is shown in Figure 1.29.

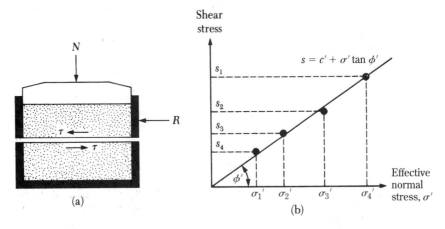

Figure 1.28   Direct shear test in sand: (a) schematic diagram of test equipment; (b) plot of test results to obtain the friction angle ϕ'

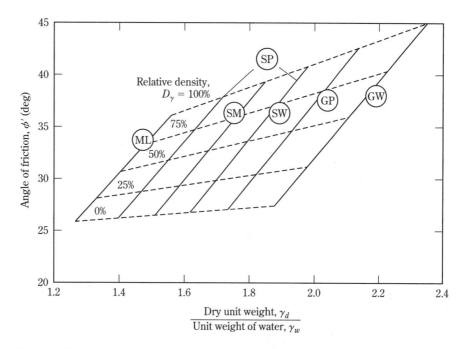

Figure 1.29 Range of relative density and corresponding range of effective stress angle of friction for coarse-grained soil (modified from U.S. Department of the Navy, 1971)

Triaxial Tests

Triaxial compression tests can be conducted on sands and clays Figure 1.30a shows a schematic diagram of the triaxial test arrangement. Essentially, the test consists of placing a soil specimen confined by a rubber membrane into a lucite chamber and then applying an all-around confining pressure (σ_3) to the specimen by means of the chamber fluid (generally, water or glycerin). An added stress ($\Delta\sigma$) can also be applied to the specimen in the axial direction to cause failure ($\Delta\sigma = \Delta\sigma_f$ at failure). Drainage from the specimen can be allowed or stopped, depending on the condition being tested. For clays, three main types of tests can be conducted with triaxial equipment (see Figure 1.31):

1. Consolidated-drained test (CD test)
2. Consolidated-undrained test (CU test)
3. Unconsolidated-undrained test (UU test)

    Consolidated-drained tests:

Step 1. Apply chamber pressure σ_3. Allow complete drainage, so that the pore water pressure ($u = u_o$) developed is zero.

Step 2. Apply a deviator stress $\Delta\sigma$ slowly. Allow drainage, so that the pore water pressure ($u = u_d$) developed through the application of $\Delta\sigma$ is zero. At failure, $\Delta\sigma = \Delta\sigma_f$; the total pore water pressure $u_f = u_o + u_d = 0$.

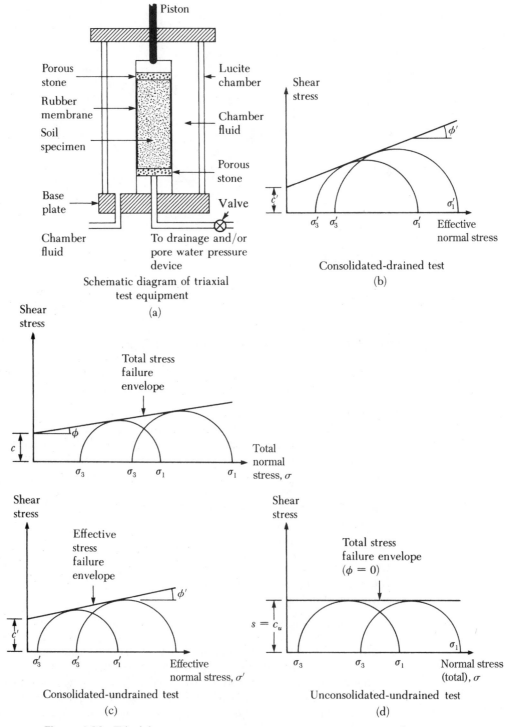

Figure 1.30 Triaxial test

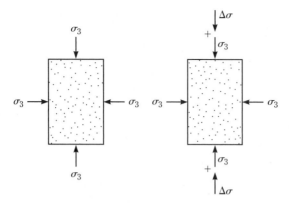

Figure 1.31 Sequence of stress application in triaxial test

So for *consolidated-drained tests,* at failure,

Major principal effective stress $= \sigma_3 + \Delta\sigma_f = \sigma_1 = \sigma_1'$

Minor principal effective stress $= \sigma_3 = \sigma_3'$

Changing σ_3 allows several tests of this type to be conducted on various clay specimens. The shear strength parameters (c' and ϕ') can now be determined by plotting Mohr's circle at failure, as shown in Figure 1.30b, and drawing a common tangent to the Mohr's circles. This is the *Mohr–Coulomb failure envelope.* (*Note:* For normally consolidated clay, $c' \approx 0$.) At failure,

$$\sigma_1' = \sigma_3' \tan^2\left(45 + \frac{\phi'}{2}\right) + 2c' \tan\left(45 + \frac{\phi'}{2}\right) \tag{1.74}$$

Consolidated-undrained tests:

Step 1. Apply chamber pressure σ_3. Allow complete drainage, so that the pore water pressure ($u = u_o$) developed is zero.

Step 2. Apply a deviator stress $\Delta\sigma$. Do not allow drainage, so that the pore water pressure $u = u_d \neq 0$. At failure, $\Delta\sigma = \Delta\sigma_f$; the pore water pressure $u_f = u_o + u_d = 0 + u_{d(f)}$.

Hence, at failure,

Major principal total stress $= \sigma_3 + \Delta\sigma_f = \sigma_1$

Minor principal total stress $= \sigma_3$

Major principal effective stress $= (\sigma_3 + \Delta\sigma_f) - u_f = \sigma_1'$

Minor principal effective stress $= \sigma_3 - u_f = \sigma_3'$

Changing σ_3 permits multiple tests of this type to be conducted on several soil specimens. The total stress Mohr's circles at failure can now be plotted, as shown in

Figure 1.30c, and then a common tangent can be drawn to define the *failure envelope*. This *total stress failure envelope* is defined by the equation

$$s = c + \sigma \tan \phi \tag{1.75}$$

where c and ϕ are the *consolidated-undrained cohesion* and *angle of friction*, respectively (*Note:* $c \approx 0$ for normally consolidated clays)

Similarly, effective stress Mohr's circles at failure can be drawn to determine the *effective stress failure envelope* (Figure 1.30c), which satisfy the relation expressed in Eq. (1.72).

Unconsolidated-undrained tests:

Step 1.   Apply chamber pressure σ_3. Do not allow drainage, so that the pore water pressure ($u = u_o$) developed through the application of σ_3 is not zero.

Step 2.   Apply a deviator stress $\Delta\sigma$. Do not allow drainage ($u = u_d \neq 0$). At failure, $\Delta\sigma = \Delta\sigma_f$; the pore water pressure $u_f = u_o + u_{d(f)}$

For *unconsolidated-undrained* triaxial tests,

$$\text{Major principal total stress} = \sigma_3 + \Delta\sigma_f = \sigma_1$$

$$\text{Minor principal total stress} = \sigma_3$$

The total stress Mohr's circle at failure can now be drawn, as shown in Figure 1.30d. For saturated clays, the value of $\sigma_1 - \sigma_3 = \Delta\sigma_f$ is a constant, irrespective of the chamber confining pressure σ_3 (also shown in Figure 1.30d). The tangent to these Mohr's circles will be a horizontal line, called the $\phi = 0$ condition. The shear stress for this condition is

$$s = c_u = \frac{\Delta\sigma_f}{2} \tag{1.76}$$

where c_u = undrained cohesion (or undrained shear strength)

The pore pressure developed in the soil specimen during the unconsolidated-undrained triaxial test is

$$u = u_a + u_d \tag{1.77}$$

The pore pressure u_a is the contribution of the hydrostatic chamber pressure σ_3. Hence,

$$u_a = B\sigma_3 \tag{1.78}$$

where B = Skempton's pore pressure parameter

Similarly, the pore parameter u_d is the result of the added axial stress $\Delta\sigma$, so

$$u_d = A\Delta\sigma \tag{1.79}$$

where A = Skempton's pore pressure parameter

However,

$$\Delta\sigma = \sigma_1 - \sigma_3 \tag{1.80}$$

Combining Eqs. (1.77), (1.78), (1.79), and (1.80) gives

$$u = u_a + u_d = B\sigma_3 + A(\sigma_1 - \sigma_3) \tag{1.81}$$

The pore water pressure parameter B in soft saturated soils is unity, so

$$u = \sigma_3 + A(\sigma_1 - \sigma_3) \tag{1.82}$$

The value of the pore water pressure parameter A at failure will vary with the type of soil. Following is a general range of the values of A at failure for various types of clayey soil encountered in nature:

Type of soil	A at failure
Sandy clays	0.5–0.7
Normally consolidated clays	0.5–1
Overconsolidated clays	−0.5–0

1.16   *Unconfined Compression Test*

The *unconfined compression test* (Figure 1.32a) is a special type of unconsolidated-undrained triaxial test in which the confining pressure $\sigma_3 = 0$, as shown in Figure 1.32b. In this test, an axial stress $\Delta\sigma$ is applied to the specimen to cause failure

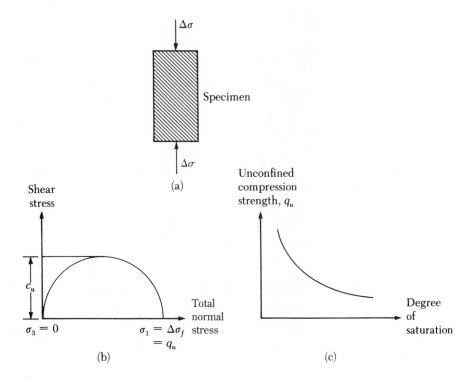

Figure 1.32 Unconfined compression test: (a) soil specimen; (b) Mohr's circle for the test; (c) variation of q_u with the degree of saturation

(i.e., $\Delta\sigma = \Delta\sigma_f$). The corresponding Mohr's circle is shown in Figure 1.32b. Note that, for this case,

$$\text{Major principal total stress} = \Delta\sigma_f = q_u$$

$$\text{Minor principal total stress} = 0$$

The axial stress at failure, $\Delta\sigma_f = q_u$, is generally referred to as the *unconfined compression strength*. The shear strength of saturated clays under this condition ($\phi = 0$), from Eq. (1.72), is

$$s = c_u = \frac{q_u}{2} \tag{1.83}$$

The unconfined compression strength can be used as an indicator of the consistency of clays.

Unconfined compression tests are sometimes conducted on unsaturated soils. With the void ratio of a soil specimen remaining constant, the unconfined compression strength rapidly decreases with the degree of saturation (Figure 1.32c). Figure 1.33 shows an unconfined compression test in progress.

Figure 1.33 Unconfined compression test in progress (courtesy of Soiltest, Inc., Lake Bluff, Illinois)

1.17 *Comments on Friction Angle, ϕ'*

Effective Stress Friction Angle of Granular Soils

In general, the direct shear test yields a higher angle of friction compared with that obtained by the triaxial test. Also, note that the failure envelope for a given soil is actually curved. The Mohr–Coulomb failure criterion defined by Eq. (1.72) is only an approximation. Because of the curved nature of the failure envelope, a soil tested at higher normal stress will yield a lower value of ϕ'. An example of this relationship is shown in Figure 1.34, which is a plot of ϕ' versus the void ratio e for Chattachoochee River sand near Atlanta, Georgia (Vesic, 1963). The friction angles shown were obtained from triaxial tests. Note that, for a given value of e, the magnitude of ϕ' is about 4° to 5° smaller when the confining pressure σ_3' is greater than about 70 kN/m² (10 lb/in²), compared with that when $\sigma_3' < 70$ kN/m².

Effective Stress Friction Angle of Cohesive Soils

Figure 1.35 shows the effective stress friction angle, ϕ', for several normally consolidated clays, obtained by conducting triaxial tests (Bjerrum and Simons, 1960). It can be

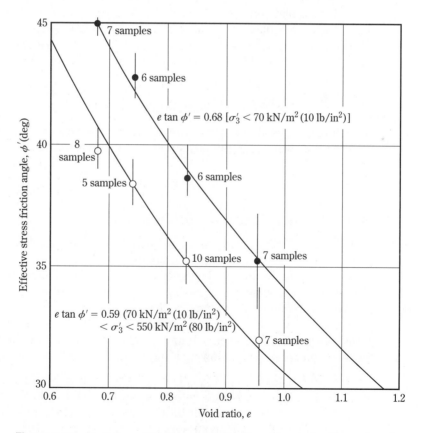

Figure 1.34 Variation of friction angle ϕ' with void ratio for Chattachoochee River sand (after Vesic, 1963)

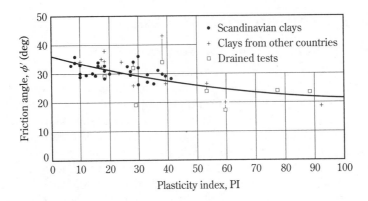

Figure 1.35 Variation of friction angle ϕ' with plasticity index for several clays (after Bjerrum and Simons, 1960)

seen from the figure that, in general, the friction angle ϕ' decreases with the increase in plasticity index. The value of ϕ' generally decreases from about 37°–38° with a plasticity index of about 10 to about 25° or less with a plasticity index of about 100. Similar results were also provided by Kenney (1959). The consolidated undrained friction angle (ϕ') of normally consolidated saturated clays generally ranges from 5°–20°.

The consolidated drained triaxial test was described in Section 1.15. Figure 1.36 shows a schematic diagram of a plot of $\Delta\sigma$ versus axial strain in a drained triaxial test for a clay. At failure, for this test, $\Delta\sigma = \Delta\sigma_f$. However, at large axial strain (i.e., the ultimate strength condition), we have the following relationships:

Major principal stress: $\sigma'_{1(\text{ult})} = \sigma_3 + \Delta\sigma_{\text{ult}}$

Minor principal stress: $\sigma'_{3(\text{ult})} = \sigma_3$

At failure (i.e., peak strength), the relationship between σ'_1 and σ'_3 is given by Eq. (1.74). However, for ultimate strength, it can be shown that

$$\sigma'_{1(\text{ult})} = \sigma'_3 \tan^2\left(45 + \frac{\phi'_r}{2}\right) \tag{1.84}$$

where ϕ'_r = residual effective stress friction angle

Figure 1.37 shows the general nature of the failure envelopes at peak strength and ultimate strength (or *residual strength*). The residual shear strength of clays is important in the evaluation of the long-term stability of new and existing slopes and the design of remedial measures. The effective stress residual friction angles ϕ'_r of clays may be substantially smaller than the effective stress peak friction angle ϕ'. Past research has shown that the clay fraction (i.e., the percent finer than 2 microns) present in a given soil, CF, and the clay mineralogy are the two primary factors that control ϕ'_r. The following is a summary of the effects of CF on ϕ'_r.

1. If CF is less than about 15%, then ϕ'_r is greater than about 25°.
2. For CF > about 50%, ϕ'_r is entirely governed by the sliding of clay minerals and may be in the range of about 10° to 15°.
3. For kaolinite, illite, and montmorillonite, ϕ'_r is about 15°, 10°, and 5°, respectively.

Illustrating these facts, Figure 1.38 shows the variation of ϕ'_r with CF for several soils (Skempton, 1985).

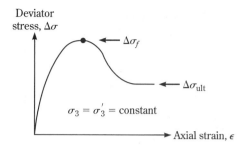

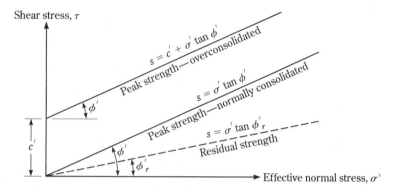

Figure 1.36 Plot of deviator stress vs. axial strain–drained triaxial test

Figure 1.37 Peak- and residual-strength envelopes for clay

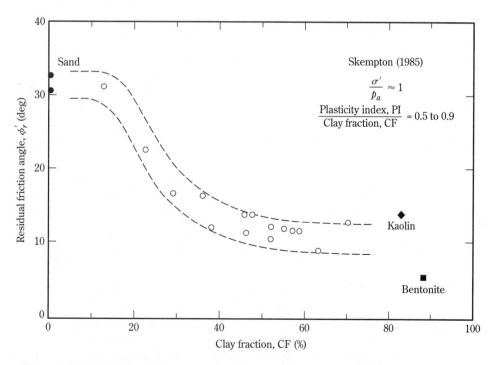

Figure 1.38 Variation of ϕ'_r with CF (*Note:* p_a = atmospheric pressure)

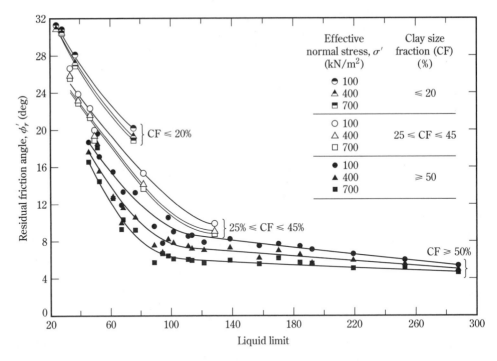

Figure 1.39 Variation of ϕ'_r with liquid limit for some clays (after Stark, 1995)

Figure 1.39 shows the variation of ϕ'_r with the liquid limit for some clays (Stark, 1995). It is important to note that

1. For a given clay, ϕ'_r decreases with the increase in liquid limit.
2. For a given liquid limit and given clay-size fractions present in the soil, the magnitude of ϕ'_r decreases with the increase in the normal effective stress, due to the curvilinear nature of the failure envelope.

1.18 *Correlations for Undrained Shear Strength, c_u*

The undrained shear strength, c_u, is an important parameter in the design of foundations. For normally consolidated clay deposits (Figure 1.40), the magnitude of c_u increases almost linearly with the increase in effective overburden pressure.

There are several empirical relations between c_u and the effective overburden pressure σ'_o in the field. Some of these relationships are summarized in Table 1.10.

1.19 *Sensitivity*

For many naturally deposited clay soils, the unconfined compression strength is much less when the soils are tested after remolding without any change in the moisture content. This property of clay soil is called *sensitivity*. The degree of sensitivity

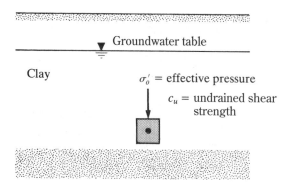

Figure 1.40 Clay deposit

Table 1.10 Empirical Equations Related to c_u and Effective Overburden Pressure

Reference	Relationship	Remarks
Skempton (1957)	$\dfrac{c_{u(\text{VST})}}{\sigma_o'} = 0.11 + 0.0037\text{PI}$ PI = plasticity index (%) $c_{u(\text{VST})}$ = undrained shear strength from vane shear test (See Chapter 3 for details of vane shear test)	For normally consolidated clay
Chandler (1988)	$\dfrac{c_{u(\text{VST})}}{\sigma_c'} = 0.11 + 0.0037\text{PI}$ σ_c' = preconsolidation pressure	Can be used for overconsolidated soil Accuracy ±25% Not valid for sensitive and fissured clays
Jamiolkowski et al. (1985)	$\dfrac{c_u}{\sigma_c'} = 0.23 \pm 0.04$	For lightly overconsolidated clays
Mesri (1989)	$\dfrac{c_u}{\sigma_o'} = 0.22$	
Bjerrum and Simons (1960)	$\dfrac{c_u}{\sigma_o'} = f(\text{LI})$ LI = liquidity index [See Eq. (1.47) for definition]	See Figure 1.41 (p. 58) for normally consolidated clays
Ladd et al. (1977)	$\dfrac{\left(\dfrac{c_u}{\sigma_o'}\right)_{\text{overconsolidated}}}{\left(\dfrac{c_u}{\sigma_o'}\right)_{\text{normally consolidated}}} = (\text{OCR})^{0.8}$ OCR = overconsolidation ratio = $\dfrac{\sigma_c'}{\sigma_o'}$	

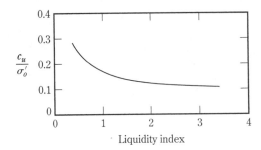

Figure 1.41 Variation of c_u/σ'_o with liquidity index (based on Bjerrum and Simons 1960)

is the ratio of the unconfined compression strength in an undisturbed state to that in a remolded state, or

$$S_t = \frac{q_{u(\text{undisturbed})}}{q_{u(\text{remolded})}} \tag{1.85}$$

The sensitivity ratio of most clays ranges from about 1 to 8; however, highly flocculent marine clay deposits may have sensitivity ratios ranging from about 10 to 80. Some clays turn to viscous liquids upon remolding, and these clays are referred to as "quick" clays. The loss of strength of clay soils from remolding is caused primarily by the destruction of the clay particle structure that was developed during the original process of sedimentation.

Problems

1.1 A moist soil has a void ratio of 0.7. The moisture content of the soil is 12%. If $G_s = 2.7$, determine the soil's (a) porosity, (b) degree of saturation, and (c) dry unit weight in kN/m^3.

1.2 For the soil described in Problem 1.1,
   a. What would be the saturated unit weight in kN/m^3?
   b. How much water, in kN/m^3, needs to be added to the soil for complete saturation?
   c. What would be the moist unit weight, in kN/m^3, when the degree of saturation is 70%?

1.3 A soil specimen has a volume of 1.75 ft^3 and a mass of 193 lb. Let $w = 15\%$ and $G_s = 2.67$. Determine the specimen's (a) void ratio, (b) porosity, (c) dry unit weight, (d) moist unit weight, and (e) degree of saturation.

1.4 A saturated soil specimen has $w = 36\%$ and $\gamma_d = 13.5 \, kN/m^3$. Determine the specimen's (a) void ratio, (b) porosity, (c) specific gravity of soil solids, and (d) saturated unit weight (in kN/m^3).

1.5 The laboratory test results on a sand are as follows: $e_{max} = 0.91$, $e_{min} = 0.48$, and $G_s = 2.67$. What would be the dry and moist unit weights of this sand (in lb/ft^3) when compacted at a moisture content of 10% to a relative density of 65%?

1.6   For a granular soil, $\gamma = 17 \text{ kN/m}^3$, $D_r = 60\%$, $w = 8\%$, and $G_s = 2.66$. If e_{min} is 0.4, what is e_{max}? What is the dry unit weight of the soil in the loosest state?

1.7   Laboratory test results on six soils are given in the following table:

<table>
<tr><th colspan="7">Sieve analysis—percent passing</th></tr>
<tr><th></th><th colspan="6">Soil</th></tr>
<tr><th>Sieve No.</th><th>A</th><th>B</th><th>C</th><th>D</th><th>E</th><th>F</th></tr>
<tr><td>4</td><td>92</td><td>100</td><td>100</td><td>95</td><td>100</td><td>100</td></tr>
<tr><td>10</td><td>48</td><td>60</td><td>98</td><td>90</td><td>91</td><td>82</td></tr>
<tr><td>40</td><td>28</td><td>41</td><td>82</td><td>79</td><td>80</td><td>74</td></tr>
<tr><td>200</td><td>13</td><td>33</td><td>72</td><td>64</td><td>30</td><td>55</td></tr>
<tr><td>Liquid limit</td><td>31</td><td>38</td><td>56</td><td>35</td><td>43</td><td>35</td></tr>
<tr><td>Plastic limit</td><td>26</td><td>25</td><td>31</td><td>26</td><td>29</td><td>21</td></tr>
</table>

Classify the soils by the AASHTO Soil Classification System and give the group indices.

1.8   Classify the soils given in Problem 1.7 by the Unified Soil Classification System. Give group symbols and group names.

1.9   The hydraulic conductivity of a sand was tested in the laboratory at a void ratio of 0.56 and was determined to be 0.14 cm/sec. Estimate the hydraulic conductivity of this sand at a void ratio of 0.79 by using Eqs. (1.26), (1.27), and (1.28).

1.10   The *in situ* hydraulic conductivity of a clay is 5.4×10^{-6} cm/sec at a void ratio of 0.92. What would be the hydraulic conductivity of the clay at a void ratio of 0.72? Use Eq. (1.30).

1.11   From the soil profile shown in Figure P1.11, determine the total stress, pore water pressure, and effective stress at *A*, *B*, *C*, and *D*.

1.12   A sandy soil ($G_s = 2.66$) has void ratios of 0.42 and 0.97 in its densest and loosest states, respectively. Estimate the range of the critical hydraulic gradient in this soil at which it might turn into quicksand.

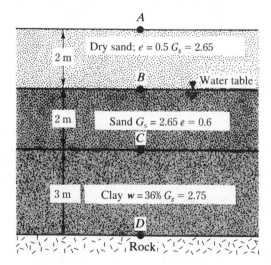

Figure P1.11

1.13 A normally consolidated clay layer 2.6 m thick has a void ratio of 1.3. Let LL = 41 and the average effective stress on the clay layer = 82 kN/m². How much consolidation settlement would the clay layer undergo if the average effective stress on it is increased to 120 kN/m² as the result of the construction of a foundation?

1.14 Assume that the clay layer in Problem 1.13 is preconsolidated. Let $\sigma'_c = 95$ kN/m² and $C_s = 1/4C_c$. Estimate the consolidation settlement.

1.15 Suppose the clay in Figure P1.11 is normally consolidated. A laboratory consolidation test on the clay gave the following results:

Pressure (kN/m²)	Void ratio
100	0.905
200	0.815

a. Calculate the average effective stress on the clay layer.
b. Determine the compression index C_c.
c. If the average effective stress on the clay layer is increased $(\sigma'_o + \Delta\sigma')$ to 115 kN/m², what would be the total consolidation settlement?

1.16 For the clay soil in Problem 1.15c, $C_v = 5.6$ mm²/min. How long will it take the clay to reach half the consolidation settlement? (*Note:* The clay layer in the field is drained on one side only.)

1.17 A clay specimen 1 in. thick (drained on top only) was tested in the laboratory. For a given load increment, the time for 60% consolidation was 6 min. How long will it take for 50% consolidation for a similar clay layer in the field that is 8 ft thick and drained on both sides?

1.18 A total of 60 mm consolidation settlement is expected in the two clay layers shown in Figure P1.18, due to a surcharge of $\Delta\sigma$. Find the duration of surcharge application at which 30 mm of total settlement would take place.

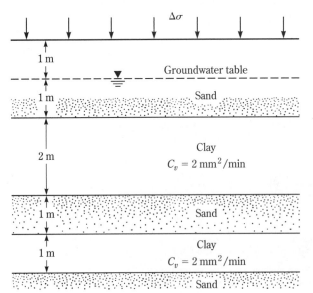

Figure P1.18

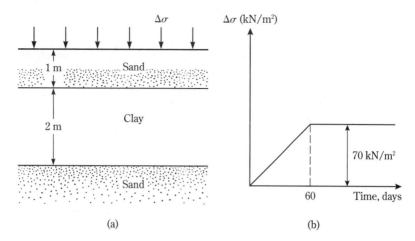

Figure P1.19

1.19 The coefficient of consolidation of a clay for a given pressure range was obtained as 8×10^{-3} mm²/sec on the basis of one-dimensional consolidation test results. In the field, there is a 2-m-thick layer of the same clay (Figure P1.19a). Based on the assumption that a uniform surcharge of 70 kN/m² was to be applied instantaneously, the total consolidation settlement was estimated to be 150 mm. However, during construction, the loading was gradual; the resulting surcharge can be approximated as shown in Figure P1.19b. Estimate the settlement at $t = 30$ and $t = 120$ days after the beginning of construction.

1.20 A direct shear test was conducted on a specimen of dry sand of area 2 in. × 2 in. The results are as follows:

Normal force (lb)	Shear force at failure (lb)
33	20.7
55.2	35.8
66.2	40.2

Draw a graph of shear stress at failure vs. normal stress, and determine the soil friction angle ϕ'.

1.21 A consolidated-drained triaxial test on a normally consolidated clay yielded the following results:

$$\text{All-around confining pressure} = \sigma_3 = 115 \text{ kN/m}^2$$
$$\text{Added axial stress at failure} = \Delta\sigma = 230 \text{ kN/m}^2$$

Determine the shear stress parameter ϕ'.

1.22 A consolidated-drained triaxial test on a normally consolidated clay yielded a friction angle ϕ' of 28°. If the all-around confining pressure during the test was 15 lb/in.², what was the major principal stress at failure?

1.23 Following are the results of two consolidated-drained triaxial tests on a clay:

$$\text{Test I: } \sigma_3 = 82.8 \text{ kN/m}^2; \sigma_{1(\text{failure})} = 329.2 \text{ kN/m}^2$$
$$\text{Test II: } \sigma_3 = 165.6 \text{ kN/m}^2; \sigma_{1(\text{failure})} = 558.6 \text{ kN/m}^2$$

Determine the shear strength parameters c' and ϕ'.

1.24 A consolidated-undrained triaxial test was conducted on a normally consolidated saturated clay. The following are the test results:

$$\sigma_3 = 13 \text{ lb/in.}^2;$$
$$\sigma_{1(\text{failure})} = 32 \text{ lb/in.}^2$$

The pore water pressure at failure, u_f, = 5.5 lb/in.^{2}
Determine c, ϕ, c', and ϕ'.

References

American Society for Testing and Materials (2000). *Annual Book of ASTM Standards,* Vol. 04.08, Conshohocken, PA.

Bjerrum, L., and Simons, N. E. (1960). "Comparison of Shear Strength Characteristics of Normally Consolidated Clay," *Proceedings, Research Conference on Shear Strength of Cohesive Soils,* ASCE, 711–726.

Casagrande, A. (1936). "Determination of the Preconsolidation Load and Its Practical Significance," *Proceedings, First International Conference on Soil Mechanics and Foundation Engineering,* Cambridge, MA, Vol. 3, pp. 60–64.

Chandler, R. J. (1988). "The *In Situ* Measurement of the Undrained Shear Strength of Clays Using the Field Vane," *STP 1014, Vane Shear Strength Testing in Soils: Field and Laboratory Studies,* ASTM, pp. 13–44.

Darcy, H. (1856). *Les Fontaines Publiques de la Ville de Dijon,* Paris.

Das, B. M. (2002). *Soil Mechanics Laboratory Manual,* 6th ed., Oxford University Press, New York.

Highway Research Board (1945). *Report of the Committee on Classification of Materials for Subgrades and Granular Type Roads,* Vol. 25, pp. 375–388.

Jamiolkowski, M., Ladd, C. C., Germaine, J. T., and Lancellotta, R. (1985). "New Developments in Field and Laboratory Testing of Soils," *Proceedings, XI International Conference on Soil Mechanics and Foundation Engineering,* San Francisco, Vol. 1, pp. 57–153.

Kenney, T. C. (1959). "Discussion," *Journal of the Soil Mechanics and Foundations Division,* American Society of Civil Engineers, Vol. 85, No. SM3, pp. 67–69.

Ladd, C. C., Foote, R., Ishihara, K., Schlosser, F., and Poulos, H. G. (1977). "Stress Deformation and Strength Characteristics," *Proceedings, Ninth International Conference on Soil Mechanics and Foundation Engineering,* Tokyo, Vol. 2, 421–494.

Mesri, G. (1989). "A Re-evaluation of $S_{u(\text{mob})} \approx 0.22\sigma_p$ Using Laboratory Shear Tests," *Canadian Geotechnical Journal,* Vol. 26, No. 1, pp. 162–164.

Olson, R. E. (1977). "Consolidation Under Time-Dependent Loading," *Journal of Geotechnical Engineering,* ASCE, Vol. 103, No. GT1, pp. 55–60.

Samarasinghe, A. M., Huang, Y. H., and Drnevich, V. P. (1982). "Permeability and Consolidation of Normally Consolidated Soils," *Journal of the Geotechnical Engineering Division,* ASCE, Vol. 108, No. GT6, 835–850.

Schmertmann, J. H. (1953). "Undisturbed Consolidation Behavior of Clay," *Transactions,* American Society of Civil Engineers, Vol. 120, p. 1201.

Sivaram, B., and Swamee, P. (1977). "A Computational Method for Consolidation Coefficient," *Soils and Foundations,* Tokyo, Vol. 17, No. 2, pp. 48–52.

Skempton, A. W. (1944). "Notes on the Compressibility of Clays," *Quarterly Journal of Geological Society,* London, Vol. C, pp. 119–135.

Skempton, A. W. (1957). "The Planning and Design of New Hong Kong Airport," *Proceedings, The Institute of Civil Engineers,* London, Vol. 7, pp. 305–307.

Skempton, A. W. (1985). "Residual Strength of Clays in Landslides, Folded Strata, and the Laboratory," *Geotechnique,* Vol. 35, No. 1, pp. 3–18.

Stark, T. D. (1995). "Measurement of Drained Residual Strength of Overconsolidated Clays," *Transportation Research Record No. 1479,* National Research Council, Washington, DC, pp. 26–34.

Tavenas, F., Jean, P., Leblond, P., and Leroueil, S. (1983). "The Permeability of Natural Soft Clays. Part II: Permeability Characteristics," *Canadian Geotechnical Journal,* Vol. 20, No. 4, pp. 645–660.

Terzaghi, K., and Peck, R. B. (1967). *Soil Mechanics in Engineering Practice,* Wiley, New York.

U.S. Department of the Navy (1971). *Design Manual—Soil Mechanics, Foundations and Earth Structures,* NAVFAC DM-7, U.S. Government Printing Office, Washington, DC.

U.S. Department of the Navy (1982). *Soil Mechanics,* NAVFAC DM7.1, U.S. Government Printing Office, Washington, DC.

Vesic, A. S. (1963). "Bearing Capacity of Deep Foundations in Sand," *Highway Research Record No. 39,* National Academy of Sciences, Washington, DC., pp. 112–154.

2

Natural Soil Deposits and Subsoil Exploration

2.1 Introduction

To design a foundation that will support a structure, an engineer must understand the types of soil deposits that will support the foundation. Moreover, foundation engineers must remember that soil at any site frequently is non-homogeneous; that is, the soil profile may vary. Soil mechanics theories involve idealized conditions, so the application of the theories to foundation engineering problems involves a judicious evaluation of site conditions and soil parameters. To do this requires some knowledge of the geological process by which the soil deposit at the site was formed, supplemented by subsurface exploration. Good professional judgment constitutes an essential part of geotechnical engineering—and it comes only with practice.

This chapter is divided into two parts. The first is a general overview of natural soil deposits generally encountered, and the second describes the general principles of subsoil exploration.

Natural Soil Deposits

2.2 Soil Origin

Most of the soils that cover the earth are formed by the weathering of various rocks. There are two general types of weathering: (1) mechanical weathering and (2) chemical weathering.

Mechanical weathering is the process by which rocks are broken into smaller and smaller pieces by physical forces, including running water, wind, ocean waves, glacier ice, frost, and expansion and contraction caused by the gain and loss of heat.

Chemical weathering is the process of chemical decomposition of the original rock. In the case of mechanical weathering, the rock breaks into smaller pieces without a change in its chemical composition. However, in chemical weathering, the original material may be changed to something entirely different. For example, the chemical weathering of feldspar can produce clay minerals. Most rock weathering is a combination of mechanical and chemical weathering.

Soil produced by the weathering of rocks can be transported by physical processes to other places. The resulting soil deposits are called *transported soils*. In contrast, some soils stay where they were formed and cover the rock surface from which they derive. These soils are referred to as *residual soils*.

Transported soils can be subdivided into three major categories based on the *transporting agent:*

1. *Alluvial,* or *fluvial:* deposited by running water
2. *Glacial:* deposited by glaciers
3. *Aeolian:* deposited by the wind

In addition to transported and residual soils, there are *peats* and *organic soils,* which derive from the decomposition of organic materials.

2.3 Residual Soil

Residual soil deposits are common in the tropics, on islands such as the Hawaiian islands, and in the southeastern United States. The nature of a residual soil deposit will generally depend on the parent rock. When hard rocks such as granite and gneiss undergo weathering, most of the materials are likely to remain in place. These soil deposits generally have a top layer of clayey or silty clay material, below which are silty or sandy soil layers. These layers in turn are generally underlain by a partially weathered rock and then sound bedrock. The depth of the sound bedrock may vary widely, even within a distance of a few meters.

In contrast to hard rocks, some chemical rocks, such as limestone, are made up chiefly of the mineral calcite ($CaCO_3$). Chalk has a large concentration of the mineral dolomite [$CaMg(CO_3)_2$]. Chemical rocks have large amounts of soluble materials, some of which are removed by groundwater, leaving behind the insoluble fraction of the rock. Residual soils that derive from chemical rocks possess a gradual transition zone to the bedrock. The residual soils derived from the weathering of limestonelike rocks are mostly gray in color. Although uniform in kind, the depth of weathering may vary greatly. The residual soils immediately above the bedrock may be normally consolidated. Large foundations with heavy loads may be susceptible to large consolidation settlements on these soils.

2.4 Alluvial Deposits

Alluvial soil deposits derive from the action of streams and rivers and can be divided into two major categories: (1) *braided-stream deposits* and (2) deposits caused by the *meandering belt of streams.*

Deposits from Braided Streams

Braided streams are high-gradient, rapidly flowing streams that are highly erosive and carry large amounts of sediment. Because of the high bed load, a minor change in the velocity of flow will cause sediments to deposit. By this process, these streams

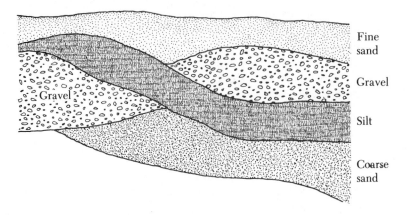

Figure 2.1 Cross section of a braided-stream deposit

may build up a complex tangle of converging and diverging channels separated by sandbars and islands.

The deposits formed from braided streams are highly irregular in stratification and have a wide range of grain sizes. Figure 2.1 shows a cross section of such a deposit. These deposits share several characteristics:

1. The grain sizes usually range from gravel to silt. Clay-size particles are generally *not* found in deposits from braided streams.
2. Although grain size varies widely, the soil in a given pocket or lens is rather uniform.
3. At any given depth, the void ratio and unit weight may vary over a wide range within a lateral distance of only a few meters. This variation can be observed during soil exploration for the construction of a foundation for a structure. The standard penetration resistance at a given depth obtained from various boreholes will be highly irregular and variable.

Alluvial deposits are present in several parts of the western United States, such as Southern California, Utah, and the basin and range sections of Nevada. Also, a large amount of sediment originally derived from the Rocky Mountain range was carried eastward to form the alluvial deposits of the Great Plains. On a smaller scale, this type of natural soil deposit, left by braided streams, can be encountered locally.

Meander Belt Deposits

The term *meander* is derived from the Greek word *maiandros,* after the Maiandros (now Menderes) River in Asia, famous for its winding course. Mature streams in a valley curve back and forth. The valley floor in which a river meanders is referred to as the *meander belt.* In a meandering river, the soil from the bank is continually eroded from the points where it is concave in shape and is deposited at points where the bank is convex in shape, as shown in Figure 2.2. These deposits are called *point bar deposits,* and they usually consist of sand and silt-size particles. Sometimes, during the process of erosion and deposition, the river abandons a meander and cuts a shorter path. The abandoned meander, when filled with water, is called an *oxbow lake.* (See Figure 2.2.)

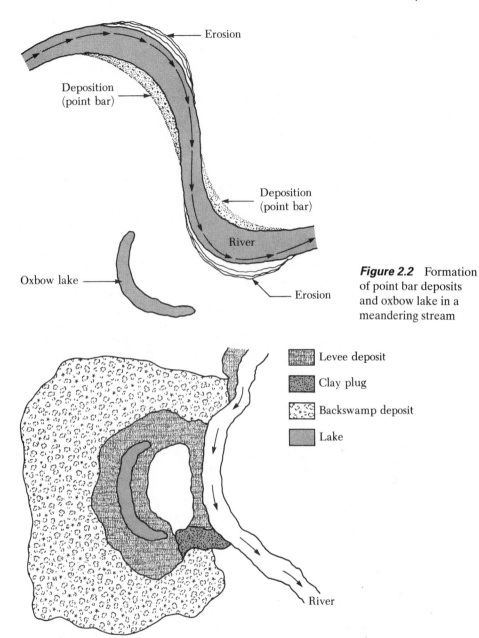

Figure 2.2 Formation of point bar deposits and oxbow lake in a meandering stream

Figure 2.3 Levee and backswamp deposit

During floods, rivers overflow low-lying areas. The sand and silt-size particles carried by the river are deposited along the banks to form ridges known as *natural levees* (Figure 2.3). Finer soil particles consisting of silts and clays are carried by the water farther onto the floodplains. These particles settle at different rates to form what is referred to as *backswamp deposits* (Figure 2.3), often highly plastic clays.

2.5	*Glacial Deposits*

During the Pleistocene Ice Age, glaciers covered large areas of the earth. The glaciers advanced and retreated with time. During their advance, the glaciers carried large amounts of sand, silt, clay, gravel, and boulders. *Drift* is a general term usually applied to the deposits laid down by glaciers. Unstratified deposits laid down by melting glaciers are referred to as *till*. The physical characteristics of till may vary from glacier to glacier.

The landforms that developed from the deposits of till are called *moraines*. A *terminal moraine* (Figure 2.4) is a ridge of till that marks the maximum limit of a glacier's advance. *Recessional moraines* are ridges of till developed behind the terminal moraine at varying distances apart. They are the result of temporary stabilization of the glacier during the recessional period. The till deposited by the glacier between the moraines is referred to as *ground moraine* (Figure 2.4). Ground moraines constitute large areas of the central United States and are called *till plains*.

The sand, silt, and gravel that are carried by the melting water from the front of a glacier are called *outwash*. In a pattern similar to that of braided-stream deposits, the melted water deposits the outwash, forming *outwash plains* (Figure 2.4), also called *glaciofluvial deposits*. The range of grain sizes present in a till varies greatly.

Glacial water also carries with it silts and clays. The water finds its way to many basins and forms lakes. The silt particles initially tend to settle to the bottom of the lake when the water is still. During the winter, when the top of the lake freezes, the suspended clay particles gradually settle to the bottom. During the summer, the snow on the lake melts. The supply of fresh water, loaded with sediments, repeats the process. As a result, the lacustrine soil formed from such a deposit has alternate layers of silt and clay. This soil is called *varved clay*. The varves are usually a few millimeters thick; however, in some instances they can be 50–100 mm (2–4 in.) thick. Varved clays can be found in the Northeast and the Pacific Northwest of the United States. They are mostly normally consolidated and may be sensitive. The hydraulic

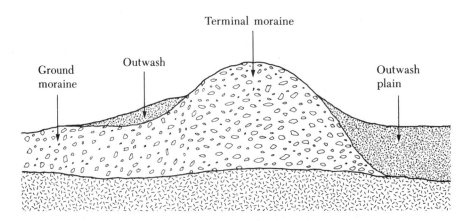

Figure 2.4 Terminal moraine, ground moraine, and outwash plain

conductivity in the vertical direction is usually several times smaller than that in the horizontal direction. The load-bearing capacity of these deposits is quite low, and significant settlement of structures with shallow foundations may be anticipated.

2.6 Aeolian Soil Deposits

Wind is also a major transporting agent leading to the formation of soil deposits. When large areas of sand lie exposed, wind can blow the sand away and redeposit it elsewhere. Deposits of windblown sand generally take the shape of *dunes* (Figure 2.5). As dunes are formed, the sand is blown over the crest by the wind. Beyond the crest, the sand particles roll down the slope. The process tends to form a *compact sand deposit* on the *windward side,* and a rather *loose deposit* on the *leeward side,* of the dune.

Dunes exist along the southern and eastern shores of Lake Michigan, the Atlantic Coast, the southern coast of California, and at various places along the coasts of Oregon and Washington. Sand dunes can also be found in the alluvial and rocky plains of the western United States. Following are some of the typical properties of *dune sand:*

1. The grain-size distribution of the sand at any particular location is surprisingly uniform. This uniformity can be attributed to the sorting action of the wind.
2. The general grain size decreases with distance from the source, because the wind carries the small particles farther than the large ones.
3. The relative density of sand deposited on the windward side of dunes may be as high as 50–65%, decreasing to about 0–15% on the leeward side.

Loess is an aeolian deposit consisting of silt and silt-size particles. The grain-size distribution of loess is rather uniform. The cohesion of loess is generally derived from a clay coating over the silt-size particles, which contributes to a stable soil structure in an unsaturated state. The cohesion may also be the result of the precipitation of chemicals leached by rainwater. Loess is a *collapsing* soil, because when the soil becomes saturated, it loses its binding strength between particles. Special precautions need to be taken for the construction of foundations over loessial deposits. There are extensive deposits of loess in the United States, mostly in the midwestern states of Iowa, Missouri, Illinois, and Nebraska and for some distance along the Mississippi River in Tennessee and Mississippi.

2.7 Organic Soil

Organic soils are usually found in low-lying areas where the water table is near or above the ground surface. The presence of a high water table helps in the growth of aquatic plants that, when decomposed, form organic soil. This type of soil deposit is usually encountered in coastal areas and in glaciated regions. Organic soils show the following characteristics:

1. Their natural moisture content may range from 200% to 300%.
2. They are highly compressible.
3. Laboratory tests have shown that, under loads, a large amount of settlement is derived from secondary consolidation.

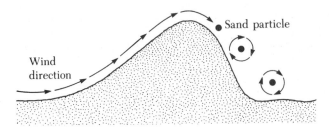

Figure 2.5   Sand dune

Subsurface Exploration

2.8   *Purpose of Subsurface Exploration*

The process of identifying the layers of deposits that underlie a proposed structure and their physical characteristics is generally referred to as *subsurface exploration*. The purpose of subsurface exploration is to obtain information that will aid the geotechnical engineer in

1. Selecting the type and depth of foundation suitable for a given structure.
2. Evaluating the load-bearing capacity of the foundation.
3. Estimating the probable settlement of a structure.
4. Determining potential foundation problems (e.g., expansive soil, collapsible soil, sanitary landfill, and so on).
5. Determining the location of the water table.
6. Predicting the lateral earth pressure for structures such as retaining walls, sheet pile bulkheads, and braced cuts.
7. Establishing construction methods for changing subsoil conditions.

Subsurface exploration may also be necessary when additions and alterations to existing structures are contemplated.

2.9   *Subsurface Exploration Program*

Subsurface exploration comprises several steps, including the collection of preliminary information, reconnaissance, and site investigation.

Collection of Preliminary Information

This step involves obtaining information regarding the type of structure to be built and its general use. For the construction of buildings, the approximate column loads and their spacing and the local building-code and basement requirements should be known. The construction of bridges requires determining the lengths of their spans and the loading on piers and abutments.

A general idea of the topography and the type of soil to be encountered near and around the proposed site can be obtained from the following sources:

1. United States Geological Survey maps.
2. State government geological survey maps.

3. United States Department of Agriculture's Soil Conservation Service county soil reports.
4. Agronomy maps published by the agriculture departments of various states.
5. Hydrological information published by the United States Corps of Engineers, including records of stream flow, information on high flood levels, tidal records, and so on.
6. Highway department soil manuals published by several states.

The information collected from these sources can be extremely helpful in planning a site investigation. In some cases, substantial savings may be realized by anticipating problems that may be encountered later in the exploration program.

Reconnaissance

The engineer should always make a visual inspection of the site to obtain information about

1. The general topography of the site, the possible existence of drainage ditches, abandoned dumps of debris, and other materials present at the site. Also, evidence of creep of slopes and deep, wide shrinkage cracks at regularly spaced intervals may be indicative of expansive soils.
2. Soil stratification from deep cuts, such as those made for the construction of nearby highways and railroads.
3. The type of vegetation at the site, which may indicate the nature of the soil. For example, a mesquite cover in central Texas may indicate the existence of expansive clays that can cause foundation problems.
4. High-water marks on nearby buildings and bridge abutments.
5. Groundwater levels, which can be determined by checking nearby wells.
6. The types of construction nearby and the existence of any cracks in walls or other problems.

The nature of the stratification and physical properties of the soil nearby can also be obtained from any available soil-exploration reports on existing structures.

Site Investigation

The site investigation phase of the exploration program consists of planning, making test boreholes, and collecting soil samples at desired intervals for subsequent observation and laboratory tests. The approximate required minimum depth of the borings should be predetermined. The depth can be changed during the drilling operation, depending on the subsoil encountered. To determine the approximate minimum depth of boring, engineers may use the rules established by the American Society of Civil Engineers (1972):

1. Determine the net increase in the effective stress, $\Delta\sigma'$, under a foundation with depth as shown in Figure 2.6. (The general equations for estimating increases in stress are given in Chapter 5.)
2. Estimate the variation of the vertical effective stress, σ'_o, with depth.
3. Determine the depth, $D = D_1$, at which the effective stress increase $\Delta\sigma'$ is equal to $(\frac{1}{10})q$ (q = estimated net stress on the foundation).

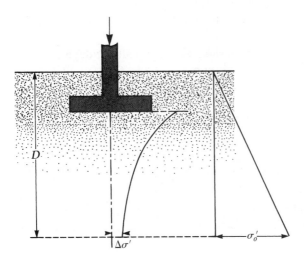

Figure 2.6 Determination of the minimum depth of boring

4. Determine the depth, $D = D_2$, at which $\Delta\sigma'/\sigma_o' = 0.05$.

5. Choose the smaller of the two depths, D_1 and D_2, just determined as the approximate minimum depth of boring required, unless bedrock is encountered.

If the preceding rules are used, the depths of boring for a building with a width of 30 m (100 ft) will be approximately the following, according to Sowers and Sowers (1970):

No. of stories	Boring depth	
1	3.5 m	(11 ft)
2	6 m	(20 ft)
3	10 m	(33 ft)
4	16 m	(53 ft)
5	24 m	(79 ft)

To determine the boring depth for hospitals and office buildings, Sowers and Sowers also use the rule

$$D_b = 3S^{0.7} \quad \text{(for light steel or narrow concrete buildings)} \quad (2.1)$$

and

$$D_b = 6S^{0.7} \quad \text{(for heavy steel or wide concrete buildings)} \quad (2.2)$$

where D_b = depth of boring, in meters
  S = number of stories

In English units, the preceding equations take the form

Table 2.1 Approximate Spacing of Boreholes

Type of project	Spacing	
	(m)	(ft)
Multistory building	10–30	30–100
One-story industrial plants	20–60	60–200
Highways	250–500	800–1600
Residential subdivision	250–500	800–1600
Dams and dikes	40–80	130–260

$$D_b \text{ (ft)} = 10S^{0.7} \quad \text{(for light steel or narrow concrete buildings)} \qquad (2.3)$$

and

$$D_b \text{ (ft)} = 20S^{0.7} \quad \text{(for heavy steel or wide concrete buildings)} \qquad (2.4)$$

When deep excavations are anticipated, the depth of boring should be at least 1.5 times the depth of excavation.

Sometimes, subsoil conditions require that the foundation load be transmitted to bedrock. The minimum depth of core boring into the bedrock is about 3 m (10 ft). If the bedrock is irregular or weathered, the core borings may have to be deeper.

There are no hard-and-fast rules for borehole spacing. Table 2.1 gives some general guidelines. Spacing can be increased or decreased, depending on the condition of the subsoil. If various soil strata are more or less uniform and predictable, fewer boreholes are needed than in nonhomogeneous soil strata.

The engineer should also take into account the ultimate cost of the structure when making decisions regarding the extent of field exploration. The exploration cost generally should be 0.1–0.5% of the cost of the structure. Soil borings can be made by several methods, including auger boring, wash boring, percussion drilling, and rotary drilling.

2.10 Exploratory Borings in the Field

Auger boring is the simplest method of making exploratory boreholes. Figure 2.7 shows two types of hand auger: the *posthole auger* and the *helical auger.* Hand augers cannot be used for advancing holes to depths exceeding 3–5 m (10–16 ft). However, they can be used for soil exploration work on some highways and small structures. *Portable power-driven helical augers* (76 mm to 305 mm in diameter) are available for making deeper boreholes. The soil samples obtained from such borings are highly disturbed. In some noncohesive soils or soils having low cohesion, the walls of the boreholes will not stand unsupported. In such circumstances, a metal pipe is used as a *casing* to prevent the soil from caving in.

When power is available, *continuous-flight augers* are probably the most common method used for advancing a borehole. The power for drilling is delivered by truck- or tractor-mounted drilling rigs. Boreholes up to about 60–70 m (200–230 ft)

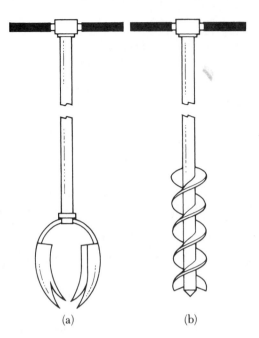

(a) (b)

Figure 2.7 Hand tools: (a) posthole auger; (b) helical auger

can easily be made by this method. Continuous-flight augers are available in sections of about 1–2 m (3–6 ft) with either a solid or hollow stem. Some of the commonly used solid–stem augers have outside diameters of 66.68 mm ($2\frac{5}{8}$ in.), 82.55 mm ($3\frac{1}{4}$ in.), 101.6 mm (4 in.), and 114.3 mm ($4\frac{1}{2}$ in.). Common commercially available hollow-stem augers have dimensions of 63.5 mm ID and 158.75 mm OD (2.5 in. × 6.25 in.), 69.85 mm ID and 177.8 OD (2.75 in. × 7 in.), 76.2 mm ID and 203.2 OD (3 in. × 8 in.), and 82.55 mm ID and 228.6 mm OD (3.25 in. × 9 in.).

The tip of the auger is attached to a cutter head (Figure 2.8). During the drilling operation (Figure 2.9), section after section of auger can be added and the hole extended downward. The flights of the augers bring the loose soil from the bottom of the hole to the surface. The driller can detect changes in the type of soil by noting changes in the speed and sound of drilling. When solid-stem augers are used, the auger must be withdrawn at regular intervals to obtain soil samples and also to conduct other operations such as standard penetration tests. Hollow-stem augers have a distinct advantage over solid-stem augers in that they do not have to be removed frequently for sampling or other tests. As shown schematically in Figure 2.10, the outside of the hollow-stem auger acts as a casing.

The hollow-stem auger system includes the following components:

Outer component: (a) hollow auger sections, (b) hollow auger cap, and (c) drive cap
Inner component: (a) pilot assembly, (b) center rod column, and (c) rod-to-cap adapter

The auger head contains replaceable carbide teeth. During drilling, if soil samples are to be collected at a certain depth, the pilot assembly and the center rod are removed. The soil sampler is then inserted through the hollow stem of the auger column.

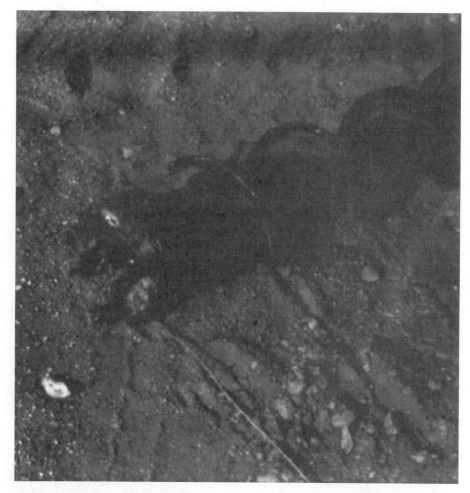

Figure 2.8 Carbide-tipped cutting head on auger flight attached with bolt (courtesy of William B. Ellis, El Paso Engineering and Testing, Inc., El Paso, Texas)

Wash boring is another method of advancing boreholes. In this method, a casing about 2–3 m (6–10 ft) long is driven into the ground. The soil inside the casing is then removed by means of a chopping bit attached to a drilling rod. Water is forced through the drilling rod and exits at a very high velocity through the holes at the bottom of the chopping bit (Figure 2.11). The water and the chopped soil particles rise in the drill hole and overflow at the top of the casing through a T connection. The washwater is collected in a container. The casing can be extended with additional pieces as the borehole progresses; however, that is not required if the borehole will stay open and not cave in. Wash borings are rarely used now in the United States and other developed countries.

Rotary drilling is a procedure by which rapidly rotating drilling bits attached to the bottom of drilling rods cut and grind the soil and advance the borehole. There are several types of drilling bit. Rotary drilling can be used in sand, clay, and rocks (unless they are badly fissured). Water or *drilling mud* is forced down the drilling

Figure 2.9 Drilling with continuous-flight augers (courtesy of Danny R.
Anderson, Danny R. Anderson Consultants, El Paso, Texas)

rods to the bits, and the return flow forces the cuttings to the surface. Boreholes with
diameters of 50–203 mm (2–8 in.) can easily be made by this technique. The drilling
mud is a slurry of water and bentonite. Generally, it is used when the soil that is encountered is likely to cave in. When soil samples are needed, the drilling rod is
raised and the drilling bit is replaced by a sampler. With the environmental drilling
applications, rotary drilling with air is becoming more common.

    Percussion drilling is an alternative method of advancing a borehole, particu-
larly through hard soil and rock. A heavy drilling bit is raised and lowered to chop
the hard soil. The chopped soil particles are brought up by the circulation of water.
Percussion drilling may require casing.

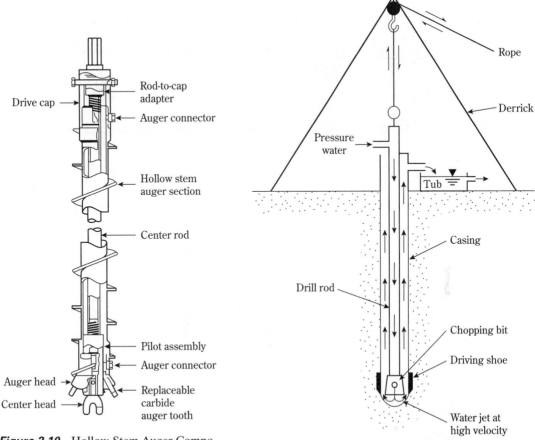

Figure 2.10 Hollow-Stem Auger Components (after ASTM, 2001)

Figure 2.11 Wash boring

2.11 *Procedures for Sampling Soil*

Two types of soil samples can be obtained during subsurface exploration: *disturbed* and *undisturbed*. Disturbed, but representative, samples can generally be used for the following types of laboratory test:

1. Grain-size analysis
2. Determination of liquid and plastic limits
3. Specific gravity of soil solids
4. Determination of organic content
5. Classification of soil

Disturbed soil samples, however, cannot be used for consolidation, hydraulic conductivity, or shear strength tests. Undisturbed soil samples must be obtained for these types of laboratory tests.

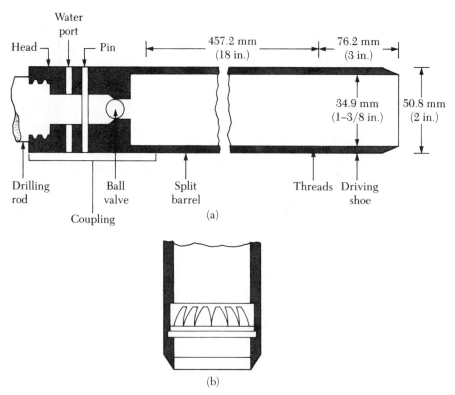

Figure 2.12   (a) Standard split-spoon sampler; (b) spring core catcher

Split-Spoon Sampling

Split-spoon samplers can be used in the field to obtain soil samples that are generally disturbed, but still representative. A section of a *standard split-spoon sampler* is shown in Figure 2.12a. The tool consists of a steel driving shoe, a steel tube that is split longitudinally in half, and a coupling at the top. The coupling connects the sampler to the drill rod. The standard split tube has an inside diameter of 34.93 mm ($1\frac{3}{8}$ in.) and an outside diameter of 50.8 mm (2 in.); however, samplers having inside and outside diameters up to 63.5 mm ($2\frac{1}{2}$ in.) and 76.2 mm (3 in.), respectively, are also available. When a borehole is extended to a predetermined depth, the drill tools are removed and the sampler is lowered to the bottom of the hole. The sampler is driven into the soil by hammer blows to the top of the drill rod. The standard weight of the hammer is 622.72 N (140 lb), and for each blow, the hammer drops a distance of 0.762 m (30 in.). The number of blows required for a spoon penetration of three 152.4-mm (6-in.) intervals are recorded. The number of blows required for the last two intervals are added to give the *standard penetration number, N,* at that depth. This number is generally referred to as the *N value* (American Society for Testing and Materials, 2001, Designation D-1586-99). The sampler is then withdrawn, and the shoe and coupling are removed. Finally, the soil sample recovered from the tube is placed in a glass bottle and transported to the laboratory. This field test is called the standard penetration test (SPT).

The degree of disturbance for a soil sample is usually expressed as

$$A_R(\%) = \frac{D_o^2 - D_i^2}{D_i^2}(100) \tag{2.5}$$

where A_R = area ratio (ratio of disturbed area to total area of soil)
   D_o = outside diameter of the sampling tube
   D_i = inside diameter of the sampling tube

When the area ratio is 10% or less, the sample generally is considered to be undisturbed. For a standard split-spoon sampler,

$$A_R(\%) = \frac{(50.8)^2 - (34.93)^2}{(34.93)^2}(100) = 111.5\%$$

Hence, these samples are highly disturbed. Split-spoon samples generally are taken at intervals of about 1.53 m (5 ft). When the material encountered in the field is sand (particularly fine sand below the water table), recovery of the sample by a split-spoon sampler may be difficult. In that case, a device such as a *spring core catcher* may have to be placed inside the split spoon (Figure 2.12b).

At this juncture, it is important to point out that several factors contribute to the variation of the standard penetration number N at a given depth for similar soil profiles. Among these factors are the SPT hammer efficiency, borehole diameter, sampling method, and rod length factor (Skempton, 1986; Seed et al., 1985). The two most common types of SPT hammers used in the field are the *safety hammer* and *donut hammer*. They are commonly dropped by a rope with *two wraps around a pulley*. Figure 2.13 shows schematic diagrams of the safety hammer and donut hammer.

On the basis of field observations, it appears reasonable to standardize the field penetration number as a function of the input driving energy and its dissipation around the sampler into the surrounding soil, or

$$N_{60} = \frac{N\eta_H\eta_B\eta_S\eta_R}{60} \tag{2.6}$$

where N_{60} = standard penetration number, corrected for field conditions
   N = measured penetration number
   η_H = hammer efficiency (%)
   η_B = correction for borehole diameter
   η_S = sampler correction
   η_R = correction for rod length

Variations of η_H, η_B, η_S, and η_R, based on recommendations by Seed et al. (1985) and Skempton (1986), are summarized in Table 2.2.

Besides compelling the geotechnical engineer to obtain soil samples, standard penetration tests provide several useful correlations. For example, the consistency of

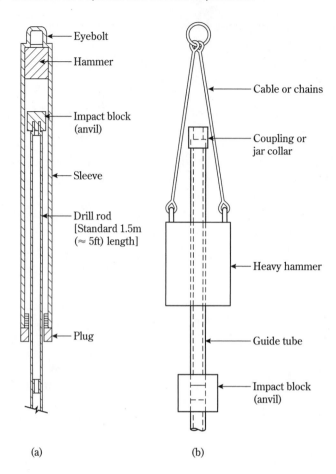

(a)

(b)

Figure 2.13   Configuration of SPT hammers: (a) safety hammer; (b) donut hammer (after Seed et al., 1985)

Table 2.2   Variations of $\eta_H, \eta_B, \eta_S,$ and η_R [Eq. (2.6)]

1. Variation of η_H

Country	Hammer type	Hammer release	η_H (%)
Japan	Donut	Free fall	78
	Donut	Rope and pulley	67
United States	Safety	Rope and pulley	60
	Donut	Rope and pulley	45
Argentina	Donut	Rope and pulley	45
China	Donut	Free fall	60
	Donut	Rope and pulley	50

2. Variation of η_B

Diameter		
mm	in.	η_B
60–120	2.4–4.7	1
150	6	1.05
200	8	1.15

4. Variation of η_R

Rod length		
m	ft	η_R
>10	>30	1.0
6–10	20–30	0.95
4–6	12–20	0.85
0–4	0–12	0.75

3. Variation of η_S

Variable	η_S
Standard sampler	1.0
With liner for dense sand and clay	0.8
With liner for loose sand	0.9

Table 2.3 Consistency of Clays and Approximate Correlation with the Standard Penetration Number, N_{60}

Standard penetration number, N_{60}	Consistency	Unconfined compression strength, q_u	
		kN/m^2	lb/ft^2
0–2	Very soft	0–25	0–500
2–5	Soft	25–50	500–1000
5–10	Medium stiff	50–100	1000–2100
10–20	Stiff	100–200	2100–4200
20–30	Very stiff	200–400	4200–8400
>30	Hard	>400	>8400

clayey soils can often be estimated from the standard penetration number, N_{60}, as shown in Table 2.3. However, correlations for clays require tests to verify that the relationships are valid for the clay deposit being examined.

The literature contains many correlations between the standard penetration number and the undrained shear strength of clay, c_u. On the basis of results of undrained triaxial tests conducted on insensitive clays, Stroud (1974) suggested that

$$c_u = KN_{60} \tag{2.7}$$

where K = constant = 3.5–6.5 kN/m^2 (0.507–0.942 lb/in^2)
   N_{60} = standard penetration number obtained from the field

The average value of K is about 4.4 kN/m^2 (0.638 lb/in^2).
   Hara et al. (1971) also suggested that

$$c_u(\text{kN/m}^2) = 29N_{60}^{0.72} \tag{2.8}$$

The overconsolidation ratio, OCR, of a natural clay deposit can also be correlated with the standard penetration number. On the basis of the regression analysis of 110 data points, Mayne and Kemper (1988) obtained the relationship

$$\text{OCR} = 0.193\left(\frac{N_{60}}{\sigma_o'}\right)^{0.689} \tag{2.9}$$

where σ_o' = effective vertical stress in MN/m^2

It is important to point out that any correlation between c_u and N_{60} is only approximate.

In granular soils, the value of N is affected by the effective overburden pressure, σ_o'. For that reason, the value of N_{60} obtained from field exploration under different effective overburden pressures should be changed to correspond to a standard value of σ'_o. That is,

$$(N_1)_{60} = C_N N_{60} \tag{2.10}$$

where $(N_1)_{60}$ = value of N_{60} corrected to a standard value of
         $\sigma'_o [100 \text{ kN/m}^2 \, (2000 \text{ lb/ft}^2)]$
       C_N = correction factor
       N_{60} = value of N obtained from field exploration [Eq. (2.6)]

In the past, a number of empirical relations were proposed for C_N. Some of the relationships are given next. The most commonly cited relationships are those of Liao and Whitman (1986) and Skempton (1986).

In the following relationships for C_N, note that σ'_o *is the effective overburden pressure and* p_a = *atmospheric pressure* $\left(\approx 100 \, kN/m^2, \, or \approx 2000 \, lb/ft^2 \right)$
Liao and Whitman's relationship (1986):

$$C_N = \left[\frac{1}{\left(\dfrac{\sigma'_o}{p_a} \right)} \right]^{0.5} \tag{2.11}$$

Skempton's relationship (1986):

$$C_N = \frac{2}{1 + \left(\dfrac{\sigma'_o}{p_a} \right)} \tag{2.12}$$

Seed et al.'s relationship (1975):

$$C_N = 1 - 1.25 \log\left(\frac{\sigma'_o}{p_a} \right) \tag{2.13}$$

Peck et al.'s relationship (1974):

$$C_N = 0.77 \log\left[\frac{20}{\left(\dfrac{\sigma'_o}{p_a} \right)} \right] \tag{2.14}$$

for $\sigma'_o \geq 25 \text{ kN//m}^2 \, (\approx 500 \text{ lb/ft}^2)$

An approximate relationship between the corrected standard penetration number and the relative density of sand is given in Table 2.4. The values are approximate primarily because the effective overburden pressure and the stress history of the soil significantly influence the N_{60} values of sand. An extensive study conducted by Marcuson and Bieganousky (1977) produced the empirical relationship

$$D_r(\%) = 11.7 + 0.76 \left(222 N_{60} + 1600 - 53\sigma'_o - 50 C_u^2 \right)^{0.5} \tag{2.15}$$

Table 2.4 Relation between the Corrected $(N_1)_{60}$
Values and the Relative Density in Sands

Standard penetration number, $(N_1)_{60}$	Approximate relative density, D_r, (%)
0–5	0–5
5–10	5–30
10–30	30–60
30–50	60–95

where D_r = relative density
$\quad\quad\ N_{60}$ = standard penetration number in the field
$\quad\quad\ \sigma'_o$ = effective overburden pressure (lb/in^2)
$\quad\quad\ C_u$ = uniformity coefficient of the sand

Kulhawy and Mayne (1990) modified the preceding relationship to consider the stress history (i.e., OCR) of soil, which can be expressed as

$$D_r(\%) = 12.2 + 0.75\left[222N_{60} + 2311 - 711\text{OCR} - 779\left(\frac{\sigma'_o}{p_a}\right) - 50C_u^2\right]^{0.5} \quad (2.16)$$

where $\text{OCR} = \dfrac{\text{preconsolidation pressure}, \sigma'_c}{\text{effective overburden pressure}, \sigma'_o}$
$\quad\quad\ p_a$ = atmospheric pressure

Cubrinovski and Ishihara (1999) also proposed a correlation between N_{60} and the relative density of sand (D_r) that can be expressed as

$$D_r(\%) = \left[\frac{N_{60}\left(0.23 + \dfrac{0.06}{D_{50}}\right)^{1.7}}{9}\left(\frac{1}{\dfrac{\sigma'_o}{p_a}}\right)\right]^{0.5}(100) \quad (2.17)$$

where p_a = atmospheric pressure (≈ 100 kN//m^2, or ≈ 2000 lb/ft^2)
$\quad\quad\ D_{50}$ = sieve size through which 50% of the soil will pass (mm)

The peak friction angle, ϕ', of granular soil has also been correlated with N_{60} or $(N_1)_{60}$ by several investigators. Some of these correlations are as follows:

1. Peck, Hanson, and Thornburn (1974) give a correlation between $(N_1)_{60}$ and ϕ' in a graphical form, which can be approximated as (Wolff, 1989)

$$\phi'(\text{deg}) = 27.1 + 0.3(N_1)_{60} - 0.00054[(N_1)_{60}]^2 \quad (2.18)$$

2. Schmertmann (1975) provided the correlation between N_{60}, σ'_o, and ϕ' shown in Figure 2.14. Mathematically, the correlation can be approximated as (Kulhawy and Mayne, 1990)

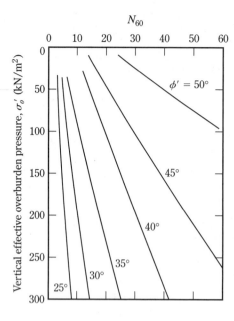

Figure 2.14   Schmertmann's (1975) correlation between N_{60}, σ'_o, and ϕ' for granular soils

$$\phi' = \tan^{-1}\left[\frac{N_{60}}{12.2 + 20.3\left(\dfrac{\sigma'_o}{p_a}\right)}\right]^{0.34} \tag{2.19}$$

where N_{60} = field standard penetration number
σ'_o = effective overburden pressure
p_a = atmospheric pressure in the same unit as σ'_o
ϕ' = soil friction angle

3. Hatanaka and Uchida (1996) provided a simple correlation between ϕ' and $(N_1)_{60}$ that can be expressed as

$$\phi' = \sqrt{20(N_1)_{60}} + 20 \tag{2.20}$$

The following qualifications should be noted when standard penetration resistance values are used in the preceding correlations to estimate soil parameters:

1. The equations are approximate.
2. Because the soil is not homogeneous, the values of N_{60} obtained from a given borehole vary widely.
3. In soil deposits that contain large boulders and gravel, standard penetration numbers may be erratic and unreliable.

Although approximate, with correct interpretation the standard penetration test provides a good evaluation of soil properties. The primary sources of error in standard penetration tests are inadequate cleaning of the borehole, careless measurement of the blow count, eccentric hammer strikes on the drill rod, and inadequate maintenance of water head in the borehole.

Scraper Bucket

When the soil deposits are sand mixed with pebbles, obtaining samples by split spoon with a spring core catcher may not be possible because the pebbles may prevent the springs from closing. In such cases, a scraper bucket may be used to obtain disturbed representative samples (Figure 2.15a). The scraper bucket has a driving point and can be attached to a drilling rod. The sampler is driven down into the soil and rotated, and the scrapings from the side fall into the bucket.

Thin-Walled Tube

Thin-walled tubes are sometimes referred to as *Shelby tubes*. They are made of seamless steel and are frequently used to obtain undisturbed clayey soils. The most common thin-walled tube samplers have outside diameters of 50.8 mm (2 in.) and 76.2 mm (3 in.). The bottom end of the tube is sharpened. The tubes can be attached to drill rods (Figure 2.15b). The drill rod with the sampler attached is lowered to the bottom of the borehole, and the sampler is pushed into the soil. The soil sample inside the tube is then pulled out. The two ends are sealed, and the sampler is sent to the laboratory for testing.

Samples obtained in this manner may be used for consolidation or shear tests. A thin-walled tube with a 50.8-mm (2-in.) outside diameter has an inside diameter of about 47.63 mm ($1\frac{7}{8}$ in.). The area ratio is

$$A_R(\%) = \frac{D_o^2 - D_i^2}{D_i^2}(100) = \frac{(50.8)^2 - (47.63)^2}{(47.63)^2}(100) = 13.75\%$$

Increasing the diameters of samples increases the cost of obtaining them.

Piston Sampler

When undisturbed soil samples are very soft or larger than 76.2 mm (3 in.) in diameter, they tend to fall out of the sampler. Piston samplers are particularly useful under such conditions. There are several types of piston sampler; however, the

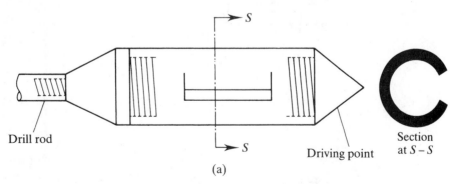

(a)

Figure 2.15 Sampling devices: (a) scraper bucket. Following page: (b) thin-walled tube; (c) and (d) piston sampler

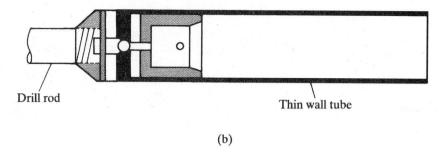

(b)

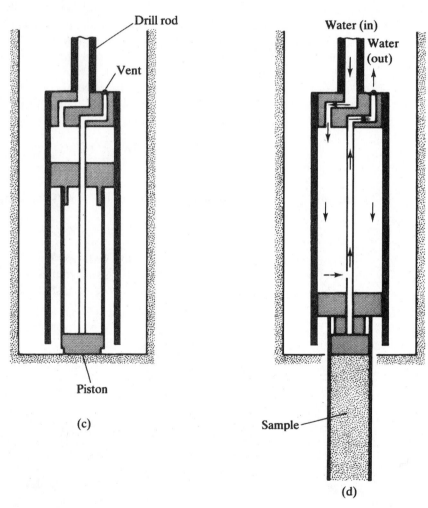

(c)

(d)

Figure 2.15 (Continued)

sampler proposed by Osterberg (1952) is the most useful. (see Figures 2.15c and 2.15d). It consists of a thin-walled tube with a piston. Initially, the piston closes the end of the tube. The sampler is lowered to the bottom of the borehole (Figure 2.15c), and the tube is pushed into the soil hydraulically, past the piston. Then the pressure is released through a hole in the piston rod (Figure 2.15d). To a large extent, the presence of the piston prevents distortion in the sample by not letting the soil squeeze into the sampling tube very fast and by not admitting excess soil. Consequently, samples obtained in this manner are less disturbed than those obtained by Shelby tubes.

2.12 Observation of Water Tables

The presence of a water table near a foundation significantly affects the foundation's load-bearing capacity and settlement, among other things. The water level will change seasonally. In many cases, establishing the highest and lowest possible levels of water during the life of a project may become necessary.

If water is encountered in a borehole during a field exploration, that fact should be recorded. In soils with high hydraulic conductivity, the level of water in a borehole will stabilize about 24 hours after completion of the boring. The depth of the water table can then be recorded by lowering a chain or tape into the borehole.

In highly impermeable layers, the water level in a borehole may not stabilize for several weeks. In such cases, if accurate water-level measurements are required, a *piezometer* can be used. A piezometer basically consists of a porous stone or a perforated pipe with a plastic standpipe attached to it. Figure 2.16 shows the general placement of a piezometer in a borehole. This procedure will allow periodic checking until the water level stabilizes.

2.13 Vane Shear Test

The *vane shear test* (ASTM D-2573) may be used during the drilling operation to determine the *in situ* undrained shear strength (c_u) of clay soils—particularly soft clays. The vane shear apparatus consists of four blades on the end of a rod, as shown in Figure 2.17. The height, H, of the vane is twice the diameter, D. The vane can be either rectangular or tapered (see Figure 2.17). The dimensions of vanes used in the field are given in Table 2.5. The vanes of the apparatus are pushed into the soil at the bottom of a borehole without disturbing the soil appreciably. Torque is applied at the top of the rod to rotate the vanes at a standard rate of 0.1°/sec. This rotation will induce failure in a soil of cylindrical shape surrounding the vanes. The maximum torque, T, applied to cause failure is measured. Note that

$$T = f(c_u, H, \text{and } D) \tag{2.21}$$

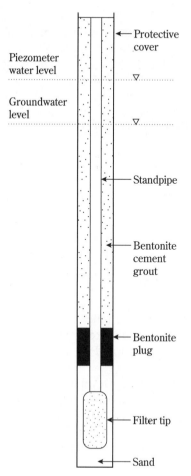

Figure 2.16 Casagrande-type piezometer (courtesy of N. Sivakugan, James Cook University, Australia)

or

$$c_u = \frac{T}{K} \tag{2.22}$$

where T is in $N \cdot m$, c_u is in kN/m^2, and

K = a constant with a magnitude depending on the dimension and shape of the vane

The constant

$$K = \left(\frac{\pi}{10^6}\right)\left(\frac{D^2 H}{2}\right)\left(1 + \frac{D}{3H}\right) \tag{2.23}$$

where D = diameter of vane in cm

H = measured height of vane in cm

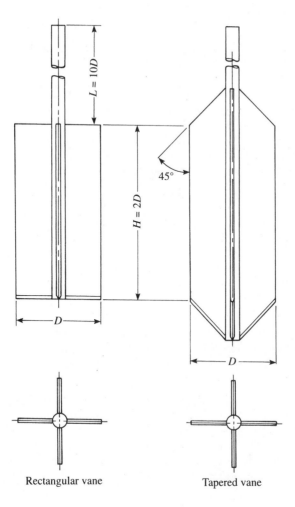

Rectangular vane

Tapered vane

Figure 2.17 Geometry of field vane (after ASTM, 2001)

If $H/D = 2$, Eq. (2.23) yields

$$K = 366 \times 10^{-8} D^3 \qquad (2.24)$$

$$\uparrow$$

$$\text{(cm)}$$

In English units, if c_u and T in Eq. (2.22) are expressed in lb/ft² and lb-ft, respectively, then

$$K = \left(\frac{\pi}{1728}\right)\left(\frac{D^2 H}{2}\right)\left(1 + \frac{D}{3H}\right) \qquad (2.25)$$

Table 2.5   Recommended Dimensions of Field Vanes[a] (after ASTM, 2001)

Casing size	Diameter, D mm (in.)	Height, H mm (in.)	Thickness of blade mm (in.)	Diameter of rod mm (in.)
AX	38.1 ($1\frac{1}{2}$)	76.2 (3)	1.6 ($\frac{1}{16}$)	12.7 ($\frac{1}{2}$)
BX	50.8 (2)	101.6 (4)	1.6 ($\frac{1}{16}$)	12.7 ($\frac{1}{2}$)
NX	63.5 ($2\frac{1}{2}$)	127.0 (5)	3.2 ($\frac{1}{8}$)	12.7 ($\frac{1}{2}$)
4 in. (101.6 mm)[b]	92.1 ($3\frac{5}{8}$)	184.1 ($7\frac{1}{4}$)	3.2 ($\frac{1}{8}$)	12.7 ($\frac{1}{2}$)

[a]The selection of a vane size is directly related to the consistency of the soil being tested; that is, the softer the soil, the larger the vane diameter should be.
[b]Inside diameter.

If $H/D = 2$, Eq. (2.25) yields

$$K = 0.0021D^3 \qquad (2.26)$$
$$\uparrow$$
$$\text{(in.)}$$

Field vane shear tests are moderately rapid and economical and are used extensively in field soil-exploration programs. The test gives good results in soft and medium-stiff clays and gives excellent results in determining the properties of sensitive clays.

Sources of significant error in the field vane shear test are poor calibration of torque measurement and damaged vanes. Other errors may be introduced if the rate of rotation of the vane is not properly controlled.

For actual design purposes, the undrained shear strength values obtained from field vane shear tests [$c_{u(\text{VST})}$] are too high, and it is recommended that they be corrected according to the equation

$$c_{u(\text{corrected})} = \lambda c_{u(\text{VST})} \qquad (2.27)$$

where λ = correction factor

Several correlations have been given previously for the correction factor λ; some more are as follows:
Bjerrum (1972):

$$\lambda = 1.7 - 0.54 \log[\text{PI}(\%)] \qquad (2.28)$$

Morris and Williams (1994):

$$\lambda = 1.18e^{-0.08(\text{PI})} + 0.57 \quad \text{(for PI} > 5) \qquad (2.29)$$

$$\lambda = 7.01e^{-0.08(\text{LL})} + 0.57 \quad \text{(where LL is in \%)} \qquad (2.30)$$

Aas et al. (1986):

$$\lambda = f\left[\frac{c_{u(\text{VST})}}{\sigma'_o}\right] \quad \text{(see Figure 2.18)} \qquad (2.31)$$

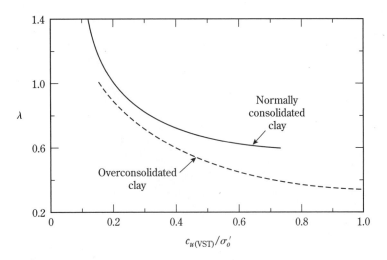

Figure 2.18 Variation of λ with $c_{u(\text{VST})}/\sigma'_o$ [see Eq. (2.31)]

The field vane shear strength can also be correlated with the preconsolidation pressure and the overconsolidation ratio of the clay. Using 343 data points, Mayne and Mitchell (1988) derived the following empirical relationship for estimating the preconsolidation pressure of a natural clay deposit:

$$\sigma'_c = 7.04[c_{u(\text{field})}]^{0.83} \tag{2.32}$$

Here, σ'_c = preconsolidation pressure (kN/m^2)
$c_{u(\text{field})}$ = field vane shear strength (kN/m^2)

Figure 2.19 shows the plot of the data points from which Mayne and Mitchell derived the foregoing relationship. The authors also showed that the OCR can be correlated with $c_{u(\text{field})}$ according to the equation

$$\text{OCR} = \beta \frac{c_{u(\text{field})}}{\sigma'_o} \tag{2.33}$$

where σ'_o = effective overburden pressure
$\beta = 22(\text{PI})^{-0.48}$ (2.34)

in which PI = plasticity index

Figure 2.20 shows the variation of β with the plasticity index.

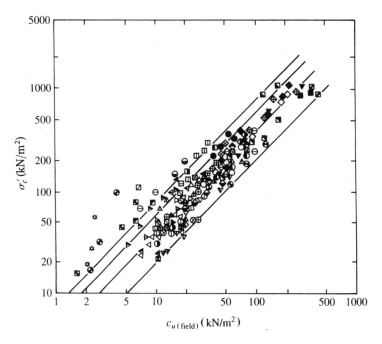

Figure 2.19   Variation of preconsolidation pressure with field vane shear strength (after Mayne and Mitchell, 1988)

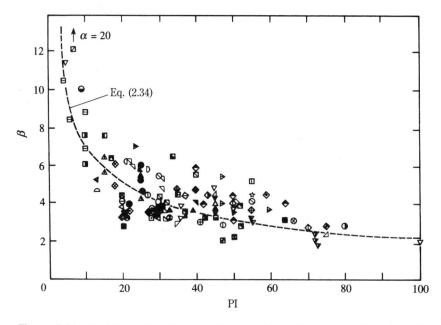

Figure 2.20   Variation of β with plasticity index (after Mayne and Mitchell, 1988)

Two other correlations for β presented in the literature are as follows:
Hansbo (1957):

$$\beta = \frac{222}{w(\%)} \tag{2.35}$$

Larsson (1980):

$$\beta = \frac{1}{0.08 + 0.0055(\text{PI})} \tag{2.36}$$

2.14 Cone Penetration Test

The cone penetration test (CPT), originally known as the Dutch cone penetration test, is a versatile sounding method that can be used to determine the materials in a soil profile and estimate their engineering properties. The test is also called the *static penetration test,* and no boreholes are necessary to perform it. In the original version, a 60° cone with a base area of 10 cm² (1.55 in.²) was pushed into the ground at a steady rate of about 20 mm/sec ($\approx$0.8 in./sec), and the resistance to penetration (called the point resistance) was measured.

The cone penetrometers in use at present measure (a) the *cone resistance* (q_c) to penetration developed by the cone, which is equal to the vertical force applied to the cone, divided by its horizontally projected area; and (b) the *frictional resistance* (f_c), which is the resistance measured by a sleeve located above the cone with the local soil surrounding it. The frictional resistance is equal to the vertical force applied to the sleeve, divided by its surface area—actually, the sum of friction and adhesion.

Generally, two types of penetrometers are used to measure q_c and f_c:

a. *Mechanical friction-cone penetrometer* (Figure 2.21). The tip of this penetrometer is connected to an inner set of rods. The tip is first advanced about 40 mm, giving the cone resistance. With further thrusting, the tip engages the friction sleeve. As the inner rod advances, the rod force is equal to the sum of the vertical force on the cone and sleeve. Subtracting the force on the cone gives the side resistance.
b. *Electric friction-cone penetrometer* (Figure 2.22). The tip of this penetrometer is attached to a string of steel rods. The tip is pushed into the ground at the rate of 20 mm/sec. Wires from the transducers are threaded through the center of the rods and continuously measure the cone and side resistances.

Figure 2.23 shows the results of penetrometer tests in a soil profile with friction measurement by a mechanical friction-cone penetrometer and an electric friction-cone penetrometer.

Several correlations that are useful in estimating the properties of soils encountered during an exploration program have been developed for the point resistance (q_c) and the friction ratio (F_r) obtained from the cone penetration tests. The friction ratio is defined as

$$F_r = \frac{\text{frictional resistance}}{\text{cone resistance}} = \frac{f_c}{q_c} \tag{2.37}$$

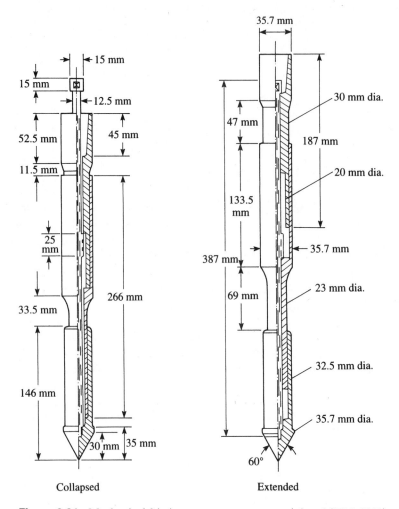

Figure 2.21 Mechanical friction-cone penetrometer (after ASTM, 2001)

As in the case of standard penetration tests, several correlations have been developed between q_c and other soil properties. Some of these correlations are presented next.

Correlation Between Relative Density (D_r) and q_o for Sand

Lancellotta (1983) and Jamiolkowski et al. (1985) showed that the relative density of *normally consolidated sand*, D_r, and q_c can be correlated according to the formula

$$D_r(\%) = A + B \log_{10}\left(\frac{q_c}{\sqrt{\sigma_o'}}\right) \tag{2.38}$$

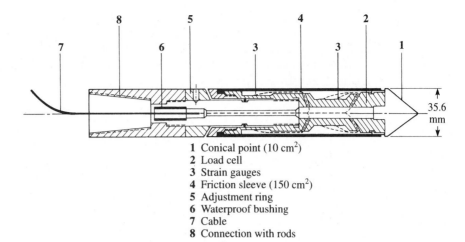

1 Conical point (10 cm²)
2 Load cell
3 Strain gauges
4 Friction sleeve (150 cm²)
5 Adjustment ring
6 Waterproof bushing
7 Cable
8 Connection with rods

Figure 2.22 Electric friction-cone penetrometer (after ASTM, 2001)

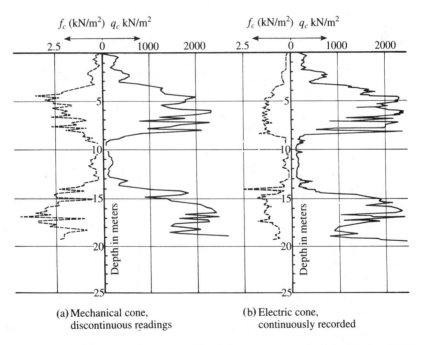

(a) Mechanical cone, discontinuous readings

(b) Electric cone, continuously recorded

Figure 2.23 Penetrometer tests with friction measurement (after Ruiter, 1971)

where A, B = constants
        σ_o' = vertical effective stress

The values of A and B are as follows:

A	B	Unit of q_c and σ_o'
−98	66	metric ton/m²

Baldi et al. (1982), and Robertson and Campanella (1983) recommended the empirical relationship shown in Figure 2.24 between vertical effective stress (σ'_o), relative density (D_r), and q_c for *normally consolidated sand.*

Kulhawy and Mayne (1990) proposed the following relationship to correlate D_r, q_c, and the vertical effective stress σ'_o:

$$D_r = \sqrt{\left[\frac{1}{305Q_c \text{OCR}^{1.8}}\right]\left[\frac{\dfrac{q_c}{p_a}}{\left(\dfrac{\sigma'_o}{p_a}\right)^{0.5}}\right]} \tag{2.39}$$

In this equation, OCR = overconsolidation ratio
$\quad\quad\quad\quad\quad\quad\quad p_a$ = atmospheric pressure
$\quad\quad\quad\quad\quad\quad\quad Q_c$ = compressibility factor

The recommended values of Q_c are as follows:

highly compressible sand = 0.91

moderately compressible sand = 1.0

low compressible sand = 1.09

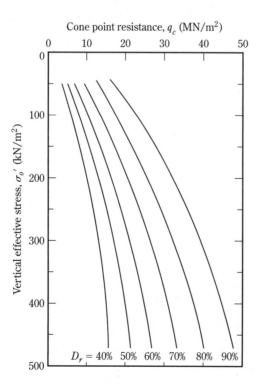

Figure 2.24 Variation of q_c, σ'_o, and D_r for normally consolidated quartz sand (based on Baldi et al., 1982, and Robertson and Campanella, 1983)

Correlation between q_c and Drained Friction Angle (ϕ') for Sand

On the basis of experimental results, Robertson and Campanella (1983) suggested the variation of D_r, σ'_o, and ϕ' for normally consolidated quartz sand. This relationship, shown in Figure 2.25, can be expressed as (Kulhawy and Mayne, 1990)

$$\phi' = \tan^{-1}\left[0.1 + 0.38 \log\left(\frac{q_c}{\sigma'_o}\right) \right] \tag{2.40}$$

Correlations of Soil Types

Robertson and Campanella (1983) provided the correlations shown in Figure 2.26 between q_c and the friction ratio [Eq. (2.37)] to identify various types of soil encountered in the field. Figure 2.27 shows a relationship between q_c/N_{60} (N_{60} = field standard penetration resistance) versus mean grain size (D_{50}) for various types of soil.

Correlations for Undrained Shear Strength (c_u), Preconsolidation Pressure (σ'_c), and Overconsolidation Ratio (OCR) for Clays

According to Mayne and Kemper (1988), in clayey soil the undrained cohesion c_u, preconsolidation pressure σ'_c, and overconsolidation ratio can be correlated via the equation

$$\left(\frac{c_u}{\sigma'_o}\right) = \left(\frac{q_c - \sigma_o}{\sigma'_o}\right)\frac{1}{N_K} \tag{2.41}$$

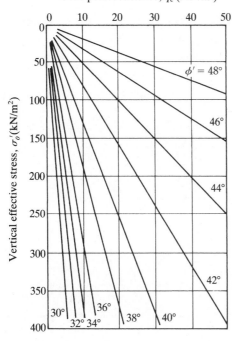

Figure 2.25 Variation of q_c with σ'_o and ϕ' in normally consolidated quartz sand (after Robertson and Campanella, 1983)

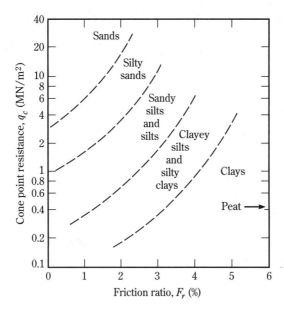

Figure 2.26 Robertson and Campanella's correlation (1983) between q_c, F_r, and the type of soil

or

$$c_u = \frac{q_c - \sigma_o}{N_K} \tag{2.41a}$$

where N_K = bearing capacity factor ($N_K = 15$ for an electric cone and $N_K = 20$ for a mechanical cone)

σ_o = *total* vertical stress

σ_o' = effective vertical stress

Consistent units of c_u, σ_o, σ_o', and q_c should be used with Eq. (2.41); that is,

$$\sigma_c' = 0.243 (q_c)^{0.96} \atop {\uparrow \qquad \qquad \uparrow \atop \text{MN/m}^2 \quad \text{MN/m}^2} \tag{2.42}$$

and

$$\text{OCR} = 0.37 \left(\frac{q_c - \sigma_o}{\sigma_o'} \right)^{1.01} \tag{2.43}$$

where σ_o and σ_o' = total and effective stress, respectively.

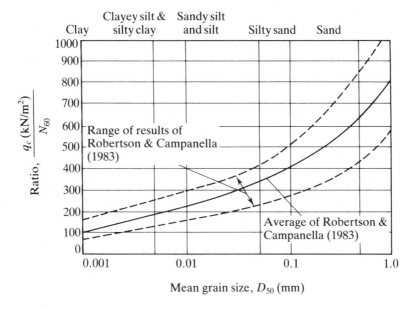

Figure 2.27 General range of variation of q_c/N_{60} for various types of soil (after Robertson and Campanella, 1983)

Pressuremeter Test (PMT)

The pressuremeter test is an *in situ* test conducted in a borehole. It was originally developed by Menard (1956) to measure the strength and deformability of soil. It has also been adopted by ASTM as Test Designation 4719. The Menard-type PMT consists essentially of a probe with three cells. The top and bottom ones are *guard cells* and the middle one is the *measuring cell*, as shown schematically in Figure 2.28a. The test is conducted in a prebored hole with a diameter that is between 1.03 and 1.2 times the nominal diameter of the probe. The probe that is most commonly used has a diameter of 58 mm and a length of 420 mm. The probe cells can be expanded by either liquid or gas. The guard cells are expanded to reduce the end-condition effect on the measuring cell, which has a volume (V_o) of 535 cm³. Following are the dimensions for the probe diameter and the diameter of the borehole, as recommended by ASTM:

Probe diameter (mm)	Borehole diameter	
	Nominal (mm)	Maximum (mm)
44	45	53
58	60	70
74	76	89

In order to conduct a test, the measuring cell volume, V_o, is measured and the probe is inserted into the borehole. Pressure is applied in increments and the new

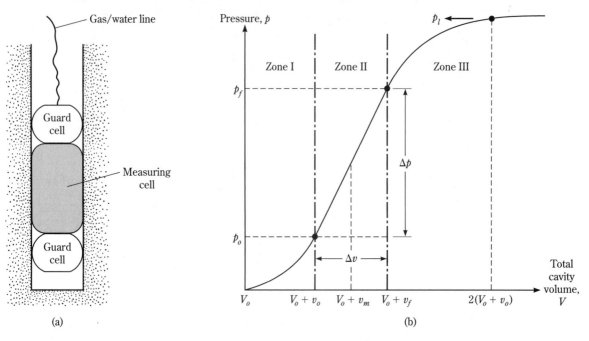

Figure 2.28 (a) Pressuremeter; (b) plot of pressure versus total cavity volume

volume of the cell is measured. The process is continued until the soil fails or until the pressure limit of the device is reached. The soil is considered to have failed when the total volume of the expanded cavity (V) is about twice the volume of the original cavity. After the completion of the test, the probe is deflated and advanced for testing at another depth.

The results of the pressuremeter test are expressed in the graphical form of pressure versus volume, as shown in Figure 2.28b. In the figure, Zone I represents the reloading portion during which the soil around the borehole is pushed back into the initial state (i.e., the state it was in before drilling). The pressure p_o represents the *in situ* total horizontal stress. Zone II represents a pseudoelastic zone in which the cell volume versus cell pressure is practically linear. The pressure p_f represents the creep, or yield, pressure. The zone marked III is the plastic zone. The pressure p_l represents the limit pressure.

The pressuremeter modulus, E_p, of the soil is determined with the use of the theory of expansion of an infinitely thick cylinder. Thus,

$$E_p = 2(1 + \mu_s)(V_o + v_m)\left(\frac{\Delta p}{\Delta v}\right) \qquad (2.44)$$

where $v_m = \dfrac{v_o + v_f}{2}$

$\Delta p = p_f - p_o$
$\Delta v = v_f - v_o$
μ_s = Poisson's ratio (which may be assumed to be 0.33)

The limit pressure p_l is usually obtained by extrapolation and not by direct measurement.

In order to overcome the difficulty of preparing the borehole to the proper size, self-boring pressuremeters (SBPMTs) have also been developed. The details concerning SBPMTs can be found in the work of Baguelin et al. (1978).

Correlations between various soil parameters and the results obtained from the pressuremeter tests have been developed by various investigators. Kulhawy and Mayne (1990) proposed that

$$\sigma_c' = 0.45 p_l \tag{2.45}$$

where σ_c' = preconsolidation pressure

On the basis of the cavity expansion theory, Baguelin et al. (1978) proposed that

$$c_u = \frac{(p_l - p_o)}{N_p} \tag{2.46}$$

where c_u = undrained shear strength of a clay

$$N_p = 1 + \ln\left(\frac{E_p}{3c_u}\right)$$

Typical values of N_p vary between 5 and 12, with an average of about 8.5. Ohya et al. (1982) (see also Kulhawy and Mayne, 1990) correlated E_p with field standard penetration numbers (N_{60}) for sand and clay as follows:

$$\text{Clay: } E_p(\text{kN/m}^2) = 1930 N_{60}^{0.63} \tag{2.47}$$

$$\text{Sand: } E_p(\text{kN/m}^2) = 908 N_{60}^{0.66} \tag{2.48}$$

2.16   *Dilatometer Test*

The use of the flat-plate dilatometer test (DMT) is relatively recent (Marchetti, 1980; Schmertmann, 1986). The equipment essentially consists of a flat plate measuring 220 mm (length) $\times$ 95 mm (width) $\times$ 14 mm (thickness) (8.66 in. $\times$ 3.74 in. $\times$ 0.55 in.). A thin, flat, circular, expandable steel membrane having a diameter of 60 mm (2.36 in.) is located flush at the center on one side of the plate (Figure 2.29a). The dilatometer probe is inserted into the ground with a cone penetrometer testing rig (Figure 2.29b). Gas and electric lines extend from the surface control box, through the penetrometer rod, and into the blade. At the required depth, high-pressure nitrogen gas is used to inflate the membrane. Two pressure readings are taken:

1. The pressure A required to "lift off" the membrane.
2. The pressure B at which the membrane expands 1.1 mm (0.4 in.) into the surrounding soil

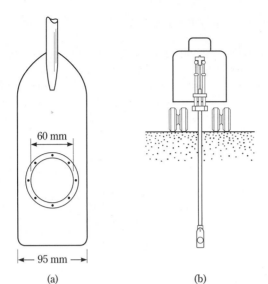

Figure 2.29 (a) Schematic diagram of a flat-plate dilatometer; (b) dilatometer probe inserted into ground

The A and B readings are corrected as follows (Schmertmann, 1986):

$$\text{Contact stress, } p_o = 1.05(A + \Delta A - Z_m) - 0.05(B - \Delta B - Z_m) \quad (2.49)$$

$$\text{Expansion stress, } p_1 = B - Z_m - \Delta B \quad (2.50)$$

where ΔA = vacuum pressure required to keep the membrane in contact with its seating

       ΔB = air pressure required inside the membrane to deflect it outward to a center expansion of 1.1 mm

       Z_m = gauge pressure deviation from zero when vented to atmospheric pressure

The test is normally conducted at depths 200 to 300 mm apart. The result of a given test is used to determine three parameters:

1. Material index, $I_D = \dfrac{p_1 - p_o}{p_o - u_o}$

2. Horizontal stress index, $K_D = \dfrac{p_o - u_o}{\sigma'_o}$

3. Dilatometer modulus, $E_D (\text{kN/m}^2) = 34.7(p_1 \text{ kN/m}^2 - p_o \text{ kN/m}^2)$

where u_o = pore water pressure

       σ'_o = *in situ* vertical effective stress

Figure 2.30 shows the results of a dilatometer test conducted in Porto Tolle, Italy (Marchetti, 1980). The subsoil consisted of recent normally consolidated delta deposits of the Po River. A thick layer of silty clay was found below a depth of 3 m (10 ft) ($c' = 0$; $\phi' \approx 28°$). The results obtained from this dilatometer test have been correlated with several soil properties (Marchetti, 1980). Some of the correlations are as follows:

a. $K_o = \left(\dfrac{K_D}{1.5}\right)^{0.47} - 0.6$ (2.51)

b. $\text{OCR} = (0.5K_D)^{1.6}$ (2.52)

c. $\dfrac{c_u}{\sigma'_o} = 0.22$ (for normally consolidated clay) (2.53)

d. $\left(\dfrac{c_u}{\sigma'_o}\right)_{\text{OC}} = \left(\dfrac{c_u}{\sigma'_o}\right)_{\text{NC}} (0.5K_D)^{1.25}$ (2.54)

e. $E_s = (1 - \mu_s^2)E_D$ (2.55)

where K_o = coefficient of at-rest earth pressure
OCR = overconsolidation ratio
OC = overconsolidated soil

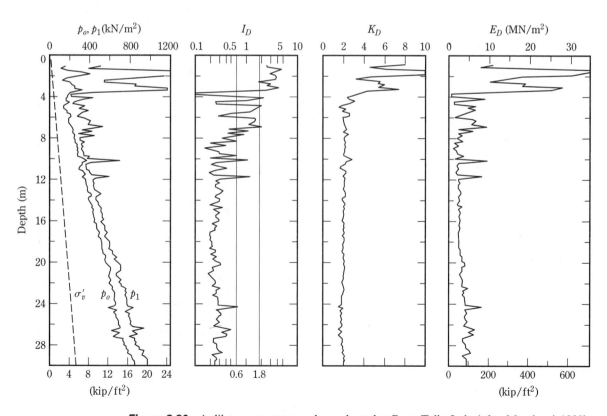

Figure 2.30 A dilatometer test result conducted at Porto Tolle, Italy (after Marchetti, 1980)

$$NC = \text{normally consolidated soil}$$
$$E_s = \text{modulus of elasticity}$$

Schmertmann (1986) also provided a correlation between the material index (I_D) and the dilatometer modulus (E_D) for a determination of the nature of the soil and its unit weight (γ). This relationship is shown in Figure 2.31.

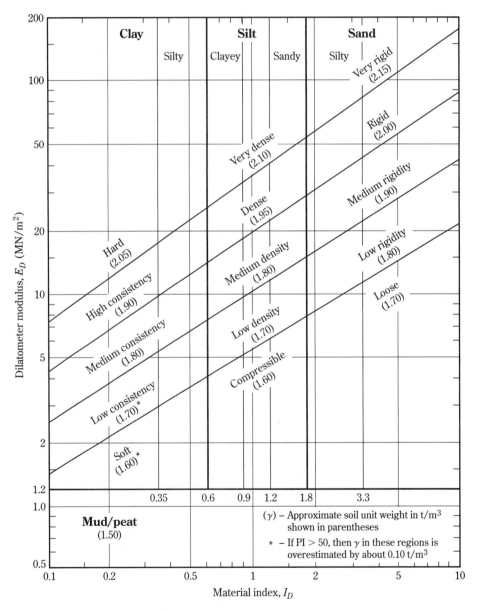

Figure 2.31 Chart for determination of soil description and unit weight (after Schmertmann, 1986); *Note:* 1 t/m³ = 9.81 kN/m³

Coring of Rocks

When a rock layer is encountered during a drilling operation, rock coring may be necessary. To core rocks, a *core barrel* is attached to a drilling rod. A *coring bit* is attached to the bottom of the barrel (Fig. 2.32). The cutting elements may be diamond, tungsten, carbide, and so on. Table 2.6 summarizes the various types of core barrel and their sizes, as well as the compatible drill rods commonly used for exploring foundations. The coring is advanced by rotary drilling. Water is circulated through the drilling rod during coring, and the cutting is washed out.

Two types of core barrel are available: the *single-tube core barrel* (Figure 2.32a) and the *double-tube core barrel* (Figure 2.32b). Rock cores obtained by single-tube core barrels can be highly disturbed and fractured because of torsion. Rock cores smaller than the BX size tend to fracture during the coring process.

When the core samples are recovered, the depth of recovery should be properly recorded for further evaluation in the laboratory. Based on the length of the rock core recovered from each run, the following quantities may be calculated for a

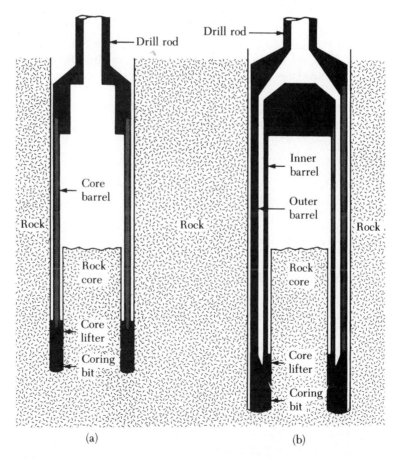

(a) (b)

Figure 2.32 Rock coring: (a) single-tube core barrel; (b) double-tube core barrel

Table 2.6   Standard Size and Designation of Casing, Core Barrel, and Compatible Drill Rod

Casing and core barrel designation	Outside diameter of core barrel bit		Drill rod designation	Outside diameter of drill rod		Diameter of borehole		Diameter of core sample	
	(mm)	(in.)		(mm)	(in.)	(mm)	(in.)	(mm)	(in.)
EX	36.51	$1\frac{7}{16}$	E	33.34	$1\frac{5}{16}$	38.1	$1\frac{1}{2}$	22.23	$\frac{7}{8}$
AX	47.63	$1\frac{7}{8}$	A	41.28	$1\frac{5}{8}$	50.8	2	28.58	$1\frac{1}{8}$
BX	58.74	$2\frac{5}{16}$	B	47.63	$1\frac{7}{8}$	63.5	$2\frac{1}{2}$	41.28	$1\frac{5}{8}$
NX	74.61	$2\frac{15}{16}$	N	60.33	$2\frac{3}{8}$	76.2	3	53.98	$2\frac{1}{8}$

general evaluation of the rock quality encountered:

$$\text{Recovery ratio} = \frac{\text{length of core recovered}}{\text{theoretical length of rock cored}} \tag{2.56}$$

Rock quality designation (RQD)

$$= \frac{\Sigma \text{ length of recovered pieces equal to or larger than 101.6 m (4 in.)}}{\text{theoretical length of rock cored}} \tag{2.57}$$

A recovery ratio of unity indicates the presence of intact rock; for highly fractured rocks, the recovery ratio may be 0.5 or smaller. Table 2.7 presents the general relationship (Deere, 1963) between the RQD and the *in situ* rock quality.

Table 2.7   Relation between *in situ* Rock Quality and RQD

RQD	Rock quality
0–0.25	Very poor
0.25–0.5	Poor
0.5–0.75	Fair
0.75–0.9	Good
0.9–1	Excellent

2.18   *Preparation of Boring Logs*

The detailed information gathered from each borehole is presented in a graphical form called the *boring log*. As a borehole is advanced downward, the driller generally should record the following information in a standard log:

1. Name and address of the drilling company
2. Driller's name
3. Job description and number
4. Number, type, and location of boring
5. Date of boring
6. Subsurface stratification, which can be obtained by visual observation of the soil brought out by auger, split-spoon sampler, and thin-walled Shelby tube sampler
7. Elevation of water table and date observed, use of casing and mud losses, and so on

8. Standard penetration resistance and the depth of SPT

9. Number, type, and depth of soil sample collected

10. In case of rock coring, type of core barrel used and, for each run, the actual length of coring, length of core recovery, and RQD

This information should never be left to memory, because doing so often results in erroneous boring logs.

After completion of the necessary laboratory tests, the geotechnical engineer prepares a finished log that includes notes from the driller's field log and the results of tests conducted in the laboratory. Figure 2.33 shows a typical boring log.

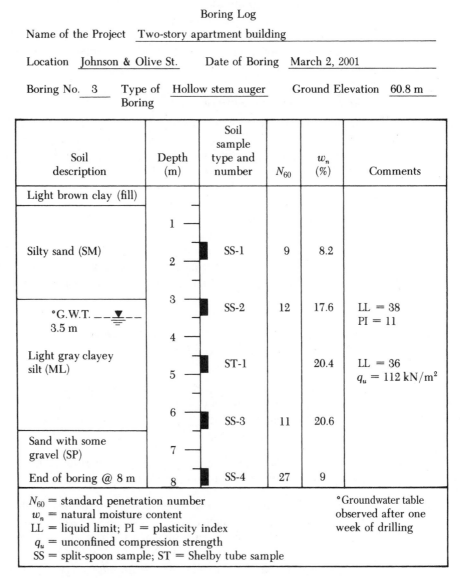

Boring Log

Name of the Project   Two-story apartment building

Location   Johnson & Olive St.   Date of Boring   March 2, 2001

Boring No.   3   Type of   Hollow stem auger   Ground Elevation   60.8 m
Boring

Soil description	Depth (m)	Soil sample type and number	N_{60}	w_n (%)	Comments
Light brown clay (fill)					
Silty sand (SM)	1 2	SS-1	9	8.2	
°G.W.T. ▽ 3.5 m	3 4	SS-2	12	17.6	LL = 38 PI = 11
Light gray clayey silt (ML)	5	ST-1		20.4	LL = 36 q_u = 112 kN/m²
	6	SS-3	11	20.6	
Sand with some gravel (SP)	7				
End of boring @ 8 m	8	SS-4	27	9	

N_{60} = standard penetration number
w_n = natural moisture content
LL = liquid limit; PI = plasticity index
q_u = unconfined compression strength
SS = split-spoon sample; ST = Shelby tube sample

°Groundwater table observed after one week of drilling

Figure 2.33   A typical boring log

These logs have to be attached to the final soil-exploration report submitted to the client. The figure also lists the classifications of the soils in the left-hand column, along with the description of each soil (based on the Unified Soil Classification System).

2.19   *Geophysical Exploration*

Several types of geophysical exploration techniques permit a rapid evaluation of subsoil characteristics. These methods also allow rapid coverage of large areas and are less expensive than conventional exploration by drilling. However, in many cases, definitive interpretation of the results is difficult. For that reason, such techniques should be used for preliminary work only. Here, we discuss three types of geophysical exploration technique: the seismic refraction survey, cross-hole seismic survey, and resistivity survey.

Seismic Refraction Survey

Seismic refraction surveys are useful in obtaining preliminary information about the thickness of the layering of various soils and the depth to rock or hard soil at a site. Refraction surveys are conducted by impacting the surface, such as at point *A* in Figure 2.34a, and observing the first arrival of the disturbance (stress waves) at several other points (e.g., *B, C, D,* ...). The impact can be created by a hammer blow or by a small explosive charge. The first arrival of disturbance waves at various points can be recorded by geophones.

The impact on the ground surface creates two types of *stress wave: P waves* (or *plane waves*) and *S waves* (or *shear waves*). *P* waves travel faster than *S* waves; hence, the first arrival of disturbance waves will be related to the velocities of the *P* waves in various layers. The velocity of *P* waves in a medium is

$$v = \sqrt{\frac{E_s}{\left(\dfrac{\gamma}{g}\right)} \frac{(1 - \mu_s)}{(1 - 2\mu_s)(1 + \mu_s)}} \tag{2.58}$$

where E_s = modulus of elasticity of the medium
  γ = unit weight of the medium
  g = acceleration due to gravity
  μ_s = Poisson's ratio

To determine the velocity v of P waves in various layers and the thicknesses of those layers, we use the following procedure:

1. Obtain the times of first arrival, $t_1, t_2, t_3, \ldots$, at various distances $x_1, x_2, x_3, \ldots$ from the point of impact.
2. Plot a graph of time t against distance x. The graph will look like the one shown in Figure 2.34b.
3. Determine the slopes of the lines *ab, bc, cd,* ... :

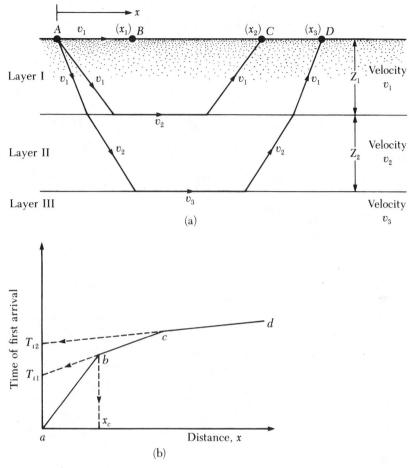

Figure 2.34 Seismic refraction survey

$$\text{Slope of } ab = \frac{1}{v_1}$$

$$\text{Slope of } bc = \frac{1}{v_2}$$

$$\text{Slope of } cd = \frac{1}{v_3}$$

Here, $v_1, v_2, v_3, \ldots$ are the *P*-wave velocities in layers I, II, III, $\ldots$, respectively (Figure 2.34a)

4. Determine the thickness of the top layer:

$$Z_1 = \frac{1}{2}\sqrt{\frac{v_2 - v_1}{v_2 + v_1}}x_c \tag{2.59}$$

The value of x_c can be obtained from the plot, as shown in Figure 2.34b.

5. Determine the thickness of the second layer:

$$Z_2 = \frac{1}{2}\left[T_{i2} - 2Z_1 \frac{\sqrt{v_3^2 - v_1^2}}{v_3 v_1} \right] \frac{v_3 v_2}{\sqrt{v_3^2 - v_2^2}} \qquad (2.60)$$

Here, T_{i2} is the time intercept of the line *cd* in Figure 2.34b, extended backwards.

(For detailed derivatives of these equations and other related information, see Dobrin, 1960, and Das, 1992).

The velocities of *P* waves in various layers indicate the types of soil or rock that are present below the ground surface. The range of the *P*-wave velocity that is generally encountered in different types of soil and rock at shallow depths is given in Table 2.8.

In analyzing the results of a refraction survey, two limitations need to be kept in mind:

1. The basic equations for the survey—that is, Eqs. (2.59) and (2.60)—are based on the assumption that the *P*-wave velocity $v_1 < v_2 < v_3 < \cdots$.
2. When a soil is saturated below the water table, the *P*-wave velocity may be deceptive. *P* waves can travel with a velocity of about 1500 m/sec (5000 ft/sec) through water. For dry, loose soils, the velocity may be well below 1500 m/sec. However, in a saturated condition, the waves will travel through water that is present in the void spaces with a velocity of about 1500 m/sec. If the presence of groundwater has not been detected, the *P*-wave velocity may be erroneously interpreted to indicate a stronger material (e.g., sandstone) than is actually present *in situ*. In general, geophysical interpretations should always be verified by the results obtained from borings.

Table 2.8 Range of *P*-Wave Velocity in Various Soils and Rocks

Type of soil or rock	P-wave velocity	
	m/sec	ft/sec
Soil		
Sand, dry silt, and fine-grained topsoil	200–1000	650–3300
Alluvium	500–2000	1650–6600
Compacted clays, clayey gravel, and dense clayey sand	1000–2500	3300–8200
Loess	250–750	800–2450
Rock		
Slate and shale	2500–5000	8200–16,400
Sandstone	1500–5000	4900–16,400
Granite	4000–6000	13,100–19,700
Sound limestone	5000–10,000	16,400–32,800

Example 2.1

The results of a refraction survey at a site are given in the following table:

Distance of geophone from the source of disturbance (m)	Time of first arrival (sec $\times$ 10^3)
2.5	11.2
5	23.3
7.5	33.5
10	42.4
15	50.9
20	57.2
25	64.4
30	68.6
35	71.1
40	72.1
50	75.5

Determine the *P*-wave velocities and the thickness of the material encountered.

Solution

Velocity

In Figure 2.35, the times of first arrival of the *P* waves are plotted against the distance of the geophone from the source of disturbance. The plot has three straight-line segments. The velocity of the top three layers can now be calculated as follows:

$$\text{Slope of segment } 0a = \frac{1}{v_1} = \frac{\text{time}}{\text{distance}} = \frac{23 \times 10^{-3}}{5.25}$$

or

$$v_1 = \frac{5.25 \times 10^3}{23} = \textbf{228 m/sec (top layer)}$$

$$\text{Slope of segment } ab = \frac{1}{v_2} = \frac{13.5 \times 10^{-3}}{11}$$

or

$$v_2 = \frac{11 \times 10^3}{13.5} = \textbf{814.8 m/sec (middle layer)}$$

$$\text{Slope of segment } bc = \frac{1}{v_3} = \frac{3.5 \times 10^{-3}}{14.75}$$

or

$$v_3 = \textbf{4214 m/sec (third layer)}$$

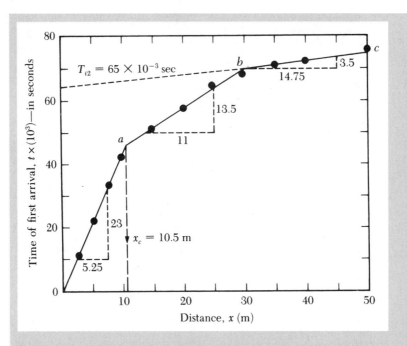

Figure 2.35 Plot of first arrival time of *P* wave vs. distance of geophone from source of disturbance

Comparing the velocities obtained here with those given in Table 2.8 indicates that the third layer is a *rock layer.*

Thickness of Layers
From Figure 2.35, $x_c = 10.5$ m, so

$$Z_1 = \frac{1}{2}\sqrt{\frac{v_2 - v_1}{v_2 + v_1}}x_c \qquad \text{[Eq. (2.59)]}$$

Thus,

$$Z_1 = \frac{1}{2}\sqrt{\frac{814.8 - 228}{814.8 + 228}} \times 10.5 = \textbf{3.94 m}$$

Again, from Eq. (2.60)

$$Z_2 = \frac{1}{2}\left[T_{i2} - \frac{2Z_1\sqrt{v_3^2 - v_1^2}}{(v_3 v_1)}\right]\frac{(v_3)(v_2)}{\sqrt{v_3^2 - v_2^2}}$$

The value of T_{i2} (from Figure 2.35) is 65×10^{-3} sec. Hence,

$$Z_2 = \frac{1}{2}\left[65 \times 10^{-3} - \frac{2(3.94)\sqrt{(4214)^2 - (228)^2}}{(4214)(228)}\right]\frac{(4214)(814.8)}{\sqrt{(4214)^2 - (814.8)^2}}$$

$$= \frac{1}{2}(0.065 - 0.0345)830.47 = \textbf{12.66 m}$$

Thus, the rock layer lies at a depth of $Z_1 + Z_2 = 3.94 + 12.66 = \textbf{16.60 m from the surface of the ground.}$ ∎

Cross-Hole Seismic Survey

The velocity of shear waves created as the result of an impact to a given layer of soil can be effectively determined by the *cross-hole seismic survey* (Stokoe and Woods, 1972). The principle of this technique is illustrated in Figure 2.36, which shows two holes drilled into the ground a distance L apart. A vertical impulse is created at the bottom of one borehole by means of an impulse rod. The shear waves thus generated are recorded by a vertically sensitive transducer. The velocity of shear waves can be calculated as

$$v_s = \frac{L}{t} \tag{2.61}$$

where t = travel time of the waves

The shear modulus G_s of the soil at the depth at which the test is taken can be determined from the relation

$$v_s = \sqrt{\frac{G_s}{(\gamma/g)}}$$

or

$$G_s = \frac{v_s^2 \gamma}{g} \tag{2.62}$$

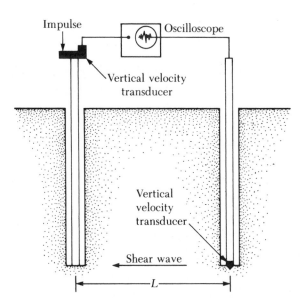

Figure 2.36 Cross-hole method of seismic survey

where v_s = velocity of shear waves
         γ = unit weight of soil
         g = acceleration due to gravity

The shear modulus is useful in the design of foundations to support vibrating machinery and the like.

Resistivity Survey

Another geophysical method for subsoil exploration is the *electrical resistivity survey*. The electrical resistivity of any conducting material having a length L and an area of cross section A can be defined as

$$\rho = \frac{RA}{L} \tag{2.63}$$

where R = electrical resistance

The unit of resistivity is the *ohm-centimeter* or *ohm-meter*. The resistivity of various soils depends primarily on their moisture content and also on the concentration of dissolved ions in them. Saturated clays have a very low resistivity; dry soils and rocks have a high resistivity. The range of resistivity generally encountered in various soils and rocks is given in Table 2.9.

The most common procedure for measuring the electrical resistivity of a soil profile makes use of four electrodes driven into the ground and spaced equally along a straight line. The procedure is generally referred to as the *Wenner method* (Figure 2.37a). The two outside electrodes are used to send an electrical current I (usually a dc current with nonpolarizing potential electrodes) into the ground. The current is typically in the range of 50–100 milliamperes. The voltage drop, V, is measured between the two inside electrodes. If the soil profile is homogeneous, its electrical resistivity is

$$\rho = \frac{2\pi dV}{I} \tag{2.64}$$

In most cases, the soil profile may consist of various layers with different resistivities, and Eq. (2.64) will yield the *apparent resistivity*. To obtain the *actual resistiv-*

Table 2.9 Representative Values of Resistivity

Material	Resistivity (ohm · m)
Sand	500–1500
Clays, saturated silt	0–100
Clayey sand	200–500
Gravel	1500–4000
Weathered rock	1500–2500
Sound rock	>5000

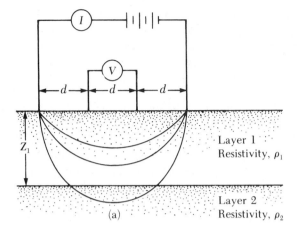

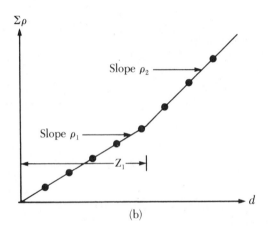

Figure 2.37 Electrical resistivity survey: (a) Wenner method; (b) empirical method for determining resistivity and thickness of each layer

ity of various layers and their thicknesses, one may use an empirical method that involves conducting tests at various electrode spacings (i.e., *d* is changed). The sum of the apparent resistivities, $\Sigma\rho$, is plotted against the spacing *d*, as shown in Figure 2.37b. The plot thus obtained has relatively straight segments, the slopes of which give the resistivity of individual layers. The thicknesses of various layers can be estimated as shown in Figure 2.37b.

The resistivity survey is particularly useful in locating gravel deposits within a fine-grained soil.

2.20 *Subsoil Exploration Report*

At the end of all soil exploration programs, the soil and rock specimens collected in the field are subject to visual observation and appropriate laboratory testing. (The basic soil tests were described in Chapter 1.) After all the required information has been compiled, a soil exploration report is prepared for use by the design office and for reference during future construction work. Although the details and sequence of information in such reports may vary to some degree, depending on the structure under consideration and the person compiling the report, each report should include the following items:

1. A description of the scope of the investigation
2. A description of the proposed structure for which the subsoil exploration has been conducted
3. A description of the location of the site, including any structures nearby, drainage conditions, the nature of vegetation on the site and surrounding it, and any other features unique to the site
4. A description of the geological setting of the site
5. Details of the field exploration—that is, number of borings, depths of borings, types of borings involved, and so on
6. A general description of the subsoil conditions, as determined from soil specimens and from related laboratory tests, standard penetration resistance and cone penetration resistance, and so on
7. A description of the water-table conditions
8. Recommendations regarding the foundation, including the type of foundation recommended, the allowable bearing pressure, and any special construction procedure that may be needed; alternative foundation design procedures should also be discussed in this portion of the report
9. Conclusions and limitations of the investigations

The following graphical presentations should be attached to the report:

1. A site location map
2. A plan view of the location of the borings with respect to the proposed structures and those nearby
3. Boring logs
4. Laboratory test results
5. Other special graphical presentations

The exploration reports should be well planned and documented, as they will help in answering questions and solving foundation problems that may arise later during design and construction.

Problems

2.1   What is the area ratio of a Shelby tube with an outside diameter of 3 in. and an inside diameter of 2.874 in.?

2.2   A soil profile is shown in Figure P2.2, along with the standard penetration numbers in the clay layer. Use Eqs. (2.8) and (2.9) to determine the variation of c_u and OCR with depth.

2.3   The following table gives the variation of the field standard penetration number (N_{60}) in a sand deposit:

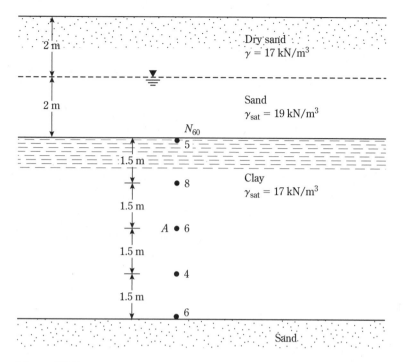

Figure P2.2

Depth (m)	N_{60}
1.5	6
3.0	8
4.5	9
6.0	8
7.5	13
9.0	14

The groundwater table is located at a depth of 6 m. The dry unit weight of sand from 0 to a depth of 6 m is 18 kN/m³, and the saturated unit weight of sand for a depth of 6 m to 12 m is 20.2 kN/m³. Use the relationship given in Eq. (2.12) to calculate the corrected standard penetration numbers.

2.4 For the soil profile described in Problem 2.3, estimate an average peak soil friction angle ϕ'. Use Eq. (2.18).

2.5 For the soil profile described in Problem 2.3, use Eq. (2.19) to determine the average drained friction angle ϕ'.

2.6 The following gives standard penetration numbers determined from a sandy soil deposit in the field:

Depth (ft)	Unit weight of soil (lb/ft³)	N_{60}
10	102	8
15	102	10
20	102	12

25	120	16
30	120	18
35	120	21
40	120	19

Using Eq. (2.19), determine the variation of the peak soil friction angle ϕ'. Estimate an average value of ϕ' for the design of a shallow foundation. (*Note*: For a depth greater than 20 ft, the unit weight of soil is 120 lb/ft³.)

2.7   Redo Problem 2.6, using Eq. (2.11) and Eq. (2.18).

2.8   The following table lists some characteristics of a soil deposit in sand:

Depth (m)	Effective overburden pressure (kN/m²)	Field standard penetration number, N_{60}
3	55	9
4.5	82	11
6	98	12

Assume the uniformity coefficient of the sand (C_u) to be 2.8 and an average overconsolidation ratio (OCR) of 2. Use Eq. (2.16) to estimate the average relative density of the sand between a depth of 3 m and 6 m.

2.9   Vane shear tests were conducted in a layer of clay. The vane dimensions were 63.5 mm(D) $\times$ 127 mm(H). At a certain depth, the torque required to cause failure was 0.051 N·m. The liquid limit of the clay was 46 and the plastic limit was 21. Estimate the undrained cohesion of the clay for use in the design by applying Bjerrum's λ relationship [Eq. (2.28)].

2.10   **a.** A vane shear test was conducted on a saturated clay. The height and diameter of the vanes were 4 in. and 2 in., respectively. During the test, the maximum torque applied was 23 lb-ft. Determine the undrained shear strength of the clay.

b. The clay soil described in Part (a) has a liquid limit of 58 and a plastic limit of 23. What would be the corrected undrained shear strength of the clay for design purposes? Use Bjerrum's relationship for λ [Eq. (2.28)].

2.11   A cone penetration test was conducted in a deposit of normally consolidated dry sand. The following table gives the results of the test:

Depth (m)	Point resistance of cone, q_c (MN/m²)
1.5	2.05
3.0	4.23
4.5	6.01
6.0	8.18
7.5	9.97
9.0	12.42

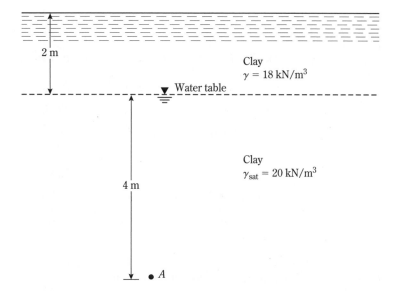

Figure P2.12

Assuming the dry unit weight of sand to be 16 kN/m^3,
a. Use Eq. (2.40) to estimate the average peak friction angle ϕ' of the sand.
b. Use Eq. (2.39) to estimate the average relative density of the sand.
Assume the sand to be moderately compressible.

2.12 In the soil profile shown in Figure P2.12, if the cone penetration resistance (q_c) at A, as determined by an electric friction-cone penetrometer, is 0.8 MN/m^2, find
a. The undrained cohesion, c_u
b. The overconsolidation ratio, OCR

2.13 In a pressuremeter test in a soft saturated clay, the measure cell volume $V_o = 535 \text{ cm}^3$, $p_o = 42.4 \text{ kN/m}^2$, $p_f = 326.5 \text{ kN/m}^2$, $v_o = 46 \text{ cm}^3$, and $v_f = 180 \text{ cm}^3$. Assuming Poisson's ratio (μ_s) to be 0.5 and using Figure 2.28, calculate the pressuremeter modulus E_p.

2.14 A dilatometer test was conducted in a clay deposit. The groundwater table was located at a depth of 3 m below the surface. At a depth of 8 m below the surface, the contact pressure (p_o) was 280 kN/m^2 and the expansion stress (p_1) was 350 kN/m^2. Determine the following:
a. Coefficient of at-rest earth pressure, K_o
b. Overconsolidation ratio, OCR
c. Modulus of elasticity, E_s
Assume σ'_o at a depth of 8 m to be 95 kN/m^2 and $\mu_s = 0.35$.

2.15 Coring of rock was required during a field exploration. The core barrel was advanced 1.5 m during the coring, and the length of the core recovered was 0.6 m. What was the recovery ratio?

2.16 The results of a refraction survey (Figure 2.34a) at a site are given in the following table:

Distance from the source of disturbance (m)	Time of first arrival of P waves (sec $\times 10^3$)
2.5	5.08
5.0	10.16
7.5	15.24
10.0	17.01
15.0	20.02
20.0	24.2
25.0	27.1
30.0	28.0
40.0	31.1
50.0	33.9

Determine the thickness and the P-wave velocity of the materials encountered.

2.17 Repeat Problem 2.16 for the following data:

Distance from the source of disturbance (ft)	Time of first arrival of P waves (sec $\times 10^3$)
25	49.06
50	81.96
75	122.8
100	148.2
150	174.2
200	202.8
250	228.6
300	256.7

References

Aas, G., Lacasse, S., Lunne, I., and Høeg, K. (1986). "Use of In Situ Tests for Foundation Design in Clay," *Proceedings, In Situ '86,* American Society of Civil Engineers, pp. 1–30.

American Society for Testing and Materials (2001). *Annual Book of ASTM Standards,* Vol. 04.08, West Conshohocken, PA.

American Society of Civil Engineers (1972). "Subsurface Investigation for Design and Construction of Foundations of Buildings," *Journal of the Soil Mechanics and Foundations Division,* American Society of Civil Engineers, Vol. 98, No. SM5, pp. 481–490.

Baguelin, F., Jézéquel, J. F., and Shields, D. H. (1978). *The Pressuremeter and Foundation Engineering,* Trans Tech Publications, Clausthal, Germany.

Baldi, G., Bellotti, R., Ghionna, V., and Jamiolkowski, M. (1982). "Design Parameters for Sands from CPT," *Proceedings, Second European Symposium on Penetration Testing,* Amsterdam, Vol. 2, pp. 425–438.

Bjerrum, L. (1972). "Embankments on Soft Ground," *Proceedings of the Specialty Conference,* American Society of Civil Engineers, Vol. 2, pp. 1–54.

Cubrinovski, M., and Ishihara, K. (1999). "Empirical Correlations between SPT N-Values and Relative Density for Sandy Soils," *Soils and Foundations,* Vol. 39, No. 5, pp. 61–92.

Das, B. M. (1992). *Principles of Soil Dynamics,* PWS Publishing Company, Boston.

Deere, D. U. (1963). "Technical Description of Rock Cores for Engineering Purposes," *Felsmechanik und Ingenieurgeologie,* Vol. 1, No. 1, pp. 16–22.

Dobrin, M. B. (1960). *Introduction to Geophysical Prospecting,* McGraw-Hill, New York.

Hansbo, S. (1957). *A New Approach to the Determination of the Shear Strength of Clay by the Fall Cone Test,* Swedish Geotechnical Institute, Report No. 114.

Hara, A., Ohata, T., and Niwa, M. (1971). "Shear Modulus and Shear Strength of Cohesive Soils," *Soils and Foundations,* Vol. 14, No. 3, pp. 1–12.

Hatanaka, M., and Uchida, A. (1996). "Empirical Correlation between Penetration Resistance and Internal Friction Angle of Sandy Soils," *Soils and Foundations,* Vol. 36, No. 4, pp. 1–10.

Jamiolkowski, M., Ladd, C. C., Germaine, J. T., and Lancellotta, R. (1985). "New Developments in Field and Laboratory Testing of Soils," *Proceedings, 11th International Conference on Soil Mechanics and Foundation Engineering,* Vol. 1, pp. 57–153.

Kulhawy, F. H., and Mayne, P. W. (1990). *Manual on Estimating Soil Properties for Foundation Design,* Electric Power Research Institute, Palo Alto, California.

Lancellotta, R. (1983). *Analisi di Affidabilità in Ingegneria Geotecnica,* Atti Istituto Scienza Construzioni, No. 625, Politecnico di Torino.

Larsson, R. (1980). "Undrained Shear Strength in Stability Calculation of Embankments and Foundations on Clay," *Canadian Geotechnical Journal,* Vol. 17, pp. 591–602.

Liao, S. S. C., and Whitman, R. V. (1986). "Overburden Correction Factors for SPT in Sand," *Journal of Geotechnical Engineering,* American Society of Civil Engineers, Vol. 112, No. 3, pp. 373–377.

Marchetti, S. (1980). "*In Situ* Test by Flat Dilatometer," *Journal of Geotechnical Engineering Division,* ASCE, Vol. 106, GT3, pp. 299–321.

Marcuson, W. F., III, and Bieganousky, W. A. (1977). "SPT and Relative Density in Coarse Sands," *Journal of Geotechnical Engineering Division,* American Society of Civil Engineers, Vol. 103, No. 11, pp. 1295–1309.

Mayne, P. W., and Kemper, J. B. (1988). "Profiling OCR in Stiff Clays by CPT and SPT," *Geotechnical Testing Journal,* ASTM, Vol. 11, No. 2, pp. 139–147.

Mayne, P. W., and Mitchell, J. K. (1988). "Profiling of Overconsolidation Ratio in Clays by Field Vane," *Canadian Geotechnical Journal,* Vol. 25, No. 1, pp. 150–158.

Menard, L. (1956). *An Apparatus for Measuring the Strength of Soils in Place,* master's thesis, University of Illinois, Urbana, Illinois.

Morris, P. M., and Williams, D. T. (1994). "Effective Stress Vane Shear Strength Correction Factor Correlations," *Canadian Geotechnical Journal,* Vol. 31, No. 3, pp. 335–342.

Ohya, S., Imai, T., and Matsubara, M. (1982). "Relationships between *N* Value by SPT and LLT Pressuremeter Results," *Proceedings, 2nd European Symposium on Penetration Testing,* Vol. 1, Amsterdam, pp. 125–130.

Osterberg, J. O. (1952). "New Piston-Type Soil Sampler," *Engineering News-Record,* April 24.

Peck, R. B., Hanson, W. E., and Thornburn, T. H. (1974). *Foundation Engineering,* 2d ed., Wiley, New York.

Robertson, P. K., and Campanella, R. G. (1983). "Interpretation of Cone Penetration Tests. Part I: Sand," *Canadian Geotechnical Journal,* Vol. 20, No. 4, pp. 718–733.

Ruiter, J. (1971). "Electric Penetrometer for Site Investigations," *Journal of the Soil Mechanics and Foundations Division,* American Society of Civil Engineers, Vol. 97, No. 2, pp. 457–472.

Schmertmann, J. H. (1975). "Measurement of *In Situ* Shear Strength," *Proceedings, Specialty Conference on* In Situ *Measurement of Soil Properties,* ASCE, Vol. 2, pp. 57–138.

Schmertmann, J. H. (1986). "Suggested Method for Performing the Flat Dilatometer Test," *Geotechnical Testing Journal,* ASTM, Vol. 9, No. 2, pp. 93–101.

Seed, H. B., Arango, I., and Chan, C. K. (1975). *Evaluation of Soil Liquefaction Potential during Earthquakes,* Report No. EERC 75-28, Earthquake Engineering Research Center, University of California, Berkeley.

Seed, H. B., Tokimatsu, K., Harder, L. F., and Chung, R. M. (1985). "Influence of SPT Procedures in Soil Liquefaction Resistance Evaluations," *Journal of Geotechnical Engineering,* ASCE, Vol. 111, No.12, pp. 1425–1445.

Skempton, A. W. (1986). "Standard Penetration Test Procedures and the Effect in Sands of Overburden Pressure, Relative Density, Particle Size, Aging and Overconsolidation," *Geotechnique,* Vol. 36, No. 3, pp. 425–447.

Sowers, G. B., and Sowers, G. F. (1970). *Introductory Soil Mechanics and Foundations,* 3d ed., Macmillan, New York.

Stokoe, K. H., and Woods, R. D. (1972). "*In Situ* Shear Wave Velocity by Cross-Hole Method," *Journal of Soil Mechanics and Foundations Division,* American Society of Civil Engineers, Vol. 98, No. SM5, pp. 443–460.

Stroud, M. (1974). "SPT in Insensitive Clays," *Proceedings, European Symposium on Penetration Testing,* Vol. 2.2, pp. 367–375.

Wolff, T. F. (1989). "Pile Capacity Prediction Using Parameter Functions," in *Predicted and Observed Axial Behavior of Piles, Results of a Pile Prediction Symposium,* sponsored by the Geotechnical Engineering Division, ASCE, Evanston, IL, June, 1989, ASCE Geotechnical Special Publication No. 23, pp. 96–106.

3

Shallow Foundations: Ultimate Bearing Capacity

Introduction

To perform satisfactorily, shallow foundations must have two main characteristics:

1. They have to be safe against overall shear failure in the soil that supports them.
2. They cannot undergo excessive displacement, or settlement. (The term *excessive* is relative, because the degree of settlement allowed for a structure depends on several considerations.)

The load per unit area of the foundation at which shear failure in soil occurs is called the *ultimate bearing capacity,* which is the subject of this chapter.

General Concept

Consider a strip foundation with a width of B resting on the surface of a dense sand or stiff cohesive soil, as shown in Figure 3.1a. Now, if a load is gradually applied to the foundation, settlement will increase. The variation of the load per unit area, with the foundation settlement on the foundation q, is also shown in Figure 3.1a. At a certain point—when the load per unit area equals q_u—a sudden failure in the soil supporting the foundation will take place, and the failure surface in the soil will extend to the ground surface. This load per unit area, q_u, is usually referred to as the *ultimate bearing capacity of the foundation.* When such sudden failure in soil takes place, it is called *general shear failure.*

If the foundation under consideration rests on sand or clayey soil of medium compaction (Figure 3.1b), an increase in the load on the foundation will also be accompanied by an increase in settlement. However, in this case the failure surface in the soil will gradually extend outward from the foundation, as shown by the solid lines in Figure 3.1b. When the load per unit area on the foundation equals $q_{u(1)}$, movement of the foundation will be accompanied by sudden jerks. A considerable movement of the foundation is then required for the failure surface in soil to extend to the ground surface (as shown by the broken lines in the figure). The load per unit area at which this happens is the *ultimate bearing capacity, q_u.* Beyond that point, an increase in load will be accompanied by a large increase in foundation settlement.

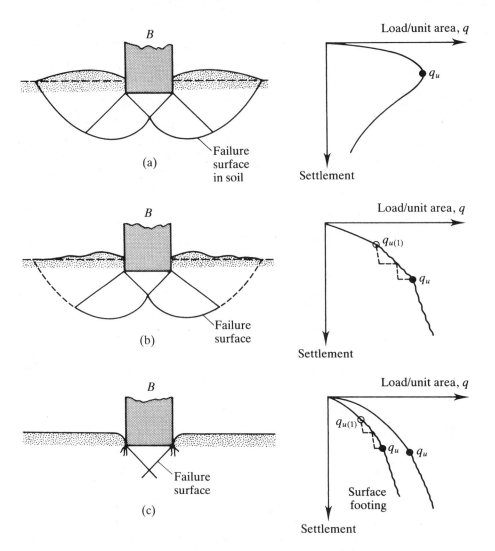

Figure 3.1 Nature of bearing capacity failure in soil: (a) general shear failure: (b) local shear failure; (c) punching shear failure (redrawn after Vesic, 1973)

The load per unit area of the foundation, $q_{u(1)}$, is referred to as the *first failure load* (Vesic, 1963). Note that a peak value of q is not realized in this type of failure, which is called the *local shear failure* in soil.

If the foundation is supported by a fairly loose soil, the load–settlement plot will be like the one in Figure 3.1c. In this case, the failure surface in soil will not extend to the ground surface. Beyond the ultimate failure load, q_u, the load–settlement plot will be steep and practically linear. This type of failure in soil is called the *punching shear failure*.

Vesic (1963) conducted several laboratory load-bearing tests on circular and rectangular plates supported by a sand at various relative densities of compaction, D_r. The variations of $q_{u(1)}/\frac{1}{2}\gamma B$ and $q_u/\frac{1}{2}\gamma B$ obtained from those tests, where B is the diameter of a circular plate or width of a rectangular plate and γ is a dry unit weight

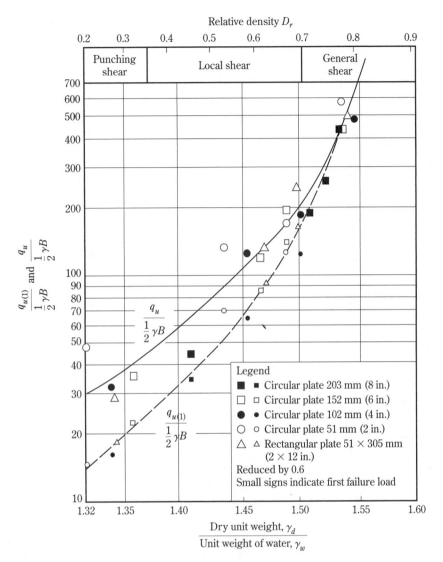

Figure 3.2 Variation of $q_{u(1)}/0.5\gamma B$ and $q_u/0.5\gamma B$ for circular and rectangular plates on the surface of a sand (adapted from Vesic, 1963)

of sand, are shown in Figure 3.2. It is important to note from this figure that, for $D_r \geqslant$ about 70%, the general shear type of failure in soil occurs.

On the basis of experimental results, Vesic (1973) proposed a relationship for the mode of bearing capacity failure of foundations resting on sands. Figure 3.3 shows this relationship, which involves the notation

D_r = relative density of sand

D_f = depth of foundation measured from the ground surface

$$B^{\star} = \frac{2BL}{B + L} \tag{3.1}$$

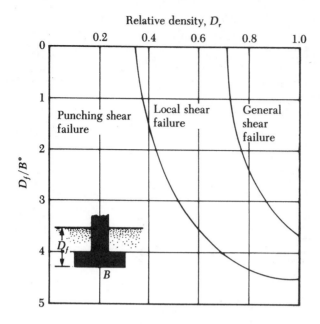

Figure 3.3 Modes of foundation failure in sand (after Vesic, 1973)

where B = width of foundation
L = length of foundation

(*Note:* L is always greater than B.)

For square foundations, $B = L$; for circular foundations, $B = L$ = diameter, so

$$B^\star = B \qquad (3.2)$$

Figure 3.4 shows the settlement S of the circular and rectangular plates on the surface of a sand at *ultimate load,* as described in Figure 3.2. The figure indicates a general range of S/B with the relative density of compaction of sand. So, in general, we can say that, for foundations at a shallow depth (i.e., small $D_f/B^\star$), the ultimate load may occur at a foundation settlement of 4–10% of B. This condition arises together with general shear failure in soil; however, in the case of local or punching shear failure, the ultimate load may occur at settlements of 15–25% of the width of the foundation (B).

3.3   *Terzaghi's Bearing Capacity Theory*

Terzaghi (1943) was the first to present a comprehensive theory for the evaluation of the ultimate bearing capacity of rough shallow foundations. According to this theory, a foundation is *shallow* if its depth, D_f (Figure 3.5), is less than or equal to its width. Later investigators, however, have suggested that foundations with D_f equal to 3–4 times their width may be defined as *shallow foundations.*

Terzaghi suggested that for a *continuous,* or *strip, foundation* (i.e., one whose width-to-length ratio approaches zero), the failure surface in soil at ultimate load may be assumed to be similar to that shown in Figure 3.5. (Note that this is the case of general shear failure, as defined in Figure 3.1a.) The effect of soil above the bottom

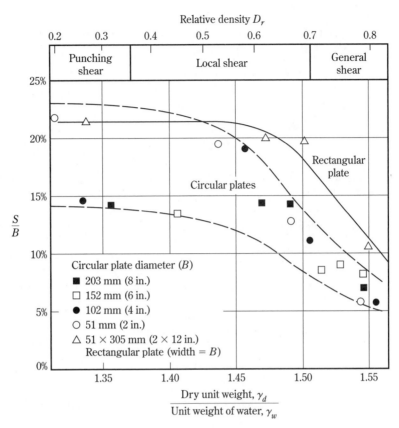

Figure 3.4 Range of settlement of circular and rectangular plates at ultimate load ($D_f/B = 0$) in sand (modified from Vesic, 1963)

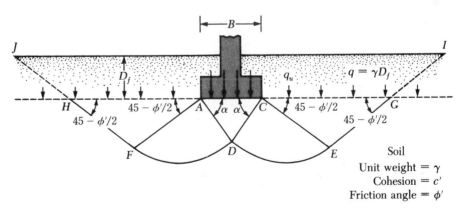

Figure 3.5 Bearing capacity failure in soil under a rough rigid continuous foundation

of the foundation may also be assumed to be replaced by an equivalent surcharge, $q = \gamma D_f$ (where γ is a unit weight of soil). The failure zone under the foundation can be separated into three parts (see Figure 3.5):

1. The *triangular zone ACD* immediately under the foundation
2. The *radial shear zones ADF* and *CDE,* with the curves *DE* and *DF* being arcs of a logarithmic spiral
3. Two triangular *Rankine passive zones AFH* and *CEG*

The angles *CAD* and *ACD* are assumed to be equal to the soil friction angle ϕ'. Note that, with the replacement of the soil above the bottom of the foundation by an equivalent surcharge q, the shear resistance of the soil along the failure surfaces *GI* and *HJ* was neglected.

Using equilibrium analysis, Terzaghi expressed the ultimate bearing capacity in the form

$$q_u = c'N_c + qN_q + \tfrac{1}{2}\gamma BN_\gamma \quad \text{(continuous foundation)} \tag{3.3}$$

where
$$\begin{aligned}
c' &= \text{cohesion of soil} \\
\gamma &= \text{unit weight of soil} \\
q &= \gamma D_f \\
N_c, N_q, N_\gamma &= \text{bearing capacity factors that are nondimensional and are func-} \\
&\quad\ \text{tions only of the soil friction angle } \phi'
\end{aligned}$$

The bearing capacity factors N_c, N_q, and N_γ are defined by

$$N_c = \cot \phi' \left[\frac{e^{2(3\pi/4 - \phi'/2)\tan \phi'}}{2 \cos^2\left(\dfrac{\pi}{4} + \dfrac{\phi'}{2}\right)} - 1 \right] = \cot \phi' (N_q - 1) \tag{3.4}$$

$$N_q = \frac{e^{2(3\pi/4 - \phi'/2)\tan \phi'}}{2 \cos^2\left(45 + \dfrac{\phi'}{2}\right)} \tag{3.5}$$

and

$$N_\gamma = \frac{1}{2}\left(\frac{K_{p\gamma}}{\cos^2 \phi'} - 1\right)\tan \phi' \tag{3.6}$$

where $K_{p\gamma}$ = passive pressure coefficient

The variations of the bearing capacity factors defined by Eqs. (3.4), (3.5), and (3.6) are given in Table 3.1.

To estimate the ultimate bearing capacity of *square* and *circular foundations,* Eq. (3.1) may be respectively modified to

$$q_u = 1.3c'N_c + qN_q + 0.4\gamma BN_\gamma \quad \text{(square foundation)} \tag{3.7}$$

Table 3.1 Terzaghi's Bearing Capacity Factors—Eqs. (3.4), (3.5), and (3.6)

ϕ'	N_c	N_q	$N_\gamma{}^a$	ϕ'	N_c	N_q	$N_\gamma{}^a$
0	5.70	1.00	0.00	26	27.09	14.21	9.84
1	6.00	1.1	0.01	27	29.24	15.90	11.60
2	6.30	1.22	0.04	28	31.61	17.81	13.70
3	6.62	1.35	0.06	29	34.24	19.98	16.18
4	6.97	1.49	0.10	30	37.16	22.46	19.13
5	7.34	1.64	0.14	31	40.41	25.28	22.65
6	7.73	1.81	0.20	32	44.04	28.52	26.87
7	8.15	2.00	0.27	33	48.09	32.23	31.94
8	8.60	2.21	0.35	34	52.64	36.50	38.04
9	9.09	2.44	0.44	35	57.75	41.44	45.41
10	9.61	2.69	0.56	36	63.53	47.16	54.36
11	10.16	2.98	0.69	37	70.01	53.80	65.27
12	10.76	3.29	0.85	38	77.50	61.55	78.61
13	11.41	3.63	1.04	39	85.97	70.61	95.03
14	12.11	4.02	1.26	40	95.66	81.27	115.31
15	12.86	4.45	1.52	41	106.81	93.85	140.51
16	13.68	4.92	1.82	42	119.67	108.75	171.99
17	14.60	5.45	2.18	43	134.58	126.50	211.56
18	15.12	6.04	2.59	44	151.95	147.74	261.60
19	16.56	6.70	3.07	45	172.28	173.28	325.34
20	17.69	7.44	3.64	46	196.22	204.19	407.11
21	18.92	8.26	4.31	47	224.55	241.80	512.84
22	20.27	9.19	5.09	48	258.28	287.85	650.67
23	21.75	10.23	6.00	49	298.71	344.63	831.99
24	23.36	11.40	7.08	50	347.50	415.14	1072.80
25	25.13	12.72	8.34				

[a]From Kumbhojkar (1993)

and

$$q_u = 1.3c'N_c + qN_q + 0.3\gamma B N_\gamma \quad \text{(circular foundation)} \qquad (3.8)$$

In Eq. (3.7), B equals the dimension of each side of the foundation; in Eq. (3.8), B equals the diameter of the foundation.

For foundations that exhibit the local shear failure mode in soils, Terzaghi suggested the following modifications to Eqs. (3.3), (3.7), and (3.8):

$$q_u = \frac{2}{3}c'N_c' + qN_q' + \tfrac{1}{2}\gamma B N_\gamma' \qquad \text{(strip foundation)} \qquad (3.9)$$

$$q_u = 0.867c'N_c' + qN_q' + 0.4\gamma B N_\gamma' \quad \text{(square foundation)} \qquad (3.10)$$

$$q_u = 0.867c'N_c' + qN_q' + 0.3\gamma B N_\gamma' \quad \text{(circular foundation)} \qquad (3.11)$$

N_c', N_q', and N_γ', the *modified bearing capacity factors,* can be calculated by using the bearing capacity factor equations (for N_c, N_q, and N_γ, respectively) by replacing ϕ' by $\overline{\phi}' = \tan^{-1}(\frac{2}{3} \tan \phi')$. The variation of N_c', N_q', and N_γ' with the soil friction angle ϕ is given in Table 3.2.

      Terzaghi's bearing capacity equations have now been modified to take into account the effects of the foundation shape (B/L), depth of embedment (D_f), and the load inclination. This is given in Section 3.7. Many design engineers, however, still use Terzaghi's equation, which provides fairly good results considering the uncertainty of the soil conditions at various sites.

3.4   *Factor of Safety*

Calculating the gross *allowable load-bearing capacity* of shallow foundations requires the application of a factor of safety (FS) to the gross ultimate bearing capacity, or

$$q_{\text{all}} = \frac{q_u}{\text{FS}} \qquad (3.12)$$

Table 3.2 Terzaghi's Modified Bearing Capacity Factors N_c', N_q', and N_γ'

ϕ'	N_c'	N_q'	N_γ'	ϕ'	N_c'	N_q'	N_γ'
0	5.70	1.00	0.00	26	15.53	6.05	2.59
1	5.90	1.07	0.005	27	16.30	6.54	2.88
2	6.10	1.14	0.02	28	17.13	7.07	3.29
3	6.30	1.22	0.04	29	18.03	7.66	3.76
4	6.51	1.30	0.055	30	18.99	8.31	4.39
5	6.74	1.39	0.074	31	20.03	9.03	4.83
6	6.97	1.49	0.10	32	21.16	9.82	5.51
7	7.22	1.59	0.128	33	22.39	10.69	6.32
8	7.47	1.70	0.16	34	23.72	11.67	7.22
9	7.74	1.82	0.20	35	25.18	12.75	8.35
10	8.02	1.94	0.24	36	26.77	13.97	9.41
11	8.32	2.08	0.30	37	28.51	15.32	10.90
12	8.63	2.22	0.35	38	30.43	16.85	12.75
13	8.96	2.38	0.42	39	32.53	18.56	14.71
14	9.31	2.55	0.48	40	34.87	20.50	17.22
15	9.67	2.73	0.57	41	37.45	22.70	19.75
16	10.06	2.92	0.67	42	40.33	25.21	22.50
17	10.47	3.13	0.76	43	43.54	28.06	26.25
18	10.90	3.36	0.88	44	47.13	31.34	30.40
19	11.36	3.61	1.03	45	51.17	35.11	36.00
20	11.85	3.88	1.12	46	55.73	39.48	41.70
21	12.37	4.17	1.35	47	60.91	44.45	49.30
22	12.92	4.48	1.55	48	66.80	50.46	59.25
23	13.51	4.82	1.74	49	73.55	57.41	71.45
24	14.14	5.20	1.97	50	81.31	65.60	85.75
25	14.80	5.60	2.25				

However, some practicing engineers prefer to use a factor of safety such that

$$\text{Net stress increase on soil} = \frac{\text{net ultimate bearing capacity}}{\text{FS}} \qquad (3.13)$$

The net ultimate bearing capacity is defined as the ultimate pressure per unit area of the foundation that can be supported by the soil in excess of the pressure caused by the surrounding soil at the foundation level. If the difference between the unit weight of concrete used in the foundation and the unit weight of soil surrounding is assumed to be negligible, then

$$q_{net(u)} = q_u - q \qquad (3.14)$$

where $q_{net(u)}$ = net ultimate bearing capacity
        $q = \gamma D_f$

So

$$q_{all(net)} = \frac{q_u - q}{\text{FS}} \qquad (3.15)$$

The factor of safety as defined by Eq. (3.15) should be at least 3 in all cases.

Example 3.1

A square foundation is 1.5 m × 1.5 m in plan. The soil supporting the foundation has a friction angle $\phi' = 20°$, and $c' = 15.2$ kN/m². The unit weight of soil, γ, is 17.8 kN/m³. Determine the allowable gross load on the foundation with a factor of safety (FS) of 4. Assume that the depth of the foundation (D_f) is 1 meter and that general shear failure occurs in soil.

Solution
From Eq. (3.7),

$$q_u = 1.3c'N_c + qN_q + 0.4\gamma BN_\gamma$$

From Table 3.1, for $\phi' = 20°$,

$$N_c = 17.69$$
$$N_q = 7.44$$
$$N_\gamma = 3.64$$

Thus,

$$q_u = (1.3)(15.2)(17.69) + (1 \times 17.8)(7.44) + (0.4)(17.8)(1.5)(3.64)$$
$$= 349.55 + 132.43 + 38.87 = 520.85 \approx 521 \text{ kN/m}^2$$

So the allowable load per unit area of the foundation is

$$q_{all} = \frac{q_u}{\text{FS}} = \frac{521}{4} = 130.25 \text{ kN/m}^2 \approx 130 \text{ kN/m}^2$$

Thus, the total allowable gross load

$$Q = (130)B^2 = (130)(1.5 \times 1.5) = \underline{292.5 \text{ kN}} \quad \blacksquare$$

Example 3.2

Repeat Example 3.1, assuming that local shear failure occurs in the soil supporting the foundation.

Solution
From Eq. (3.10),

$$q_u = 0.867c'N_c' + qN_q' + 0.4\gamma BN_\gamma'$$

From Table 3.2, for $\phi' = 20°$,

$$N_c' = 11.85$$
$$N_q' = 3.88$$
$$N_\gamma' = 1.12$$

So

$$q_u = (0.867)(15.2)(11.85) + (1 \times 17.8)(3.88) + (0.4)(17.8)(1.5)(1.12)$$

$$= 156.2 + 69.1 + 12.0 = 237.3 \text{ kN/m}^2$$

$$q_{\text{all}} = \frac{237.3}{4} = 59.3 \text{ kN/m}^2$$

Thus,

$$\text{Allowable gross load} = Q = (q_{\text{all}})(B^2) = (59.3)(1.5^2) = \underline{133.4 \text{ kN}} \quad \blacksquare$$

3.5 *Modification of Bearing Capacity Equations for Water Table*

Equations (3.3) and (3.7) through (3.11) give the ultimate bearing capacity, based on the assumption that the water table is located well below the foundation. However, if the water table is close to the foundation, some modifications of the bearing capacity equations will be necessary. (See Figure 3.6.)

Case I. If the water table is located so that $0 \le D_1 \le D_f$, the factor q in the bearing capacity equations takes the form

$$q = \text{effective surcharge} = D_1\gamma + D_2(\gamma_{\text{sat}} - \gamma_w) \tag{3.16}$$

where γ_{sat} = saturated unit weight of soil
γ_w = unit weight of water

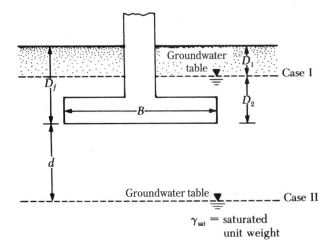

Groundwater table

D_1

Case I

D_f

D_2

B

d

Groundwater table

Case II

γ_{sat} = saturated unit weight

Figure 3.6 Modification of bearing capacity equations for water table

Also, the value of γ in the last term of the equations has to be replaced by $\gamma' = \gamma_{sat} - \gamma_w$.

Case II. For a water table located so that $0 \leq d \leq B$,

$$q = \gamma D_f \tag{3.17}$$

In this case, the factor γ in the last term of the bearing capacity equations must be replaced by the factor

$$\bar{\gamma} = \gamma' + \frac{d}{B}(\gamma - \gamma') \tag{3.18}$$

The preceding modifications are based on the assumption that there is no seepage force in the soil.

Case III. When the water table is located so that $d \geq B$, the water will have no effect on the ultimate bearing capacity.

3.6 Case History: Ultimate Bearing Capacity in Saturated Clay

Brand et al. (1972) reported field-test results of soil explorations on small foundations in soft Bangkok clay (a deposit of marine clay) in Rangsit, Thailand. The results are shown in Figure 3.7. Because of the sensitivity of the clay, the laboratory test results for c_u (unconfined compression and unconsolidated undrained triaxial) were rather scattered; however, better results were obtained for the

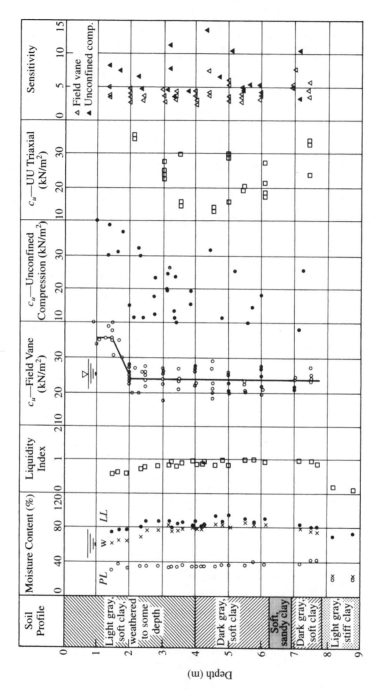

Figure 3.7 Results of soil exploration in soft Bangkok clay at Rangsit, Thailand (redrawn after Brand et al., 1972)

variation of c_u with depth from field vane shear tests, which showed that the average variations of the undrained cohesion were as follows:

Depth (m)	C_u (kN/m²)
0–1.5	≈35
1.5–2	Decreasing linearly from 35 to 24
2–8	≈24

Five small square foundations were tested for ultimate bearing capacity. The sizes of the foundations were 0.6 m × 0.6 m, 0.675 m × 0.675 m, 0.75 m × 0.75 m, 0.9 m × 0.9 m, and 1.05 m × 1.05 m. The depth of the bottom of the foundations was 1.5 m, measured from the ground surface. The load–settlement plots obtained from the bearing capacity tests are shown in Figure 3.8.

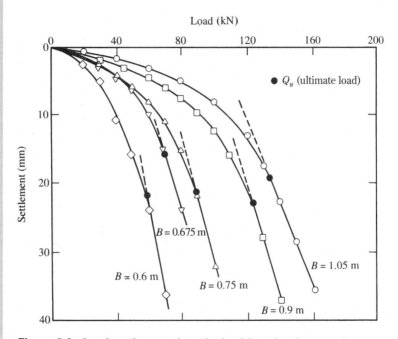

Figure 3.8 Load–settlement plots obtained from bearing capacity tests

Analysis of the Field-Test Results

The ultimate loads, Q_u, obtained from each test are also shown in Figure 3.8. The ultimate load is defined as the point where the load displacement becomes practically linear. The failure in soil below the foundation is of the local shear type. Hence, we may apply Eq. (3.10):

$$q_u = 0.867c_u N_c' + qN_q' + 0.4\gamma BN_\gamma'$$

For $\phi = 0$, $c = c_u$ and, from Table 3.2, $N_c' = 5.7$, $N_q' = 1$, and $N_\gamma' = 0$. Thus, for $\phi = 0$,

$$q_u = 4.94c_u + q \tag{3.19}$$

Assuming that the unit weight of soil is about 18.5 kN/m^3, it follows that $q \approx D_f \gamma = (1.5)(18.5) = 27.75 \text{ kN/m}^2$. We can then assume average values of c_u: For depths of 1.5 m to 2.0 m, $c_u \approx (35 + 24)/2 = 29.5 \text{ kN/m}^2$; for depths greater than 2.0 m, $c_u \approx 24 \text{ kN/m}^2$. If we further assume that the undrained cohesion of clay at depth $\leq B$ below the foundation controls the ultimate bearing capacity, then

$$c_{u(\text{average})} \approx \frac{(29.5)(2.0 - 1.5) + (24)[B - (2.0 - 1.5)]}{B} \quad (3.20)$$

The value of $c_{u(\text{average})}$ obtained for each foundation needs to be corrected in view of Eq. (2.19). Table 3.3 presents the details of other calculations and a comparison of the theoretical and field ultimate bearing capacities.

Table 3.3 Comparison of Theoretical and Field Ultimate Bearing Capacities

B (m)	$c_{u(\text{average})}$[a] (kN/m²)	Plasticity index[b]	Correction factor, λ[c]	$c_{u(\text{corrected})}$[d] (kN/m²)	$q_{u(\text{theory})}$[e] (kN/m²)	$Q_{u(\text{field})}$[f] (kN)	$q_{u(\text{field})}$[g] (kN/m²)
0.6	28.58	40	0.84	24.01	146.4	60	166.6
0.675	28.07	40	0.84	23.58	144.2	71	155.8
0.75	27.67	40	0.84	23.24	142.6	90	160
0.9	27.06	40	0.84	22.73	140.0	124	153
1.05	26.62	40	0.84	22.36	138.2	140	127

[a] Eq. (3.20)
[b] From Figure 3.7
[c] From Eq. (2.28) [$\lambda = 1.7 - 0.54 \log PI$; Bjerrum (1972)]
[d] Eq. (2.27)
[e] Eq. (3.19)
[f] Figure 3.8
[g] $Q_{u(\text{field})}/B^2$

Note that the ultimate bearing capacities obtained from the field are about 10% higher than those obtained from theory. One reason for such a difference is that the ratio D_f/B for the field tests varies from 1.5 to 2.5. The increase in the bearing capacity due to the depth of embedment has not been accounted for in Eq. (3.20).

3.7   *The General Bearing Capacity Equation*

The ultimate bearing capacity equations (3.3), (3.7), and (3.8) are for continuous, square, and circular foundations only; they do not address the case of rectangular foundations ($0 < B/L < 1$). Also, the equations do not take into account the shearing resistance along the failure surface in soil above the bottom of the foundation (the portion of the failure surface marked as *GI* and *HJ* in Figure 3.5). In addition, the load on the foundation may be inclined. To account for all these shortcomings, Meyerhof (1963) suggested the following form of the general bearing capacity equation:

$$q_u = c'N_cF_{cs}F_{cd}F_{ci} + qN_qF_{qs}F_{qd}F_{qi} + \frac{1}{2}\gamma BN_\gamma F_{\gamma s}F_{\gamma d}F_{\gamma i} \qquad (3.21)$$

In this equation, c' = cohesion

q = effective stress at the level of the bottom of the foundation

γ = unit weight of soil

B = width of foundation (= diameter for a circular foundation)

$F_{cs}, F_{qs}, F_{\gamma s}$ = shape factors

$F_{cd}, F_{qd}, F_{\gamma d}$ = depth factors

$F_{ci}, F_{qi}, F_{\gamma i}$ = load inclination factors

N_c, N_q, N_γ = bearing capacity factors

The equations for determining the various factors given in Eq. (3.21) are described briefly in the sections that follow. Note that the original equation for ultimate bearing capacity is derived only for the plane-strain case (i.e., for continuous foundations). The shape, depth, and load inclination factors are empirical factors based on experimental data.

Bearing Capacity Factors

The basic nature of the failure surface in soil suggested by Terzaghi now appears to have been borne out by laboratory and field studies of bearing capacity (Vesic, 1973). However, the angle α shown in Figure 3.5 is closer to $45 + \phi'/2$ than to ϕ'. If this change is accepted, the values of N_c, N_q, and N_γ for a given soil friction angle will also change from those given in Table 3.1. With $\alpha = 45 + \phi'/2$, it can be shown that

$$N_q = \tan^2\left(45 + \frac{\phi'}{2}\right)e^{\pi \tan \phi'} \qquad (3.22)$$

and

$$N_c = (N_q - 1)\cot \phi' \qquad (3.23)$$

Equation (3.23) for N_c was originally derived by Prandtl (1921), and Eq. (3.22) for N_q was presented by Reissner (1924). Caquot and Kerisel (1953) and Vesic (1973) gave the relation for N_γ as

$$N_\gamma = 2(N_q + 1)\tan \phi' \qquad (3.24)$$

Table 3.4 shows the variation of the preceding bearing capacity factors with soil friction angles.

Shape Factors   The equations for the shape factors F_{cs}, F_{qs}, and $F_{\gamma s}$ were recommended by De Beer (1970) and are

$$F_{cs} = 1 + \left(\frac{B}{L}\right)\left(\frac{N_q}{N_c}\right) \tag{3.25}$$

$$F_{qs} = 1 + \left(\frac{B}{L}\right)\tan \phi' \tag{3.26}$$

and

$$F_{\gamma s} = 1 - 0.4\left(\frac{B}{L}\right) \tag{3.27}$$

where L = length of the foundation $(L > B)$

Table 3.4   Bearing Capacity Factors

ϕ'	N_c	N_q	N_γ	ϕ'	N_c	N_q	N_γ
0	5.14	1.00	0.00	26	22.25	11.85	12.54
1	5.38	1.09	0.07	27	23.94	13.20	14.47
2	5.63	1.20	0.15	28	25.80	14.72	16.72
3	5.90	1.31	0.24	29	27.86	16.44	19.34
4	6.19	1.43	0.34	30	30.14	18.40	22.40
5	6.49	1.57	0.45	31	32.67	20.63	25.99
6	6.81	1.72	0.57	32	35.49	23.18	30.22
7	7.16	1.88	0.71	33	38.64	26.09	35.19
8	7.53	2.06	0.86	34	42.16	29.44	41.06
9	7.92	2.25	1.03	35	46.12	33.30	48.03
10	8.35	2.47	1.22	36	50.59	37.75	56.31
11	8.80	2.71	1.44	37	55.63	42.92	66.19
12	9.28	2.97	1.69	38	61.35	48.93	78.03
13	9.81	3.26	1.97	39	67.87	55.96	92.25
14	10.37	3.59	2.29	40	75.31	64.20	109.41
15	10.98	3.94	2.65	41	83.86	73.90	130.22
16	11.63	4.34	3.06	42	93.71	85.38	155.55
17	12.34	4.77	3.53	43	105.11	99.02	186.54
18	13.10	5.26	4.07	44	118.37	115.31	224.64
19	13.93	5.80	4.68	45	133.88	134.88	271.76
20	14.83	6.40	5.39	46	152.10	158.51	330.35
21	15.82	7.07	6.20	47	173.64	187.21	403.67
22	16.88	7.82	7.13	48	199.26	222.31	496.01
23	18.05	8.66	8.20	49	229.93	265.51	613.16
24	19.32	9.60	9.44	50	266.89	319.07	762.89
25	20.72	10.66	10.88				

The shape factors are empirical relations based on extensive laboratory tests.

Depth Factors   Hansen (1970) proposed the following equations for the depth factors:

$$F_{cd} = 1 + 0.4\left(\frac{D_f}{B}\right) \tag{3.28}$$

$$F_{qd} = 1 + 2\tan\phi'(1 - \sin\phi')^2\frac{D_f}{B} \tag{3.29}$$

$$F_{\gamma d} = 1 \tag{3.30}$$

Equations (3.28) and (3.29) are valid for $D_f/B \leq 1$. For a depth-of-embedment-to-foundation-width ratio greater than unity $(D_f/B > 1)$, the equations have to be modified to

$$F_{cd} = 1 + (0.4)\tan^{-1}\left(\frac{D_f}{B}\right) \tag{3.31}$$

$$F_{qd} = 1 + 2\tan\phi'(1 - \sin\phi')^2\tan^{-1}\left(\frac{D_f}{B}\right) \tag{3.32}$$

and

$$F_{\gamma d} = 1 \tag{3.33}$$

respectively. The factor $\tan^{-1}(D_f/B)$ is in radians in Eqs. (3.31) and (3.32).

Inclination Factors   Meyerhof (1963) and Hanna and Meyerhof (1981) suggested the following inclination factors for use in Eq. (3.21):

$$F_{ci} = F_{qi} = \left(1 - \frac{\beta°}{90°}\right)^2 \tag{3.34}$$

$$F_{\gamma i} = \left(1 - \frac{\beta}{\phi'}\right)^2 \tag{3.35}$$

Here, β = inclination of the load on the foundation with respect to the vertical.

3.8   *Meyerhof's Bearing Capacity, Shape, Depth, and Inclination Factors*

In most solutions, presented in this text, the bearing capacity, shape, depth, and inclination factors presented in Section 3.7 will be used. However, many geotechnical engineers employ the various factors recommended by Meyerhof (1963) for use in Eq. (3.21). Table 3.5 is a summary of those factors.

Table 3.5 Meyerhof's Bearing Capacity, Shape, Depth, and Inclination Factors [Eq. (3.21)]

Factor	Relationship
Bearing capacity	
N_c	Equation (3.23)
N_q	Equation (3.22)
N_γ	$(N_q - 1) \tan(1.4\,\phi')$; see Table 3.6
Shape	
For $\phi = 0$,	
$\quad F_{cs}$	$1 + 0.2\,(B/L)$
$\quad F_{qs} = F_{\gamma s}$	1
For $\phi' \geq 10°$,	
$\quad F_{cs}$	$1 + 0.2\,(B/L)\tan^2(45 + \phi'/2)$
$\quad F_{qs} = F_{\gamma s}$	$1 + 0.1\,(B/L)\tan^2(45 + \phi'/2)$
Depth	
For $\phi = 0$,	
$\quad F_{cd}$	$1 + 0.2\,(D_f/B)$
$\quad F_{qd} = F_{\gamma d}$	1
For $\phi' \geq 10°$	
$\quad F_{cd}$	$1 + 0.2\,(D_f/B)\tan(45 + \phi'/2)$
$\quad F_{qd} = F_{\gamma d}$	$1 + 0.1\,(D_f/B)\tan(45 + \phi'/2)$
Inclination	
$F_{ci} = F_{qi}$	Equation (3.34)
$F_{\gamma i}$	Equation (3.35)

Table 3.6 Meyerhof's Bearing Capacity Factor $N_\gamma = (N_q - 1)\tan(1.4\,\phi')$

ϕ'	N_γ	ϕ'	N_γ	ϕ'	N_γ	ϕ'	N_γ
0	0.00	14	0.92	28	11.19	42	139.32
1	0.002	15	1.13	29	13.24	43	171.14
2	0.01	16	1.38	30	15.67	44	211.41
3	0.02	17	1.66	31	18.56	45	262.74
4	0.04	18	2.00	32	22.02	46	328.73
5	0.07	19	2.40	33	26.17	47	414.32
6	0.11	20	2.87	34	31.15	48	526.44
7	0.15	21	3.42	35	37.15	49	674.91
8	0.21	22	4.07	36	44.43	50	873.84
9	0.28	23	4.82	37	53.27	51	1143.93
10	0.37	24	5.72	38	64.07	52	1516.05
11	0.47	25	6.77	39	77.33	53	2037.26
12	0.60	26	8.00	40	93.69		
13	0.74	27	9.46	41	113.99		

Example 3.3

A square foundation $(B \times B)$ has to be constructed as shown in Figure 3.9. Assume that $\gamma = 105 \text{ lb/ft}^3$, $\gamma_{\text{sat}} = 118 \text{ lb/ft}^3$, $D_f = 4 \text{ ft}$, and $D_1 = 2 \text{ ft}$. The gross allowable load, Q_{all}, with FS $= 3$ is 150,000 lb. The standard penetration resistance, N_{60} values are as follows:

Depth (ft)	N_{60} (blow/ft)
5	4
10	6
15	6
20	10
25	5

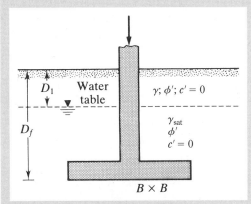

Figure 3.9 A square foundation

Determine the size of the footing. Use Eq. (3.21).

Solution

From Eqs. (2.10) and (2.11),

$$(N_1)_{60} = N_{60}\left(\frac{p_a}{\sigma'_o}\right)^{0.5} \tag{a}$$

Combining Eqs. (2.20) and (a) yields

$$\phi' = \left[20N_{60}\left(\frac{p_a}{\sigma'_o}\right)^{0.5}\right]^{0.5} + 20 \tag{b}$$

Thus, $p_a \approx 2000 \text{ lb/ft}^2$. Now the following table can be prepared:

Depth (ft)	N_{60}	σ'_o(lb/ft²)	ϕ' (deg) [Eq. (b)]
5	4	$2 \times 105 + 3(118 - 62.4) = 376.8$	33.6
10	6	$376.8 + 5(118 - 62.4) = 654.8$	34.5
15	6	$654.8 + 5(118 - 62.4) = 932.8$	33.3
20	10	$932.8 + 5(118 - 62.4) = 1210.8$	36.0
25	5	$1210.8 + 5(118 - 62.4) = 1488.8$	31.6

Average $\phi' = 33.8° \approx 34°$

Next, we have

$$q_{all} = \frac{Q_{all}}{B^2} = \frac{150,000}{B^2} \text{ lb/ft}^2 \qquad \text{(c)}$$

From Eq. (3.21) (with $c' = 0$), we obtain

$$q_{all} = \frac{q_u}{\text{FS}} = \frac{1}{3}\left(qN_q F_{qs} F_{qd} + \frac{1}{2}\gamma' B N_\gamma F_{\gamma s} F_{\gamma d} \right)$$

For $\phi' = 34°$, from Table 3.4, $N_q = 29.44$ and $N_\gamma = 41.06$. Hence,

$$F_{qs} = 1 + \frac{B}{L}\tan\phi' = 1 + \tan 34 = 1.67$$

$$F_{\gamma s} = 1 - 0.4\left(\frac{B}{L}\right) = 1 - 0.4 = 0.6$$

$$F_{qd} = 1 + 2\tan\phi'(1 - \sin\phi')^2\frac{D_f}{B} = 1 + 2\tan 34(1 - \sin 34)^2\frac{4}{B} = 1 + \frac{1.05}{B}$$

$$F_{\gamma d} = 1$$

and

$$q = (2)(105) + 2(118 - 62.4) = 321.2 \text{ lb/ft}^2$$

So

$$\begin{aligned}
q_{all} = \frac{1}{3}&\left[(321.2)(29.44)(1.67)\left(1 + \frac{1.05}{B}\right) \right.\\
&\left. + \left(\frac{1}{2}\right)(118 - 62.4)(B)(41.06)(0.6)(1) \right] \qquad \text{(d)}\\
= 5263.9 &+ \frac{5527.1}{B} + 228.3B
\end{aligned}$$

Combining Eqs. (c) and (d) results in

$$\frac{150,000}{B^2} = 5263.9 + \frac{5527.1}{B} + 228.3B$$

By trial and error, we find that $B \approx 4.5$ ft. ∎

Example 3.4

A square column foundation has to carry a gross allowable total mass of 15,290 kg. The depth of the foundation is 0.7 m. The load is inclined at an angle of 20° to the vertical. (See Figure 3.10.) Determine the width of the foundation, B. Use Eq. (3.21) and a factor of safety of 3.

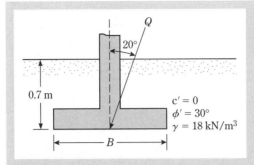

Figure 3.10 A square column foundation

Solution

With $c' = 0$, the ultimate bearing capacity becomes

$$q_u = qN_qF_{qs}F_{qd}F_{qi} + \frac{1}{2}\gamma BN_\gamma F_{\gamma s}F_{\gamma d}F_{\gamma i} \qquad (3.21)$$

with

$$q = (0.7)(18) = 12.6 \text{ kN/m}^2$$

and

$$\gamma = 18 \text{ kN/m}^3$$

From Table 3.4, for $\phi' = 30°$,

$$N_q = 18.4$$

$$N_\gamma = 22.4$$

$$F_{qs} = 1 + \left(\frac{B}{L}\right)\tan \phi' = 1 + 0.577 = 1.577$$

$$F_{\gamma s} = 1 - 0.4\left(\frac{B}{L}\right) = 0.6$$

$$F_{qd} = 1 + 2\tan \phi'(1 - \sin \phi')^2 \frac{D_f}{B} = 1 + \frac{(0.289)(0.7)}{B} = 1 + \frac{0.202}{B}$$

$$F_{\gamma d} = 1$$

$$F_{qi} = \left(1 - \frac{\beta°}{90°}\right)^2 = \left(1 - \frac{20}{90}\right)^2 = 0.605$$

and

$$F_{\gamma i} = \left(1 - \frac{\beta°}{\phi'}\right)^2 = \left(1 - \frac{20}{30}\right)^2 = 0.11$$

Hence,

$$q_u = (12.6)(18.4)(1.577)\left(1 + \frac{0.202}{B}\right)(0.605)$$

$$+ (0.5)(18)(B)(22.4)(0.6)(1)(0.11)$$

$$= 221.2 + \frac{44.68}{B} + 13.3B \tag{a}$$

Thus,

$$q_{all} = \frac{q_u}{3} = 73.73 + \frac{14.89}{B} + 4.43\,B \tag{b}$$

We are given that Q = total allowable load = $q_{all} \times B^2$, or

$$q_{all} = \frac{15,290 \times 9.81}{B^2}\,\text{N/m}^2 \approx \frac{150}{B^2}\,\text{kN/m}^2 \tag{c}$$

Equating the right-hand sides of Eqs. (b) and (c) yields

$$\frac{150}{B^2} = 73.73 + \frac{14.89}{B} + 4.43B$$

By trial and error, we obtain $B \approx$ **1.3 m** ∎

3.9 *Effect Of Soil Compressibility*

In Section 3.3, Eqs. (3.3), (3.7), and (3.8), which apply to the case of general shear failure, were modified to Eqs. (3.9), (3.10), and (3.11) to take into account the change of failure mode in soil (i.e., local shear failure). The change of failure mode is due to soil compressibility, to account for which Vesic (1973) proposed the following modification of Eq. (3.21):

$$q_u = c'N_c F_{cs}F_{cd}F_{cc} + qN_q F_{qs}F_{qd}F_{qc} + \tfrac{1}{2}\gamma B N_\gamma F_{\gamma s}F_{\gamma d}F_{\gamma c} \tag{3.36}$$

In this equation, F_{cc}, F_{qc}, and $F_{\gamma c}$ are soil compressibility factors.

The soil compressibility factors were derived by Vesic (1973) by analogy to the expansion of cavities. According to that theory, in order to calculate F_{cc}, F_{qc}, and $F_{\gamma c}$, the following steps should be taken:

1. Calculate the *rigidity index, I_r,* of the soil at a depth approximately $B/2$ below the bottom of the foundation, or

$$I_r = \frac{G_s}{c' + q'\tan\phi'} \tag{3.37}$$

where G_s = shear modulus of the soil

q = effective overburden pressure at a depth of $D_f + B/2$

2. The critical rigidity index, $I_{r(cr)}$, can be expressed as

$$I_{r(cr)} = \frac{1}{2}\left\{ \exp\left[\left(3.30 - 0.45\frac{B}{L} \right) \cot\left(45 - \frac{\phi'}{2} \right) \right] \right\}$$ (3.38)

The variations of $I_{r(cr)}$ for $B/L = 0$ and $B/L = 1$ are given in Table 3.7.

3. If $I_r \geqslant I_{r(cr)}$, then

$$F_{cc} = F_{qc} = F_{\gamma c} = 1$$

However, if $I_r < I_{r(cr)}$, then

$$F_{\gamma c} = F_{qc} = \exp\left\{ \left(-4.4 + 0.6\frac{B}{L} \right)\tan\phi' + \left[\frac{(3.07 \sin\phi')(\log 2I_r)}{1 + \sin\phi'} \right] \right\}$$ (3.39)

Figure 3.11 shows the variation of $F_{\gamma c} = F_{qc}$ [see Eq. (3.39)] with ϕ' and I_r. For $\phi = 0$,

$$F_{cc} = 0.32 + 0.12\frac{B}{L} + 0.60 \log I_r$$ (3.40)

Table 3.7 Variation of $I_{r(cr)}$ with ϕ' and B/L[1]

	$I_{r(cr)}$	
ϕ' (deg)	$\dfrac{B}{L} = 0$	$\dfrac{B}{L} = 1$
0	13	8
5	18	11
10	25	15
15	37	20
20	55	30
25	89	44
30	152	70
35	283	120
40	592	225
45	1442	482
50	4330	1258

[1] After Vesic (1973)

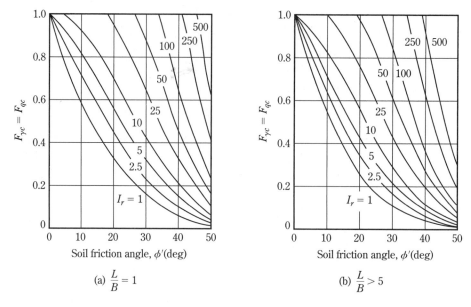

Figure 3.11 Variation of $F_{\gamma c} = F_{qc}$ with I_r and ϕ'

For $\phi' > 0$,

$$F_{cc} = F_{qc} - \frac{1 - F_{qc}}{N_q \tan \phi'} \qquad (3.41)$$

Example 3.5

For a shallow foundation, $B = 0.6$ m, $L = 1.2$ m, and $D_f = 0.6$ m. The known soil characteristics are as follows:

    Soil: $\phi' = 25°$

           $c' = 48$ kN/m²

           $\gamma = 18$ kN/m³

           Modulus of elasticity, $E_s = 620$ kN/m²

           Poisson's ratio, $\mu_s = 0.3$

Calculate the ultimate bearing capacity.

Solution   From Eq. (3.37),

$$I_r = \frac{G_s}{c' + q' \tan \phi'}$$

However,

$$G = \frac{E_s}{2(1 + \mu_s)}$$

So

$$I_r = \frac{E_s}{2(1 + \mu_s)[c' + q'\tan\phi']}$$

Now,

$$q' = \gamma\left(D_f + \frac{B}{2}\right) = 18\left(0.6 + \frac{0.6}{2}\right) = 16.2 \text{ kN/m}^2$$

Thus,

$$I_r = \frac{620}{2(1 + 0.3)[48 + 16.2\tan 25]} = 4.29$$

From Eq. (3.38),

$$I_{r(cr)} = \frac{1}{2}\left\{\exp\left[\left(3.3 - 0.45\frac{B}{L}\right)\cot\left(45 - \frac{\phi'}{2}\right)\right]\right\}$$

$$= \frac{1}{2}\left\{\exp\left[\left(3.3 - 0.45\frac{0.6}{1.2}\right)\cot\left(45 - \frac{25}{2}\right)\right]\right\} = 62.46$$

Since $I_{r(cr)} > I_r$, we use Eqs. (3.39) and (3.41) to obtain

$$F_{\gamma c} = F_{qc} = \exp\left\{\left(-4.4 + 0.6\frac{B}{L}\right)\tan\phi' + \left[\frac{(3.07\sin\phi')\log(2I_r)}{1 + \sin\phi'}\right]\right\}$$

$$= \exp\left\{\left(-4.4 + 0.6\frac{0.6}{1.2}\right)\tan 25\right.$$

$$\left. + \left[\frac{(3.07\sin 25)\log(2 \times 4.29)}{1 + \sin 25}\right]\right\} = 0.347$$

and

$$F_{cc} = F_{qc} - \frac{1 - F_{qc}}{N_q\tan\phi'}$$

For $\phi' = 25°$, $N_q = 10.66$ (see Table 3.4); therefore,

$$F_{cc} = 0.347 - \frac{1 - 0.347}{10.66\tan 25} = 0.216$$

Now, from Eq. (3.36),

$$q_u = c'N_cF_{cs}F_{cd}F_{cc} + qN_qF_{qs}F_{qd}F_{qc} + \tfrac{1}{2}\gamma BN_\gamma F_{\gamma s}F_{\gamma d}F_{\gamma c}$$

From Table 3.4, for $\phi' = 25°$, $N_c = 20.72$, $N_q = 10.66$, and $N_\gamma = 10.88$. Consequently,

$$F_{cs} = 1 + \left(\frac{N_q}{N_c}\right)\left(\frac{B}{L}\right) = 1 + \left(\frac{10.66}{20.72}\right)\left(\frac{0.6}{1.2}\right) = 1.257$$

$$F_{qs} = 1 + \frac{B}{L}\tan\phi' = 1 + \frac{0.6}{1.2}\tan 25 = 1.233$$

$$F_{\gamma s} = 1 - 0.4\frac{B}{L} = 1 - 0.4\frac{0.6}{1.2} = 0.8$$

$$F_{cd} = 1 + 0.4\left(\frac{D_f}{B}\right) = 1 + 0.4\left(\frac{0.6}{0.6}\right) = 1.4$$

$$F_{qd} = 1 + 2\tan\phi'(1 - \sin\phi')^2\left(\frac{D_f}{B}\right)$$

$$= 1 + 2\tan 25\,(1 - \sin 25)^2\left(\frac{0.6}{0.6}\right) = 1.311$$

and

$$F_{\gamma d} = 1$$

Thus,

$$q_u = (48)(20.72)(1.257)(1.4)(0.216) + (0.6 \times 18)(10.66)(1.233)(1.311)$$
$$(0.347) + (\tfrac{1}{2})(18)(0.6)(10.88)(0.8)(1)(0.347) = \textbf{459 kN/m}^2 \quad \blacksquare$$

3.10   *Eccentrically Loaded Foundations*

In several instances, as with the base of a retaining wall, foundations are subjected to moments in addition to the vertical load, as shown in Figure 3.12a. In such cases, the distribution of pressure by the foundation on the soil is not uniform. The nominal distribution of pressure is

$$q_{max} = \frac{Q}{BL} + \frac{6M}{B^2 L} \tag{3.42}$$

and

$$q_{min} = \frac{Q}{BL} - \frac{6M}{B^2 L} \tag{3.43}$$

where Q = total vertical load
        M = moment on the foundation

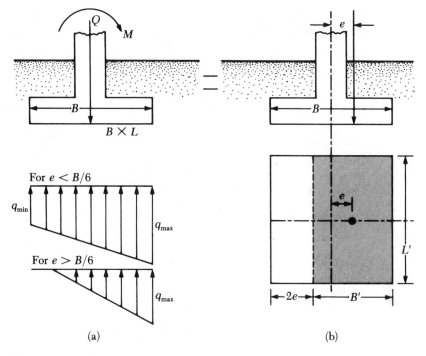

Figure 3.12 Eccentrically loaded foundations

Figure 3.12b shows a force system equivalent to that shown in Figure 3.12a. The distance

$$e = \frac{M}{Q} \tag{3.44}$$

is the eccentricity. Substituting Eq. (3.44) into Eqs. (3.42) and (3.43) gives

$$q_{max} = \frac{Q}{BL}\left(1 + \frac{6e}{B}\right) \tag{3.45}$$

and

$$q_{min} = \frac{Q}{BL}\left(1 - \frac{6e}{B}\right) \tag{3.46}$$

Note that, in these equations, when the eccentricity e becomes $B/6$, q_{min} is zero. For $e > B/6$, q_{min} will be negative, which means that tension will develop. Because soil cannot take any tension, there will then be a separation between the foundation and the soil underlying it. The nature of the pressure distribution on the soil will be as shown in Figure 3.12a. The value of q_{max} is then

$$q_{max} = \frac{4Q}{3L(B - 2e)} \tag{3.47}$$

The exact distribution of pressure is difficult to estimate.

The factor of safety for such types of loading against bearing capacity failure can be evaluated by using the procedure suggested by Meyerhof (1953), which is generally referred to as the *effective area* method. The following is Meyerhof's step-by-step procedure for determining the ultimate load that the soil can support and the factor of safety against bearing capacity failure:

1. Determine the effective dimensions of the foundation:

$$B' = \text{effective width} = B - 2e$$

$$L' = \text{effective length} = L$$

Note that if the eccentricity were in the direction of the length of the foundation, the value of L' would be equal to $L - 2e$. The value of B' would equal B. The smaller of the two dimensions (i.e., L' and B') is the effective width of the foundation.

2. Use Eq. (3.21) for the ultimate bearing capacity:

$$q'_u = c'N_cF_{cs}F_{cd}F_{ci} + qN_qF_{qs}F_{qd}F_{qi} + \tfrac{1}{2}\gamma B'N_\gamma F_{\gamma s}F_{\gamma d}F_{\gamma i} \tag{3.48}$$

To evaluate F_{cs}, F_{qs}, and $F_{\gamma s}$, use Eqs. (3.25) through (3.27) with *effective length* and *effective width* dimensions instead of L and B, respectively. To determine F_{cd}, F_{qd}, and $F_{\gamma d}$, use Eqs. (3.28) through (3.33). Do not replace B with B'.

3. The total ultimate load that the foundation can sustain is

$$Q_{ult} = q'_u \overbrace{(B')(L')}^{A} \tag{3.49}$$

where $A' = \text{effective area}$

4. The factor of safety against bearing capacity failure is

$$\text{FS} = \frac{Q_{ult}}{Q} \tag{3.50}$$

5. Check the factor of safety against q_{max}, or $\text{FS} = q'_u/q_{max}$.

Foundations with Two-Way Eccentricity

Consider a situation in which a foundation is subjected to a vertical ultimate load Q_{ult} and a moment M, as shown in Figures 3.13a and b. For this case, the components of the moment M about the x- and y-axes can be determined as M_x and M_y, respectively. (See Figure 3.13.) This condition is equivalent to a load Q_{ult} placed eccentrically on the foundation with $x = e_B$ and $y = e_L$ (Figure 3.13d). Note that

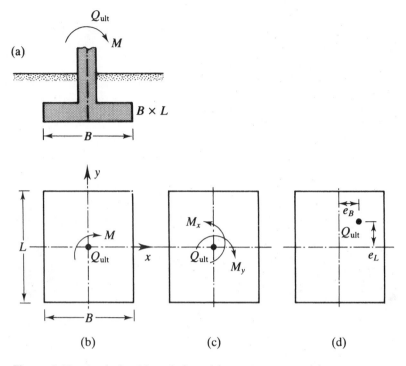

Figure 3.13 Analysis of foundation with two-way eccentricity

$$e_B = \frac{M_y}{Q_{ult}} \qquad (3.51)$$

and

$$e_L = \frac{M_x}{Q_{ult}} \qquad (3.52)$$

If Q_{ult} is needed, it can be obtained from Eq. (3.49); that is,

$$Q_{ult} = q'_u A'$$

where, from Eq. (3.48),

$$q'_u = c' N_c F_{cs} F_{cd} F_{ci} + q N_q F_{qs} F_{qd} F_{qi} + \tfrac{1}{2} \gamma B' N_\gamma F_{\gamma s} F_{\gamma d} F_{\gamma i}$$

and

$$A' = \text{effective area} = B'L'$$

As before, to evaluate F_{cs}, F_{qs}, and $F_{\gamma s}$ [Eqs. (3.25) through (3.27)], we use the effective length L' and effective width B' instead of L and B, respectively. To calculate F_{cd}, F_{qd}, and $F_{\gamma d}$, we use Eqs. (3.28) through (3.33); however, we do not replace B with B'. In determining the effective area A', effective width B', and effective length L', five possible cases may arise (Highter and Anders, 1985).

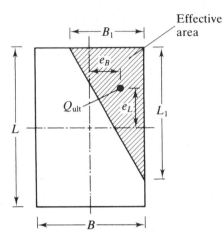

Figure 3.14 Effective area for the case of $e_L/L \geq \frac{1}{6}$ and $e_B/B \geq \frac{1}{6}$

Case I. $e_L/L \geq \frac{1}{6}$ and $e_B/B \geq \frac{1}{6}$. The effective area for this condition is shown in Figure 3.14, or

$$A' = \tfrac{1}{2}B_1L_1 \tag{3.53}$$

where

$$B_1 = B\left(1.5 - \frac{3e_B}{B}\right) \tag{3.54}$$

and

$$L_1 = L\left(1.5 - \frac{3e_L}{L}\right) \tag{3.55}$$

The effective length L' is the larger of the two dimensions B_1 and L_1. So the effective width is

$$B' = \frac{A'}{L'} \tag{3.56}$$

Case II. $e_L/L < 0.5$ and $0 < e_B/B < \frac{1}{6}$. The effective area for this case, shown in Figure 3.15a, is

$$A' = \frac{1}{2}(L_1 + L_2)B \tag{3.57}$$

The magnitudes of L_1 and L_2 can be determined from Figure 3.15b. The effective width is

$$B' = \frac{A'}{L_1 \text{ or } L_2 \quad \text{(whichever is larger)}} \tag{3.58}$$

The effective length is

$$L' = L_1 \text{ or } L_2 \quad \text{(whichever is larger)} \tag{3.59}$$

Case III. $e_L/L < \frac{1}{6}$ and $0 < e_B/B < 0.5$. The effective area, shown in Figure 3.16a, is

$$A' = \frac{1}{2}(B_1 + B_2)L \tag{3.60}$$

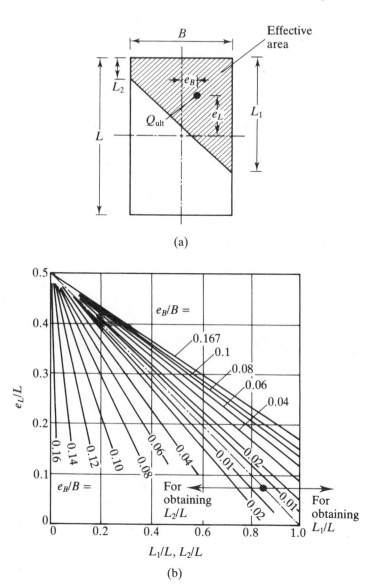

Figure 3.15 Effective area for the case of $e_L/L < 0.5$ and $0 < e_B/B < \frac{1}{6}$ (after Higher and Anders, 1985)

The effective width is

$$B' = \frac{A'}{L} \tag{3.61}$$

The effective length is

$$L' = L \tag{3.62}$$

The magnitudes of B_1 and B_2 can be determined from Figure 3.16b.

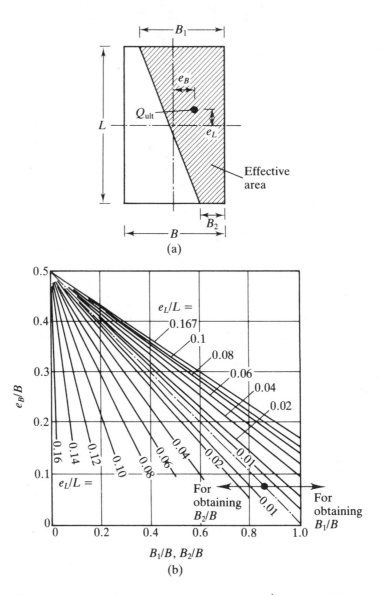

Figure 3.16   Effective area for the case of $e_L/L < \frac{1}{6}$ and $0 < e_B/B < 0.5$ (after Highter and Anders, 1985)

Case IV. $e_L/L < \frac{1}{6}$ and $e_B/B < \frac{1}{6}$. Figure 3.17a shows the effective area for this case. The ratio B_2/B, and thus B_2, can be determined by using the e_L/L curves that slope upward. Similarly, the ratio L_2/L, and thus L_2, can be determined by using the e_L/L curves that slope downward. The effective area is then

$$A' = L_2 B + \tfrac{1}{2}(B + B_2)(L - L_2) \qquad (3.63)$$

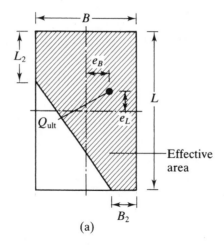

(a)

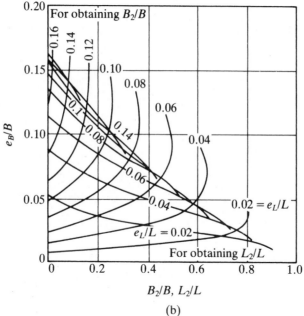

$B_2/B, L_2/L$

(b)

Figure 3.17 Effective area for the case of $e_L/L < \frac{1}{6}$ and $e_B/B < \frac{1}{6}$ (after Highter and Anders, 1985)

The effective width is

$$B' = \frac{A'}{L} \tag{3.64}$$

The effective length is

$$L' = L \tag{3.65}$$

Case V. (Circular Foundation) In the case of circular foundations under eccentric loading (Figure 3.18a), the eccentricity is always one way. The effective area A' and

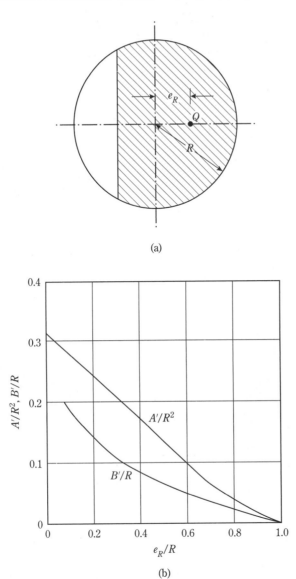

Figure 3.18 Effective area for circular foundation (after Highter and Anders, 1985)

the effective width B' for a circular foundation are given in a nondimensional (unitless) form in Figure 3.18b. Hence, the effective length

$$L' = \frac{A'}{B'}$$

Example 3.6

A continuous foundation is shown in Figure 3.19. If the load eccentricity is 0.5 ft, determine the ultimate load, Q_{ult}, per unit length of the foundation.

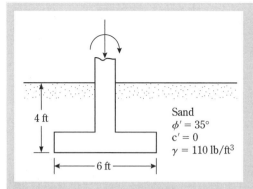

Figure 3.19 A continuous foundation with load eccentricity

Solution

For $c' = 0$, Eq. (3.48) gives

$$q_u' = qN_qF_{qs}F_{qd}F_{qi} + \frac{1}{2}\gamma B'N_\gamma F_{\gamma s}F_{\gamma d}F_{\gamma i}$$

where $q = (110)(4) = 440 \text{ lb/ft}^2$

For $\phi' = 35°$, from Table 3.4, $N_q = 33.3$ and $N_\gamma = 48.03$. Also,

$$B' = 6 - (2)(0.5) = 5 \text{ ft}$$

Because the foundation in question is a strip foundation, B'/L' is zero. Hence, $F_{qs} = 1$, $F_{\gamma s} = 1$,

$$F_{qi} = F_{\gamma i} = 1$$

$$F_{qd} = 1 + 2\tan\phi'(1 - \sin\phi')^2\frac{D_f}{B} = 1 + 0.255\left(\frac{4}{6}\right) = 1.17$$

$$F_{\gamma d} = 1$$

and

$$q_u' = (440)(33.3)(1)(1.17)(1) + \left(\frac{1}{2}\right)(110)(5)(48.03)(1)(1)(1) = 30{,}351 \text{ lb/ft}^2$$

Consequently,

$$Q_{\text{ult}} = (B')(1)(q_u') = (5)(1)(30{,}351) = 151{,}755 \text{ lb/ft} = \textbf{75.88 ton/ft} \quad \blacksquare$$

Example 3.7

A square foundation is shown in Figure 3.20, with $e_L = 0.3$ m and $e_B = 0.15$ m. Assume two-way eccentricity, and determine the ultimate load, Q_{ult}.

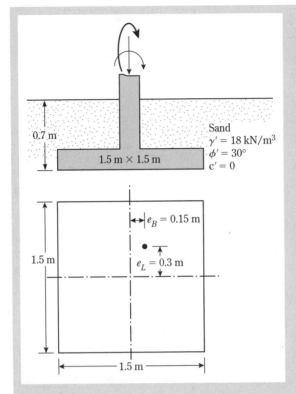

Figure 3.20 An eccentrically loaded foundation

Solution

We have

$$\frac{e_L}{L} = \frac{0.3}{1.5} = 0.2$$

and

$$\frac{e_B}{B} = \frac{0.15}{1.5} = 0.1$$

This case is similar to that shown in Figure 3.15a. From Figure 3.15b, for $e_L/L = 0.2$ and $e_B/B = 0.1$,

$$\frac{L_1}{L} \approx 0.85; \qquad L_1 = (0.85)(1.5) = 1.275 \text{ m}$$

and

$$\frac{L_2}{L} \approx 0.21; \qquad L_2 = (0.21)(1.5) = 0.315 \text{ m}$$

From Eq. (3.57),

$$A' = \tfrac{1}{2}(L_1 + L_2)B = \tfrac{1}{2}(1.275 + 0.315)(1.5) = 1.193 \text{ m}^2$$

From Eq. (3.59),

$$L' = L_1 = 1.275 \text{ m}$$

From Eq. (3.58),

$$B' = \frac{A'}{L'} = \frac{1.193}{1.275} = 0.936 \text{ m}$$

Note from Eq. (3.48) with $c' = 0$,

$$q'_u = qN_qF_{qs}F_{qd}F_{qi} + \frac{1}{2}\gamma B'N_\gamma F_{\gamma s}F_{\gamma d}F_{\gamma i}$$

where $q = (0.7)(18) = 12.6 \text{ kN/m}^2$

For $\phi' = 30°$, from Table 3.4, $N_q = 18.4$ and $N_\gamma = 22.4$. Thus,

$$F_{qs} = 1 + \left(\frac{B'}{L'}\right)\tan\phi' = 1 + \left(\frac{0.936}{1.275}\right)\tan 30° = 1.424$$

$$F_{\gamma s} = 1 - 0.4\left(\frac{B'}{L'}\right) = 1 - 0.4\left(\frac{0.936}{1.275}\right) = 0.706$$

$$F_{qd} = 1 + 2\tan\phi'(1 - \sin\phi')^2\frac{D_f}{B} = 1 + \frac{(0.289)(0.7)}{1.5} = 1.135$$

and

$$F_{\gamma d} = 1$$

So

$$\begin{aligned}Q_{ult} = A'q'_u &= A'(qN_qF_{qs}F_{qd} + \frac{1}{2}\gamma B'N_\gamma F_{\gamma s}F_{\gamma d}) \\ &= (1.193)[(12.6)(18.4)(1.424)(1.135) \\ &\quad + (0.5)(18)(0.936)(22.4)(0.706)(1)] \approx 606 \text{ kN} \quad\blacksquare\end{aligned}$$

Problems

3.1 For a continuous foundation, the following are given:
   a. $B = 4$ ft, $D_f = 3$ ft, $\gamma = 110$ lb/ft³, $\phi' = 25°$, $c' = 600$ lb/ft²
   b. $B = 2$ m, $D_f = 1$ m, $\gamma = 17$ kN/m³, $\phi' = 30°$, $c' = 0$
   Use Terzaghi's equation and a factor of safety of 4 to determine the gross allowable vertical load-bearing capacity. Assume that general shear failure occurs in soil.

3.2 A square column foundation is 2 m × 2 m in plan. Let $D_f = 1.5$ m, $\gamma = 16.5$ kN/m³, $\phi' = 36°$, and $c' = 0$. Assuming general shear failure in soil, use Terzaghi's equation and a factor of safety of 3 to determine the gross allowable vertical load the column could carry.

3.3 Redo Problem 3.1, using Eq. (3.21) and the bearing capacity, shape, and depth factors given in Section 3.7.

3.4 Redo Problem 3.2, using Eq. (3.21) and the bearing capacity, shape, and depth factors given in Section 3.7.

3.5   For the column foundation shown in Figure P3.5, determine the gross allowable load Q_{all}. Use FS = 4, Eq. (3.21), and other factors given in Section 3.7.

3.6   A column foundation is shown in Figure P3.6. Using Eq. (3.21) and the bearing capacity, shape, and depth factors given in Section 3.7, determine the net allowable load [see Eq. (3.15)] the foundation can carry. Use FS = 3.

3.7   Solve Problem 3.6, using Eq. (3.21) and Meyerhof's bearing capacity, shape, and depth factors (Section 3.8). Use FS = 3.

3.8   For a square foundation, $D_f = 2$ m, $\gamma = 16.5$ kN/m^3, $\phi' = 30°$, $c' = 0$, gross allowable load $Q_{all} = 3330$ kN, and FS = 4. Determine the size of the foundation. Use Eq. (3.21) and the bearing capacity, shape, and depth factors given in Section 3.7.

3.9   A foundation measuring 8 ft $\times$ 8 ft has to be constructed in a granular soil deposit. For this foundation, $D_f = 5$ ft and $\gamma = 110$ lb/ft^3. Following are the results of a standard penetration test in that soil:

Depth (ft)	Field standard penetration number, N_{60}
5	11
10	14
15	16
20	21
25	24

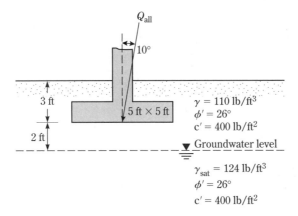

Figure P3.5

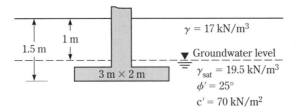

Figure P3.6

a. Use Eq. (2.19) to estimate an average friction angle ϕ' for the soil.
b. From Eq. (3.21), estimate the gross allowable load the foundation can carry. Use the bearing capacity, shape, and depth factors given in Section 3.7. Also, use FS = 4.

3.10 For the design of a shallow foundation, the following characteristics apply:

Soil: $\phi' = 25°$
$c' = 50 \text{ kN/m}^2$
Unit weight, $\gamma = 17 \text{ kN/m}^3$
Modulus of elasticity, $E_s = 1020 \text{ kN/m}^2$
Poisson's ratio, $\mu_s = 0.35$

Foundation: $L = 1.5 \text{ m}$
$B = 1 \text{ m}$
$D_f = 1 \text{ m}$

Calculate the ultimate bearing capacity. Use Eq. (3.36) and the bearing capacity, shape, and depth factors given in Section 3.7.

3.11 An eccentrically loaded foundation is shown in Figure P3.11. Determine the ultimate load Q_u that the foundation can carry. Use Meyerhof's bearing capacity, shape, and depth factors. (See Section 3.8.)

3.12 An eccentrically loaded foundation is shown in Figure P3.12. Use FS = 4, and determine the allowable load that the foundation can carry. Use the bearing capacity, shape, and depth factors given in Section 3.7.

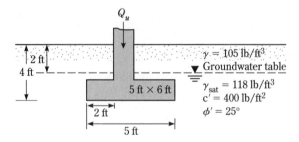

Figure P3.11

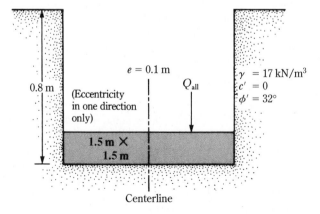

Figure P3.12

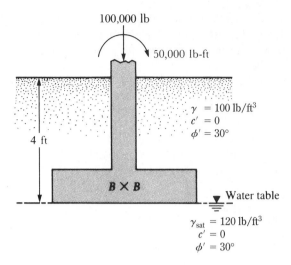

Figure P3.13

3.13   A square footing is shown in Figure P3.13. Use FS = 6, and determine the size of the footing. Use bearing capacity, shape, and depth factors given in Section 3.7.

3.14   The shallow foundation measures shown in Figure 3.13 are 4 ft × 6 ft and is subjected to a centric load and a moment. If e_B = 0.4 ft, e_L = 1.2 ft, and the depth of the foundation is 3 ft, determine the allowable load the foundation can carry. Use a factor of safety of 4. For the soil, we are told that unit weight γ = 115 lb/ft³, friction angle ϕ' = 35°, and cohesion c' = 0. Use the bearing capacity, shape, and depth factors given in Section 3.7.

References

Bjerrum, L. (1972). "Embankments on Soft Ground." *Proceedings of the Specialty Conference, American Society of Civil Engineers,* Vol. 2, pp. 1–54.

Brand, E. W., Muktabhant, C., and Taechanthummarak, A. (1972). "Load Test on Small Foundations in Soft Clay," *Proceedings, Specialty Conference on Performance of Earth and Earth-Supported Structures, American Society of Civil Engineers,* Vol. 1, Part 2, pp. 903–928.

Caquot, A., and Kerisel, J. (1953). "Sur le terme de surface dans le calcul des fondations en milieu pulverulent," *Proceedings, Third International Conference on Soil Mechanics and Foundation Engineering,* Zürich, Vol. I, pp. 336–337.

De Beer, E. E. (1970). "Experimental Determination of the Shape Factors and Bearing Capacity Factors of Sand," *Geotechnique,* Vol. 20, No. 4, pp. 387–411.

Hanna, A. M., and Meyerhof, G. G. (1981). "Experimental Evaluation of Bearing Capacity of Footings Subjected to Inclined Loads," *Canadian Geotechnical Journal,* Vol. 18, No. 4, pp. 599–603.

Hansen, J. B. (1970). *A Revised and Extended Formula for Bearing Capacity,* Bulletin 28, Danish Geotechnical Institute, Copenhagen.

Highter, W. H., and Anders, J. C. (1985). "Dimensioning Footings Subjected to Eccentric Loads," *Journal of Geotechnical Engineering*, American Society of Civil Engineers, Vol. 111, No. GT5, pp. 659–665.

Kumbhojkar, A. S. (1993). "Numerical Evaluation of Terzaghi's N_γ," *Journal of Geotechnical Engineering*, American Society of Civil Engineers, Vol. 119, No. 3, pp. 598–607.

Meyerhof, G. G. (1953). "The Bearing Capacity of Foundations Under Eccentric and Inclined Loads," *Proceedings, Third International Conference on Soil Mechanics and Foundation Engineering*, Zürich, Vol. 1, pp. 440–445.

Meyerhof, G. G. (1963). "Some Recent Research on the Bearing Capacity of Foundations," *Canadian Geotechnical Journal*, Vol. 1, No. 1, pp. 16–26.

Prandtl, L. (1921). "Über die Eindringungsfestigkeit (Härte) plastischer Baustoffe und die Festigkeit von Schneiden," *Zeitschrift für angewandte Mathematik und Mechanik*, Vol. 1, No. 1, pp. 15–20.

Reissner, H. (1924). "Zum Erddruckproblem," *Proceedings, First International Congress of Applied Mechanics*, Delft, pp. 295–311.

Terzaghi, K. (1943). *Theoretical Soil Mechanics*, Wiley, New York.

Vesic, A. S. (1963). "Bearing Capacity of Deep Foundations in Sand," *Highway Research Record* No. 39, National Academy of Sciences, pp. 112–153.

Vesic, A. S. (1973). "Analysis of Ultimate Loads of Shallow Foundations," *Journal of the Soil Mechanics and Foundations Division*, American Society of Civil Engineers, Vol. 99, No. SM1, pp. 45–73.

4

Ultimate Bearing Capacity of Shallow Foundations: Special Cases

4.1 Introduction

The ultimate bearing capacity problems described in Chapter 3 assume that the soil supporting the foundation is homogeneous and extends to a great depth below the bottom of the foundation. They also assume that the ground surface is horizontal. However, that is not true in all cases: It is possible to encounter a rigid layer at a shallow depth, or the soil may be layered and have different shear strength parameters. In some instances, it may be necessary to construct foundations on or near a slope. This chapter discusses bearing capacity problems relating to these special cases, as well as the seismic bearing capacity of shallow foundations.

4.2 Foundation Supported by a Soil With a Rigid Base at Shallow Depth

Figure 4.1(a) shows a shallow, rough *continuous* foundation supported by a soil that extends to a great depth. Neglecting the depth factor, for vertical loading Eq. (3.21) will take the form

$$q_u = c'N_c + qN_q + \frac{1}{2}\gamma B N_\gamma \tag{4.1}$$

The general approach for obtaining expressions for N_c, N_q, and N_γ was outlined in Chapter 3. The extent of the failure zone in soil, D, at ultimate load obtained in the derivation of N_c and N_q by Prandtl (1921) and Reissner (1924) is given in Figure 4.1(b). Similarly, the magnitude of D obtained by Lundgren and Mortensen (1953) in evaluating N_γ is given in the figure.

Now, if a rigid, rough base is located at a depth of $H < D$ below the bottom of the foundation, full development of the failure surface in soil will be restricted. In such a case, the soil failure zone and the development of slip lines at ultimate load

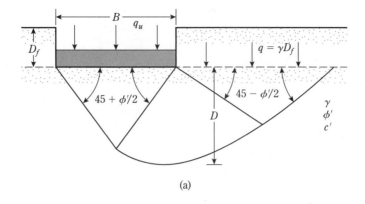

(a)

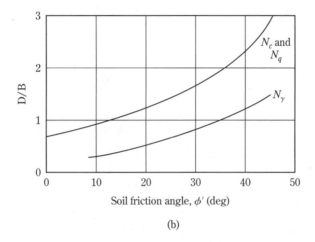

(b)

Figure 4.1 (a) Failure surface under a rough continuous foundation; (b) variation of D/B with soil friction angle ϕ'

will be as shown in Figure 4.2. Mandel and Salencon (1972) determined the bearing capacity factors applicable to this case by numerical integration, using the theory of plasticity. According to their theory, the ultimate bearing capacity of a rough continuous foundation with a rigid, rough base located at a shallow depth can be given by the relation

$$q_u = c'N_c^* + qN_q^* + \frac{1}{2}\gamma B N_\gamma^* \qquad (4.2)$$

where $N_c^*, N_q^*; N_\gamma^*$ = modified bearing capacity factors
B = width of foundation
γ = unit weight of soil

Note that, for $H \geq D$, $N_c^* = N_c$, $N_q^* = N_q$, and $N_\gamma^* = N_\gamma$ (Lundgren and Mortensen, 1953). The variations of N_c^*, N_q^*, and N_γ^* with H/B and the soil friction angle ϕ' are given in Figures 4.3, 4.4, and 4.5, respectively.

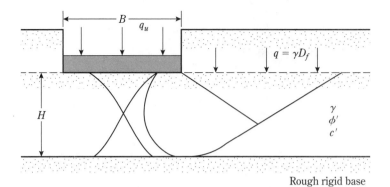

Figure 4.2 Failure surface under a rough, continuous foundation with a rigid, rough base located at a shallow depth

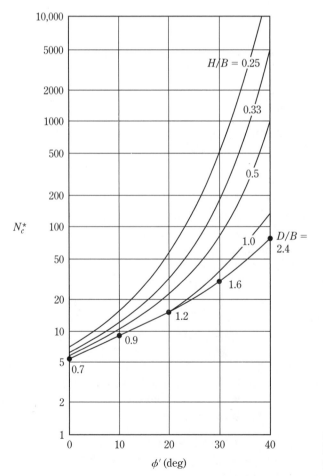

Figure 4.3 Mandel and Salencon's bearing capacity factor N_c^* [Eq. (4.2)]

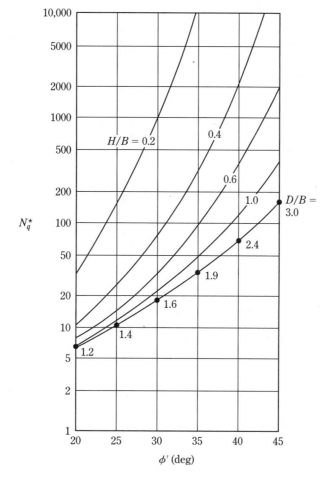

Figure 4.4 Mandel and Salencon's bearing capacity factor N_q^* [Eq. (4.2)]

Neglecting the depth factors, the ultimate bearing capacity of rough circular and rectangular foundations on a sand layer ($c' = 0$) with a rough, rigid base located at a shallow depth can be given as

$$q_u = qN_q^*F_{qs}^* + \frac{1}{2}\gamma BN_\gamma^*F_{\gamma s}^*$$

(4.3)

where $F_{qs}^*, F_{\gamma s}^* = $ modified shape factors

The shape factors F_{qs}^* and $F_{\gamma s}^*$ are functions of H/B and ϕ'. On the basis of the work of Meyerhof and Chaplin (1953), and simplifying the assumption that, in radial planes, the stresses and shear zones are identical to those in transverse planes, Meyerhof (1974) proposed that

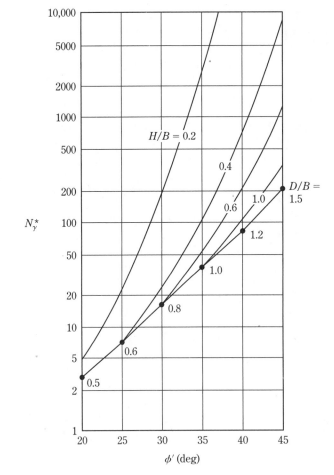

Figure 4.5 Mandel and Salencon's bearing capacity factor N_γ^* [Eq. (4.2)]

$$F_{qs}^* \approx 1 - m_1\left(\frac{B}{L}\right) \tag{4.4}$$

and

$$F_{\gamma s}^* \approx 1 - m_2\left(\frac{B}{L}\right) \tag{4.5}$$

where L = length of the foundation

The variations of m_1 and m_2 with H/B and ϕ' are shown in Figure 4.6.

For saturated clay (i.e., under the undrained condition, or $\phi = 0$), Eq. (4.2) will simplify to the form

$$q_u = c_u N_c^* + q \tag{4.6}$$

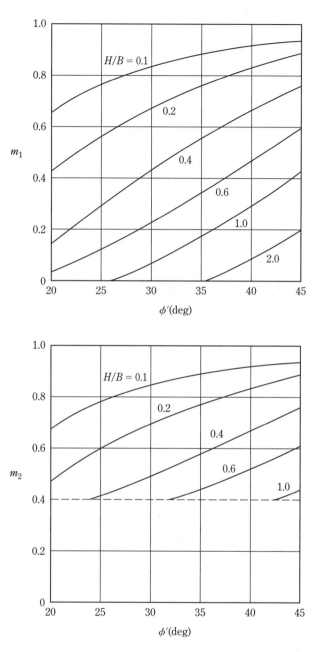

Figure 4.6 Variation of m_1 and m_2 with H/B and ϕ' (Meyerhof)

Mandel and Salencon (1972) performed calculations to evaluate N_c^* for *continuous foundations.* Similarly, Buisman (1940) gave the following relationship for obtaining the ultimate bearing capacity of square foundations:

$$q_{u(\text{square})} = \left(\pi + 2 + \frac{B}{2H} - \frac{\sqrt{2}}{2}\right)c_u + q \qquad \left(\text{for } \frac{B}{2H} - \frac{\sqrt{2}}{2} \geq 0\right) \qquad (4.7)$$

Table 4.1 Values of N_c^* for Continuous and Square Foundations ($\phi = 0$)

$\dfrac{B}{H}$	N_c^*	
	Square[a]	Continuous[b]
2	5.43	5.24
3	5.93	5.71
4	6.44	6.22
5	6.94	6.68
6	7.43	7.20
8	8.43	8.17
10	9.43	9.05

[a] Buisman's analysis (1940)
[b] Mandel and Salencon's analysis (1972)

In this equation, c_u is the undrained shear strength.

Equation (4.7) can be rewritten as

$$q_{u(\text{square})} = 5.14 \underbrace{\left(1 + \frac{0.5\dfrac{B}{H} - 0.707}{5.14} \right)}_{N_{c(\text{square})}^*} c_u + q \tag{4.8}$$

Table 4.1 gives the values of N_c^* for continuous and square foundations.

Example 4.1

A square foundation measuring 1 m × 1 m is constructed on a layer of sand. We are given that $D_f = 0.75$ m, $\gamma = 17.3$ kN/m³, $\phi' = 35°$, and $c' = 0$. A rock layer is located at a depth of 0.6 m below the bottom of the foundation. Using a factor of safety of 4, determine the gross allowable load the foundation can carry.

Solution
From Eq. (4.3),

$$q_u = qN_q^* F_{qs}^* + \frac{1}{2}\gamma B N_\gamma^* F_{\gamma s}^*$$

and we also have

$$q = 17.3 \times 0.75 = 12.98 \text{ kN/m}^2$$

For $\phi' = 35°$, $H/B = 0.6/1 = 0.6$, $N_q^* \approx 90$ (Figure 4.4), and $N_\gamma^* \approx 50$ (Figure 4.5), and we have

$$F_{qs}^* = 1 - m_1(B/L)$$

From Figure 4.6(a), for $\phi' = 35°$, $H/B = 0.6$, and the value of $m_1 = 0.34$, so

$$F_{qs}^* = 1 - (0.34)(1/1) = 0.66$$

Similarly,

$$F_{\gamma s}^* = 1 - m_2(B/L)$$

From Figure 4.6(b), $m_2 = 0.45$, so

$$F_{\gamma s}^* = 1 - (0.45)(1/1) = 0.55$$

Hence,

$$q_u = (12.98)(90)(0.66) + (1/2)(17.3)(1)(50)(0.55) = 1009 \text{ kN/m}^2$$

and

$$Q_{\text{all}} = \frac{q_u B^2}{\text{FS}} = \frac{(1009)(1 \times 1)}{4} = \textbf{252.25 kN} \qquad \blacksquare$$

4.3   *Bearing Capacity of Layered Soils: Stronger Soil Underlain by Weaker Soil*

The bearing capacity equations presented in Chapter 3 involve cases in which the soil supporting the foundation is homogeneous and extends to a considerable depth. The cohesion, angle of friction, and unit weight of soil were assumed to remain constant for the bearing capacity analysis. However, in practice, layered soil profiles are often encountered. In such instances, the failure surface at ultimate load may extend through two or more soil layers, and a determination of the ultimate bearing capacity in layered soils can be made in only a limited number of cases. This section features the procedure for estimating the bearing capacity for layered soils proposed by Meyerhof and Hanna (1978) and Meyerhof (1974).

Figure 4.7 shows a shallow continuous foundation supported by a *stronger soil layer,* underlain by a weaker soil that extends to a great depth. For the two soil layers, the physical parameters are as follows:

Layer	Unit weight	Soil friction angle	Cohesion
Top	γ_1	ϕ_1'	c_1'
Bottom	γ_2	ϕ_2'	c_2'

At ultimate load per unit area (q_u), the failure surface in soil will be as shown in the figure. If the depth H is relatively small compared with the foundation width B, a punching shear failure will occur in the top soil layer, followed by a general shear failure in the bottom soil layer. This is shown in Figure 4.7(a). However, if the depth H is relatively large, then the failure surface will be completely located in the top soil

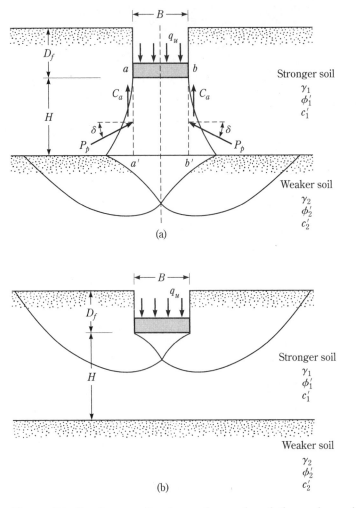

Figure 4.7 Bearing capacity of a continuous foundation on layered soil

layer, which is the upper limit for the ultimate bearing capacity. This is shown in Figure 4.7b.

   The ultimate bearing capacity for this problem, as shown in Figure 4.7a, can be given as

$$q_u = q_b + \frac{2(C_a + P_p \sin \delta)}{B} - \gamma_1 H \tag{4.9}$$

where B = width of the foundation
    C_a = adhesive force
    P_p = passive force per unit length of the faces aa' and bb'
    q_b = bearing capacity of the bottom soil layer
    δ = inclination of the passive force P_p with the horizontal

Note that, in Eq. (4.7),

$$C_a = c'_a H \tag{4.8}$$

where c'_a = adhesion

Equation (4.7) can be simplified to the form

$$q_u = q_b + \frac{2c'_a H}{B} + \gamma_1 H^2 \left(1 + \frac{2D_f}{H}\right)\frac{K_{pH} \tan \delta}{B} - \gamma_1 H \tag{4.9}$$

where K_{pH} = horizontal component of passive earth pressure coefficient

However, let

$$K_{pH} \tan \delta = K_s \tan \phi'_1 \tag{4.10}$$

where K_s = punching shear coefficient

Then

$$q_u = q_b + \frac{2c'_a H}{B} + \gamma_1 H^2 \left(1 + \frac{2D_f}{H}\right)\frac{K_s \tan \phi'_1}{B} - \gamma_1 H \tag{4.11}$$

The punching shear coefficient, K_s, is a function of q_2/q_1 and ϕ'_1, or, specifically,

$$K_s = f\left(\frac{q_2}{q_1}, \phi'_1\right)$$

Note that q_1 and q_2 are the ultimate bearing capacities of a continuous foundation of width B under vertical load on the surfaces of homogeneous thick beds of upper and lower soil, or

$$q_1 = c'_1 N_{c(1)} + \tfrac{1}{2}\gamma_1 B N_{\gamma(1)} \tag{4.12}$$

and

$$q_2 = c'_2 N_{c(2)} + \tfrac{1}{2}\gamma_2 B N_{\gamma(2)} \tag{4.13}$$

where $N_{c(1)}, N_{\gamma(1)}$ = bearing capacity factors for friction angle ϕ'_1 (Table 3.4)
$N_{c(2)}, N_{\gamma(2)}$ = bearing capacity factors for friction angle ϕ'_2 (Table 3.4)

Observe that, for the top layer to be a stronger soil, q_2/q_1 should be less than unity.

The variation of K_s with q_2/q_1 and ϕ'_1 is shown in Figure 4.8. The variation of c'_a/c'_1 with q_2/q_1 is shown in Figure 4.9. If the height H is relatively large, then the failure surface in soil will be completely located in the stronger upper-soil layer (Figure 4.7b). For this case,

$$q_u = q_t = c'_1 N_{c(1)} + q N_{q(1)} + \tfrac{1}{2}\gamma_1 B N_{\gamma(1)} \tag{4.14}$$

where $N_{q(1)}$ = bearing capacity factor for $\phi' = \phi'_1$ (Table 3.4) and $q = \gamma_1 D_f$

Combining Eqs. (4.11) and (4.14) yields

$$q_u = q_b + \frac{2c'_a H}{B} + \gamma_1 H^2 \left(1 + \frac{2D_f}{H}\right)\frac{K_s \tan \phi'_1}{B} - \gamma_1 H \le q_t \tag{4.15}$$

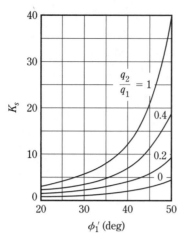

Figure 4.8 Meyerhof and Hanna's punching shear coefficient K_s

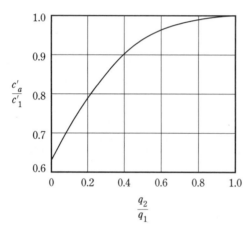

Figure 4.9 Variation of c_a'/c_1' with q_2/q_1 based on the theory of Meyerhof and Hanna (1978)

For rectangular foundations, the preceding equation can be extended to the form

$$q_u = q_b + \left(1 + \frac{B}{L}\right)\left(\frac{2c_a'H}{B}\right)$$
$$+ \gamma_1 H^2 \left(1 + \frac{B}{L}\right)\left(1 + \frac{2D_f}{H}\right)\left(\frac{K_s \tan \phi_1'}{B}\right) - \gamma_1 H \leq q_t \tag{4.16}$$

where

$$q_b = c_2' N_{c(2)} F_{cs(2)} + \gamma_1 (D_f + H) N_{q(2)} F_{qs(2)} + \frac{1}{2} \gamma_2 B N_{\gamma(2)} F_{\gamma s(2)} \tag{4.17}$$

and

$$q_t = c_1' N_{c(1)} F_{cs(1)} + \gamma_1 D_f N_{q(1)} F_{qs(1)} + \frac{1}{2} \gamma_1 B N_{\gamma(1)} F_{ys(1)} \tag{4.18}$$

in which $F_{cs(1)}, F_{qs(1)}, F_{ys(1)}$ = shape factors with respect to top soil layer (Eqs. 3.25 through 3.27)

$F_{cs(2)}, F_{qs(2)}, F_{ys(2)}$ = shape factors with respect to bottom soil layer (Eqs. 3.25 through 3.27)

Special Cases

1. *Top layer is strong sand and bottom layer is saturated soft clay* ($\phi_2 = 0$). From Eqs. (4.16), (4.17), and (4.18),

$$q_b = \left(1 + 0.2 \frac{B}{L}\right) 5.14 c_2 + \gamma_1 (D_f + H) \tag{4.19}$$

and

$$q_t = \gamma_1 D_f N_{q(1)} F_{qs(1)} + \tfrac{1}{2} \gamma_1 B N_{\gamma(1)} F_{ys(1)} \tag{4.20}$$

Hence,

$$q_u = \left(1 + 0.2 \frac{B}{L}\right) 5.14 c_2 + \gamma_1 H^2 \left(1 + \frac{B}{L}\right)\left(1 + \frac{2D_f}{H}\right) \frac{K_s \tan \phi_1'}{B}$$

$$+ \gamma_1 D_f \le \gamma_1 D_f N_{q(1)} F_{qs(1)} + \frac{1}{2} \gamma_1 B N_{\gamma(1)} F_{ys(1)} \tag{4.21}$$

where c_2 = undrained cohesion

For a determination of K_s from Figure 4.8,

$$\frac{q_2}{q_1} = \frac{c_2 N_{c(2)}}{\frac{1}{2} \gamma_1 B N_{\gamma(1)}} = \frac{5.14 c_2}{0.5 \gamma_1 B N_{\gamma(1)}} \tag{4.22}$$

2. *Top layer is stronger sand and bottom layer is weaker sand* ($c_1' = 0, c_2' = 0$). The ultimate bearing capacity can be given as

$$q_u = \left[\gamma_1 (D_f + H) N_{q(2)} F_{qs(2)} + \frac{1}{2} \gamma_2 B N_{\gamma(2)} F_{ys(2)} \right]$$

$$+ \gamma_1 H^2 \left(1 + \frac{B}{L}\right)\left(1 + \frac{2D_f}{H}\right) \frac{K_s \tan \phi_1'}{B} - \gamma_1 H \le q_t \tag{4.23}$$

where

$$q_t = \gamma_1 D_f N_{q(1)} F_{qs(1)} + \frac{1}{2} \gamma_1 B N_{\gamma(1)} F_{\gamma s(1)} \tag{4.24}$$

Then

$$\frac{q_2}{q_1} = \frac{\frac{1}{2}\gamma_2 B N_{\gamma(2)}}{\frac{1}{2}\gamma_1 B N_{\gamma(1)}} = \frac{\gamma_2 N_{\gamma(2)}}{\gamma_1 N_{\gamma(1)}} \tag{4.25}$$

3. *Top layer is stronger saturated clay* ($\phi_1 = 0$) *and bottom layer is weaker saturated clay* ($\phi_2 = 0$). The ultimate bearing capacity can be given as

$$q_u = \left(1 + 0.2\frac{B}{L}\right)5.14c_2 + \left(1 + \frac{B}{L}\right)\left(\frac{2c_aH}{B}\right) + \gamma_1 D_f \leq q_t \tag{4.26}$$

where

$$q_t = \left(1 + 0.2\frac{B}{L}\right)5.14c_1 + \gamma_1 D_f \tag{4.27}$$

and c_1 and c_2 are undrained cohesions. For this case,

$$\frac{q_2}{q_1} = \frac{5.14c_2}{5.14c_1} = \frac{c_2}{c_1} \tag{4.28}$$

Example 4.2

A foundation 1.5 m $\times$ 1 m is located at a depth D_f of 1 m in a stronger clay. A softer clay layer is located at a depth H of 1 m, measured from the bottom of the foundation. For the top clay layer,

$$\text{Undrained shear strength} = 120 \text{ kN/m}^2$$
$$\text{Unit weight} = 16.8 \text{ kN/m}^3$$

and for the bottom clay layer,

$$\text{Undrained shear strength} = 48 \text{ kN/m}^2$$
$$\text{Unit weight} = 16.2 \text{ kN/m}^3$$

Determine the gross allowable load for the foundation with an FS of 4.

Solution
For this problem, Eqs. (4.26), (4.27), and (4.28) will apply, or

$$q_u = \left(1 + 0.2\frac{B}{L}\right)5.14c_2 + \left(1 + \frac{B}{L}\right)\left(\frac{2c_aH}{B}\right) + \gamma_1 D_f$$

$$\leq \left(1 + 0.2\frac{B}{L}\right)5.14c_1 + \gamma_1 D_f$$

We are given the following data:

$$B = 1\,\text{m} \qquad H = 1\,\text{m} \qquad D_f = 1\,\text{m}$$
$$L = 1.5\,\text{m} \qquad \gamma_1 = 16.8\,\text{kN/m}^3$$

From Figure 4.9 for $c_2/c_1 = 48/120 = 0.4$, the value of $c_a/c_1 \approx 0.9$, so

$$c_a = (0.9)(120) = 108\,\text{kN/m}^2$$

and

$$q_u = \left[1 + (0.2)\left(\frac{1}{1.5}\right)\right](5.14)(48) + \left(1 + \frac{1}{1.5}\right)\left[\frac{(2)(108)(1)}{1}\right] + (16.8)(1)$$

$$= 279.6 + 360 + 16.8 = 656.4\,\text{kN/m}^2$$

As a check, we have, from Eq. (4.27),

$$q_t = \left[1 + (0.2)\left(\frac{1}{1.5}\right)\right](5.14)(120) + (16.8)(1)$$

$$= 699 + 16.8 = 715.8\,\text{kN/m}^2$$

Thus, $q_u = 656.4\,\text{kN/m}^2$ (i.e., the smaller of the two values just calculated), and

$$q_{\text{all}} = \frac{q_u}{\text{FS}} = \frac{656.4}{4} = 164.1\,\text{kN/m}^2$$

The total allowable load is therefore

$$(q_{\text{all}})(1 \times 1.5) = \textbf{246.15 kN} \qquad\blacksquare$$

Example 4.3

In Figure 4.7, assume that the top layer is sand and the bottom layer is soft saturated clay. We are given the following data:

For the sand, $\gamma_1 = 117\,\text{lb/ft}^3$; $\phi_1' = 40°$

For the soft clay (bottom layer), $c_2 = 400\,\text{lb/ft}^2$; $\phi_2 = 0$

For the foundation, $B = 3\,\text{ft}$; $D_f = 3\,\text{ft}$; $L = 4.5\,\text{ft}$; $H = 4\,\text{ft}$

Determine the gross ultimate bearing capacity of the foundation.

Solution
For this case, Eqs. (4.21) and (4.22) apply. For $\phi_1' = 40°$, from Table 3.4, $N_\gamma = 109.41$ and

$$\frac{q_2}{q_1} = \frac{c_2 N_{c(2)}}{0.5\gamma_1 B N_{\gamma(1)}} = \frac{(400)(5.14)}{(0.5)(117)(3)(109.41)} = 0.107$$

From Figure 4.8, for $c_2 N_{c(2)}/0.5\gamma_1 BN_{\gamma(1)} = 0.107$ and $\phi'_1 = 40°$, the value of $K_s \approx 2.5$. Equation (4.21) then gives

$$q_u = \left[1 + (0.2)\left(\frac{B}{L}\right)\right]5.14c_2 + \left(1 + \frac{B}{L}\right)\gamma_1 H^2\left(1 + \frac{2D_f}{H}\right)K_s\frac{\tan \phi'_1}{B} + \gamma_1 D_f$$

$$= \left[1 + (0.2)\left(\frac{3}{4.5}\right)\right](5.14)(400) + \left(1 + \frac{3}{4.5}\right)(117)(4)^2$$

$$\times \left[1 + \frac{(2)(3)}{4}\right](2.5)\frac{\tan 40}{3} + (117)(3)$$

$$= 2330 + 5454 + 351 = 8135 \text{ lb/ft}^2$$

Again, from Eq. (4.21),

$$q_t = \gamma_1 D_f N_{q(1)}F_{qs(1)} + \tfrac{1}{2}\gamma_1 BN_{\gamma(1)}F_{\gamma s(1)}$$

From Table 3.4, for $\phi'_1 = 40°$, $N_\gamma = 109.4$ and $N_q = 64.20$.
From Eqs. (3.26) and (3.27),

$$F_{qs(1)} = \left(1 + \frac{B}{L}\right)\tan \phi'_1 = \left(1 + \frac{3}{4.5}\right)\tan 40 = 1.4$$

and

$$F_{\gamma s(1)} = 1 - 0.4\frac{B}{L} = 1 - (0.4)\left(\frac{3}{4.5}\right) = 0.733$$

so that

$$q_t = (117)(3)(64.20)(1.4) + (\tfrac{1}{2})(117)(3)(109.4)(0.733) = 45,622 \text{ lb/ft}^2$$

Hence,

$$q_u = \mathbf{8135 \text{ lb/ft}^2} \qquad \blacksquare$$

4.4   *Bearing Capacity of Foundations on Top of a Slope*

In some instances, shallow foundations need to be constructed on top of a slope. In Figure 4.10, the height of the slope is H, and the slope makes an angle β with the horizontal. The edge of the foundation is located at a distance b from the top of the slope. At ultimate load, q_u, the failure surface will be as shown in the figure.

Meyerhof (1957) developed the following theoretical relation for the ultimate bearing capacity for *continuous foundations*:

$$q_u = c'N_{cq} + \tfrac{1}{2}\gamma BN_{\gamma q} \qquad (4.29)$$

For purely granular soil, $c' = 0$, thus,

$$q_u = \tfrac{1}{2}\gamma BN_{\gamma q} \qquad (4.30)$$

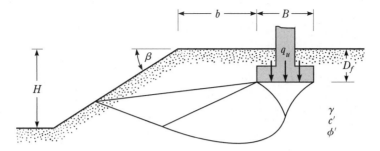

Figure 4.10 Shallow foundation on top of a slope

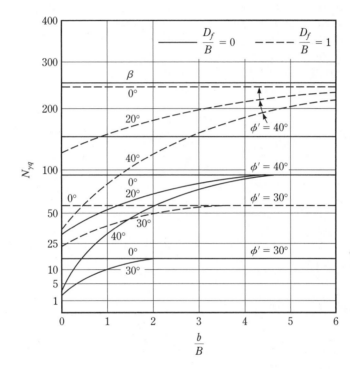

Figure 4.11 Meyerhof's bearing capacity factor $N_{\gamma q}$ for granular soil ($c' = 0$)

Again, for purely cohesive soil, $\phi = 0$ (the undrained condition); hence,

$$q_u = cN_{cq} \tag{4.31}$$

where c = undrained cohesion.

The variations of $N_{\gamma q}$ and N_{cq} defined by Eqs. (4.30) and (4.31) are shown in Figures 4.11 and 4.12, respectively. In using N_{cq} in Eq. (4.31) as given in Figure 4.12, the following points need to be kept in mind:

1. The term

$$N_s = \frac{\gamma H}{c} \tag{4.32}$$

is defined as the stability number.

2. If $B < H$, use the curves for $N_s = 0$.
3. If $B \geq H$, use the curves for the calculated stability number N_s.

Stress Characteristics Solution for Granular Soil Slopes

For slopes in granular soils, the ultimate bearing capacity of a continuous foundation can be given by Eq. (4.30), or

$$q_u = \frac{1}{2}\gamma B N_{\gamma q}$$

On the basis of the method of stress characteristics, Graham, Andrews, and Shields (1988) provided a solution for the bearing capacity factor $N_{\gamma q}$ for a shallow continuous foundation on the top of a slope in *granular soil.* Figure 4.13 shows the schematics of the failure zone in the soil for embedment (D_f/B) and setback (b/B) assumed for those authors' analysis. The variations of $N_{\gamma q}$ obtained by this method are shown in Figures 4.14, 4.15, and 4.16.

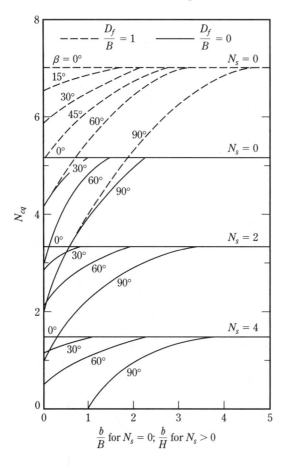

Figure 4.12 Meyerhof's bearing capacity factor N_{cq} for purely cohesive soil

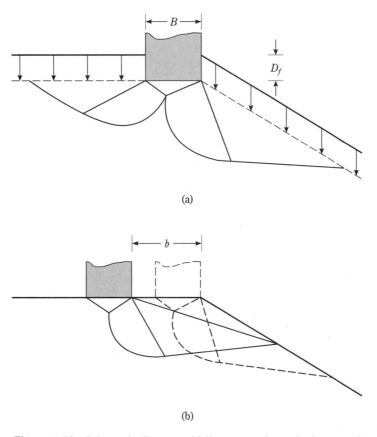

(a)

(b)

Figure 4.13 Schematic diagram of failure zones for embedment and setback: (a) $D_f/B > 0$; (b) $b/B > 0$

Example 4.4

In Figure 4.10, for a shallow continuous foundation in a clay, the following data are given: $B = 1.2$ m; $D_f = 1.2$ m; $b = 0.8$ m; $H = 6.2$ m; $\beta = 30°$; unit weight of soil $= 17.5$ kN/m^3; $\phi = 0$; and $c = 50$ kN/m^2. Determine the gross allowable bearing capacity with a factor of safety FS $= 4$.

Solution
Since $B < H$, we will assume the stability number $N_s = 0$. From Eq. (4.31),

$$q_u = cN_{cq}$$

We are given that

$$\frac{D_f}{B} = \frac{1.2}{1.2} = 1$$

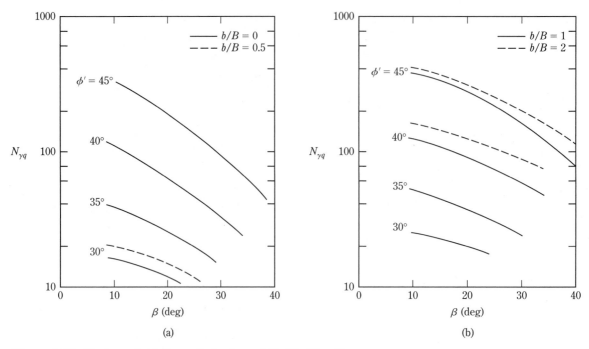

Figure 4.14 Graham et al.'s theoretical values of $N_{\gamma q}$ $(D_f/B = 0)$

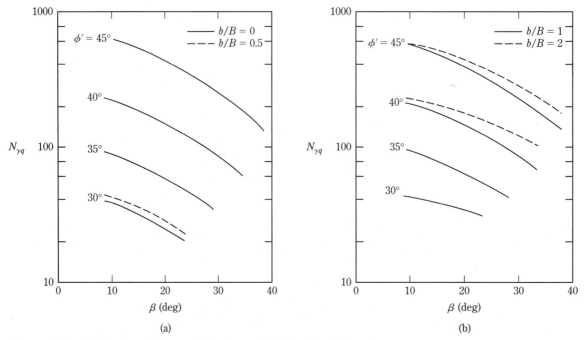

Figure 4.15 Graham et al.'s theoretical values of $N_{\gamma q}(D_f/B = 0.5)$

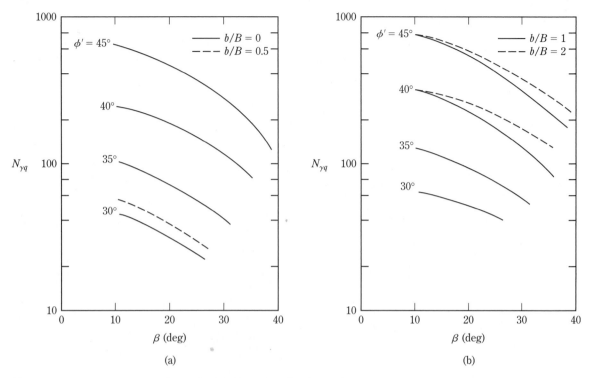

Figure 4.16 Graham et al.'s theoretical values of $N_{\gamma q} (D_f/B = 1)$

and

$$\frac{b}{B} = \frac{0.8}{1.2} = 0.75$$

For $\beta = 30°$, $D_f/B = 1$ and $b/B = 0.75$, Figure 4.12 gives $N_{cq} = 6.3$. Hence,

$$q_u = (50)(6.3) = 315 \text{ kN/m}^2$$

and

$$q_{\text{all}} = \frac{q_u}{\text{FS}} = \frac{315}{4} = \textbf{78.8 kN/m}^2 \qquad \blacksquare$$

Example 4.5

Figure 4.17 shows a continuous foundation on a slope of a granular soil. Estimate the ultimate bearing capacity by

  a. Meyerhof's method
  b. the stress characteristic solution method

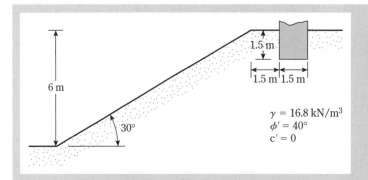

Figure 4.17   Foundation on a granular slope

Solution

a. For granular soil $(c' = 0)$, from Eq. (4.30),

$$q_u = \frac{1}{2}\gamma B N_{\gamma q}$$

We are given that $b/B = 1.5/1.5 = 1$, $D_f/B = 1.5/1.5 = 1$, $\phi' = 40°$, and $\beta = 30°$.

From Figure 4.11, $N_{\gamma q} \approx 120$. So

$$q_u = \frac{1}{2}(16.8)(1.5)(120) = \mathbf{1512\ kN/m^2}$$

b. We have

$$q_u = \frac{1}{2}\gamma B N_{\gamma q}$$

From Figure 4.16b, $N_{\gamma q} \approx 110$. Hence,

$$q_u = \frac{1}{2}(16.8)(1.5)(110) = \mathbf{1386\ kN/m^2} \qquad ∎$$

Problems

4.1   A rectangular foundation is shown in Figure P4.1. Determine the gross allowable load the foundation can carry, given that $B = 3$ ft, $L = 6$ ft, $D_f = 3$ ft, $H = 2$ ft, $\phi' = 40°$, $c' = 0$, and $\gamma = 115$ lb/ft³. Use FS = 4.

4.2   Repeat Problem 4.1 with the following data: $B = 1.5$ m, $L = 1.5$ m, $D_f = 1$ m, $H = 0.6$ m, $\phi' = 35°$, $c' = 0$, and $\gamma = 15$ kN/m³. Use FS = 3.

4.3   A square foundation measuring 4 ft × 4 ft is supported by a saturated clay layer of limited depth underlain by a rock layer. Given that $D_f = 3$ ft,

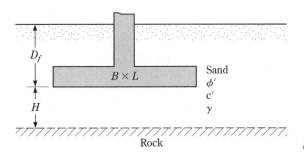

Figure P4.1

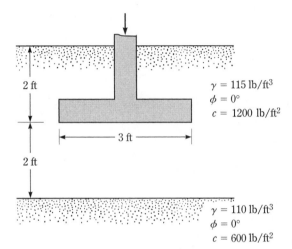

Figure P4.4

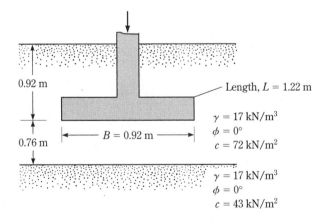

Figure P4.5

$H = 2$ ft, $c_u = 2400$ lb/ft^2, and $\gamma = 120$ lb/ft^3, estimate the ultimate bearing capacity of the foundation.

4.4 A strip foundation in a two-layered clay is shown in Figure P4.4. Find the gross allowable bearing capacity. Use a factor of safety of 3.

4.5 Find the gross ultimate load that the footing shown in Figure P4.5 can carry.

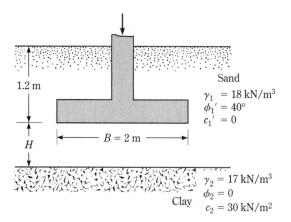

Sand
$\gamma_1 = 18 \text{ kN/m}^3$
$\phi_1' = 40°$
$c_1' = 0$

$B = 2 \text{ m}$

H

$\gamma_2 = 17 \text{ kN/m}^3$
$\phi_2 = 0$
Clay $c_2 = 30 \text{ kN/m}^2$

Figure P4.6

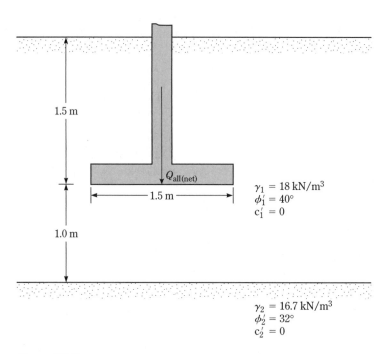

1.5 m

$Q_{\text{all(net)}}$
1.5 m

$\gamma_1 = 18 \text{ kN/m}^3$
$\phi_1' = 40°$
$c_1' = 0$

1.0 m

$\gamma_2 = 16.7 \text{ kN/m}^3$
$\phi_2' = 32°$
$c_2' = 0$

Figure P4.7

4.6 Figure P4.6 shows a continuous foundation.

    a. If $H = 1.5$ m, determine the ultimate bearing capacity q_u.

    b. At what minimum value of H/B will the clay layer not have any effect on the ultimate bearing capacity of the foundation?

4.7 A square foundation on a layered sand is shown in Figure P4.7. Determine the net allowable load that the foundation can support. Use FS = 4.

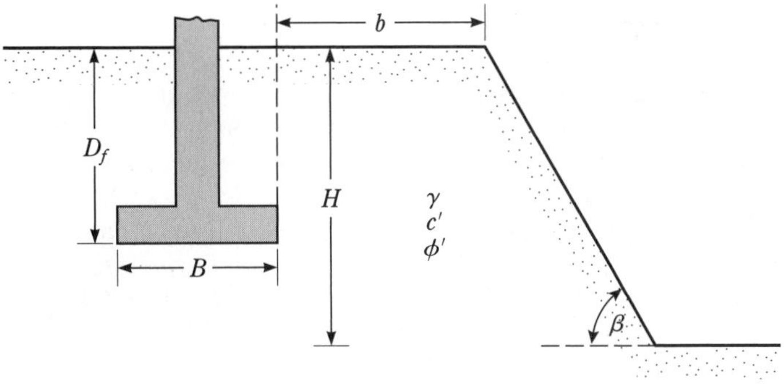

Figure P4.10

4.8   A continuous foundation with a width of 1 m is located on a slope made of clay soil. Refer to Figure 4.10, and let $D_f = 1$ m, $H = 4$ m, $b = 2$ m, $\gamma = 16.8$ kN/m³, $c = 68$ kN/m³, $\phi = 0$, and $\beta = 60°$. Using Eq. (4.31),
  a. Determine the allowable bearing capacity of the foundation. (Let FS = 3.)
  b. Plot a graph of the ultimate bearing capacity q_u if b is changed from 0 to 6 m.

4.9   A continuous foundation is to be constructed near a slope made of granular soil. (See Figure 4.10). If $B = 4$ ft, $b = 6$ ft, $H = 15$ ft, $D_f = 4$ ft, $\beta = 30°$, $\phi' = 40°$, and $\gamma = 110$ lb/ft³, estimate the allowable bearing capacity of the foundation. Use Eq. (4.30) and FS = 4.

4.10 Figure P4.10 shows a shallow strip foundation on the top of a slope. We are given the following data:

$$
\begin{aligned}
Slope\ (sand): \quad &\beta = 15° \\
&c' = 0 \\
&\phi' = 40° \\
&\gamma = 15\ \text{kN/m}^3 \\
Foundation: \quad &B = 1.5\ \text{m} \\
&D_f = 0.75\ \text{m} \\
&b = 1.5\ \text{m}
\end{aligned}
$$

Estimate the allowable bearing capacity. Use the stress characteristics solution and a factor of safety of 4.

4.11 Repeat Problem 4.10 with the following data:

$$
\begin{aligned}
Slope\ (sand): \quad &\beta = 20° \\
&c' = 0 \\
&\phi' = 30° \\
&\gamma = 112\ \text{lb/ft}^3 \\
Foundation: \quad &B = 4\ \text{ft} \\
&D_f = 4\ \text{ft} \\
&b = 4\ \text{ft}
\end{aligned}
$$

Use a factor of safety of 3.

References

Buisman, A. S. K. (1940). *Grondmechanica,* Waltman, Delft, the Netherlands.

Graham, J., Andrews, M., and Shields, D. H. (1988). "Stress Characteristics for Shallow Footings in Cohesionless Slopes," *Canadian Geotechnical Journal,* Vol. 25, No. 2, pp. 238–249.

Lundgren, H., and Mortensen, K. (1953). "Determination by the Theory of Plasticity on the Bearing Capacity of Continuous Footings on Sand," *Proceedings, Third International Conference on Soil Mechanics and Foundation Engineering,* Zurich, Vol. 1, pp. 409–412.

Mandel, J., and Salencon, J. (1972). "Force portante d'un sol sur une assise rigide (étude théorique)," *Geotechnique,* Vol. 22, No. 1, pp. 79–93.

Meyerhof, G. G. (1957). "The Ultimate Bearing Capacity of Foundations on Slopes," *Proceedings, Fourth International Conference on Soil Mechanics and Foundation Engineering,* London, Vol. 1, pp. 384–387.

Meyerhof, G. G. (1974). "Ultimate Bearing Capacity of Footings on Sand Layer Overlying Clay," *Canadian Geotechnical Journal,* Vol. 11, No. 2, pp. 224–229.

Meyerhof, G. G., and Chaplin, T. K. (1953). "The Compression and Bearing Capacity of Cohesive Soils," *British Journal of Applied Physics,* Vol. 4, pp. 20–26.

Meyerhof, G. G., and Hanna, A. M. (1978). "Ultimate Bearing Capacity of Foundations on Layered Soil under Inclined Load," *Canadian Geotechnical Journal,* Vol. 15, No. 4, pp. 565–572.

Prandtl, L. (1921). "Über die Eindringungsfestigkeit (Härte) plastischer Baustoffe und die Festigkeit von Schneiden," *Zeitschrift für angewandte Mathematik und Mechanik,* Vol. 1, No. 1, pp. 15–20.

Reissner, H. (1924). "Zum Erddruckproblem," *Proceedings, First International Congress of Applied Mechanics,* Delft, the Netherlands, pp. 295–311.

5

Shallow Foundations: Allowable Bearing Capacity and Settlement

Introduction

It was mentioned in Chapter 3 that, in many cases, the allowable settlement of a shallow foundation may control the allowable bearing capacity. The allowable settlement itself may be controlled by local building codes. Thus, the allowable bearing capacity will be the smaller of the following two conditions:

$$q_{\text{all}} = \begin{cases} \dfrac{q_u}{\text{FS}} \\ \text{or} \\ q_{\text{allowable settlement}} \end{cases}$$

The settlement of a foundation can be divided into two major categories: (a) elastic, or immediate, settlement and (b) consolidation settlement. Immediate, or elastic, settlement of a foundation takes place during or immediately after the construction of the structure. Consolidation settlement occurs over time. Pore water is extruded from the void spaces of saturated clayey soils submerged in water. The total settlement of a foundation is the sum of the elastic settlement and the consolidation settlement.

Consolidation settlement comprises two phases: *primary* and *secondary.* The fundamentals of primary consolidation settlement were explained in detail in Chapter 1. Secondary consolidation settlement occurs after the completion of primary consolidation caused by slippage and reorientation of soil particles under a sustained load. Primary consolidation settlement is more significant than secondary settlement in inorganic clays and silty soils. However, in organic soils, secondary consolidation settlement is more significant.

For the calculation of foundation settlement (both elastic and consolidation), it is required that we estimate the vertical stress increase in the soil mass due to the net load applied on the foundation. Hence, this chapter is divided into the following four parts:

1. Procedure for calculation of vertical stress increase
2. Settlement calculation (elastic and consolidation)
3. Allowable bearing capacity based on elastic settlement
4. Foundation with soil reinforcement.

Vertical Stress Increase in a Soil Mass Caused by Foundation Load

5.2   Stress Due To a Concentrated Load

In 1885, Boussinesq developed the mathematical relationships for determining the normal and shear stresses at any point inside *homogeneous, elastic,* and *isotropic* mediums due to a *concentrated point load* located at the surface, as shown in Figure 5.1. According to his analysis, the *vertical stress increase* at point A caused by a point load of magnitude P is given by

$$\Delta\sigma = \frac{3P}{2\pi z^2 \left[1 + \left(\dfrac{r}{z}\right)^2\right]^{5/2}} \tag{5.1}$$

where $r = \sqrt{x^2 + y^2}$
           x, y, z = coordinates of the point A

Note that Eq. (5.1) is not a function of Poisson's ratio of the soil.

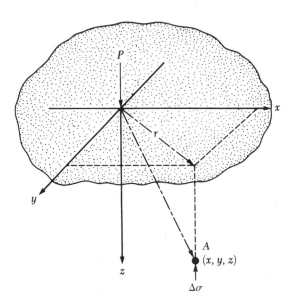

Figure 5.1   Vertical stress at a point A caused by a point load on the surface

5.3   *Stress Due to a Circularly Loaded Area*

The Boussinesq equation (5.1) can also be used to determine the vertical stress below the center of a flexible circularly loaded area, as shown in Figure 5.2. Let the radius of the loaded area be $B/2$, and let q_o be the uniformly distributed load per unit area. To determine the stress increase at a point A, located at a depth z below the center of the circular area, consider an elemental area on the circle. The load on this elemental area may be taken to be a point load and expressed as $q_o r \, d\theta \, dr$. The stress increase at A caused by this load can be determined from Eq. (5.1) as

$$d\sigma = \frac{3(q_o r \, d\theta \, dr)}{2\pi z^2 \left[1 + \left(\dfrac{r}{z} \right)^2 \right]^{5/2}} \tag{5.2}$$

The total increase in stress caused by the entire loaded area may be obtained by integrating Eq. (5.2), or

$$\Delta\sigma = \int d\sigma = \int_{\theta=0}^{\theta=2\pi} \int_{r=0}^{r=B/2} \frac{3(q_o r \, d\theta \, dr)}{2\pi z^2 \left[1 + \left(\dfrac{r}{z} \right)^2 \right]^{5/2}}$$

$$= q_o \left\{ 1 - \frac{1}{\left[1 + \left(\dfrac{B}{2z} \right)^2 \right]^{3/2}} \right\} \tag{5.3}$$

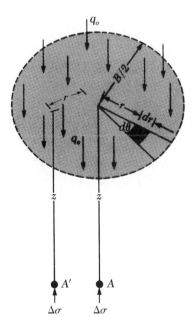

Figure 5.2   Increase in pressure under a uniformly loaded flexible circular area

Table 5.1   Variation of $\Delta\sigma/q_o$ for a Uniformly Loaded Flexible Circular Area

z/(B/2)	r/(B/2)					
	0	**0.2**	**0.4**	**0.6**	**0.8**	**1.0**
0	1.000	1.000	1.000	1.000	1.000	1.000
0.1	0.999	0.999	0.998	0.996	0.976	0.484
0.2	0.992	0.991	0.987	0.970	0.890	0.468
0.3	0.976	0.973	0.963	0.922	0.793	0.451
0.4	0.949	0.943	0.920	0.860	0.712	0.435
0.5	0.911	0.902	0.869	0.796	0.646	0.417
0.6	0.864	0.852	0.814	0.732	0.591	0.400
0.7	0.811	0.798	0.756	0.674	0.545	0.367
0.8	0.756	0.743	0.699	0.619	0.504	0.366
0.9	0.701	0.688	0.644	0.570	0.467	0.348
1.0	0.646	0.633	0.591	0.525	0.434	0.332
1.2	0.546	0.535	0.501	0.447	0.377	0.300
1.5	0.424	0.416	0.392	0.355	0.308	0.256
2.0	0.286	0.286	0.268	0.248	0.224	0.196
2.5	0.200	0.197	0.191	0.180	0.167	0.151
3.0	0.146	0.145	0.141	0.135	0.127	0.118
4.0	0.087	0.086	0.085	0.082	0.080	0.075

Similar integrations could be performed to obtain the vertical stress increase at A', located a distance r from the center of the loaded area at a depth z (Ahlvin and Ulery, 1962). Table 5.1 gives the variation of $\Delta\sigma/q_o$ with $r/(B/2)$ and $z/(B/2)$ [for $0 \leqslant r/(B/2) \leqslant 1$]. Note that the variation of $\Delta\sigma/q_o$ with depth at $r/(B/2) = 0$ can be obtained from Eq. (5.3).

5.4   *Stress below a Rectangular Area*

The integration technique of Boussinesq's equation also allows the vertical stress at any point A below the corner of a flexible rectangular loaded area to be evaluated. (See Figure 5.3.) To do so, consider an elementary area $dA = dx\,dy$ on the flexible loaded area. If the load per unit area is q_o, the total load on the elemental area is

$$dP = q_o\,dx\,dy \tag{5.4}$$

This elemental load, dP, may be treated as a point load. The increase in vertical stress at point A caused by dP may be evaluated by using Eq. (5.1). Note, however, the need to substitute $dP = q_o\,dx\,dy$ for P and $x^2 + y^2$ for r^2 in that equation. Thus,

$$\text{The stress increase at } A \text{ caused by } dP = \frac{3q_o\,(dx\,dy)z^3}{2\pi(x^2 + y^2 + z^2)^{5/2}}$$

The total stress increase $\Delta\sigma$ caused by the entire loaded area at point A may now be obtained by integrating the preceding equation:

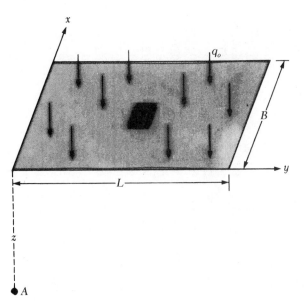

Figure 5.3 Determination of stress below the corner of a flexible rectangular loaded area

$$\Delta\sigma = \int_{y=0}^{L}\int_{x=0}^{B}\frac{3q_o\,(dx\,dy)z^3}{2\pi(x^2 + y^2 + z^2)^{5/2}} = q_o I \tag{5.5}$$

Here,

$$I = \text{influence factor} = \frac{1}{4\pi}\left(\frac{2mn\sqrt{m^2 + n^2 + 1}}{m^2 + n^2 + m^2 n^2 + 1}\cdot\frac{m^2 + n^2 + 2}{m^2 + n^2 + 1}\right.$$

$$\left. + \tan^{-1}\frac{2mn\sqrt{m^2 + n^2 + 1}}{m^2 + n^2 + 1 - m^2 n^2}\right) \tag{5.6}$$

When $m^2 + n^2 + 1 < m^2 n^2$, the argument of $\tan^{-1}$ becomes negative. In that case,

$$I = \text{influence factor} = \frac{1}{4\pi}\left[\frac{2mn\sqrt{m^2 + n^2 + 1}}{m^2 + n^2 + m^2 n^2 + 1}\cdot\frac{m^2 + n^2 + 2}{m^2 + n^2 + 1}\right.$$

$$\left. + \tan^{-1}\left(\pi - \frac{2mn\sqrt{m^2 + n^2 + 1}}{m^2 + n^2 + 1 - m^2 n^2}\right)\right] \tag{5.6a}$$

where $m = \dfrac{B}{z}$ $\tag{5.7}$

and $n = \dfrac{L}{z}$ $\tag{5.8}$

The variations of the influence values with m and n are given in Table 5.2.

The stress increase at any point below a rectangular loaded area can also be found by using Eq. (5.5) in conjunction with Figure 5.4. To determine the stress at a

Table 5.2 Variation of Influence Value I [Eq. (5.6)][a]

m	n											
	0.1	0.2	0.3	0.4	0.5	0.6	0.7	0.8	0.9	1.0	1.2	1.4
0.1	0.00470	0.00917	0.01323	0.01678	0.01978	0.02223	0.02420	0.02576	0.02698	0.02794	0.02926	0.03007
0.2	0.00917	0.01790	0.02585	0.03280	0.03866	0.04348	0.04735	0.05042	0.05283	0.05471	0.05733	0.05894
0.3	0.01323	0.02585	0.03735	0.04742	0.05593	0.06294	0.06858	0.07308	0.07661	0.07938	0.08323	0.08561
0.4	0.01678	0.03280	0.04742	0.06024	0.07111	0.08009	0.08734	0.09314	0.09770	0.10129	0.10631	0.10941
0.5	0.01978	0.03866	0.05593	0.07111	0.08403	0.09473	0.10340	0.11035	0.11584	0.12018	0.12626	0.13003
0.6	0.02223	0.04348	0.06294	0.08009	0.09473	0.10688	0.11679	0.12474	0.13105	0.13605	0.14309	0.14749
0.7	0.02420	0.04735	0.06858	0.08734	0.10340	0.11679	0.12772	0.13653	0.14356	0.14914	0.15703	0.16199
0.8	0.02576	0.05042	0.07308	0.09314	0.11035	0.12474	0.13653	0.14607	0.15371	0.15978	0.16843	0.17389
0.9	0.02698	0.05283	0.07661	0.09770	0.11584	0.13105	0.14356	0.15371	0.16185	0.16835	0.17766	0.18357
1.0	0.02794	0.05471	0.07938	0.10129	0.12018	0.13605	0.14914	0.15978	0.16835	0.17522	0.18508	0.19139
1.2	0.02926	0.05733	0.08323	0.10631	0.12626	0.14309	0.15703	0.16843	0.17766	0.18508	0.19584	0.20278
1.4	0.03007	0.05894	0.08561	0.10941	0.13003	0.14749	0.16199	0.17389	0.18357	0.19139	0.20278	0.21020
1.6	0.03058	0.05994	0.08709	0.11135	0.13241	0.15028	0.16515	0.17739	0.18737	0.19546	0.20731	0.21510
1.8	0.03090	0.06058	0.08804	0.11260	0.13395	0.15207	0.16720	0.17967	0.18986	0.19814	0.21032	0.21836
2.0	0.03111	0.06100	0.08867	0.11342	0.13496	0.15326	0.16856	0.18119	0.19152	0.19994	0.21235	0.22058
2.5	0.03138	0.06155	0.08948	0.11450	0.13628	0.15483	0.17036	0.18321	0.19375	0.20236	0.21512	0.22364
3.0	0.03150	0.06178	0.08982	0.11495	0.13684	0.15550	0.17113	0.18407	0.19470	0.20341	0.21633	0.22499
4.0	0.03158	0.06194	0.09007	0.11527	0.13724	0.15598	0.17168	0.18469	0.19540	0.20417	0.21722	0.22600
5.0	0.03160	0.06199	0.09014	0.11537	0.13737	0.15612	0.17185	0.18488	0.19561	0.20440	0.21749	0.22632
6.0	0.03161	0.06201	0.09017	0.11541	0.13741	0.15617	0.17191	0.18496	0.19569	0.20449	0.21760	0.22644
8.0	0.03162	0.06202	0.09018	0.11543	0.13744	0.15621	0.17195	0.18500	0.19574	0.20455	0.21767	0.22652
10.0	0.03162	0.06202	0.09019	0.11544	0.13745	0.15622	0.17196	0.18502	0.19576	0.20457	0.21769	0.22654
∞	0.03162	0.06202	0.09019	0.11544	0.13745	0.15623	0.17197	0.18502	0.19577	0.20458	0.21770	0.22656

Table 5.2 (Continued)

m	|					n						
	|	1.6	1.8	2.0	2.5	3.0	4.0	5.0	6.0	8.0	10.0	∞
0.1	|	0.03058	0.03090	0.03111	0.03138	0.03150	0.03158	0.03160	0.03161	0.03162	0.03162	0.03162
0.2	|	0.05994	0.06058	0.06100	0.06155	0.06178	0.06194	0.06199	0.06201	0.06202	0.06202	0.06202
0.3	|	0.08709	0.08804	0.08867	0.08948	0.08982	0.09007	0.09014	0.09017	0.09018	0.09019	0.09019
0.4	|	0.11135	0.11260	0.11342	0.11450	0.11495	0.11527	0.11537	0.11541	0.11543	0.11544	0.11544
0.5	|	0.13241	0.13395	0.13496	0.13628	0.13684	0.13724	0.13737	0.13741	0.13744	0.13745	0.13745
0.6	|	0.15028	0.15207	0.15326	0.15483	0.15550	0.15598	0.15612	0.15617	0.15621	0.15622	0.15623
0.7	|	0.16515	0.16720	0.16856	0.17036	0.17113	0.17168	0.17185	0.17191	0.17195	0.17196	0.17197
0.8	|	0.17739	0.17967	0.18119	0.18321	0.18407	0.18469	0.18488	0.18496	0.18500	0.18502	0.18502
0.9	|	0.18737	0.18986	0.19152	0.19375	0.19470	0.19540	0.19561	0.19569	0.19574	0.19576	0.19577
1.0	|	0.19546	0.19814	0.19994	0.20236	0.20341	0.20417	0.20440	0.20449	0.20455	0.20457	0.20458
1.2	|	0.20731	0.21032	0.21235	0.21512	0.21633	0.21722	0.21749	0.21760	0.21767	0.21769	0.21770
1.4	|	0.21510	0.21836	0.22058	0.22364	0.22499	0.22600	0.22632	0.22644	0.22652	0.22654	0.22656
1.6	|	0.22025	0.22372	0.22610	0.22940	0.23088	0.23200	0.23236	0.23249	0.23258	0.23261	0.23263
1.8	|	0.22372	0.22736	0.22986	0.23334	0.23495	0.23617	0.23656	0.23671	0.23681	0.23684	0.23686
2.0	|	0.22610	0.22986	0.23247	0.23614	0.23782	0.23912	0.23954	0.23970	0.23981	0.23985	0.23987
2.5	|	0.22940	0.23334	0.23614	0.24010	0.24196	0.24344	0.24392	0.24412	0.24425	0.24429	0.24432
3.0	|	0.23088	0.23495	0.23782	0.24196	0.24394	0.24554	0.24608	0.24630	0.24646	0.24650	0.24654
4.0	|	0.23200	0.23617	0.23912	0.24344	0.24554	0.24729	0.24791	0.24817	0.24836	0.24842	0.24846
5.0	|	0.23236	0.23656	0.23954	0.24392	0.24608	0.24791	0.24857	0.24885	0.24907	0.24914	0.24919
6.0	|	0.23249	0.23671	0.23970	0.24412	0.24630	0.24817	0.24885	0.24916	0.24939	0.24946	0.24952
8.0	|	0.23258	0.23681	0.23981	0.24425	0.24646	0.24836	0.24907	0.24939	0.24964	0.24973	0.24980
10.0	|	0.23261	0.23684	0.23985	0.24429	0.24650	0.24842	0.24914	0.24946	0.24973	0.24981	0.24989
∞	|	0.23263	0.23686	0.23987	0.24432	0.24654	0.24846	0.24919	0.24952	0.24980	0.24989	0.25000

[a] After Newmark, 1935.

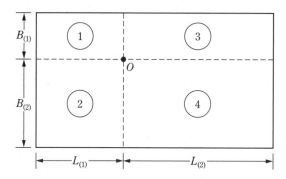

Figure 5.4 Stress below any point of a loaded flexible rectangular area

depth z below point O, divide the loaded area into four rectangles, with O the corner common to each. Then use Eq. (5.5) to calculate the increase in stress at a depth z below O caused by each rectangular area. The total stress increase caused by the entire loaded area may now be expressed as

$$\Delta\sigma = q_o (I_1 + I_2 + I_3 + I_4) \tag{5.9}$$

where $I_1, I_2, I_3,$ and I_4 = the influence values of rectangles 1, 2, 3, and 4, respectively

In most cases, the vertical stress below the center of a rectangular area is of importance. This can be given by the relationship

$$\Delta\sigma = q_o I_c \tag{5.10}$$

where

$$I_c = \frac{2}{\pi}\left[\frac{m_1 n_1}{\sqrt{1 + m_1^2 + n_1^2}} \frac{1 + m_1^2 + 2n_1^2}{(1 + n_1^2)(m_1^2 + n_1^2)}\right]$$

$$+ \sin^{-1}\frac{m_1}{\sqrt{m_1^2 + n_1^2}\sqrt{1 + n_1^2}} \tag{5.11}$$

$$m_1 = \frac{L}{B} \tag{5.12}$$

$$n_1 = \frac{z}{\left(\dfrac{B}{2}\right)} \tag{5.13}$$

The variation of I_c with m_1 and n_1 is given in Table 5.3.

Foundation engineers often use an approximate method to determine the increase in stress with depth caused by the construction of a foundation. The method is referred to as the *2:1 method.* (See Figure 5.5.) According to this method, the increase in stress at depth z is

$$\Delta\sigma = \frac{q_0 \times B \times L}{(B + z)(L + z)} \tag{5.14}$$

Note that Eq. (5.14) is based on the assumption that the stress from the foundation spreads out along lines with a *vertical-to-horizontal slope of 2:1.*

Table 5.3 Variation of I_c with m_1 and n_1

	m_1									
n_i	1	2	3	4	5	6	7	8	9	10
0.20	0.994	0.997	0.997	0.997	0.997	0.997	0.997	0.997	0.997	0.997
0.40	0.960	0.976	0.977	0.977	0.977	0.977	0.977	0.977	0.977	0.977
0.60	0.892	0.932	0.936	0.936	0.937	0.937	0.937	0.937	0.937	0.937
0.80	0.800	0.870	0.878	0.880	0.881	0.881	0.881	0.881	0.881	0.881
1.00	0.701	0.800	0.814	0.817	0.818	0.818	0.818	0.818	0.818	0.818
1.20	0.606	0.727	0.748	0.753	0.754	0.755	0.755	0.755	0.755	0.755
1.40	0.522	0.658	0.685	0.692	0.694	0.695	0.695	0.696	0.696	0.696
1.60	0.449	0.593	0.627	0.636	0.639	0.640	0.641	0.641	0.641	0.642
1.80	0.388	0.534	0.573	0.585	0.590	0.591	0.592	0.592	0.593	0.593
2.00	0.336	0.481	0.525	0.540	0.545	0.547	0.548	0.549	0.549	0.549
3.00	0.179	0.293	0.348	0.373	0.384	0.389	0.392	0.393	0.394	0.395
4.00	0.108	0.190	0.241	0.269	0.285	0.293	0.298	0.301	0.302	0.303
5.00	0.072	0.131	0.174	0.202	0.219	0.229	0.236	0.240	0.242	0.244
6.00	0.051	0.095	0.130	0.155	0.172	0.184	0.192	0.197	0.200	0.202
7.00	0.038	0.072	0.100	0.122	0.139	0.150	0.158	0.164	0.168	0.171
8.00	0.029	0.056	0.079	0.098	0.113	0.125	0.133	0.139	0.144	0.147
9.00	0.023	0.045	0.064	0.081	0.094	0.105	0.113	0.119	0.124	0.128
10.00	0.019	0.037	0.053	0.067	0.079	0.089	0.097	0.103	0.108	0.112

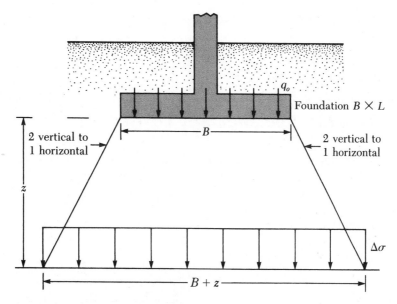

Figure 5.5 2:1 method of finding stress increase under a foundation

---

Example 5.1

A flexible rectangular area 2.5 m × 5 m is located on the surface of the ground and is loaded with $q_o = 145$ kN/m². Determine the stress increase caused by this loading at a depth of 6.25 m below the center of the rectangular area. Use Eq. (5.5).

Solution
From Figure 5.4,

$$B_{(1)} = \frac{2.5 \text{ m}}{2} = 1.25 \text{ m}$$

and

$$L_{(1)} = \frac{5}{2} = 2.5 \text{ m}$$

From Eqs. (5.7) and (5.8),

$$m_{(1)} = \frac{B_1}{z} = \frac{1.25}{6.25} = 0.2$$

and

$$n_{(1)} = \frac{L_1}{z} = \frac{2.5}{6.25} = 0.4$$

From Table 5.2, for $m_{(1)} = 0.20$ and $n_{(1)} = 0.4$, $I_{(1)} = 0.0328$. Also, note that $I_{(1)} = I_{(2)} = I_{(3)} = I_{(4)}$. Thus,

$$\Delta\sigma = q_o[4I_{(1)}] = (145)(4)(0.0328) = \mathbf{19.024 \text{ kN/m}^2} \qquad \blacksquare$$

5.5   *Average Vertical Stress Increase Due to a Rectangularly Loaded Area*

In Section 5.4, the vertical stress increase below the corner of a uniformly loaded rectangular area was given as

$$\Delta\sigma = q_o I$$

In many cases, one must find the average stress increase, $\Delta\sigma_{av}$, below the corner of a uniformly loaded rectangular area with limits of $z = 0$ to $z = H$, as shown in Figure 5.6. This can be evaluated as

$$\Delta\sigma_{av} = \frac{1}{H}\int_0^H (q_o I)\,dz = q_o I_a \qquad (5.15)$$

where $I_a = f(m, n)$ $\qquad\qquad\qquad\qquad\qquad\qquad\qquad$ (5.16)

$\qquad\quad m = \dfrac{B}{H}$ $\qquad\qquad\qquad\qquad\qquad\qquad\qquad\quad$ (5.17)

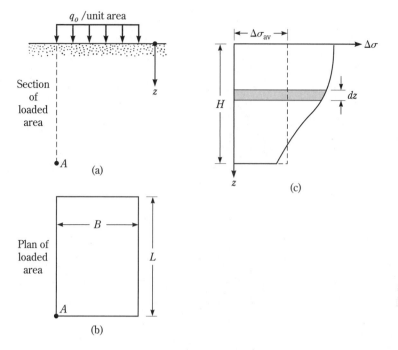

Figure 5.6 Average vertical stress increase due to a rectangularly loaded flexible area

and

$$n = \frac{L}{H} \tag{5.18}$$

The variation of I_a with m and n is shown in Figure 5.7, as proposed by Griffiths (1984).

In estimating the consolidation settlement under a foundation, it may be required to determine the average vertical stress increase in only a given layer—that is, between $z = H_1$ and $z = H_2$, as shown in Figure 5.8. This can be done as (Griffiths, 1984)

$$\Delta\sigma_{av(H_2/H_1)} = q_o \left[\frac{H_2 I_{a(H_2)} - H_1 I_{a(H_1)}}{H_2 - H_1} \right] \tag{5.19}$$

where $\Delta\sigma_{av(H_2/H_1)}$ = average stress increase immediately below the corner of a uniformly loaded rectangular area between depths $z = H_1$ and $z = H_2$

$$I_{a(H_2)} = I_a \text{ for } z = 0 \text{ to } z = H_2 = f\left(m = \frac{B}{H_2}, n = \frac{L}{H_2} \right)$$

and

$$I_{a(H_1)} = I_a \text{ for } z = 0 \text{ to } z = H_1 = f\left(m = \frac{B}{H_1}, n = \frac{L}{H_1} \right)$$

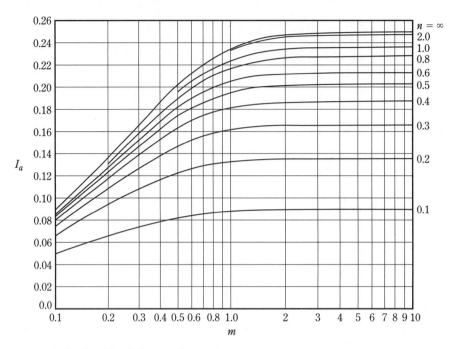

Figure 5.7   Griffiths' influence factor I_a

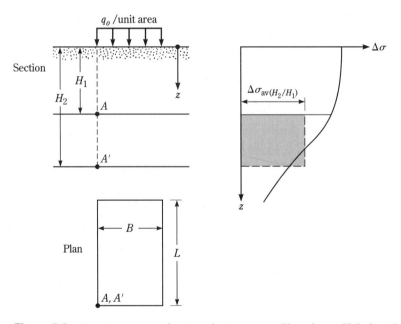

Figure 5.8   Average pressure increase between $z = H_1$ and $z = H_2$ below the corner of a uniformly loaded rectangular area

Example 5.2

Determine the *average* stress increase below the center of the loaded area in Figure 5.9 between $z = 9$ ft and $z = 15$ ft (i.e., between points A and A').

Solution

The loaded area can be divided into four rectangular areas, each measuring 4.5 ft $\times$ 4.5 ft ($L \times B$). From Eq. (5.19), the average stress increase (between the required depths) below the corner of each rectangular area can be given as

$$\Delta\sigma_{av(H_2/H_1)} = q_o\left[\frac{H_2 I_{a(H_2)} - H_1 I_{a(H_1)}}{H_2 - H_1}\right] = 1000\left[\frac{(15)I_{a(H_2)} - (9)I_{a(H_1)}}{15 - 9}\right]$$

For $I_{a(H_2)}$,

$$m = \frac{B}{H_2} = \frac{4.5}{15} = 0.3$$

and

$$n = \frac{L}{H_2} = \frac{4.5}{15} = 0.3$$

From Figure 5.7, for $m = 0.3$ and $n = 0.3$, $I_{a(H_2)} = 0.136$. For $I_{a(H_1)}$,

$$m = \frac{B}{H_1} = \frac{4.5}{9} = 0.5$$

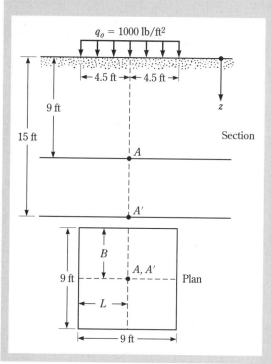

Figure 5.9 Determination of average increase in stress below a rectangular area

and

$$n = \frac{L}{H_1} = \frac{4.5}{9} = 0.5$$

Again from Figure 5.7, $I_{a(H_1)} = 0.175$, so

$$\Delta\sigma_{\mathrm{av}(H_2/H_1)} = 1000\left[\frac{(15)(0.136) - (9)(0.175)}{15 - 9}\right] = 77.5 \text{ lb/ft}^2$$

The stress increase between $z = 9$ ft and $z = 15$ ft below the center of the loaded area is equal to

$$4\Delta\sigma_{\mathrm{av}(H_2/H_1)} = (4)(77.5) = \mathbf{310 \text{ lb/ft}^2} \qquad \blacksquare$$

5.6   *Stress Increase under an Embankment*

Figure 5.10 shows the cross section of an embankment of height H. For this two-dimensional loading condition, the vertical stress increase may be expressed as

$$\Delta\sigma = \frac{q_o}{\pi}\left[\left(\frac{B_1 + B_2}{B_2}\right)(\alpha_1 + \alpha_2) - \frac{B_1}{B_2}(\alpha_2)\right] \qquad (5.20)$$

where $q_o = \gamma H$
  γ = unit weight of the embankment soil
  H = height of the embankment

$$\alpha_1 = \tan^{-1}\left(\frac{B_1 + B_2}{z}\right) - \tan^{-1}\left(\frac{B_1}{z}\right) \qquad (5.21)$$

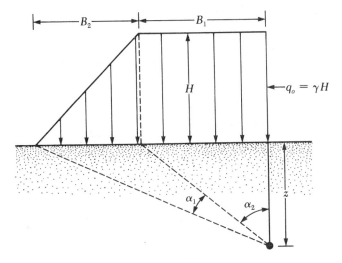

Figure 5.10   Embankment loading

$$\alpha_2 = \tan^{-1}\left(\frac{B_1}{z}\right) \tag{5.22}$$

(Note that α_1 and α_2 are in radians.)

For a detailed derivation of Eq. (5.20), see Das (1997). A simplified form of the equation is

$$\Delta\sigma = q_o I' \tag{5.23}$$

where $I' = $ a function of B_1/z and B_2/z

The variation of I' with B_1/z and B_2/z is shown in Figure 5.11. An application of this diagram is given in Example 5.3.

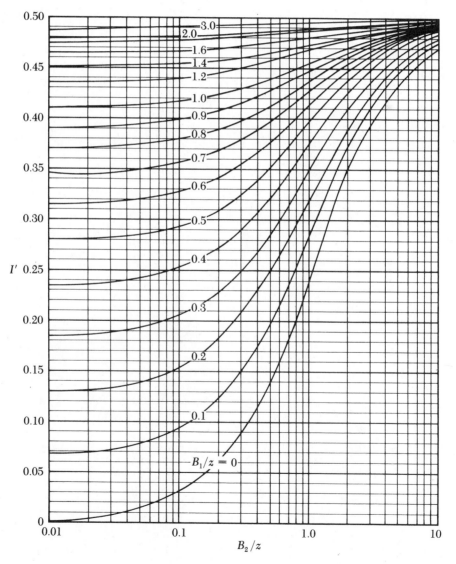

Figure 5.11 Influence value I' for embankment loading (after Osterberg, 1957)

Example 5.3

An embankment is shown in Figure 5.12a. Determine the stress increase under the embankment at points A_1 and A_2.

Solution
We have

$$\gamma H = (17.5)(7) = 122.5 \text{ kN/m}^2$$

Stress Increase at A_1
The left side of Figure 5.12b indicates that $B_1 = 2.5$ m and $B_2 = 14$ m, so

$$\frac{B_1}{z} = \frac{2.5}{5} = 0.5$$

and

$$\frac{B_2}{z} = \frac{14}{5} = 2.8$$

According to Figure 5.11, in this case $I' = 0.445$. Because the two sides in Figure 5.12b are symmetrical, the value of I' for the right side will also be 0.445, so

$$\Delta\sigma = \Delta\sigma_{(1)} + \Delta\sigma_{(2)} = q_0[I'_{(\text{left side})} + I'_{(\text{right side})}]$$

$$= 122.5[0.445 + 0.445] = \mathbf{109.03 \text{ kN/m}^2}$$

Stress Increase at A_2
In Figure 5.12c, for the left side, $B_2 = 5$ m and $B_1 = 0$, so

$$\frac{B_2}{z} = \frac{5}{5} = 1$$

and

$$\frac{B_1}{z} = \frac{0}{5} = 0$$

According to Figure 5.11, for these values of B_2/z and B_1/z, $I' = 0.25$; hence,

$$\Delta\sigma_{(1)} = 43.75(0.25) = 10.94 \text{ kN/m}^2$$

For the middle section,

$$\frac{B_2}{z} = \frac{14}{5} = 2.8$$

and

$$\frac{B_1}{z} = \frac{14}{5} = 2.8$$

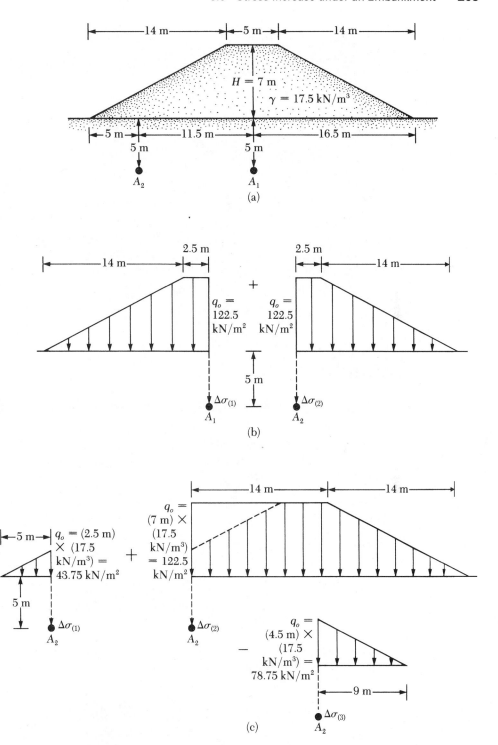

Figure 5.12 Stress increase due to embankment loading

Thus, $I' = 0.495$, so

$$\Delta\sigma_{(2)} = 0.495(122.5) = 60.64 \text{ kN/m}^2$$

For the right side,

$$\frac{B_2}{z} = \frac{9}{5} = 1.8$$

$$\frac{B_1}{z} = \frac{0}{5} = 0$$

and $I' = 0.335$, so

$$\Delta\sigma_{(3)} = (78.75)(0.335) = 26.38 \text{ kN/m}^2$$

The total stress increase at point A_2 is

$$\Delta\sigma = \Delta\sigma_{(1)} + \Delta\sigma_{(2)} - \Delta\sigma_{(3)} = 10.94 + 60.64 - 26.38 = \mathbf{45.2 \text{ kN/m}^2} \quad \blacksquare$$

Elastic Settlement

5.7   Elastic Settlement Based on the Theory of Elasticity

The elastic settlement of a shallow foundation can be estimated by using the theory of elasticity. From Hooke's law, as applied to Figure 5.13, we obtain

$$S_e = \int_0^H \varepsilon_z dz = \frac{1}{E_s} \int_0^H (\Delta\sigma_z - \mu_s\Delta\sigma_x - \mu_s\Delta\sigma_y)dz \tag{5.24}$$

where
$$\begin{aligned}
S_e &= \text{elastic settlement} \\
E_s &= \text{modulus of elasticity of soil} \\
H &= \text{thickness of the soil layer} \\
\mu_s &= \text{Poisson's ratio of the soil} \\
\Delta\sigma_x, \Delta\sigma_y, \Delta\sigma_z &= \text{stress increase due to the net applied foundation load in} \\
&\quad \text{the } x, y, \text{ and } z \text{ directions, respectively}
\end{aligned}$$

Theoretically, if the foundation is perfectly flexible (see Figure 5.14 and Bowles, 1987), the settlement may be expressed as

$$S_e = q_o(\alpha B')\frac{1 - \mu_s^2}{E_s}I_s I_f \tag{5.25}$$

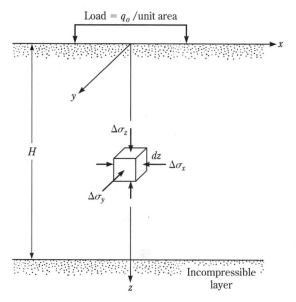

Figure 5.13 Elastic settlement of shallow foundation

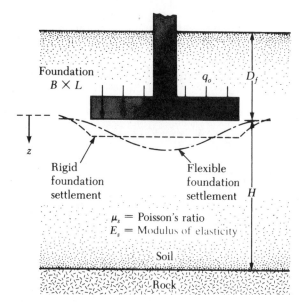

Figure 5.14 Elastic settlement of flexible and rigid foundations

where q_o = net applied pressure on the foundation
μ_s = Poisson's ratio of soil
E_s = average modulus of elasticity of the soil under the foundation, measured from $z = 0$ to about $z = 4B$

$B' = B/2$ for center of foundation
$\quad = B$ for corner of foundation

$$I_s = \text{shape factor (Steinbrenner, 1934)}$$

$$= F_1 + \frac{1 - 2\mu_s}{1 - \mu_s} F_2 \tag{5.26}$$

$$F_1 = \frac{1}{\pi}(A_0 + A_1) \tag{5.27}$$

$$F_2 = \frac{n'}{2\pi} \tan^{-1} A_2 \tag{5.28}$$

$$A_0 = m' \ln \frac{\left(1 + \sqrt{m'^2 + 1}\right)\sqrt{m'^2 + n'^2}}{m'\left(1 + \sqrt{m'^2 + n'^2 + 1}\right)} \tag{5.29}$$

$$A_1 = \ln \frac{\left(m' + \sqrt{m'^2 + 1}\right)\sqrt{1 + n'^2}}{m' + \sqrt{m'^2 + n'^2 + 1}} \tag{5.30}$$

$$A_2 = \frac{m'}{n'\sqrt{m'^2 + n'^2 + 1}} \tag{5.31}$$

$$I_f = \text{depth factor (Fox, 1948)} = f\left(\frac{D_f}{B}, \mu_s, \text{ and } \frac{L}{B}\right) \tag{5.32}$$

$\alpha =$ a factor that depends on the location on the foundation where settlement is being calculated

To calculate settlement at the *center* of the foundation, we use

$$\alpha = 4$$

$$m' = \frac{L}{B}$$

and

$$n' = \frac{H}{\left(\dfrac{B}{2}\right)}$$

To calculate settlement at a *corner* of the foundation,

$$\alpha = 1$$

$$m' = \frac{L}{B}$$

and

$$n' = \frac{H}{B}$$

The variations of F_1 and F_2 [see Eqs. (5.27) and (5.28)] with m' and n' are given in Table 5.4. Also, the variation of I_f with D_f/B and μ_s is given in Table 5.5. Note that *when $D_f = 0$, the value of $I_f = 1$ in all cases.*

The elastic settlement of a *rigid foundation* can be estimated as

$$S_{e(\text{rigid})} \approx 0.93 S_{e(\text{flexible, center})} \tag{5.33}$$

Table 5.4 Variation of F_1 and F_2 with m' and n'[a]

n'	m'									
	1.0	**1.2**	**1.4**	**1.6**	**1.8**	**2.0**	**2.5**	**3.0**	**3.5**	**4.0**
0.5										
F_1	0.049	0.046	0.044	0.042	0.041	0.040	0.038	0.038	0.037	0.037
F_2	0.074	0.077	0.080	0.081	0.083	0.084	0.085	0.086	0.087	0.087
0.8										
F_1	0.104	0.100	0.096	0.093	0.091	0.089	0.086	0.084	0.083	0.082
F_2	0.083	0.090	0.095	0.098	0.101	0.103	0.107	0.109	0.110	0.111
1.0										
F_1	0.142	0.138	0.134	0.130	0.127	0.125	0.121	0.118	0.116	0.115
F_2	0.083	0.091	0.098	0.102	0.106	0.109	0.114	0.117	0.119	0.120
2.0										
F_1	0.285	0.290	0.292	0.292	0.291	0.289	0.284	0.279	0.275	0.271
F_2	0.064	0.074	0.083	0.090	0.097	0.102	0.114	0.121	0.127	0.131
4.0										
F_1	0.408	0.431	0.448	0.460	0.469	0.476	0.484	0.487	0.486	0.484
F_2	0.037	0.044	0.051	0.057	0.063	0.069	0.082	0.093	0.102	0.110
6.0										
F_1	0.457	0.489	0.514	0.534	0.550	0.563	0.585	0.598	0.606	0.609
F_2	0.026	0.031	0.036	0.040	0.045	0.050	0.060	0.070	0.079	0.087
8.0										
F_1	0.482	0.519	0.549	0.573	0.594	0.611	0.643	0.664	0.678	0.688
F_2	0.020	0.023	0.027	0.031	0.035	0.038	0.047	0.055	0.063	0.071
10.0										
F_1	0.498	0.537	0.570	0.597	0.621	0.641	0.679	0.707	0.726	0.740
F_2	0.016	0.019	0.022	0.025	0.028	0.031	0.038	0.046	0.052	0.059
12.0										
F_1	0.508	0.550	0.585	0.614	0.639	0.661	0.704	0.736	0.760	0.777
F_2	0.013	0.016	0.018	0.021	0.024	0.026	0.032	0.038	0.044	0.050
100.0										
F_1	0.555	0.605	0.649	0.688	0.722	0.753	0.819	0.872	0.918	0.956
F_2	0.002	0.002	0.002	0.003	0.003	0.003	0.004	0.005	0.006	0.006
1,000.0										
F_1	0.560	0.612	0.657	0.697	0.733	0.765	0.833	0.890	0.938	0.979
F_2	0.000	0.000	0.000	0.000	0.000	0.000	0.000	0.000	0.001	0.001

[a] After Bowles (1987)   (Continued)

Table 5.4   (Continued)

n'	4.5	5.0	6.0	7.0	8.0	9.0	10.0	25.0	50.0	100.0
						m'				
0.5										
F_1	0.036	0.036	0.036	0.036	0.036	0.036	0.036	0.036	0.036	0.036
F_2	0.087	0.087	0.088	0.088	0.088	0.088	0.088	0.088	0.088	0.088
0.8										
F_1	0.081	0.081	0.080	0.080	0.080	0.079	0.079	0.079	0.079	0.079
F_2	0.112	0.112	0.113	0.113	0.113	0.113	0.114	0.114	0.114	0.114
1.0										
F_1	0.114	0.113	0.112	0.112	0.112	0.111	0.111	0.110	0.110	0.110
F_2	0.121	0.122	0.123	0.123	0.124	0.124	0.124	0.125	0.125	0.125
2.0										
F_1	0.269	0.267	0.264	0.262	0.261	0.260	0.259	0.257	0.256	0.256
F_2	0.134	0.136	0.139	0.141	0.143	0.144	0.145	0.147	0.147	0.148
4.0										
F_1	0.482	0.479	0.474	0.470	0.466	0.464	0.462	0.453	0.451	0.451
F_2	0.116	0.121	0.129	0.135	0.139	0.142	0.145	0.154	0.155	0.156
6.0										
F_1	0.611	0.610	0.608	0.604	0.601	0.598	0.595	0.579	0.576	0.575
F_2	0.094	0.101	0.111	0.120	0.126	0.131	0.135	0.153	0.157	0.157
8.0										
F_1	0.694	0.697	0.700	0.700	0.698	0.695	0.692	0.672	0.666	0.665
F_2	0.077	0.084	0.095	0.104	0.112	0.118	0.124	0.151	0.156	0.158
10.0										
F_1	0.750	0.758	0.766	0.770	0.770	0.770	0.768	0.745	0.738	0.735
F_2	0.065	0.071	0.082	0.091	0.099	0.106	0.112	0.147	0.156	0.158
12.0										
F_1	0.791	0.801	0.815	0.823	0.826	0.828	0.828	0.806	0.796	0.793
F_2	0.056	0.061	0.071	0.080	0.088	0.095	0.102	0.143	0.154	0.158
100.0										
F_1	0.990	1.020	1.072	1.114	1.150	1.182	1.209	1.408	1.489	1.499
F_2	0.007	0.008	0.010	0.011	0.013	0.014	0.016	0.039	0.071	0.113
1,000.0										
F_1	1.016	1.049	1.106	1.154	1.196	1.233	1.266	1.548	1.752	1.941
F_2	0.001	0.001	0.001	0.001	0.001	0.001	0.002	0.004	0.008	0.016

Due to the nonhomogeneous nature of soil deposits, the magnitude of E_s may vary with depth. For that reason, Bowles (1987) recommended using a weighted average of E_s in Eq. (5.25), or

$$E_s = \frac{\sum E_{s(i)} \Delta z}{\bar{z}} \tag{5.34}$$

where $E_{s(i)}$ = soil modulus of elasticity within a depth Δz
$\bar{z}$ = H or $5B$, whichever is smaller

Table 5.5 Fox depth factor[a]

D_f/B	L/B						
	1.0	1.2	1.4	1.6	1.8	2.0	5.0
Poisson's Ratio = 0.00 = μ_s							
0.05	0.950	0.954	0.957	0.959	0.961	0.963	0.973
0.10	0.904	0.911	0.917	0.922	0.925	0.928	0.948
0.20	0.825	0.838	0.847	0.855	0.862	0.867	0.903
0.40	0.710	0.727	0.740	0.752	0.761	0.769	0.827
0.60	0.635	0.652	0.666	0.678	0.689	0.698	0.769
0.80	0.585	0.600	0.614	0.626	0.637	0.646	0.723
1.00	0.549	0.563	0.576	0.587	0.598	0.607	0.686
2.00	0.468	0.476	0.484	0.492	0.499	0.506	0.577
Poisson's Ratio = 0.10 = μ_s							
0.05	0.958	0.962	0.965	0.967	0.968	0.970	0.978
0.10	0.919	0.926	0.930	0.934	0.938	0.940	0.957
0.20	0.848	0.859	0.868	0.875	0.881	0.886	0.917
0.40	0.739	0.755	0.768	0.779	0.788	0.795	0.848
0.60	0.665	0.682	0.696	0.708	0.718	0.727	0.793
0.80	0.615	0.630	0.644	0.656	0.667	0.676	0.749
1.00	0.579	0.593	0.606	0.618	0.628	0.637	0.714
2.00	0.496	0.505	0.513	0.521	0.528	0.535	0.606
Poisson's Ratio = 0.30 = μ_s							
0.05	0.979	0.981	0.982	0.983	0.984	0.985	0.990
0.10	0.954	0.958	0.962	0.964	0.966	0.968	0.977
0.20	0.902	0.911	0.917	0.923	0.927	0.930	0.951
0.40	0.808	0.823	0.834	0.843	0.851	0.857	0.899
0.60	0.738	0.754	0.767	0.778	0.788	0.796	0.852
0.80	0.687	0.703	0.716	0.728	0.738	0.747	0.813
1.00	0.650	0.665	0.678	0.689	0.700	0.709	0.780
2.00	0.562	0.571	0.580	0.588	0.596	0.603	0.675
Poisson's Ratio = 0.40 = μ_s							
0.05	0.989	0.990	0.991	0.992	0.992	0.993	0.995
0.10	0.973	0.976	0.978	0.980	0.981	0.982	0.988
0.20	0.932	0.940	0.945	0.949	0.952	0.955	0.970
0.40	0.848	0.862	0.872	0.881	0.887	0.893	0.927
0.60	0.779	0.795	0.808	0.819	0.828	0.836	0.886
0.80	0.727	0.743	0.757	0.769	0.779	0.788	0.849
1.00	0.689	0.704	0.718	0.730	0.740	0.749	0.818
2.00	0.596	0.606	0.615	0.624	0.632	0.640	0.714
Poisson's Ratio = 0.50 = μ_s							
0.05	0.997	0.997	0.998	0.998	0.998	0.998	0.999
0.10	0.988	0.990	0.991	0.992	0.993	0.993	0.996
0.20	0.960	0.966	0.969	0.972	0.974	0.976	0.985
0.40	0.886	0.899	0.908	0.916	0.922	0.926	0.953
0.60	0.818	0.834	0.847	0.857	0.866	0.873	0.917
0.80	0.764	0.781	0.795	0.807	0.817	0.826	0.883
1.00	0.723	0.740	0.754	0.766	0.777	0.786	0.852
2.00	0.622	0.633	0.643	0.653	0.662	0.670	0.747

[a]Fox (1948), after Bowles (1987)

Example 5.4

A rigid shallow foundation 1 m × 2 m is shown in Figure 5.15. Calculate the elastic settlement at the center of the foundation.

Solution

We are given that $B = 1$ m and $L = 2$ m. Note that $\bar{z} = 5$ m $= 5B$. From Eq. (5.34),

$$E_s = \frac{\Sigma E_{s(i)} \Delta z}{\bar{z}}$$

$$= \frac{(10{,}000)(2) + (8{,}000)(1) + (12{,}000)(2)}{5} = 10{,}400 \text{ kN/m}^2$$

For the *center of the foundation,*

$$\alpha = 4$$

$$m' = \frac{L}{B} = \frac{2}{1} = 2$$

and

$$n' = \frac{H}{\left(\dfrac{B}{2}\right)} = \frac{5}{\left(\dfrac{1}{2}\right)} = 10$$

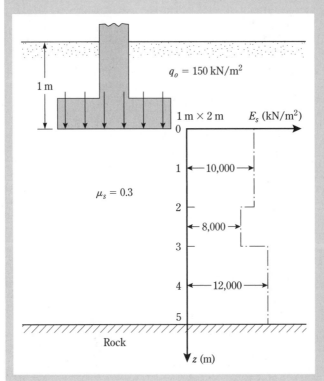

Figure 5.15   Elastic settlement below the center of a foundation

From Table 5.4, $F_1 = 0.641$ and $F_2 = 0.031$. From Eq. (5.26),

$$I_s = F_1 + \frac{2 - \mu_s}{1 - \mu_s}F_2$$

$$= 0.641 + \frac{2 - 0.3}{1 - 0.3}(0.031) = 0.716$$

Again, $D_f/B = 1/1 = 1$, $L/B = 2$, and $\mu_s = 0.3$. From Table 5.5, $I_f = 0.709$. Hence,

$$S_{e(\text{flexible})} = q_o(\alpha B')\frac{1 - \mu_s^2}{E_s}I_s I_f$$

$$= (150)\left(4 \times \frac{1}{2}\right)\left(\frac{1 - 0.3^2}{10,400}\right)(0.716)(0.709) = 0.0133 \text{ m} = 13.3 \text{ mm}$$

Since the foundation is rigid, from Eq. (5.33) we obtain

$$S_{e(\text{rigid})} = (0.93)(13.3) = \textbf{12.9 mm} \qquad \blacksquare$$

5.8   *Improved Equation for Elastic Settlement*

In 1999, Mayne and Poulos presented an improved formula for calculating the elastic settlement of foundations. The formula takes into account the rigidity of the foundation, the depth of embedment of the foundation, the increase in the modulus of elasticity of the soil with depth, and the location of rigid layers at a limited depth. To use Mayne and Poulos's equation, one needs to determine the equivalent diameter B_e of a rectangular foundation, or

$$B_e = \sqrt{\frac{4BL}{\pi}} \qquad (5.35)$$

where B = width of foundation
$\qquad L$ = length of foundation

For circular foundations,

$$B_e = B \qquad (5.36)$$

where B = diameter of foundation

Figure 5.16 shows a foundation with an equivalent diameter B_e located at a depth D_f below the ground surface. Let the thickness of the foundation be t and the modulus of elasticity of the foundation material be E_f. A rigid layer is located at a

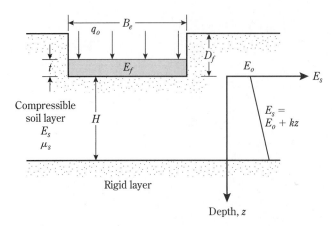

Figure 5.16   Improved equation for calculating elastic settlement: general parameters

depth H below the bottom of the foundation. The modulus of elasticity of the compressible soil layer can be given as

$$E_s = E_o + kz \tag{5.37}$$

With the preceding parameters defined, the elastic settlement below the center of the foundation is

$$S_e = \frac{q_o B_e I_G I_F I_E}{E_o}\left(1 - \mu_s^2\right) \tag{5.38}$$

where I_G = influence factor for the variation of E_s with depth

$$= f\left(\beta = \frac{E_o}{kB_e}, \frac{H}{B_e}\right)$$

I_F = foundation rigidity correction factor
I_E = foundation embedment correction factor

Figure 5.17 shows the variation of I_G with $\beta = E_o/kB_e$ and H/B_e. The foundation rigidity correction factor can be expressed as

$$I_F = \frac{\pi}{4} + \frac{1}{4.6 + 10\left(\dfrac{E_f}{E_o + \dfrac{B_e}{2}k}\right)\left(\dfrac{2t}{B_e}\right)^3} \tag{5.39}$$

Similarly, the embedment correction factor is

$$I_E = 1 - \frac{1}{3.5 \exp(1.22\mu_s - 0.4)\left(\dfrac{B_e}{D_f} + 1.6\right)} \tag{5.40}$$

Figures 5.18 and 5.19 show the variation of I_F and I_E with terms expressed in Eqs. (5.39) and (5.40).

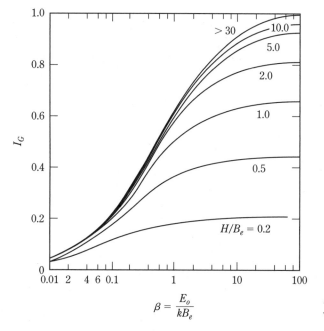

$$\beta = \frac{E_o}{kB_e}$$

Figure 5.17 Variation of I_G with β

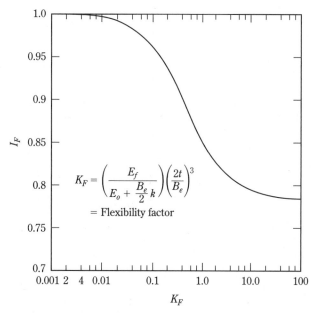

$$K_F = \left(\frac{E_f}{E_o + \frac{B_e}{2} k}\right)\left(\frac{2t}{B_e}\right)^3$$
$$= \text{Flexibility factor}$$

Figure 5.18 Variation of rigidity correction factor I_F with flexibility factor K_F. [Eq. (5.39)]

Example 5.5

For a shallow foundation supported by a silty clay, as shown in Figure 5.16,

$$\text{Length} = L = 1.5 \text{ m}$$
$$\text{Width} = B = 1 \text{ m}$$

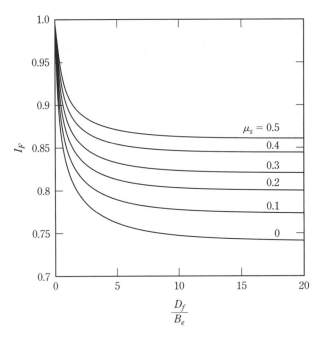

Figure 5.19 Variation of embedment correction factor I_E with D_f/B_e [Eq (5.40)]

Depth of foundation $= D_f = 1 \text{ m}$

Thickness of foundation $= t = 0.23 \text{ m}$

Load per unit area $= q_o = 190 \text{ kN/m}^2$

$E_f = 15 \times 10^6 \text{ kN/m}^2$

The silty clay soil has the following properties:

$$H = 2 \text{ m}$$

$$\mu_s = 0.3$$

$$E_o = 9000 \text{ kN/m}^2$$

$$k = 500 \text{ kN/m}^2/\text{m}$$

Estimate the elastic settlement of the foundation.

Solution

From Eq. (5.35), the equivalent diameter is

$$B_e = \sqrt{\frac{4BL}{\pi}} = \sqrt{\frac{(4)(1.5)(1)}{\pi}} = 1.38 \text{ m}$$

so

$$\beta = \frac{E_o}{kB_e} = \frac{9000}{(500)(1.38)} = 13.04$$

and

$$\frac{H}{B_e} = \frac{2}{1.38} = 1.45$$

From Figure 5.17, for $\beta = 13.04$ and $H/B_e = 1.45$, the value of $I_G \approx 0.74$. From Eq. (5.39),

$$I_F = \frac{\pi}{4} + \frac{1}{4.6 + 10\left(\dfrac{E_f}{E_o + \dfrac{B_e}{2}k}\right)\left(\dfrac{2t}{B_e}\right)^3}$$

$$= \frac{\pi}{4} + \frac{1}{4.6 + 10\left[\dfrac{15 \times 10^6}{9000 + \left(\dfrac{1.38}{2}\right)(500)}\right]\left[\dfrac{(2)(0.23)}{1.38}\right]^3} = 0.787$$

From Eq. (5.40),

$$I_E = 1 - \frac{1}{3.5\,\exp(1.22\mu_s - 0.4)\left(\dfrac{B_e}{D_f} + 1.6\right)}$$

$$= 1 - \frac{1}{3.5\,\exp[(1.22)(0.3) - 0.4]\left(\dfrac{1.38}{1} + 1.6\right)} = 0.907$$

From Eq. (5.38),

$$S_e = \frac{q_o B_e I_G I_F I_E}{E_o}(1 - \mu_s^2)$$

so, with $q_o = 190$ kN/m^2, it follows that

$$S_e = \frac{(190)(1.38)(0.74)(0.787)(0.907)}{9000}(1 - 0.3^2) = 0.014 \text{ m} \approx \textbf{14 mm} \quad \blacksquare$$

5.9   *Settlement of Sandy Soil: Use of Strain Influence Factor*

The settlement of granular soils can also be evaluated by the use of a semiempirical *strain influence factor* proposed by Schmertmann and Hartman (1978). According to this method, the settlement is

$$S_e = C_1 C_2 (\bar{q} - q) \sum_0^{z_2} \frac{I_z}{E_s} \Delta z \qquad (5.41)$$

where I_z = strain influence factor
C_1 = a correction factor for the depth of foundation embedment = $1 - 0.5$ $[q/(\bar{q} - q)]$
C_2 = a correction factor to account for creep in soil
= $1 + 0.2 \log$ (time in years/0.1)
$\bar{q}$ = stress at the level of the foundation
$q = \gamma D_f$

The variation of the strain influence factor with depth below the foundation is shown in Figure 5.20a. Note that, for square or circular foundations,

$$I_z = 0.1 \quad \text{at } z = 0$$

$$I_z = 0.5 \quad \text{at } z = z_1 = 0.5B$$

and

$$I_z = 0 \quad \text{at } z = z_2 = 2B$$

Similarly, for foundations with $L/B \geqslant 10$,

$$I_z = 0.2 \quad \text{at } z = 0$$

$$I_z = 0.5 \quad \text{at } z = z_1 = B$$

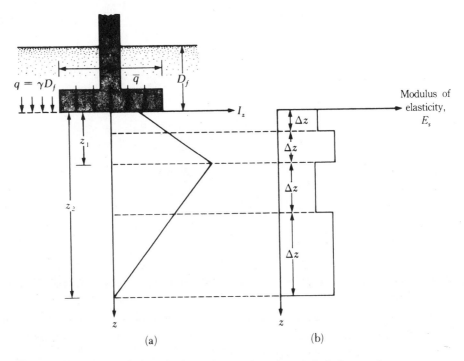

Figure 5.20 Calculation of elastic settlement by using strain influence factor

and

$$I_z = 0 \qquad \text{at } z = z_2 = 4B$$

where B = width of the foundation and L = length of the foundation

Values of L/B between 1 and 10 can be interpolated.

      The use of Eq. (5.41) first requires an evaluation of the approximate variation of the modulus of elasticity with depth (Figure 5.20). This evaluation can be made by using the standard penetration numbers or cone penetration resistances. (See Chapter 2.) The soil can be divided into several layers to a depth of $z = z_2$, and the elastic settlement of each layer can be estimated. The sum of the settlement of all layers equals S_e.

A Case History of the Calculation of S_e Using the Strain Influence Factor

Schmertmann (1970) provided a case history of a rectangular foundation (a Belgian bridge pier) having $L = 23$ m and $B = 2.6$ m and being supported by a granular soil deposit. For this foundation, we may assume that $L/B \approx 10$ for plotting the strain influence factor diagram. Figure 5.21 shows the details of the foundation, along with

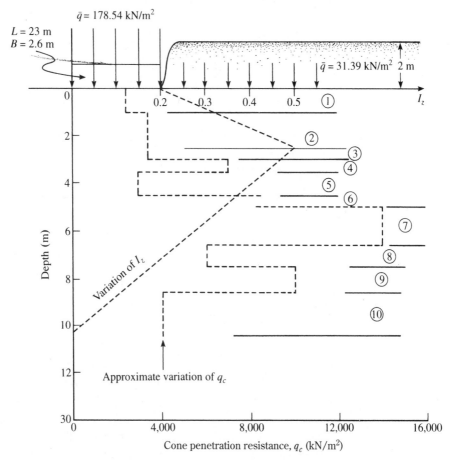

Figure 5.21 Variation of I_z and q_c below the foundation

the approximate variation of the cone penetration resistance, q_c, with depth. For this foundation [see Eq. (5.41)], note that

$$\bar{q} = 178.54 \text{ kN/m}^2$$

$$q = 31.39 \text{ kN/m}^2$$

$$C_1 = 1 - 0.5\frac{1}{\bar{q} - q} = 1 - (0.5)\left(\frac{31.39}{178.54 - 31.39}\right) = 0.893$$

and

$$C_2 = 1 + 0.2 \log\left(\frac{t \text{ yr}}{0.1}\right)$$

For $t = 5$ yr,

$$C_2 = 1 + 0.2 \log\left(\frac{5}{0.1}\right) = 1.34$$

The following table shows the calculation of $\sum_0^{z_2}(I_z/E_s)\,\Delta z$ in conjunction with Figure 5.21:

Layer	Δz (m)	q_c (kN/m²)	E_s^a (kN/m²)	z to the center of the layer (m)	I_z at the center of the layer	$(I_z/E_s)\,\Delta z$ (m²/kN)
1	1	2,450	8,575	0.5	0.258	3.00×10^{-5}
2	1.6	3,430	12,005	1.8	0.408	5.43×10^{-5}
3	0.4	3,430	12,005	2.8	0.487	1.62×10^{-5}
4	0.5	6,870	24,045	3.25	0.458	0.95×10^{-5}
5	1.0	2,950	10,325	4.0	0.410	3.97×10^{-5}
6	0.5	8,340	29,190	4.75	0.362	0.62×10^{-5}
7	1.5	14,000	49,000	5.75	0.298	0.91×10^{-5}
8	1	6,000	21,000	7.0	0.247	1.17×10^{-5}
9	1	10,000	35,000	8.0	0.154	0.44×10^{-5}
10	1.9	4,000	14,000	9.45	0.062	0.84×10^{-5}
	$\Sigma\ 10.4 \text{ m} = 4B$					$\Sigma\ 18.95 \times 10^{-5}$

$^a E_s \approx 3.5q_c$

Hence, the immediate settlement is calculated as

$$S_e = C_1 C_2 (\bar{q} - q) \Sigma \frac{I_z}{E_s} \Delta z$$

$$= (0.893)(1.34)(178.54 - 31.39)(18.95 \times 10^{-5})$$

$$= 0.03336 \text{ m} \approx 33 \text{ mm}$$

After five years, the actual *maximum* settlement observed for the foundation was about 39 mm.

Elastic Settlement of Sandy Soils: Use of Burland and Burbidge's Method

Burland and Burbidge (1985) proposed a method of calculating the elastic settlement of sandy soil using the *field standard penetration number, N_{60}*. (See Chapter 2.) The method can be summarized as follows:

1. **Variation of Standard Penetration Number with Depth**
   Obtain the field penetration numbers (N_{60}) with depth at the location of the foundation. The following adjustments of N_{60} may be necessary, depending on the field conditions:
   For gravel or sandy gravel,

   $$N_{60(a)} \approx 1.25 \, N_{60} \tag{5.42}$$

   For fine sand or silty sand below the groundwater table and $N_{60} > 15$,

   $$N_{60(a)} \approx 15 + 0.5(N_{60} - 15) \tag{5.43}$$

   where $N_{60(a)}$ = adjusted N_{60} value.

2. **Determination of Depth of Stress Influence (z')**
   In determining the depth of stress influence, the following three cases may arise:

   Case I. If N_{60} [or $N_{60(a)}$] is approximately constant with depth, calculate z' from

   $$\frac{z'}{B_R} = 1.4\left(\frac{B}{B_R}\right)^{0.75} \tag{5.44}$$

   where B_R = reference width $\begin{cases} = 1 \text{ ft (if } B \text{ is in ft)} \\ = 0.3 \text{ m (if } B \text{ is in m)} \end{cases}$
   B = width of the actual foundation

   Case II. If N_{60} [or $N_{60(a)}$] is increasing with depth, use Eq. (5.44) to calculate z'.

   Case III. If N_{60} [or $N_{60(a)}$] is decreasing with depth, calculate $z' = 2B$ and z' = distance from the bottom of the foundation to the bottom of the soft soil layer (z''). Use $z' = 2B$ or $z' = z''$ (whichever is smaller).

3. **Calculation of Elastic Settlement S_e**

   The elastic settlement of the foundation, S_e, can be calculated from

   $$\frac{S_e}{B_R} = \alpha_1 \alpha_2 \alpha_3 \left[\frac{1.25\left(\dfrac{L}{B}\right)}{0.25 + \left(\dfrac{L}{B}\right)}\right]^2 \left(\frac{B}{B_R}\right)\left(\frac{q'}{p_a}\right) \tag{5.45}$$

   where α_1 = a constant
   α_2 = compressibility index

α_3 = correction for the depth of influence
p_a = atmospheric pressure = 100 kN/m^2 ($\approx$2000 lb/ft^2)
L = length of the foundation

For *normally consolidated* sand,

$$\alpha_1 = 0.14 \tag{5.46}$$

and

$$\alpha_2 = \frac{1.71}{[\overline{N}_{60} \text{ or } \overline{N}_{60(a)}]^{1.4}} \tag{5.47}$$

where $\overline{N}_{60}$ or $\overline{N}_{60(a)}$ = average value of N_{60} or $N_{60(a)}$ in the depth of stress influence

$$\alpha_3 = \frac{z''}{z'}\left(2 - \frac{z''}{z'}\right) \leq 1 \tag{5.48}$$

and

$$q' = q_o \tag{5.49}$$

in which q_o = net applied stress at the level of the foundation (i.e., the stress at the level of the foundation minus the overburden pressure)

For *overconsolidated sand with* $q_o \leq \sigma'_c$, the preconsolidation pressure,

$$\alpha_1 = 0.047 \tag{5.50}$$

and

$$\alpha_2 = \frac{0.57}{[\overline{N}_{60} \text{ or } \overline{N}_{60(a)}]^{1.4}} \tag{5.51}$$

For α_3, use Eq. (5.48):

$$q' = q_o \tag{5.52}$$

For *overconsolidated sand with* $q_o > \sigma'_c$,

$$\alpha_1 = 0.14 \tag{5.53}$$

For α_2, use Eq. (5.51), and for α_3, use Eq. (5.48). Finally, use

$$q' = q_o - 0.67\sigma'_c \tag{5.54}$$

5.11 Range of Material Parameters for Computing Elastic Settlement

Sections 5.7 and 5.8 presented the equations for calculating the elastic settlement of foundations. Those equations contain the elastic parameters, such as E_s and μ_s. If the laboratory test results for these parameters are not available, certain real-

Table 5.6 Elastic Parameters of Various Soils

Type of soil	Modulus of elasticity, E_s		Poisson's ratio, μ_s
	MN/m²	lb/in²	
Loose sand	10.5–24.0	1500–3500	0.20–0.40
Medium dense sand	17.25–27.60	2500–4000	0.25–0.40
Dense sand	34.50–55.20	5000–8000	0.30–0.45
Silty sand	10.35–17.25	1500–2500	0.20–0.40
Sand and gravel	69.00–172.50	10,000–25,000	0.15–0.35
Soft clay	4.1–20.7	600–3000	
Medium clay	20.7–41.4	3000–6000	0.20–0.50
Stiff clay	41.4–96.6	6000–14,000	

istic assumptions have to be made. Table 5.6 shows the approximate ranges of the elastic parameters for various soils.

Several investigators have correlated the values of the modulus of elasticity, E_s, with the field standard penetration number, N_{60} and the cone penetration resistance, q_c. Mitchell and Gardner (1975) compiled a list of these correlations. Schmertmann (1970) indicated that the modulus of elasticity of sand may be given by

$$\frac{E_s}{p_a} = 8N_{60} \tag{5.55}$$

where N_{60} = standard penetration resistance (see Chapter 2)
p_a = atmospheric pressure ≈ 100 kN/m² (≈2000 lb/ft²)

Similarly,

$$E_s = 2q_c \tag{5.56}$$

where q_c = static cone penetration resistance

Schmertmann and Hartman (1978) further suggested that the following correlations may be used with the strain influence factors described in Section 5.9:

$$E_s = 2.5q_c \quad \text{(for square and circular foundations)} \tag{5.57}$$

and

$$E_s = 3.5q_c \quad \text{(for strip foundations)} \tag{5.58}$$

[*Note:* Any consistent set of units may be used in Eqs. (5.56)–(5.58).]
The modulus of elasticity of normally consolidated clays may be estimated as

$$E_s = 250c_u \text{ to } 500c_u \tag{5.59}$$

and for overconsolidated clays as

$$E_s = 750c_u \text{ to } 1000c_u \tag{5.60}$$

where c_u = undrained cohesion of clay soil

5.12 Seismic Bearing Capacity and Settlement in Granular Soil

In some instances, shallow foundations may fail during seismic events. Published studies relating to the bearing capacity of shallow foundations in such instances are rare. In 1993, however, Richards et al. developed a seismic bearing capacity theory that we shall examine in this section. The theory is not yet supported by field data.

Figure 5.22 shows a failure surface in soil assumed for the subsequent analysis, under static conditions. Similarly, Figure 5.23 shows the failure surface under earthquake conditions. Note that, in the two figures,

$$\alpha_A, \alpha_{AE} = \text{inclination angles for active pressure conditions}$$

and

$$\alpha_P, \alpha_{PE} = \text{inclination angles for passive pressure conditions}$$

According to this theory, the ultimate bearing capacities for *continuous foundations* in granular soil are

$$q_u = qN_q + \tfrac{1}{2}\gamma BN_\gamma \ (static\ conditions) \tag{5.61}$$

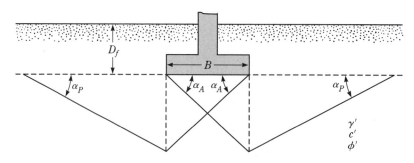

Figure 5.22 Failure surface in soil for static bearing capacity analysis. (*Note:* $\alpha_A = 45 + \phi'/2$ and $\alpha_p = 45 - \phi'/2$)

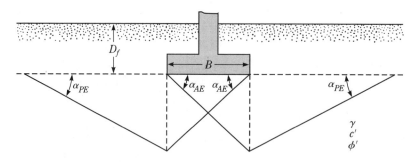

Figure 5.23 Failure surface in soil for seismic bearing capacity analysis

and

$$q_{uE} = qN_{qE} + \tfrac{1}{2}\gamma BN_{\gamma E} \, (earthquake \; conditions) \tag{5.62}$$

where $N_q, N_\gamma, N_{qE}, N_{\gamma E}$ = bearing capacity factors
$$q = \gamma D_f$$

Note that

$$N_q \text{ and } N_\gamma = f(\phi')$$

and

$$N_{qE} \text{ and } N_{\gamma E} = f(\phi', \tan\theta)$$

where $\tan\theta = \dfrac{k_h}{1 - k_v}$

with k_h = horizontal coefficient of acceleration due to an earthquake
k_v = vertical coefficient of acceleration due to an earthquake

The variations of N_q and N_γ with ϕ' are shown in Figure 5.24. Figure 5.25 shows the variations of $N_{\gamma E}/N_\gamma$ and N_{qE}/N_q with $\tan\theta$ and the soil friction angle ϕ'.

Under static conditions, bearing capacity failure can lead to a substantial sudden downward movement of the foundation. However, bearing capacity–related settlement in an earthquake takes place when the ratio $k_h/(1 - k_v)$ reaches the critical

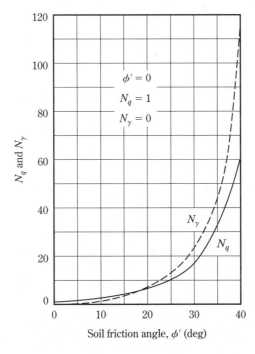

Figure 5.24 Variation of N_q and N_γ based on failure surface assumed in Figure 5.22

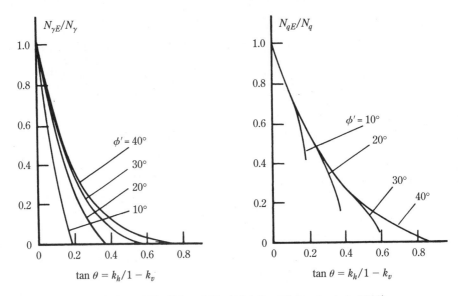

Figure 5.25   Variation of $N_{\gamma E}/N_{\gamma}$ and N_{qE}/N_q (after Richards et al., 1993)

value $(k_h/1 - k_v)^*$. If $k_v = 0$, then $(k_h/1 - k_v)^*$ becomes equal to k_h^*. Figure 5.26 shows the variation of k_h^* (for $k_v = 0$ and $c' = 0$ and in granular soil) with the factor of safety (FS) applied to the ultimate static bearing capacity [Eq. 5.61], with ϕ', and with D_f/B.

The settlement of a strip foundation due to an earthquake can be estimated (Richards et al., 1993) as

$$S_{Eq}(m) = 0.174\frac{V^2}{Ag}\left|\frac{k_h^*}{A}\right|^{-4} \tan \alpha_{AE} \qquad (5.63)$$

where V = peak velocity for the design earthquake (m/sec)
   A = acceleration coefficient for the design earthquake
   g = acceleration due to gravity (9.18 m/sec²)

The values of k_h^* and α_{AE} can be obtained from Figures 5.26 and 5.27, respectively.

Example 5.6

A strip foundation is to be constructed on a sandy soil with $B = 4$ ft, $D_f = 3$ ft, $\gamma = 110$ lb/ft³, and $\phi' = 30°$.

a. Determine the gross ultimate bearing capacity q_{uE}. Assume that $k_v = 0$ and $k_h = 0.176$.

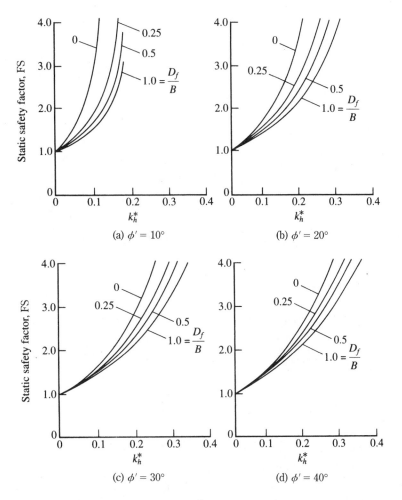

Figure 5.26   Critical acceleration k_h^* for $c' = 0$ (after Richards et al., 1993)

b. If the design earthquake parameters are $V = 1.3$ ft/sec and $A = 0.32$, determine the seismic settlement of the foundation. Use FS $= 3$ to obtain the static allowable bearing capacity.

Solution

a. From Figure 5.24, for $\phi' = 30°$, $N_q = 16.51$ and $N_\gamma = 23.76$. Also,

$$\tan \theta = \frac{k_h}{1 - k_v} = 0.176$$

For $\tan \theta = 0.176$, Figure 5.25 gives

$$\frac{N_{\gamma E}}{N_\gamma} = 0.4 \qquad \text{and} \qquad \frac{N_{qE}}{N_q} = 0.6$$

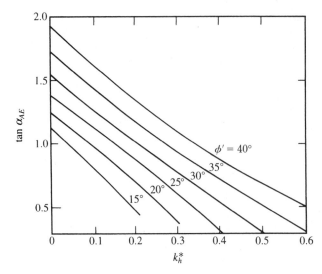

Figure 5.27 Variation of $\tan \alpha_{AE}$ with k_h^* and soil friction angle ϕ' (after Richards et al., 1993)

Thus,

$$N_{\gamma E} = (0.4)(23.76) = 9.5$$

$$N_{qE} = (0.6)(16.51) = 9.91$$

and

$$q_{uE} = qN_{qE} + \tfrac{1}{2}\gamma BN_{\gamma E}$$

$$= (3 \times 110)(9.91) + (\tfrac{1}{2})(110)(4)(9.5) \approx 5360 \ \text{lb/ft}^2$$

b. For the foundation,

$$\frac{D_f}{B} = \frac{3}{4} = 0.75$$

From Figure 5.26(c), for $\phi' = 30°$, FS = 3, and $D_f/B = 0.75$, the value of $k_h^* = 0.26$. Also, from Figure 5.27, for $k_h^* = 0.26$ and $\phi' = 30°$, the value of $\tan \alpha_{AE} = 0.88$. From Eq. (5.63), we have

$$S_{Eq} = 0.174 \left| \frac{k_h^*}{A} \right|^{-4} \tan \alpha_{AE} \left(\frac{V^2}{Ag} \right) (\text{meters})$$

with

$$V = 1.3 \ \text{ft} \approx 0.4 \ \text{m}$$

it follows that

$$S_{Eq} = 0.174 \frac{(0.4)^2}{(0.32)(9.81)} \left| \frac{0.26}{0.32} \right|^{-4} (0.88) = 0.0179 \ \text{m} = \textbf{0.7 in.} \quad \blacksquare$$

Primary Consolidation Settlement

5.13 ## Primary Consolidation Settlement Relationships

As mentioned before, consolidation settlement occurs over time in saturated clayey soils subjected to an increased load caused by construction of the foundation. (See Figure 5.28.) On the basis of the one-dimensional consolidation settlement equations given in Chapter 1, we write

$$S_{c(p)} = \int \varepsilon_z dz$$

where ε_z = vertical strain

$$= \frac{\Delta e}{1 + e_o}$$

Δe = change of void ratio

$$= f(\sigma'_o, \sigma'_c, \text{ and } \Delta\sigma')$$

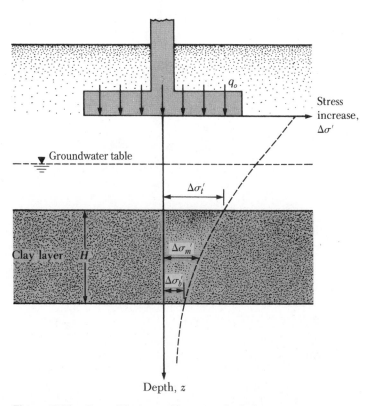

Figure 5.28 Consolidation settlement calculation

So

$$S_{c(p)} = \frac{C_c H_c}{1 + e_o} \log \frac{\sigma'_o + \Delta \sigma'_{av}}{\sigma'_o} \qquad \text{(for normally consolidated clays)} \qquad (1.53)$$

$$S_{c(p)} = \frac{C_s H_c}{1 + e_o} \log \frac{\sigma'_o + \Delta \sigma'_{av}}{\sigma'_o} \qquad \text{(for overconsolidated clays with } \sigma'_o + \Delta \sigma'_{av} < \sigma'_c) \qquad (1.55)$$

$$S_{c(p)} = \frac{C_s H_c}{1 + e_o} \log \frac{\sigma'_c}{\sigma'_o} + \frac{C_c H_c}{1 + e_o} \log \frac{\sigma'_o + \Delta \sigma'_{av}}{\sigma'_c} \qquad \text{(for overconsolidated clays with } \sigma'_o < \sigma'_c < \sigma'_o + \Delta \sigma'_{av}) \qquad (1.57)$$

where σ'_o = average effective pressure on the clay layer before the construction of the foundation

$\Delta \sigma'_{av}$ = average increase in effective pressure on the clay layer caused by the construction of the foundation

σ'_c = preconsolidation pressure

e_o = initial void ratio of the clay layer

C_c = compression index

C_s = swelling index

H_c = thickness of the clay layer

The procedures for determining the compression and swelling indexes were discussed in Chapter 1.

Note that the increase in effective pressure, $\Delta \sigma'$, on the clay layer is not constant with depth: The magnitude of $\Delta \sigma'$ will decrease with the increase in depth measured from the bottom of the foundation. However, the average increase in pressure may be approximated by

$$\Delta \sigma'_{av} = \frac{1}{6}(\Delta \sigma'_t + 4 \Delta \sigma'_m + \Delta \sigma'_b) \qquad (5.64)$$

where $\Delta \sigma'_t$, $\Delta \sigma'_m$, and $\Delta \sigma'_b$ are, respectively, the effective pressure increases at the *top, middle,* and *bottom* of the clay layer that are caused by the construction of the foundation.

The method of determining the pressure increase caused by various types of foundation load is discussed in Sections 5.2–5.6. $\Delta \sigma'_{av}$ can also be directly obtained from the method presented in Section 5.5.

Example 5.7

A plan of a foundation 1 m × 2 m is shown in Figure 5.29. Estimate the consolidation settlement of the foundation.

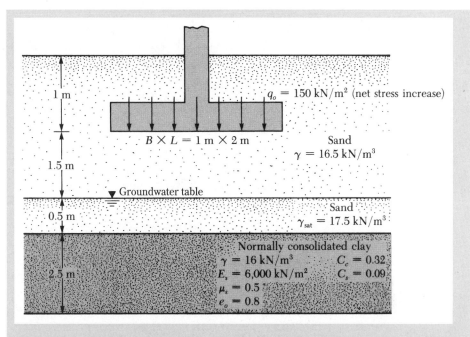

Figure 5.29 Calculation of primary consolidation settlement for a foundation

Solution
The clay is normally consolidated. Thus,

$$S_c = \frac{C_c H_c}{1 + e_o} \log \frac{\sigma'_o + \Delta\sigma'_{av}}{\sigma'_o}$$

so

$$\sigma'_o = (2.5)(16.5) + (0.5)(17.5 - 9.81) + (1.25)(16 - 9.81)$$

$$= 41.25 + 3.85 + 7.74 = 52.84 \text{ kN/m}^2$$

From Eq. (5.64),

$$\Delta\sigma'_{av} = \tfrac{1}{6}(\Delta\sigma'_t + 4\Delta\sigma'_m + \Delta\sigma'_b)$$

Now the following table can be prepared (*Note:* $L = 2$ m; $B = 1$ m):

$m_1 = L/B$	z(m)	$z/(B/2) = n_1$	$I_c{}^a$	$\Delta\sigma = q_o I_c{}^b$
2	2	4	0.190	$28.5 = \Delta\sigma'_t$
2	$2 + 2.5/2 = 3.25$	6.5	≈ 0.085	$12.75 = \Delta\sigma'_m$
2	$2 + 2.5 = 4.5$	9	0.045	$6.75 = \Delta\sigma'_b$

[a] Table 5.3
[b] Eq. (5.10)

Now,

$$\Delta\sigma'_{av} = \tfrac{1}{6}(28.5 + 4 \times 12.75 + 6.75) = 14.38 \text{ kN/m}^2$$

so

$$S_{c(p)} = \frac{(0.32)(2.5)}{1 + 0.8}\log\left(\frac{52.84 + 14.38}{52.84}\right) = 0.0465 = \textbf{46.5 mm} \qquad \blacksquare$$

5.14 *Three-Dimensional Effect on Primary Consolidation Settlement*

The consolidation settlement calculation presented in the preceding section is based on Eqs. (1.53), (1.55), and (1.57). These equations, as shown in Chapter 1, are in turn based on one-dimensional laboratory consolidation tests. The underlying assumption is that the increase in pore water pressure, Δu, immediately after application of the load equals the increase in stress, $\Delta\sigma$, at any depth. In this case,

$$S_{c(p)-\text{oed}} = \int \frac{\Delta e}{1 + e_o}\, dz = \int m_v \Delta\sigma'_{(1)} dz$$

where $S_{c(p)-\text{oed}}$ = consolidated settlement calculated by using Eqs. (1.53), (1.55), and (1.57)

$\Delta\sigma'_{(1)}$ = effective vertical stress increase

m_v = volume coefficient of compressibility (see Chapter 1)

In the field, however, when a load is applied over a limited area on the ground surface, such an assumption will not be correct. Consider the case of a circular foundation on a clay layer, as shown in Figure 5.30. The vertical and the horizontal stress increases at a point in the layer immediately below the center of the foundation are $\Delta\sigma_{(1)}$ and $\Delta\sigma_{(3)}$, respectively. For a saturated clay, the pore water pressure increase at that depth (see Chapter 1) is

$$\Delta u = \Delta\sigma_{(3)} + A[\Delta\sigma_{(1)} - \Delta\sigma_{(3)}] \qquad (5.65)$$

where A = pore water pressure parameter

For this case,

$$S_{c(p)} = \int m_v \Delta u\, dz = \int (m_v)\{\Delta\sigma_{(3)} + A[\Delta\sigma_{(1)} - \Delta\sigma_{(3)}]\}\, dz \qquad (5.66)$$

Thus, we can write

$$K_{\text{cir}} = \frac{S_{c(p)}}{S_{c(p)-\text{oed}}} = \frac{\displaystyle\int_0^{H_c} m_v \Delta u\, dz}{\displaystyle\int_0^{H_c} m_v \Delta\sigma'_{(1)} dz} = A + (1 - A)\left[\frac{\displaystyle\int_0^{H_c} \Delta\sigma'_{(3)} dz}{\displaystyle\int_0^{H_c} \Delta\sigma'_{(1)} dz}\right] \qquad (5.67)$$

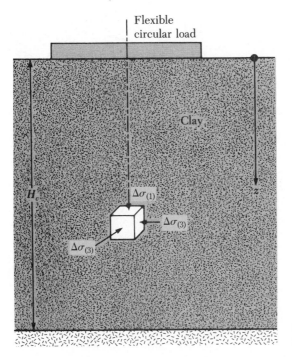

Figure 5.30 Circular foundation on a clay layer

where K_{cir} = settlement ratio for circular foundations

The settlement ratio for a continuous foundation, K_{str}, can be determined in a manner similar to that for a circular foundation. The variation of K_{cir} and K_{str} with A and H_c/B is given in Figure 5.31. (*Note:* B = diameter of a circular foundation, and B = width of a continuous foundation.)

The preceding technique is generally referred to as the *Skempton–Bjerrum modification* (1957) for a consolidation settlement calculation.

Leonards (1976) examined the correction factor K_{cr} for a three-dimensional consolidation effect in the field for a circular foundation located over *overconsolidated clay.* Referring to Figure 5.30, we have

$$S_{c(p)} = K_{cr(OC)} \, S_{c(p)-oed} \tag{5.68}$$

where $\quad K_{cr(OC)} = f\left(OCR, \dfrac{B}{H_c}\right) \tag{5.69}$

in which $\quad$ OCR = overconsolidation ratio = $\dfrac{\sigma_c'}{\sigma_o'} \tag{5.70}$

where $\quad \sigma_c'$ = preconsolidation pressure
$\quad \sigma_o'$ = present

The interpolated values of $K_{cr(OC)}$ from Leonard's 1976 work are given in Table 5.7.

The procedure for using the aforementioned modification factors is demonstrated in Example 5.8.

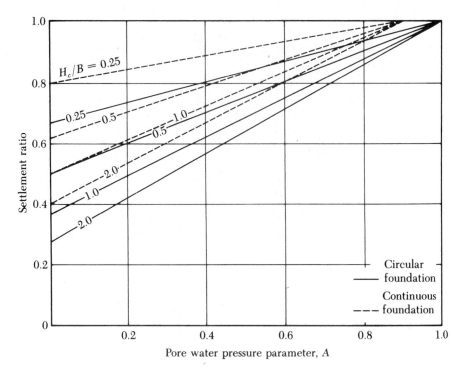

Figure 5.31   Settlement ratios for circular (K_{cir}) and continuous (K_{str}) foundations

Table 5.7   Variation of $K_{cr(OC)}$ with OCR and B/H_c

OCR	$K_{cr(OC)}$		
	$B/H_c = 4.0$	$B/H_c = 1.0$	$B/H_c = 0.2$
1	1	1	1
2	0.986	0.957	0.929
3	0.972	0.914	0.842
4	0.964	0.871	0.771
5	0.950	0.829	0.707
6	0.943	0.800	0.643
7	0.929	0.757	0.586
8	0.914	0.729	0.529
9	0.900	0.700	0.493
10	0.886	0.671	0.457
11	0.871	0.643	0.429
12	0.864	0.629	0.414
13	0.857	0.614	0.400
14	0.850	0.607	0.386
15	0.843	0.600	0.371
16	0.843	0.600	0.357

Example 5.8

Assume in Example 5.7 that the pore water pressure parameter $A = 0.6$. Taking into account the three-dimensional effect, estimate the primary consolidation settlement.

Solution

Assuming that the 2:1 method of stress increase (see Figure 5.5) holds good, the area of distribution of stress at the top of the clay layer will have dimensions

$$B' = \text{width} = B + z = 1 + (1.5 + 0.5) = 3 \text{ m}$$

and

$$L' = \text{width} = L + z = 2 + (1.5 + 0.5) = 4 \text{ m}$$

The diameter of an equivalent circular area, B_{eq}, can be given as

$$\frac{\pi}{4} B_{eq}^2 = B'L'$$

so that

$$B_{eq} = \sqrt{\frac{4B'L'}{\pi}} = \sqrt{\frac{(4)(3)(4)}{\pi}} = 3.91 \text{ m}$$

Also,

$$\frac{H_c}{B_{eq}} = \frac{2.5}{3.91} = 0.64$$

From Figure 5.31, for $A = 0.6$ and $H_c/B_{eq} = 0.64$, the magnitude of $K_{cr} \approx 0.78$. Hence,

$$S_{e(p)} = K_{cr}S_{e(p)-oed} = (0.78)(46.5) \approx \mathbf{36.3 \text{ mm}}$$ ∎

5.15 *Consolidation Settlement: A Case History*

In predicting the consolidation settlement and the time rate of settlement of a foundation under actual field conditions, an engineer has to make several simplifying assumptions pertaining to, among other things, the compression index, coefficient of consolidation, preconsolidation pressure, drainage conditions, and thickness of the clay layer. Further, the soil layering is not always uniform with ideal properties; hence, the field performance may deviate from the prediction, requiring adjustments during construction. The following case history on consolidation, as reported by Schnabel (1972), illustrates this reality.

Figure 5.32 shows the subsoil conditions for the construction of a school building in Waldorf, Maryland. Upper Pleistocene sand and gravel soils are underlain by deposits of very loose fine silty sand, soft silty clay, and clayey silt. The softer surface layers are underlain by various layers of stiff-to-firm silty clay, clayey silt, and sandy

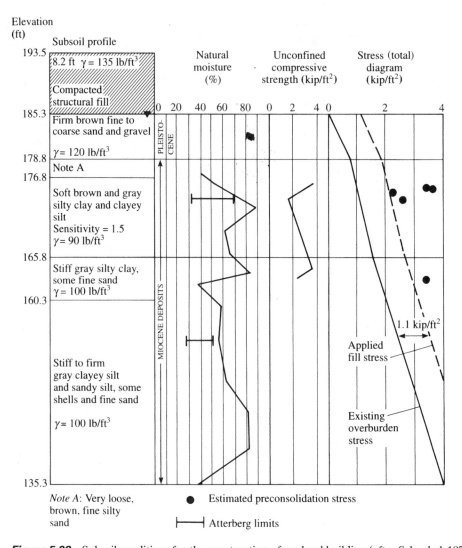

Figure 5.32   Subsoil conditions for the construction of a school building (after Schnabel, 1972)

silt to a depth of 50 ft ($\approx$15 m). Before construction of the building began, a compacted fill having a thickness of 8–10 ft ($\approx$2.4 m–3 m) was placed on the ground surface. This fill initiated the consolidation settlement in the soft silty clay and clayey silt.

To predict the time rate of settlement, based on the laboratory test results, the engineers made the following approximations:

a. The preconsolidation pressure, σ'_c, was 1600 to 2800 lb/ft^2 *in excess* of the existing overburden pressure.

b. The swell index, C_s, was 0.01 to 0.03.

c. For the more compressive layers, $C_v \approx 0.36$ ft^2/day; for the stiffer soil layers, $C_v \approx 3.1$ ft^2/day.

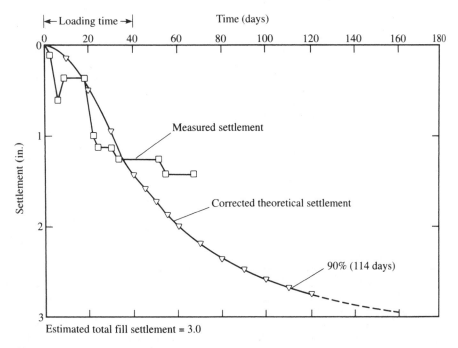

Figure 5.33 Comparison of measured and predicted consolidation settlement with time (after Schnabel, 1972)

The total consolidation settlement was estimated to be about 3 in. (76 mm). Under double drainage conditions, 90% settlement was expected to occur in 114 days.

Figure 5.33 shows a comparison of the measured and predicted settlement with time. The plot indicates that

a. $\dfrac{S_{c(p)\text{observed}}}{S_{c(p)\text{estimated}}} \approx 0.47$

b. Ninety percent of the settlement occurred in about 70 days; hence, $t_{90(\text{observed})}/t_{90(\text{estimated})} \approx 0.58$.

The relatively rapid settlement in the field was believed to be due to the presence of a fine sand layer within the Miocene deposits.

Secondary Consolidation Settlement

5.16 Settlement Due to Secondary Consolidation

At the end of primary consolidation (i.e., after the complete dissipation of excess pore water pressure) some settlement is observed that is due to the plastic adjustment of soil fabrics. This stage of consolidation is called *secondary consolidation*. A plot of deformation against the logarithm of time during secondary consolidation is

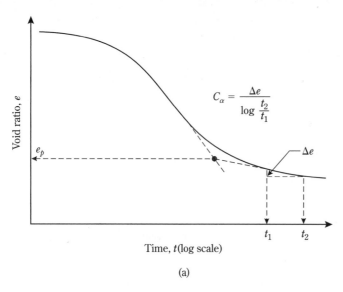

$$C_\alpha = \frac{\Delta e}{\log \frac{t_2}{t_1}}$$

(a)

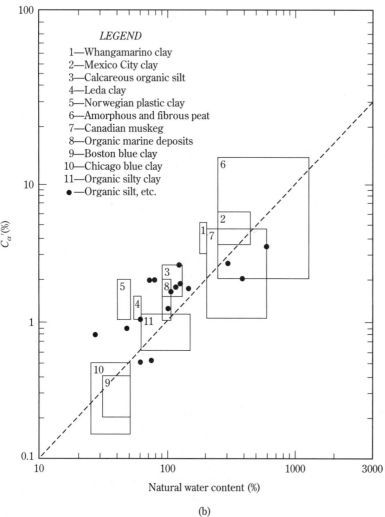

LEGEND

1—Whangamarino clay
2—Mexico City clay
3—Calcareous organic silt
4—Leda clay
5—Norwegian plastic clay
6—Amorphous and fibrous peat
7—Canadian muskeg
8—Organic marine deposits
9—Boston blue clay
10—Chicago blue clay
11—Organic silty clay
●—Organic silt, etc.

(b)

Figure 5.34 (a) Variation of e with log t under a given load increment, and definition of secondary compression index; (b) C'_α for natural soil deposits (after Mesri, 1973)

practically linear as shown in Figure 5.34a. From the figure, the secondary compression index can be defined as

$$C_\alpha = \frac{\Delta e}{\log t_2 - \log t_1} = \frac{\Delta e}{\log (t_2/t_1)} \tag{5.71}$$

where C_α = secondary compression index
   Δe = change of void ratio
   t_1, t_2 = time

The magnitude of the secondary consolidation can be calculated as

$$S_{c(s)} = C'_\alpha H_c \log(t_2/t_1) \tag{5.72}$$

where $C'_\alpha = C_\alpha/(1 + e_p)$ $\hspace{2cm}$ (5.73)
   e_p = void ratio at the end of primary consolidation (see Figure 5.33)
   H_c = thickness of clay layer

The general magnitudes of C'_α as observed in various natural deposits are given in Figure 5.34b.

Secondary consolidation settlement is more important in the case of all organic and highly compressible inorganic soils. In overconsolidated inorganic clays, the secondary compression index is very small and of less practical significance.

There are several factors that might affect the magnitude of secondary consolidation, some of which are not yet very clearly understood (Mesri, 1973). The ratio of secondary to primary compression for a given thickness of soil layer is dependent on the ratio of the stress increment, $\Delta\sigma'$, to the initial effective overburden stress, σ'_o. For small $\Delta\sigma'/\sigma'_o$ ratios, the secondary-to-primary compression ratio is larger.

Allowable Bearing Capacity

5.17 Allowable Bearing Pressure in Sand Based on Settlement Consideration

Meyerhof (1956) proposed a correlation for the *net allowable bearing pressure* for foundations with the corrected standard penetration resistance, $(N_1)_{60}$. The net pressure has been defined as

$$q_{net(all)} = q_{all} - \gamma D_f$$

According to Meyerhof's theory, for 25 mm (1 in.) of estimated maximum settlement,

$$q_{net(all)}(kN/m^2) = 11.98 \, (N_1)_{60} \qquad (\text{for } B \leqslant 1.22 \text{ m}) \tag{5.74}$$

and

$$q_{net(all)}(kN/m^2) = 7.99(N_1)_{60}\left(\frac{3.28B + 1}{3.28B}\right)^2 \quad \text{(for } B > 1.22 \text{ m)} \quad (5.75)$$

where $(N_1)_{60}$ = corrected standard penetration number

Note that in Eqs. (5.74) and (5.75) B is in meters.
In English units,

$$q_{net(all)}(kip/ft^2) = \frac{(N_1)_{60}}{4} \quad \text{(for } B \leq 4 \text{ ft)} \quad (5.76)$$

and

$$q_{net(all)}(kip/ft^2) = \frac{(N_1)_{60}}{6}\left(\frac{B + 1}{B}\right)^2 \quad \text{(for } B > 4 \text{ ft)} \quad (5.77)$$

Since the time that Meyerhof proposed his original correlation, researchers have observed that its results are rather conservative. Later, Meyerhof (1965) suggested that the net allowable bearing pressure should be increased by about 50%. Bowles (1977) proposed that the modified form of the bearing pressure equations be expressed as

$$q_{net(all)}(kN/m^2) = 19.16(N_1)_{60}F_d\left(\frac{S_e}{25}\right) \quad \text{(for } B \leq 1.22 \text{ m)} \quad (5.78)$$

and

$$q_{net(all)}(kN/m^2) = 11.98(N_1)_{60}\left(\frac{3.28B + 1}{3.28B}\right)^2 F_d\left(\frac{S_e}{25}\right) \quad \text{(for } B > 1.22 \text{ m)} \quad (5.79)$$

where F_d = depth factor = $1 + 0.33(D_f/B) \leq 1.33$ $\quad\quad\quad$ (5.80)
S_e = tolerable settlement, in mm

Again, the unit of B is meters.
In English units,

$$q_{net(all)}(kip/ft^2) = \frac{(N_1)_{60}}{2.5}F_dS_e \quad \text{(for } B \leq 4 \text{ ft)} \quad (5.81)$$

and

$$q_{net(all)}(kip/ft^2) = \frac{(N_1)_{60}}{4}\left(\frac{B + 1}{B}\right)^2 F_dS_e \quad \text{(for } B > 4 \text{ ft)} \quad (5.82)$$

where F_d is given by Eq. (5.80)
S_e = tolerable settlement, in in.

Meyerhof (1956) also prepared the following empirical relations, based on the cone penetration resistance q_c, for the net allowable bearing capacity of foundations:

$$q_{\text{net(all)}} = \frac{q_c}{15} \quad \text{(for } B \leq 1.22 \text{ m and settlement of 25 mm)} \tag{5.83}$$

$$q_{\text{net(all)}} = \frac{q_c}{25}\left(\frac{3.28B + 1}{3.28B}\right)^2 \quad \text{(for } B > 1.22 \text{ m and settlement of 25 mm)} \tag{5.84}$$

Note that in Eqs. (5.83) and (5.84) the unit of B is meters and the units of $q_{\text{net(all)}}$ and q_c are kN/m^2.

In English units,

$$q_{\text{net(all)}}(\text{lb/ft}^2) = \frac{q_c\,(\text{lb/ft}^2)}{15} \quad \text{(for } B \leq 4 \text{ ft and settlement of 1 in.)} \tag{5.85}$$

and

$$q_{\text{net(all)}}(\text{lb/ft}^2) = \frac{q_c\,(\text{lb/ft}^2)}{25}\left(\frac{B + 1}{B}\right)^2 \quad \text{(for } B > 4 \text{ ft and settlement of 1 in.)} \tag{5.86}$$

Note that in Eqs. (5.85) and (5.86), the unit of B is feet.

The basic philosophy behind the development of these correlations is that, if the maximum settlement is no more than 25 mm (1 in.) for any foundation, the differential settlement would be no more than 19 mm (0.75 in.). These are probably the allowable limits for most building foundation designs.

5.18 *Field Load Test*

The ultimate load-bearing capacity of a foundation, as well as the allowable bearing capacity based on tolerable settlement considerations, can be effectively determined from the field load test, generally referred to as the *plate load test* (ASTM, 2000; Test Designation D-1194-94). The plates that are used for tests in the field are usually made of steel and are 25 mm (1 in.) thick and 150 mm to 762 mm (6 in. to 30 in.) in diameter. Occasionally, square plates that are 305 mm × 305 mm (12 in. × 12 in.) are also used.

To conduct a plate load test, a hole is excavated with a minimum diameter of $4B$ (B is the diameter of the test plate) to a depth of D_f, the depth of the proposed foundation. The plate is placed at the center of the hole, and a load that is about one-fourth to one-fifth of the estimated ultimate load is applied to the plate in steps by means of a jack. A schematic diagram of the test arrangement is shown in Figure 5.35a. During each step of the application of the load, the settlement of the plate is observed on dial gauges. At least one hour is allowed to elapse between each application. The test should be conducted until failure, or at least until the plate has gone through 25 mm (1 in.) of settlement. Figure 5.35b shows the nature of the load–settlement curve obtained from such tests, from which the ultimate load per unit area can be determined. Figure 5.36 shows a plate load test conducted in the field.

For tests in clay,

$$q_{u(F)} = q_{u(P)} \tag{5.87}$$

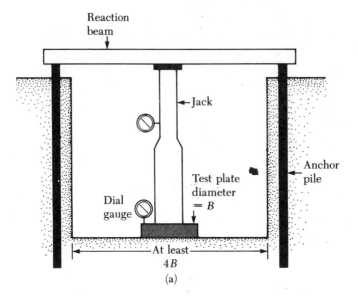

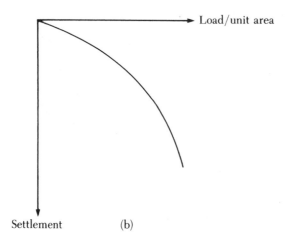

Figure 5.35 Plate load test: (a) test arrangement; (b) nature of load–settlement curve

where $q_{u(F)}$ = ultimate bearing capacity of the proposed foundation
$q_{u(P)}$ = ultimate bearing capacity of the test plate

Equation (5.87) implies that the ultimate bearing capacity in clay is virtually independent of the size of the plate.
For tests in sandy soils,

$$q_{u(F)} = q_{u(P)} \frac{B_F}{B_P} \tag{5.88}$$

where B_F = width of the foundation
B_P = width of the test plate

Figure 5.36 Plate load test in the field

The allowable bearing capacity of a foundation, based on settlement considerations and for a given intensity of load, q_o, is

$$S_F = S_P \frac{B_F}{B_P} \quad \text{(for clayey soil)} \tag{5.89}$$

and

$$S_F = S_P \left(\frac{2B_F}{B_F + B_P} \right)^2 \quad \text{(for sandy soil)} \tag{5.90}$$

The preceding relationship is based on the work of Terzaghi and Peck (1967).

Case Studies to Verify Eq. (5.90)

D'Appolonia et al. (1970) compiled several field-test results in sandy soils to establish the applicability of Eq. (5.90), and these are summarized in Figure 5.37. On the basis of the test results, it can be said that Eq. (5.90) is a fairly good approximation.

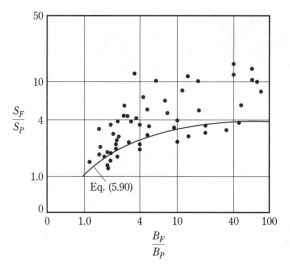

Figure 5.37   Comparison of field-test results with Eq. (5.90) (after D'Appolonia et al., 1970)

Example 5.9

A shallow square foundation for a column is to be constructed on sand. The foundation must carry a net vertical mass of 102,000 kg. The standard penetration numbers (N_{60}) obtained from exploration are given in Figure 5.38. Assume that the depth of the foundation will be 1.5 m and the tolerable settlement is 25 mm. Determine the size of the foundation.

Solution
The standard penetration numbers N_{60} need to be corrected by using Liao and Whitman's relationship [Eq. (2.11)]. This is done in the following table:

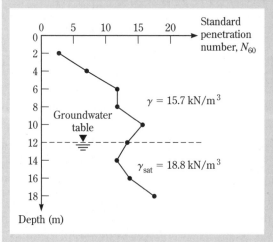

Figure 5.38   Variation of N_{60} with depth

Depth (m)	N_{60}	σ_o' (kN/m²)	$(N_1)_{60}^a$
2	3	31.4	5
4	7	62.8	9
6	12	94.2	12
8	12	125.6	11
10	16	157.0	13
12	13	188.4	9
14	12	206.4	8
16	14	224.36	9
18	18	242.34	12

[a] Rounded off

From the table, it appears than an average value of about 10 for $(N_1)_{60}$ would be appropriate. Using Eq. (5.79), we obtain

$$q_{net(all)} = 11.98(N_1)_{60}\left(\frac{3.28B + 1}{3.28B}\right)^2 F_d\left(\frac{S_e}{25}\right)$$

The allowable $S_e = 25$ mm and $(N_1)_{60} = 10$, so

$$q_{net(all)} = 119.8\left(\frac{3.28B + 1}{3.28B}\right)^2 F_d$$

Also, given $Q_o = \dfrac{102{,}000 \text{ kg} \times 9.81}{1000} \approx 1000$ kN, the following table can be prepared for trial calculations:

B (m)	F_d^a	$q_{net(all)}$ (kN/m²)	$Q_o = q_{net(all)} \times B^2$ (kN)
2	1.248	197.24	788.96
2.25	1.22	187.19	947.65
2.3	1.215	185.46	981.1
2.4	1.206	182.29	1050.0
2.5	1.198	179.45	1121.56

[a] $D_f = 1.5$ m

Because the required Q_o is 1000 kN, B will be approximately equal to **2.4 m**. ∎

5.19 *Presumptive Bearing Capacity*

Several building codes (e.g., the Uniform Building Code, Chicago Building Code, and New York City Building Code) specify the allowable bearing capacity of foundations on various types of soil. For minor construction, they often provide fairly acceptable guidelines. However, these bearing capacity values are based primarily on the *visual* classification of near-surface soils and generally do not take into consideration factors such as the stress history of the soil, the location of the water table, the depth of the foundation, and the tolerable settlement. So, for large construction projects, the codes' presumptive values should be used only as guides.

5.20 *Tolerable Settlement of Buildings*

In most instances of construction, the subsoil is not homogeneous and the load carried by various shallow foundations of a given structure can vary widely. As a result, it is reasonable to expect varying degrees of settlement in different parts of a given building. The *differential settlement* of the parts of a building can lead to damage of the superstructure. Hence, it is important to define certain parameters that quantify differential settlement and to develop limiting values for those parameters in order that the resulting structures be safe. Burland and Worth (1970) summarized the important parameters relating to differential settlement.

Figure 5.39 shows a structure in which various foundations, at *A, B, C, D,* and *E,* have gone through some settlement. The settlement at *A* is *AA′*, at *B* is *BB′*, etc. Based on this figure, the definitions of the various parameters are as follows:

$$S_T = \text{total settlement of a given point}$$

$$\Delta S_T = \text{difference in total settlement between any two points}$$

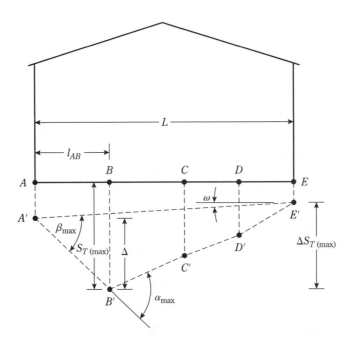

Figure 5.39 Definition of parameters for differential settlement

α = gradient between two successive points

β = angular distortion = $\dfrac{\Delta S_{T(ij)}}{l_{ij}}$

(*Note*: l_{ij} = distance between points i and j)

ω = tilt

Δ = relative deflection (i.e. movement from a straight line joining two reference points)

$\dfrac{\Delta}{L}$ = deflection ratio

Since the 1950s, various researchers and building codes have recommended allowable values for the preceding parameters. A summary of several of these recommendations is presented next.

In 1956, Skempton and McDonald proposed the following limiting values for maximum settlement and maximum angular distortion, to be used for building purposes:

Maximum settlement, $S_{T(max)}$
In sand	32 mm
In clay	45 mm

Maximum differential settlement, $\Delta S_{T(max)}$
Isolated foundations in sand	51 mm
Isolated foundations in clay	76 mm
Raft in sand	51–76 mm
Raft in clay	76–127 mm

Maximum angular distortion, β_{max} 1/300

On the basis of experience, Polshin and Tokar (1957) suggested the following allowable deflection ratios for buildings as a function of L/H, the ratio of the length to the height of a building:

$$\Delta/L = 0.0003 \text{ for } L/H \leqslant 2$$

$$\Delta/L = 0.001 \text{ for } L/H = 8$$

The 1955 Soviet Code of Practice gives the following allowable values:

Type of building	L/H	Δ/L
Multistory buildings and civil dwellings	$\leqslant 3$	0.0003 (for sand) 0.0004 (for clay)
	$\geqslant 5$	0.0005 (for sand) 0.0007 (for clay)
One-story mills		0.001 (for sand and clay)

Bjerrum (1963) recommended the following limiting angular distortion, β_{max} for various structures:

Category of potential damage	β_{max}
Safe limit for flexible brick wall $(L/H > 4)$	1/150
Danger of structural damage to most buildings	1/150
Cracking of panel and brick walls	1/150
Visible tilting of high rigid buildings	1/250
First cracking of panel walls	1/300
Safe limit for no cracking of building	1/500
Danger to frames with diagonals	1/600

If the maximum allowable values of β_{max} are known, the magnitude of the allowable $S_{T(max)}$ can be calculated with the use of the foregoing correlations.

The European Committee for Standardization has also provided limiting values for serviceability and the maximum accepted foundation movements. (See Table 5.8.)

Table 5.8   Recommendations of European Committee for Standardization on Differential Settlement Parameters

Item	Parameter	Magnitude	Comments
Limiting values for	S_T	25 mm	Isolated shallow foundation
serviceability		50 mm	Raft foundation
(European Committee	ΔS_T	5 mm	Frames with rigid cladding
for Standardization,		10 mm	Frames with flexible cladding
1994a)		20 mm	Open frames
	β	1/500	—
Maximum acceptable	S_T	50	Isolated shallow foundation
foundation movement	ΔS_T	20	Isolated shallow foundation
(European Committee	β	$\approx 1/500$	—
for Standardization, 1994b)			

Problems

5.1   A flexible circular area on the surface of a clay layer has a diameter of 2 m and is subjected to a uniformly distributed load of 100 kN/m². Determine the stress increase, $\Delta\sigma$, in the soil mass at points located at depths of 1.5 m and 3 m below the center of the loaded area.

5.2   Determine the stress increase, $\Delta\sigma$, at A and B in Figure P5.2 due to the loaded area.

5.3   As in Problem 5.2, determine the stress increase at depths of 2 m, 4 m, and 6 m below the center of the flexible loaded area shown in Figure P5.2. Use Eq. (5.10).

5.4   In the flexible rectangular area shown in Figure 5.4, $B_{(1)} = 3.3$ ft, $B_{(2)} = 6.6$ ft, $L_{(1)} = 6.6$ ft, and $L_{(2)} = 9.9$ ft. If the area is subjected to a uniform load of 2250 lb/ft², determine the stress increase at a depth of 33 ft immediately below point O.

5.5   A square column foundation is shown in Figure P5.5. Determine the average increase in pressure in the clay layer below the center of the foundation

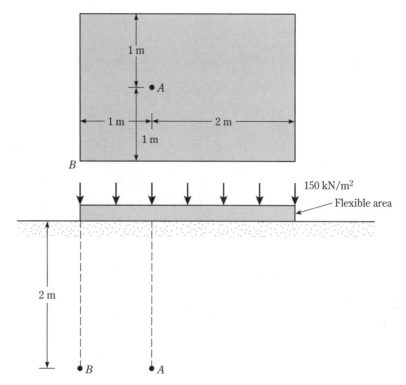

Figure P5.2

      a. by using Eqs. (5.10) and (5.64);
      b. by using the 2:1 method and Eq. (5.64)

5.6   Solve Problem 5.5, using Eq. (5.15).

5.7   Figure P5.7 shows an embankment load on a silty clay layer of soil. Determine the stress increase at points A, B, and C, located at a depth of 5 m below the ground surface.

5.8   A planned flexible load area (see Figure P5.8) is to be 2 m × 3.2 m and carries a uniformly distributed load of 210 kN/m². Estimate the elastic settlement below the center of the loaded area. Assume that $D_f = 1.6$ m and $H = \infty$. Use Eq. (5.25).

5.9   Redo Problem 5.8, assuming that $D_f = 1.2$ m and $H = 4$ m.

5.10   Figure 5.14 shows a foundation 10 ft × 6.25 ft resting on a sand deposit. The net load per unit area at the level of the foundation, q_o, is 3000 lb/ft². For the sand, $\mu_s = 0.3$, $E_s = 3200$ lb/in.², $D_f = 2.5$ ft, and $H = 32$ ft. Assume that the foundation is rigid, and determine the elastic settlement that the foundation would undergo. Use Eqs. (5.25) and (5.33).

5.11   Repeat Problem 5.10 for a foundation of size = 1.8 m × 1.8 m and with $q_o = 190$ kN/m², $D_f = 1$ m, and $H = 15$ m and soil conditions of $\mu_s = 0.4$, $E_s = 15,400$ kN/m², and $\gamma = 17$ kN/m³.

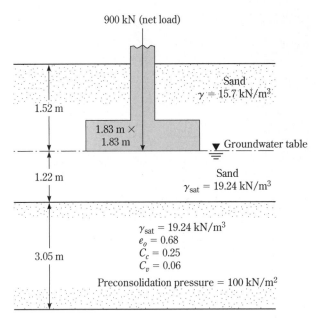

900 kN (net load)

Sand
$\gamma = 15.7 \text{ kN/m}^3$

1.52 m

1.83 m ×
1.83 m

▼ Groundwater table

1.22 m

Sand
$\gamma_{sat} = 19.24 \text{ kN/m}^3$

3.05 m

$\gamma_{sat} = 19.24 \text{ kN/m}^3$
$e_o = 0.68$
$C_c = 0.25$
$C_v = 0.06$

Preconsolidation pressure = 100 kN/m^2

Figure P5.5

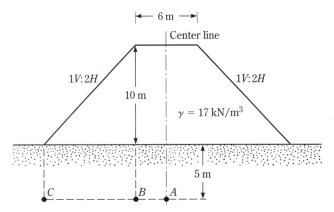

|← 6 m →|

Center line

1V:2H

1V:2H

10 m

$\gamma = 17 \text{ kN/m}^3$

5 m

C

B

A

Figure P5.7

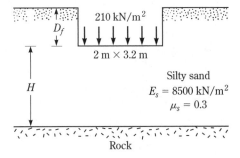

210 kN/m^2

D_f

2 m × 3.2 m

Silty sand
$E_s = 8500 \text{ kN/m}^2$
$\mu_s = 0.3$

H

Rock

Figure P5.8

5.12 For a shallow foundation supported by a silty clay, as shown in Figure 5.16, the following are given:

$$\text{Length} = L = 2 \text{ m}$$
$$\text{Width} = B = 1 \text{ m}$$
$$\text{Depth of foundation} = D_f = 1 \text{ m}$$
$$\text{Thickness of foundation} = t = 0.23 \text{ m}$$
$$\text{Load per unit area} = q_o = 190 \text{ kN/m}^2$$
$$E_f = 15 \times 10^6 \text{ kN/m}^2$$

The silty clay soil has the following properties:

$$H = 2 \text{ m}$$
$$\mu_S = 0.4$$
$$E_o = 9000 \text{ kN/m}^2$$
$$k = 500 \text{ kN/m}^2/\text{m}$$

Using Eq. (5.38), estimate the elastic settlement of the foundation.

5.13 A plan calls for a square foundation measuring 3 m × 3 m, supported by a layer of sand. (See Figure 5.16.) Let $D_f = 1.5$ m, $t = 0.25$ m, $E_o = 16,000$ kN/m^2, $k = 400$ kN/m^2/m, $\mu_S = 0.3$, $H = 20$ m, $E_f = 15 \times 10^6$ kN/m^2, and $q_o = 150$ kN/m^2, and calculate the elastic settlement. Use Eq. (5.38)

5.14 Solve Problem 5.10 with Eq. (5.41). For the correction factor C_2, use a time of 5 yr for creep, and for the unit weight of soil, use $\gamma = 115$ lb/ft^3. Assume an I_z, plot the same as that for a square foundation.

5.15 Solve Problem 5.11 with Eq. (5.41). For the correction factor C_2, use a time of 5 yr for creep.

5.16 A continuous foundation on a layer of sand is shown in Figure P5.16, along with the variation of the modulus of elasticity of the soil, E_s. Assuming that

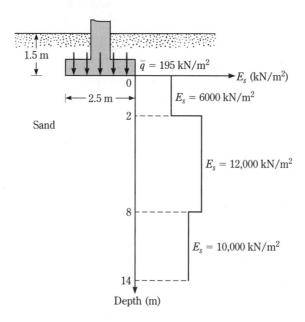

Figure P5.16

$\gamma = 18 \text{ kN/m}^3$, and assuming a creep time of 10 years for the correction factor C_2, calculate the elastic settlement of the foundation, using the strain influence factor.

5.17 Following are the average values of cone penetration resistance in a granular soil deposit:

Depth (m)	Cone penetration resistance, q_c (MN/m²)
2	1.73
4	3.6
6	4.9
8	6.8
10	8.7
15	13

Assume that $\gamma = 16.5 \text{ kN/m}^3$, and estimate the seismic ultimate bearing capacity (q_{uE}) for a continuous foundation with $B = 1.5 \text{ m}$, $D_f = 1.0 \text{ m}$, $k_h = 0.2$, and $k_v = 0$. Use Eqs. (2.40), (5.61), and (5.62).

5.18 In Problem 5.17, if the design earthquake parameters are $V = 0.35 \text{ m/sec}$ and $A = 0.3$, determine the seismic settlement of the foundation. Assume that FS = 4 for use in obtaining the static allowable bearing capacity.

5.19 Estimate the consolidation settlement of the clay layer shown in Figure P5.5, using the results of Part (a) of Problem 5.5.

5.20 Estimate the consolidation settlement of the clay layer shown in Figure P5.5, using the results of Part (b) of Problem 5.5.

5.21 Following are the results of standard penetration tests in a granular soil deposit:

Depth (ft)	Field standard penetration number, N_{60}
5	11
10	10
15	12
20	9
25	14

a. Assume that $\gamma = 115 \text{ lb/ft}^3$, and use Skempton's relation given in Eq. (2.12) to obtain corrected standard penetration numbers.

b. What will be the net allowable bearing capacity of a foundation planned to be 4 ft × 4 ft? Let $D_f = 3 \text{ ft}$ and allowable settlement = 1 in., and use the relationships presented in Section 5.17.

References

Ahlvin, R. G., and Ulery, H. H. (1962). *Tabulated Values of Determining the Composite Pattern of Stresses, Strains, and Deflections beneath a Uniform Load on a Homogeneous Half Space.* Highway Research Board Bulletin 342, pp. 1–13.

American Society for Testing and Materials (2000). *Annual Book of ASTM Standards,* Vol. 04.08, West Conshohocken, PA.

Bjerrum, L. (1963). "Allowable Settlement of Structures," *Proceedings, European Conference on Soil Mechanics and Foundation Engineering,* Wiesbaden, Germany, Vol. III, pp. 135–137.

Boussinesq, J. (1883). *Application des Potentials á L'Étude de L'Équilibre et du Mouvement des Solides Élastiques,* Gauthier-Villars, Paris.

Bowles, J. E. (1987). "Elastic Foundation Settlement on Sand Deposits," *Journal of Geotechnical Engineering,* ASCE, Vol. 113, No. 8, pp. 846–860.

Bowles, J. E. (1977). *Foundation Analysis and Design,* 2d ed., McGraw-Hill, New York.

Burland, J. B., and Burbidge, M. C. (1985). "Settlement of Foundations on Sand and Gravel," *Proceedings, Institute of Civil Engineers,* Part I, Vol. 7, pp. 1325–1381.

Burland, J. B., and Worth, C. P. (1974). "Allowable and Differential Settlement of Structures Including Damage and Soil–Structure Interaction," *Proceedings, Conference on Settlement of Structures.* Cambridge University, England, pp. 611–654.

D'Appolonia, D. J., D'Appolonia, E., and Brissettee, R. F. (1970). "Settlement of Spread Footings on Sand: Closure," *Journal of the Soil Mechanics and Foundation Engineering Division,* ASCE, Vol. 96, No. 2, pp. 754–762.

Das, B. (1997). *Advanced Soil Mechanics,* 2d ed., Taylor and Francis, Washington, DC.

European Committee for Standardization (1994a). *Basis of Design and Actions on Structures,* Eurocode 1, Brussels, Belgium.

European Committee for Standardization (1994b). *Geotechnical Design, General Rules—Part 1,* Eurocode 7, Brussels, Belgium.

Fox, E. N. (1948). "The Mean Elastic Settlement of a Uniformly Loaded Area at a Depth below the Ground Surface," *Proceedings, 2nd International Conference on Soil Mechanics and Foundation Engineering,* Rotterdam, Vol. 1, pp. 129–132.

Griffiths, D. V. (1984). "A Chart for Estimating the Average Vertical Stress Increase in an Elastic Foundation below a Uniformly Loaded Rectangular Area," *Canadian Geotechnical Journal,* Vol. 21, No. 4, 710–713.

Leonards, G. A. (1976). *Estimating Consolidation Settlement of Shallow Foundations on Overconsolidated Clay,* Special Report No. 163, Transportation Research Board, Washington, DC., pp. 13–16.

Liao, S. S. C., and Whitman, R. V. (1986). "Overburden Correction Factors for SPT in Sand," *Journal of Geotechnical Engineering,* American Society of Civil Engineers, Vol. 112, No. 3, pp. 373–377.

Mayne, P. W., and Poulos, H. G. (1999). "Approximate Displacement Influence Factors for Elastic Shallow Foundations," *Journal of Geotechnical and Geoenvironmental Engineering,* ASCE, Vol. 125, No. 6, 453–460.

Mesri, G. (1973). "Coefficient of Secondary Compression," *Journal of the Soil Mechanics and Foundations Division, American Society of Civil Engineers,* Vol. 99, No. SM1, 122–137.

Meyerhof, G. G. (1956). "Penetration Tests and Bearing Capacity of Cohesionless Soils," *Journal of the Soil Mechanics and Foundations Division,* American Society of Civil Engineers, Vol. 82, No. SM1, pp. 1–19.

Meyerhof, G. G. (1965). "Shallow Foundations," *Journal of the Soil Mechanics and Foundations Division,* American Society of Civil Engineers, Vol. 91, No. SM2, pp. 21–31.

Mitchell, J. K., and Gardner, W. S. (1975). "*In Situ* Measurement of Volume Change Characteristics," *Proceedings, Specialty Conference,* American Society of Civil Engineers, Vol. 2, pp. 279–345.

Newmark, N. M. (1935). *Simplified Computation of Vertical Pressure in Elastic Foundation,* Circular 24, University of Illinois Engineering Experiment Station, Urbana, IL.

Osterberg, J. O. (1957). "Influence Values for Vertical Stresses in Semi-Infinite Mass Due to Embankment Loading," *Proceedings, Fourth International Conference on Soil Mechanics and Foundation Engineering,* London, Vol. 1, pp. 393–396.

Polshin, D. E., and Tokar, R. A. (1957). "Maximum Allowable Nonuniform Settlement of Structures," *Proceedings, Fourth International Conference on Soil Mechanics and Foundation Engineering,* London, Vol. 1, pp. 402–405.

Richards, R., Jr., Elms, D. G., and Budhu, M. (1993). "Seismic Bearing Capacity and Settlement of Foundations," *Journal of Geotechnical Engineering,* American Society of Civil Engineers, Vol. 119, No. 4, pp. 662–674.

Schmertmann, J. H. (1970). "Static Cone to Compute Settlement Over Sand," *Journal of the Soil Mechanics and Foundations Division,* American Society of Civil Engineers, Vol. 96, No. SM3, pp. 1011–1043.

Schmertmann, J. H., and Hartman, J. P. (1978). "Improved Strain Influence Factor Diagrams," *Journal of the Geotechnical Engineering Division,* American Society of Civil Engineers, Vol. 104, No. GT8, pp. 113–1135.

Schnabel, J. J. (1972). "Foundation Construction on Compacted Structural Fill in the Washington, D.C., Area." *Proceedings, Specialty Conference on Performance of Earth and Earth-Supported Structures,* American Society of Civil Engineers, Vol. 1, Part 2, pp. 1019–1036.

Skempton, A. W., and Bjerrum, L. (1957). " A Contribution to Settlement Analysis of Foundations in Clay," *Geotechnique,* London, Vol 7, p. 178.

Skempton, A. W., and McDonald, D. M. (1956). "The Allowable Settlement of Buildings," *Proceedings of Institute of Civil Engineers,* Vol. 5, Part III, p. 727.

Steinbrenner, W. (1934). "Tafeln zur Setzungsberechnung," *Die Strasse,* Vol. 1, pp. 121–124.

Terzaghi, K., and Peck, R. B. (1967). *Soil Mechanics in Engineering Practice,* 2d ed., Wiley, New York.

6

Mat Foundations

6.1 Introduction

Under normal conditions, square and rectangular footings such as those described in Chapters 3 and 4 are economical for supporting columns and walls. However, under certain circumstances, it may be desirable to construct a footing that supports a line of two or more columns. These footings are referred to as *combined footings*. When more than one line of columns is supported by a concrete slab, it is called a *mat foundation*. Combined footings can be classified generally under the following categories:

a. Rectangular combined footing
b. Trapezoidal combined footing
c. Strap footing

Mat foundations are generally used with soil that has a low bearing capacity. A brief overview of the principles of combined footings is given in Section 6.2, followed by a more detailed discussion on mat foundations.

6.2 Combined Footings

Rectangular Combined Footing

In several instances, the load to be carried by a column and the soil bearing capacity are such that the standard spread footing design will require extension of the column foundation beyond the property line. In such a case, two or more columns can be supported on a single rectangular foundation, as shown in Figure 6.1. If the net allowable soil pressure is known, the size of the foundation ($B \times L$) can be determined in the following manner:

a. Determine the area of the foundation

$$A = \frac{Q_1 + Q_2}{q_{\text{net(all)}}} \tag{6.1}$$

where Q_1, Q_2 = column loads
$q_{\text{net(all)}}$ = net allowable soil bearing capacity

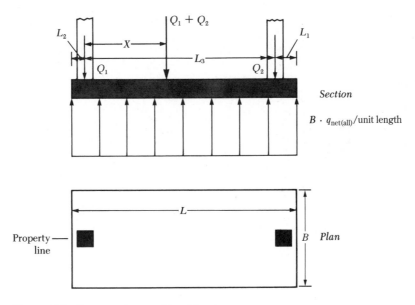

Figure 6.1   Rectangular combined footing

b. Determine the location of the resultant of the column loads. From Figure 6.1,

$$X = \frac{Q_2 L_3}{Q_1 + Q_2} \tag{6.2}$$

c. For a uniform distribution of soil pressure under the foundation, the resultant of the column loads should pass through the centroid of the foundation. Thus,

$$L = 2(L_2 + X) \tag{6.3}$$

where L = length of the foundation

d. Once the length L is determined, the value of L_1 can be obtained as follows:

$$L_1 = L - L_2 - L_3 \tag{6.4}$$

Note that the magnitude of L_2 will be known and depends on the location of the property line.

e. The width of the foundation is then

$$B = \frac{A}{L} \tag{6.5}$$

Trapezoidal Combined Footing

Trapezoidal combined footing (see Figure 6.2) is sometimes used as an isolated spread foundation of columns carrying large loads where space is tight. The size of

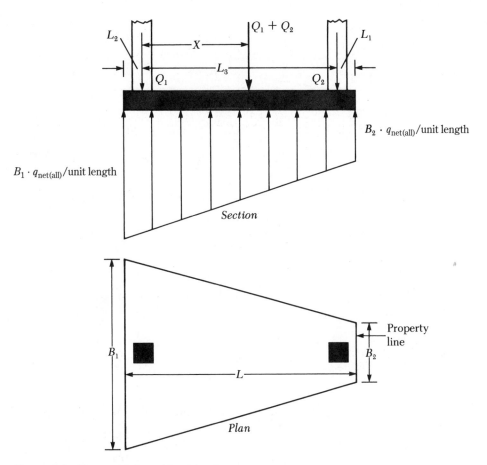

Figure 6.2 Trapezoidal combined footing

the foundation that will uniformly distribute pressure on the soil can be obtained in the following manner:

a. If the net allowable soil pressure is known, determine the area of the foundation:

$$A = \frac{Q_1 + Q_2}{q_{net(all)}}$$

From Figure 6.2,

$$A = \frac{B_1 + B_2}{2}L \tag{6.6}$$

b. Determine the location of the resultant for the column loads:

$$X = \frac{Q_2 L_3}{Q_1 + Q_2}$$

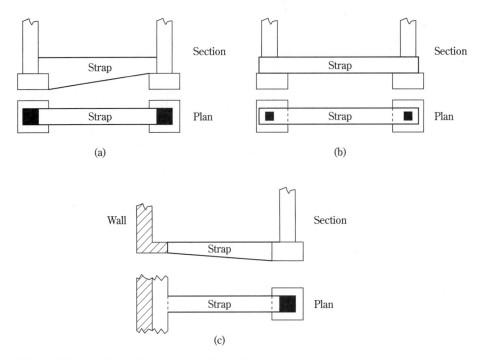

Figure 6.3 Cantilever footing—use of strap beam

c. From the property of a trapezoid,

$$X + L_2 = \left(\frac{B_1 + 2B_2}{B_1 + B_2}\right)\frac{L}{3} \tag{6.7}$$

With known values of A, L, X, and L_2, solve Eqs. (6.6) and (6.7) to obtain B_1 and B_2. Note that, for a trapezoid,

$$\frac{L}{3} < X + L_2 < \frac{L}{2}$$

Cantilever Footing

Cantilever footing construction uses a *strap beam* to connect an eccentrically loaded column foundation to the foundation of an interior column. (See Figure 6.3). Cantilever footings may be used in place of trapezoidal or rectangular combined footings when the allowable soil bearing capacity is high and the distances between the columns are large.

6.3 Common Types of Mat Foundations

The mat foundation, which is sometimes referred to as a *raft foundation,* is a combined footing that may cover the entire area under a structure supporting several columns and walls. Mat foundations are sometimes preferred for soils that have low load-bearing capacities, but that will have to support high column or wall loads. Under some conditions, spread footings would have to cover more than half the building area, and mat foundations might be more economical. Several types of mat foundations are used currently. Some of the common ones are shown schematically in Figure 6.4 and include the following:

1. Flat plate (Figure 6.4a). The mat is of uniform thickness.
2. Flat plate thickened under columns (Figure 6.4b).
3. Beams and slab (Figure 6.4c). The beams run both ways, and the columns are located at the intersection of the beams.
4. Flat plates with pedestals (Figure 6.4d).
5. Slab with basement walls as a part of the mat (Figure 6.4e). The walls act as stiffeners for the mat.

Mats may be supported by piles, which help reduce the settlement of a structure built over highly compressible soil. Where the water table is high, mats are often placed over piles to control buoyancy. Figure 6.5 shows the difference between the depth D_f and the width B of isolated foundations and mat foundations.

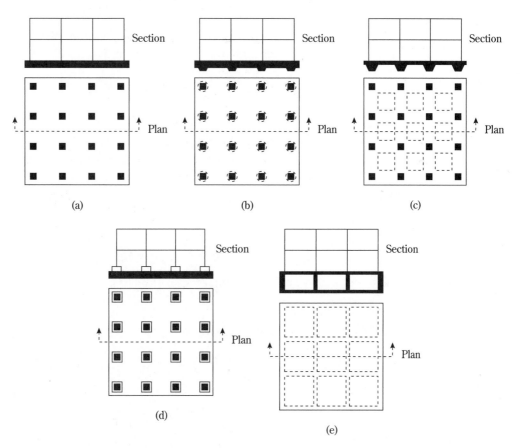

Figure 6.4 Common types of mat foundation

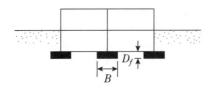

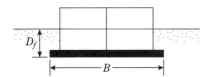

Figure 6.5   Comparison of isolated foundation and mat foundation (B = width, D_f = depth)

Bearing Capacity of Mat Foundations

The *gross ultimate bearing capacity* of a mat foundation can be determined by the same equation used for shallow foundations (see Section 3.7), or

$$q_u = c'N_cF_{cs}F_{cd}F_{ci} + qN_qF_{qs}F_{qd}F_{qi} + \tfrac{1}{2}\gamma B N_\gamma F_{\gamma s}F_{\gamma d}F_{\gamma i} \qquad (3.21)$$

(Chapter 3 gives the proper values of the bearing capacity factors, as well as the shape depth, and load inclination factors.) The term B in Eq. (3.21) is the smallest dimension of the mat. The *net ultimate capacity* of a mat foundation is

$$q_{\text{net}(u)} = q_u - q \qquad (3.14)$$

A suitable factor of safety should be used to calculate the net *allowable* bearing capacity. For rafts on clay, the factor of safety should not be less than 3 under dead load or maximum live load. However, under the most extreme conditions, the factor of safety should be at least 1.75 to 2. For rafts constructed over sand, a factor of safety of 3 should normally be used. Under most working conditions, the factor of safety against bearing capacity failure of rafts on sand is very large.

For saturated clays with $\phi = 0$ and a vertical loading condition, Eq. (3.21) gives

$$q_u = c_u N_c F_{cs} F_{cd} + q \qquad (6.8)$$

where c_u = undrained cohesion

(*Note:* $N_c = 5.14$, $N_q = 1$, and $N_\gamma = 0$.)

From Eqs. (3.25) and (3.28), for $\phi = 0$,

$$F_{cs} = 1 + \frac{B}{L}\left(\frac{N_q}{N_c}\right) = 1 + \left(\frac{B}{L}\right)\left(\frac{1}{5.14}\right) = 1 + \frac{0.195B}{L}$$

and

$$F_{cd} = 1 + 0.4\left(\frac{D_f}{B}\right)$$

Substitution of the preceding shape and depth factors into Eq. (6.8) yields

$$q_u = 5.14c_u\left(1 + \frac{0.195B}{L}\right)\left(1 + 0.4\frac{D_f}{B}\right) + q \qquad (6.9)$$

Hence, the net ultimate bearing capacity is

$$q_{net(u)} = q_u - q = 5.14c_u\left(1 + \frac{0.195B}{L}\right)\left(1 + 0.4\frac{D_f}{B}\right) \qquad (6.10)$$

For FS = 3, the net allowable soil bearing capacity becomes

$$q_{net(all)} = \frac{q_{u(net)}}{FS} = 1.713c_u\left(1 + \frac{0.195B}{L}\right)\left(1 + 0.4\frac{D_f}{B}\right) \qquad (6.11)$$

The net allowable bearing capacity for mats constructed over granular soil deposits can be adequately determined from the standard penetration resistance numbers. From Eq. (5.79), for shallow foundations,

$$q_{net(all)}(kN/m^2) = 11.98(N_1)_{60}\left(\frac{3.28B + 1}{3.28B}\right)^2 F_d\left(\frac{S_e}{25}\right)$$

where $(N_1)_{60}$ = corrected standard penetration resistance
   B = width (m)
   $F_d = 1 + 0.33(D_f/B) \le 1.33$
   S_e = settlement, in mm

When the width B is large, the preceding equation can be approximated (assuming that $3.28B + 1 \approx 3.28B$) as

$$\begin{aligned} q_{net(all)}(kN/m^2) &\approx 11.98(N_1)_{60}F_d\left(\frac{S_e}{25}\right) \\ &= 11.98(N_1)_{60}\left[1 + 0.33\left(\frac{D_f}{B}\right)\right]\left[\frac{S_e(mm)}{25}\right] \\ &\le 15.93(N_1)_{60}\left[\frac{S_e(mm)}{25}\right] \end{aligned} \qquad (6.12)$$

In English units, Eq. (6.12) may be expressed as

$$\begin{aligned} q_{net(all)}(kip/ft^2) &= 0.25(N_1)_{60}\left[1 + 0.33\left(\frac{D_f}{B}\right)\right][S_e(in.)] \\ &\le 0.33(N_1)_{60}[S_e(in.)] \end{aligned} \qquad (6.13)$$

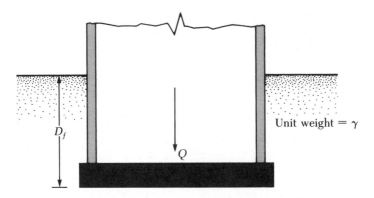

Figure 6.6 Definition of net pressure on soil caused by a mat foundation

Note that Eq. (6.13) could have been derived from Eqs. (5.80) and (5.82).

Note that the original equations (5.79) and (5.82) were for a settlement of 25 mm (1 in.), with a differential settlement of about 19 mm (0.75 in.). However, the width of the raft foundations are larger than those of the isolated spread footings. As shown in Table 5.3, the depth of significant stress increase in the soil below a foundation depends on the width of the foundation. Hence, for a raft foundation, the depth of the zone of influence is likely to be much larger than that of a spread footing. Thus, the loose soil pockets under a raft may be more evenly distributed, resulting in a smaller differential settlement. Accordingly, the customary assumption is that, for a maximum raft settlement of 50 mm (2 in.), the differential settlement would be 19 mm (0.75 in.). Using this logic and conservatively assuming that $F_d = 1$, we can respectively approximate Eqs. (6.12) and (6.13) as

$$q_{\text{net(all)}}(\text{kN/m}^2) \approx 23.96(N_1)_{60} \tag{6.14}$$

and

$$q_{\text{net(all)}}(\text{kip/ft}^2) = 0.5(N_1)_{60} \tag{6.15}$$

The net pressure applied on a foundation (see Figure 6.6) may be expressed as

$$q = \frac{Q}{A} - \gamma D_f \tag{6.16}$$

where Q = dead weight of the structure and the live load
A = area of the raft

In all cases, q should be less than or equal to $q_{\text{net(all)}}$.

Example 6.1

Determine the net ultimate bearing capacity of a mat foundation measuring 15 m × 10 m on a saturated clay with $c_u = 95$ kN/m², $\phi = 0$, and $D_f = 2$ m.

Solution
From Eq. (6.10),

$$q_{net(u)} = 5.14c_u\left[1 + \left(\frac{0.195B}{L}\right)\right]\left[1 + 0.4\frac{D_f}{B}\right]$$

$$= (5.14)(95)\left[1 + \left(\frac{0.195 \times 10}{15}\right)\right]\left[1 + \left(\frac{0.4 \times 2}{10}\right)\right]$$

$$= \textbf{595.9 kN/m}^2 \qquad\blacksquare$$

Example 6.2

What will be the net allowable bearing capacity of a mat foundation with dimensions of 15 m × 10 m constructed over a sand deposit? Here, $D_f = 2$ m, the allowable settlement is 25 mm, and the corrected average penetration number $(N_1)_{60} = 10$.

Solution
From Eq. (6.12),

$$q_{net(all)} = 11.98(N_1)_{60}\left[1 + 0.33\left(\frac{D_f}{B}\right)\right]\left(\frac{S_e}{25}\right) \leq 15.93(N_1)_{60}\left(\frac{S_e}{25}\right)$$

or

$$q_{net(all)} = (11.98)(10)\left[1 + \frac{0.33 \times 2}{10}\right]\left(\frac{25}{25}\right) = \textbf{127.7 kN/m}^2 \qquad\blacksquare$$

6.5 *Differential Settlement of Mats*

In 1988, the American Concrete Institute Committee 336 suggested a method for calculating the differential settlement of mat foundations. According to this method, the rigidity factor K_r is calculated as

$$K_r = \frac{E'I_b}{E_s B^3} \qquad (6.17)$$

where E' = modulus of elasticity of the material used in the structure
E_s = modulus of elasticity of the soil
B = width of foundation
I_b = moment of inertia of the structure per unit length at right angles to B

The term $E'I_b$ can be expressed as

$$E'I_b = E'\left(I_F + \sum I_{b'} + \sum \frac{ah^3}{12}\right) \qquad (6.18)$$

where $\qquad E'I_b$ = flexural rigidity of the superstructure and foundation per
unit length at right angles to B

$\Sigma E'I_b'$ = flexural rigidity of the framed members at right angles to B

$\Sigma(E'ah^3/12)$ = flexural rigidity of the shear walls

a = shear wall thickness

h = shear wall height

$E'I_F$ = flexibility of the foundation

Based on the value of K_r, the ratio (δ) of the differential settlement to the total settlement can be estimated in the following manner:

1. If $K_r > 0.5$, it can be treated as a rigid mat, and $\delta = 0$.
2. If $K_r = 0.5$, then $\delta \approx 0.1$.
3. If $K_r = 0$, then $\delta = 0.35$ for square mats ($B/L = 1$) and $\delta = 0.5$ for long foundations ($B/L = 0$).

6.6 Field Settlement Observations for Mat Foundations

Several field settlement observations for mat foundations are currently available in the literature. In this section, we compare the observed settlements for some mat foundations constructed over granular soil deposits with those obtained from Eqs. (6.12) and (6.13).

Meyerhof (1965) compiled the observed maximum settlements for mat foundations constructed on sand and gravel, as listed in Table 6.1. In Eq. (6.12), if the depth factor, $1 + 0.33(D_f/B)$, is assumed to be approximately unity, then

$$S_e(\text{mm}) \approx 25\frac{q_{\text{net(all)}}}{11.98(N_1)_{60}} \qquad (6.19)$$

From the values of $q_{\text{net(all)}}$ and $(N_1)_{60}$ given in Columns 6 and 5, respectively, of Table 6.1, the magnitudes of S_e were calculated and are given in Column 8.

Column 9 of Table 6.1 gives the ratios of calculated to measured values of S_e. These ratios vary from about 0.83 to 3.5. Thus, calculating the net allowable bearing capacity with the use of Eq. (6.12) or (6.13) will yield safe and conservative values.

Stuart and Graham (1975) reported the case history of the 13-story Ashby Institute building of Queens University, Belfast, Ireland, the construction of which began in August 1960. The building was supported by a mat foundation 54.9 m (180 ft) $\times$ 19.8 m (65 ft). Figure 6.7a shows a schematic cross section of the building. The nature of the subsoil, along with the field standard penetration resistance values at the south end of the building, is shown in Figure 6.7b. The base of the mat was constructed about 6.1 m (20 ft) below the ground surface.

The average corrected standard penetration number $(N_1)_{60}$ between the bottom of the mat and a depth of about $B/2$ below the mat is about 17. The engineers estimate that $q_{\text{net(all)}}$ was about 161 kN/m^2 (3360 lb/ft^2). From Eq. (6.12),

$$S_e(\text{mm}) = \frac{25q_{\text{net(all)}}}{11.98(N_1)_{60}\left[1 + 0.33\left(\dfrac{D_f}{B}\right)\right]} \qquad (6.20)$$

Table 6.1 Settlement of Mat Foundations on Sand and Gravel (based on Meyerhof, 1965)

Case No. (1)	Structure (2)	Reference (3)	B m (ft) (4)	Average $(N_1)_{60}$ (5)	$q_{net(all)}$ kN/m² (kip/ft²) (6)	Observed maximum settlement, S_e mm (in.) (7)	Calculated maximum settlement, S_e mm (in.) (8)	$\dfrac{\text{calculated } S_e}{\text{observed } S_e}$ (9)
1	T. Edison São Paulo, Brazil	Rios and Silva (1948)	18.29 (60)	15	229.8 (4.8)	15.24 (0.6)	31.97 (1.26)	2.1
2	Banco do Brazil São Paulo, Brazil	Rios and Silva (1948); Vargas (1961)	22.86 (75)	18	239.4 (5.0)	27.94 (1.1)	27.75 (1.09)	0.99
3	Iparanga São Paulo, Brazil	Vargas (1948)	9.14 (30)	9	304.4 (6.4)	35.56 (1.4)	70.58 (2.78)	1.99
4	C.B.I., Esplanda São Paulo, Brazil	Vargas (1961)	14.63 (48)	22	383.0 (8.0)	27.94 (1.1)	36.33 (1.43)	1.3
5	Riscala São Paulo, Brazil	Vargas (1948)	3.96 (13)	20	229.8 (4.8)	12.7 (0.5)	23.98 (0.94)	1.89
6	Thyssen Düsseldorf, Germany	Schultze (1962)	22.55 (74)	25	239.4 (5)	24.13 (0.95)	19.98 (0.79)	0.83
7	Ministry Düsseldorf, Germany	Schultze (1962)	15.85 (52)	20	220.2 (4.6)	20.32 (0.8)	22.98 (0.9)	1.13
8	Chimney Cologne, Germany	Schultze (1962)	20.42 (67)	10	172.4 (3.6)	10.16 (0.4)	35.98 (1.42)	3.54

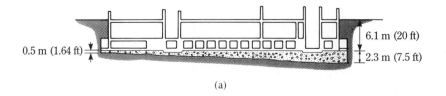

(a)

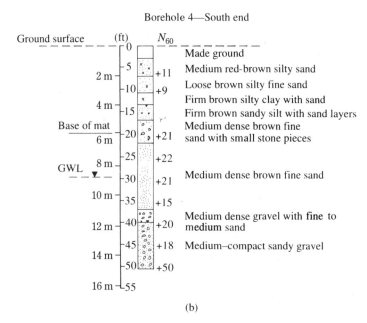

(b)

Figure 6.7   Ashby Institute Building of Queens University as reported by Stuart and Graham (1975): (a) cross section of building; (b) subsoil conditions at south end

Substituting the appropriate values into the preceding equation yields

$$S_e = \frac{(25)(161)}{(11.98)(17)\left[1 + 0.33\left(\dfrac{6.1}{19.8}\right)\right]} = 17.9 \text{ mm (0.7 cm)}$$

The construction of the building was completed in February 1964. Figure 6.8 shows the mean settlement of the mat at the south end from 1960 to 1972. In the latter year (eight years after completion of the building), the mean settlement was about 14 mm (0.55 in.). Thus, the estimated settlement of 17.9 mm (0.7 in.) was about 30% higher than that actually observed.

6.7   Compensated Foundation

Figure 6.6 and Eq. (6.16) indicate that the net pressure increase in the soil under a mat foundation can be reduced by increasing the depth D_f of the mat. This approach is generally referred to as the *compensated foundation design* and is extremely use-

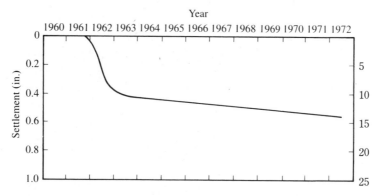

Figure 6.8 Mean settlement at the south end of the mat foundation, as reported by Stuart and Graham (1975)

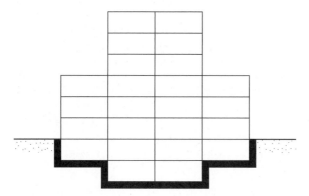

Figure 6.9 Compensated foundation

ful when structures are to be built on very soft clays. In this design, a deeper basement is made below the higher portion of the superstructure, so that the net pressure increase in soil at any depth is relatively uniform. (See Figure 6.9.) From Eq. (6.16) and Figure 6.6, the net average applied pressure on soil is

$$q = \frac{Q}{A} - \gamma D_f$$

For no increase in the net pressure on soil below a mat foundation, q should be zero. Thus,

$$D_f = \frac{Q}{A\gamma} \tag{6.21}$$

This relation for D_f is usually referred to as the depth of a *fully compensated foundation.*

The factor of safety against bearing capacity failure for partially compensated foundations (i.e., $D_f < Q/A\gamma$) may be given as

$$\text{FS} = \frac{q_{\text{net}(u)}}{q} = \frac{q_{\text{net}(u)}}{\dfrac{Q}{A} - \gamma D_f} \tag{6.22}$$

For saturated clays, the factor of safety against bearing capacity failure can thus be obtained by substituting Eq. (6.10) into Eq. (6.22):

$$FS = \frac{5.14c_u\left(1 + \dfrac{0.195B}{L}\right)\left(1 + 0.4\dfrac{D_f}{B}\right)}{\dfrac{Q}{A} - \gamma D_f} \tag{6.23}$$

Example 6.3

The mat shown in Figure 6.6 has dimensions of 60 ft × 100 ft. The total dead and live load on the mat is 25×10^3 kip. The mat is placed over a saturated clay having a unit weight of 120 lb/ft^3 and c_u = 2800 lb/ft^2. Given that D_f = 5 ft, determine the factor of safety against bearing capacity failure.

Solution
From Eq. (6.23), the factor of safety

$$FS = \frac{5.14c_u\left(1 + \dfrac{0.195B}{L}\right)\left(1 + 0.4\dfrac{D_f}{B}\right)}{\dfrac{Q}{A} - \gamma D_f}$$

We are given that c_u = 2800 lb/ft^2, D_f = 5 ft, B = 60 ft, L = 100 ft, and γ = 120 lb/ft^3. Hence,

$$FS = \frac{(5.14)(2800)\left[1 + \dfrac{(0.195)(60)}{100}\right]\left[1 + 0.4\left(\dfrac{5}{60}\right)\right]}{\left(\dfrac{25 \times 10^6\, lb}{60 \times 100}\right) - (120)(5)} = \textbf{4.66} \qquad \blacksquare$$

Example 6.4

Consider a planned mat foundation 30 m × 40 m, as shown in Figure 6.10. The total dead load plus the live load on the raft is 200 MN. Estimate the consolidation settlement at the center of the foundation.

Solution
We are given that $Q = 200 \times 10^3$ kN, so the net load per unit area is

$$q = \frac{Q}{A} - \gamma D_f = \frac{200 \times 10^3}{30 \times 40} - (15.7)(2) = 166.67 - 31.4 = 135.27 \text{ kN/m}^2$$

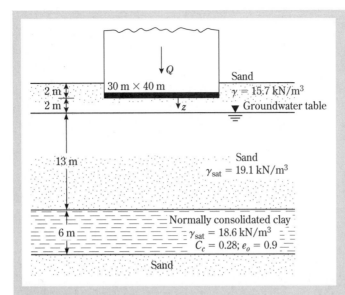

Figure 6.10 Consolidation settlement under a mat foundation

From Eq. (5.64), the average pressure increase on the clay layer below the center of the foundation is

$$\Delta\sigma'_{av} = \tfrac{1}{6}(\Delta\sigma'_t + 4\Delta\sigma'_m + \Delta\sigma'_b)$$

The values of $\Delta\sigma'_t$, $\Delta\sigma'_m$, and $\Delta\sigma'_b$ can be determined by referring to Table 5.3. At the *top of the clay layer,*

$$n_1 = \frac{z}{(B/2)} = \frac{15}{(30/2)} = 1.0$$

and

$$m_1 = \frac{L}{B} = \frac{40}{30} = 1.33$$

So, for $n_1 = 1.0$ and $m_1 = 1.33$,

$$\frac{\Delta\sigma'_t}{q} = 0.75$$

and

$$\Delta\sigma'_t = (0.75)(135.27) = 101.45 \text{ kN/m}^2$$

Similarly, for the *middle of the clay layer,*

$$n_1 = \frac{z}{(B/2)} = \frac{18}{(30/2)} = 1.2$$

and

$$m_1 = \frac{L}{B} = 1.33$$

So $\Delta\sigma'_m/q = 0.66$,
and

$$\Delta\sigma'_m = 89.3 \text{ kN/m}^2$$

At the *bottom of the clay layer,*

$$n_1 = \frac{z}{(B/2)} = \frac{21}{(30/2)} = 1.4$$

and

$$m_1 = \frac{L}{B} = 1.33$$

So $\Delta\sigma'_b/q = 0.58$,
and

$$\Delta\sigma'_b = 78.46 \text{ kN/m}^2$$

Hence,

$$\Delta\sigma'_{av} = \tfrac{1}{6}[101.45 + 4(89.3) + 78.46] = 89.5 \text{ kN/m}^2$$

From Eq. (1.53), the consolidation settlement

$$S_{c(p)} = \frac{C_c H_c}{1 + e_o} \log \frac{\sigma'_o + \Delta\sigma'_{av}}{\sigma'_o}$$

so that,

$$\sigma'_o = 4(15.7) + 13(19.1 - 9.81) + \tfrac{6}{2}(18.6 - 9.81) = 209.94 \text{ kN/m}^2$$

Hence,

$$S_{c(p)} = \frac{(0.28)(6 \times 1000)}{1 - 0.9}\log\left(\frac{209.94 + 89.5}{209.94}\right) = \mathbf{136.4 \text{ mm}} \quad \blacksquare$$

6.8 *Structural Design of Mat Foundations*

The structural design of mat foundations can be carried out by two conventional methods: the conventional rigid method and the approximate flexible method. Finite-difference and finite-element methods can also be used, but this section covers only the basic concepts of the first two design methods.

Conventional Rigid Method

The *conventional rigid method* of mat foundation design can be explained step by step with reference to Figure 6.11:

1. Figure 6.11a shows mat dimensions of $L \times B$ and column loads of $Q_1, Q_2,$ $Q_3, \dots.$ Calculate the total column load as

$$Q = Q_1 + Q_2 + Q_3 + \cdots \tag{6.24}$$

2. Determine the pressure on the soil, q, below the mat at points $A, B, C, D, \dots,$ by using the equation

$$q = \frac{Q}{A} \pm \frac{M_y x}{I_y} \pm \frac{M_x y}{I_x} \tag{6.25}$$

where $A = BL$
$I_x = (1/12)BL^3$ = moment of inertia about the x-axis
$I_y = (1/12)LB^3$ = moment of inertia about the y-axis
M_x = moment of the column loads about the x-axis = Qe_y
M_y = moment of the column loads about the y-axis = Qe_x

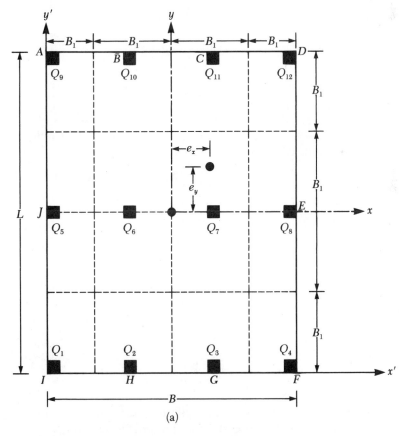

(a)

Figure 6.11 Conventional rigid mat foundation design

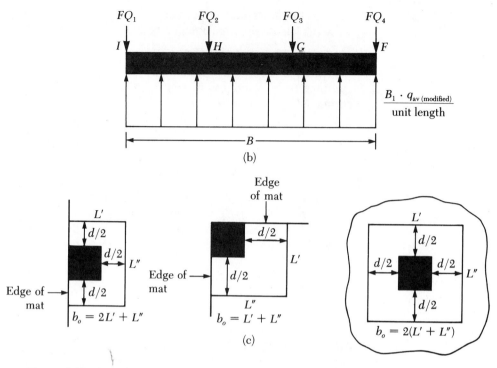

Figure 6.11 (continued)

The load eccentricities, e_x and e_y, in the x and y directions can be determined by using (x', y') coordinates:

$$x' = \frac{Q_1 x'_1 + Q_2 x'_2 + Q_3 x'_3 + \cdots}{Q}$$ (6.26)

and

$$e_x = x' - \frac{B}{2}$$ (6.27)

Similarly,

$$y' = \frac{Q_1 y'_1 + Q_2 y'_2 + Q_3 y'_3 + \cdots}{Q}$$ (6.28)

and

$$e_y = y' - \frac{L}{2}$$ (6.29)

3. Compare the values of the soil pressures determined in Step 2 with the net allowable soil pressure to determine whether $q \leq q_{\text{all(net)}}$.
4. Divide the mat into several strips in the x and y directions. (See Figure 6.11). Let the width of any strip be B_1.

5. Draw the shear, *V*, and the moment, *M*, diagrams for each individual strip (in the *x* and *y* directions). For example, the average soil pressure of the bottom strip in the *x* direction of Figure 6.11a is

$$q_{av} \approx \frac{q_I + q_F}{2} \qquad (6.30)$$

where q_I and q_F = soil pressures at points *I* and *F*, as determined from Step 2.

The total soil reaction is equal to $q_{av}B_1B$. Now obtain the total column load on the strip as $Q_1 + Q_2 + Q_3 + Q_4$. The sum of the column loads on the strip will not equal $q_{av}B_1B$, because the shear between the adjacent strips has not been taken into account. For this reason, the soil reaction and the column loads need to be adjusted, or

$$\text{Average load} = \frac{q_{av}B_1B + (Q_1 + Q_2 + Q_3 + Q_4)}{2} \qquad (6.31)$$

Now, the modified average soil reaction becomes

$$q_{av(\text{modified})} = q_{av}\left(\frac{\text{average load}}{q_{av}B_1B}\right) \qquad (6.32)$$

and the column load modification factor is

$$F = \frac{\text{average load}}{Q_1 + Q_2 + Q_3 + Q_4} \qquad (6.33)$$

So the modified column loads are FQ_1, FQ_2, FQ_3, and FQ_4. This modified loading on the strip under consideration is shown in Figure 6.11b. The shear and the moment diagram for this strip can now be drawn, and the procedure is repeated in the *x* and *y* directions for all strips.

6. Determine the effective depth *d* of the mat by checking for diagonal tension shear near various columns. According to ACI Code 318-95 (Section 11.12.2.1c, American Concrete Institute, 1995), for the critical section,

$$U = b_o d[\phi(0.34)\sqrt{f_c'}] \qquad (6.34)$$

where   *U* = factored column load (MN), or (column load) × (load factor)
ϕ = reduction factor = 0.85
f_c' = compressive strength of concrete at 28 days (MN/m²)

The units of b_o and *d* in Eq. (6.34) are in meters. In English units, Eq. (6.34) may be expressed as

$$U = b_o d(4\phi\sqrt{f_c'}) \qquad (6.35)$$

where   *U* is in lb, b_o and *d* are in in., and f_c' is in lb/in²

The expression for b_o in terms of *d*, which depends on the location of the column with respect to the plan of the mat, can be obtained from Figure 6.11c.

7. From the moment diagrams of all strips *in one direction* (*x* or *y*), obtain the *maximum* positive and negative moments per unit width (i.e., $M' = M/B_1$).

8. Determine the areas of steel per unit width for positive and negative reinforcement in the *x* and *y* directions. We have

$$M_u = (M')(\text{load factor}) = \phi A_s f_y \left(d - \frac{a}{2} \right) \tag{6.36}$$

and

$$a = \frac{A_s f_y}{0.85 f'_c b} \tag{6.37}$$

where A_s = area of steel per unit width
  f_y = yield stress of reinforcement in tension
  M_u = factored moment
  $\phi = 0.9$ = reduction factor

Examples 6.5 and 6.6 illustrate the use of the conventional rigid method of mat foundation design.

Approximate Flexible Method

In the conventional rigid method of design, the mat is assumed to be infinitely rigid. Also, the soil pressure is distributed in a straight line, and the centroid of the soil pressure is coincident with the line of action of the resultant column loads. (See Figure 6.12a.) In the *approximate flexible method* of design, the soil is assumed to be equivalent to an infinite number of elastic springs, as shown in Figure 6.12b. This assumption is sometimes referred to as the *Winkler foundation*. The elastic constant of these assumed springs is referred to as the *coefficient of subgrade reaction, k*.

To understand the fundamental concepts behind flexible foundation design, consider a beam of width B_1 having infinite length, as shown in Figure 6.12c. The beam is subjected to a single concentrated load Q. From the fundamentals of mechanics of materials,

$$M = E_F I_F \frac{d^2 z}{dx^2} \tag{6.38}$$

where M = moment at any section
  E_F = modulus of elasticity of foundation material
  I_F = moment of inertia of the cross section of the beam = $\left(\frac{1}{12}\right) B_1 h^3$ (see Figure 6.12c).

However,

$$\frac{dM}{dx} = \text{shear force} = V$$

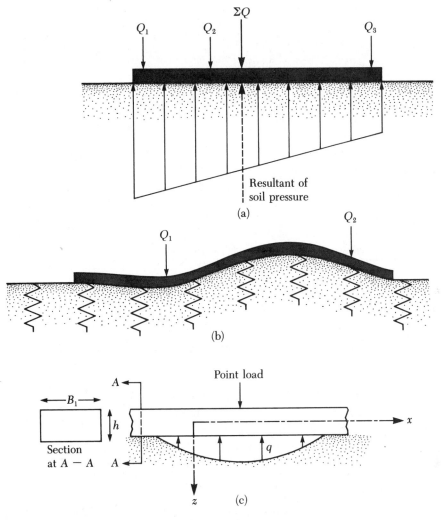

Figure 6.12 (a) Principles of design by conventional rigid method; (b) principles of approximate flexible method; (c) derivation of Eq. (6.42) for beams on elastic foundation

and

$$\frac{dV}{dx} = q = \text{soil reaction}$$

Hence,

$$\frac{d^2M}{dx^2} = q \tag{6.39}$$

Combining Eqs. (6.38) and (6.39) yields

$$E_F I_F \frac{d^4z}{dx^4} = q \tag{6.40}$$

However, the soil reaction is

$$q = -zk'$$

where z = deflection
$k' = kB_1$
k = coefficient of subgrade reaction $(kN/m^3$ or $lb/in^3)$

So

$$E_F I_F = \frac{d^4z}{dx^4} = -zkB_1 \tag{6.41}$$

Solving Eq. (6.41) yields

$$z = e^{-\alpha x}(A' \cos \beta x + A'' \sin \beta x) \tag{6.42}$$

where A' and A'' are constants and

$$\beta = \sqrt[4]{\frac{B_1 k}{4 E_F I_F}} \tag{6.43}$$

The unit of the term β, as defined by the preceding equation, is $(length)^{-1}$. This parameter is very important in determining whether a mat foundation should be designed by the conventional rigid method or the approximate flexible method. According to the American Concrete Institute Committee 336 (1988), mats should be designed by the conventional rigid method if the spacing of columns in a strip is less than $1.75/\beta$. If the spacing of columns is larger than $1.75/\beta$, the approximate flexible method may be used.

To perform the analysis for the structural design of a flexible mat, one must know the principles involved in evaluating the coefficient of subgrade reaction, k. Before proceeding with the discussion of the approximate flexible design method, let us discuss this coefficient in more detail.

If a foundation of width B (see Figure 6.13) is subjected to a load per unit area of q, it will undergo a settlement Δ. The coefficient of subgrade modulus can be defined as

$$k = \frac{q}{\Delta} \tag{6.44}$$

The unit of k is kN/m^3 (or lb/in^3). The value of the coefficient of subgrade reaction is not a constant for a given soil, but rather depends on several factors, such as the length L and width B of the foundation and also the depth of embedment of the foundation. A comprehensive study by Terzaghi (1955) of the parameters affecting the coefficient of subgrade reaction indicated that the value of the coefficient decreases with the width of the foundation. In the field, load tests can be carried out by means of square plates measuring 0.3 m $\times$ 0.3 m (1 ft $\times$ 1 ft), and values of k can

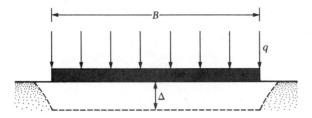

Figure 6.13 Definition of coefficient of subgrade reaction, k

be calculated. The value of k can be related to large foundations measuring $B \times B$ in the following ways:

Foundations on Sandy Soils For foundations on sandy soils,

$$k = k_{0.3}\left(\frac{B + 0.3}{2B}\right)^2$$

(6.45)

where $k_{0.3}$ and k = coefficients of subgrade reaction of foundations measuring 0.3 m $\times$ 0.3 m and B (m) $\times$ B (m), respectively (unit is kN/m^3)

In English units, Eq. (6.45) may be expressed as

$$k = k_1\left(\frac{B + 1}{2B}\right)^2$$

(6.46)

where k_1 and k = coefficients of subgrade reaction of foundations measuring 1 ft $\times$ 1 ft and B (ft) $\times$ B (ft), respectively (unit is lb/in^3)

Foundations on Clays For foundations on clays,

$$k\,(\text{kN/m}^3) = k_{0.3}(\text{kN/m}^3)\left[\frac{0.3\ (\text{m})}{B\,(\text{m})}\right]$$

(6.47)

The definition of k in Eq. (6.47) is the same as in Eq. (6.45).
   In English units,

$$k\,(\text{lb/in}^3) = k_1\,(\text{lb/in}^3)\left[\frac{1\ (\text{ft})}{B\,(\text{ft})}\right]$$

(6.48)

The definitions of k and k_1 are the same as in Eq. (6.46).

For rectangular foundations having dimensions of $B \times L$ (for similar soil and q),

$$k = \frac{k_{(B \times B)}\left(1 + 0.5\dfrac{B}{L}\right)}{1.5} \tag{6.49}$$

where k = coefficient of subgrade modulus of the rectangular foundation $(L \times B)$

$k_{(B \times B)}$ = coefficient of subgrade modulus of a square foundation having dimension of $B \times B$

Equation (6.49) indicates that the value of k for a very long foundation with a width B is approximately $0.67k_{(B \times B)}$.

The modulus of elasticity of granular soils increases with depth. Because the settlement of a foundation depends on the modulus of elasticity, the value of k increases with the depth of the foundation.

Table 6.2 provides typical ranges of values for the coefficient of subgrade reaction, k_1, for sandy and clayey soils.

For long beams, Vesic (1961) proposed an equation for estimating subgrade reaction, namely,

$$k' = Bk = 0.65 \sqrt[12]{\frac{E_s B^4}{E_F I_F}} \frac{E_s}{1 - \mu_s^2}$$

or

$$k = 0.65 \sqrt[12]{\frac{E_s B^4}{E_F I_F}} \frac{E_s}{B(1 - \mu_s^2)} \tag{6.50}$$

where E_s = modulus of elasticity of soil

B = foundation width

E_F = modulus of elasticity of foundation material

I_F = moment of inertia of the cross section of the foundation

μ_s = Poisson's ratio of soil

For most practical purposes, Eq. (6.50) can be approximated as

$$k = \frac{E_s}{B(1 - \mu_s^2)} \tag{6.51}$$

Now that we have discussed the coefficient of subgrade reaction, we will proceed with the discussion of the approximate flexible method of designing mat foundations. This method, as proposed by the American Concrete Institute Committee 336 (1988), is described step by step. The use of the design procedure, which is based primarily on the theory of plates, allows the effects (i.e., moment, shear, and deflec-

Table 6.2 Typical Subgrade Reaction Values, k_1

Soil type	k_1	
	MN/m³	lb/in.³
Dry or moist sand:		
Loose	8–25	30–90
Medium	25–125	90–450
Dense	125–375	450–1350
Saturated sand:		
Loose	10–15	35–55
Medium	35–40	125–145
Dense	130–150	475–550
Clay:		
Stiff	10–25	40–90
Very stiff	25–50	90–185
Hard	>50	>185

tion) of a concentrated column load in the area surrounding it to be evaluated. If the zones of influence of two or more columns overlap, superposition can be employed to obtain the net moment, shear, and deflection at any point. The method is as follows:

1. Assume a thickness h for the mat, according to Step 6 of the conventional rigid method. (*Note:* h is the *total* thickness of the mat.)
2. Determine the flexural ridigity R of the mat as given by the formula

$$R = \frac{E_F h^3}{12(1 - \mu_F^2)} \tag{6.52}$$

where E_F = modulus of elasticity of foundation material
        μ_F = Poisson's ratio of foundation material

3. Determine the radius of effective stiffness—that is,

$$L' = \sqrt[4]{\frac{R}{k}} \tag{6.53}$$

where k = coefficient of subgrade reaction

The zone of influence of any column load will be on the order of 3 to 4 L'.

4. Determine the moment (in polar coordinates at a point) caused by a column load (see Figure 6.14a). The formulas to use are

$$M_r = \text{radial moment} = -\frac{Q}{4}\left[A_1 - \frac{(1 - \mu_F)A_2}{\frac{r}{L'}}\right] \tag{6.54}$$

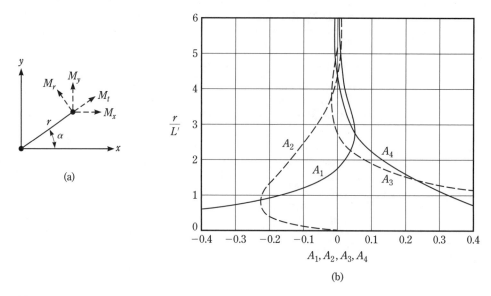

Figure 6.14 Approximate flexible method of mat design

and

$$M_t = \text{tangential moment} = -\frac{Q}{4}\left[\mu_F A_1 + \frac{(1 - \mu_F)A_2}{\dfrac{r}{L'}}\right] \qquad (6.55)$$

where r = radial distance from the column load
            Q = column load
       A_1, A_2 = functions of r/L'

The variations of A_1 and A_2 with r/L' are shown in Figure 6.14b. (For details see Hetenyi, 1946.)

In the Cartesian coordinate system (see Figure 6.14a),

$$M_x = M_t \sin^2\alpha + M_r \cos^2\alpha \qquad (6.56)$$

and

$$M_y = M_t \cos^2\alpha + M_r \sin^2\alpha \qquad (6.57)$$

5. For the unit width of the mat, determine the shear force V caused by a column load:

$$V = \frac{Q}{4L'}A_3 \qquad (6.58)$$

The variation of A_3 with r/L' is shown in Figure 6.14b.

6. If the edge of the mat is located in the zone of influence of a column, determine the moment and shear along the edge. (Assume that the mat is continuous.) Moment and shear opposite in sign to those determined are applied at the edges to satisfy the known conditions.

7. The deflection at any point is given by

$$\delta = \frac{QL'^2}{4R}A_4 \tag{6.59}$$

The variation of A_4 is presented in Figure 6.14.

Example 6.5

The plan of a mat foundation with column loads is shown in Figure 6.15. Use Eq. (6.25) to calculate the soil pressures at points *A, B, C, D, E, F, G, H, I, J, K, L, M,* and *N*. The size of the mat is 76 ft × 96 ft, all columns are 24 in. × 24 in. in section,

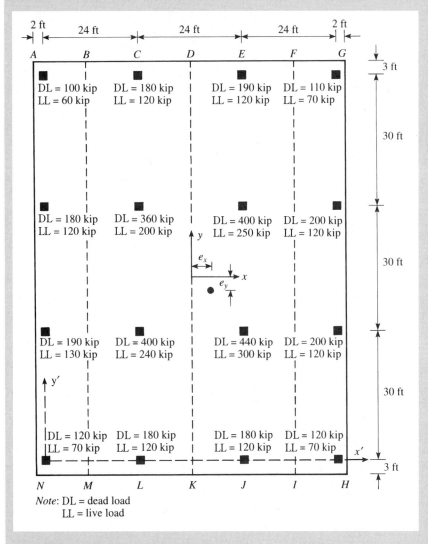

Figure 6.15 Plan of a mat foundation

and $q_{all(net)} = 1.5 \text{ kip/ft}^2$. Verify that the soil pressures are less than the net allowable bearing capacity.

Solution

From Figure 6.15,

$$
\begin{aligned}
\text{Column dead load } (DL) &= 100 + 180 + 190 + 110 + 180 + 360 + 400 + 200 \\
&\quad + 190 + 400 + 440 + 200 + 120 + 180 + 180 + 120 \\
&= 3550 \text{ kip}
\end{aligned}
$$

and

$$
\begin{aligned}
\text{Column live load } (LL) &= 60 + 120 + 120 + 70 + 120 + 200 + 250 + 120 \\
&\quad + 130 + 240 + 300 + 120 + 70 + 120 + 120 + 70 \\
&= 2230 \text{ kip}
\end{aligned}
$$

So

$$\text{Service load} = 3550 + 2230 = 5780 \text{ kip}$$

According to ACI 318-95 (Section 9.2), the factored load $U = (1.4)$ (Dead load) $+ (1.7)$ (Live load). So

$$\text{Factored load} = (1.4)(3550) + (1.7)(2230) = 8761 \text{ kip}$$

The moments of inertia of the foundation are

$$I_x = \tfrac{1}{12}(76)(96)^3 = 5603 \times 10^3 \text{ ft}^4$$

and

$$I_y = \tfrac{1}{12}(96)(76)^3 = 3512 \times 10^3 \text{ ft}^4$$

Also,

$$\Sigma M_{y'} = 0$$

so

$$
\begin{aligned}
5780x' &= (24)(300 + 560 + 640 + 300) + (48)(310 + 650 + 740 + 300) \\
&\quad + (72)(180 + 320 + 320 + 190)
\end{aligned}
$$

or

$$x' = 36.664 \text{ ft}$$

and

$$e_x = 36.664 - 36.0 = 0.664 \text{ ft}$$

Similarly,

$$\Sigma M_{x'} = 0$$

so

$$
\begin{aligned}
5780y' &= (30)(320 + 640 + 740 + 320) + (60)(300 + 560 + 650 + 320) \\
&\quad + (90)(160 + 300 + 310 + 180)
\end{aligned}
$$

or

$$y' = 44.273 \text{ ft}$$

and

$$e_y = 44.273 - \tfrac{90}{2} = -0.727 \text{ ft}$$

The moments caused by eccentricity are

$$M_x = Qe_y = (8761)(0.727) = 6369 \text{ kip-ft}$$

and

$$M_y = Qe_x = (8761)(0.664) = 5817 \text{ kip-ft}$$

From Eq. (6.25),

$$q = \frac{Q}{A} \pm \frac{M_y x}{I_y} \pm \frac{M_x y}{I_x}$$

$$= \frac{8761}{(76)(96)} \pm \frac{(5817)(x)}{3512 \times 10^3} \pm \frac{(6369)(y)}{5603 \times 10^3}$$

or

$$q = 1.20 \pm 0.0017x \pm 0.0011y \ (\text{kip/ft}^2)$$

Now the following table can be prepared:

Point	$\frac{Q}{A}$ (kip/ft²)	x (ft)	$\pm 0.0017x$ (ft)	y (ft)	$\pm 0.0011y$ (ft)	q(kip/ft²)
A	1.2	−38	−0.065	48	−0.053	**1.082**
B	1.2	−24	−0.041	48	−0.053	**1.106**
C	1.2	−12	−0.020	48	−0.053	**1.127**
D	1.2	0	0.0	48	−0.053	**1.147**
E	1.2	12	0.020	48	−0.053	**1.167**
F	1.2	24	0.041	48	−0.053	**1.188**
G	1.2	38	0.065	48	−0.053	**1.212**
H	1.2	38	0.065	−48	0.053	**1.318**
I	1.2	24	0.041	−48	0.053	**1.294**
J	1.2	12	0.020	−48	0.053	**1.273**
K	1.2	0	0.0	−48	0.053	**1.253**
L	1.2	−12	−0.020	−48	0.053	**1.233**
M	1.2	−24	−0.041	−48	0.053	**1.212**
N	1.2	−38	−0.065	−48	0.053	**1.188**

The soil pressures at all points are less than the given value of $q_{\text{all(net)}} = 1.5 \text{ kip/ft}^2$.

■

Example 6.6

Use the results of Example 6.5 and the conventional rigid method.

a. Determine the thickness of the slab.
b. Divide the mat into four strips (that is, *ABMN, BCDKLM, DEFIJK,* and *FGHI*), and determine the average soil reactions at the ends of each strip.
c. Determine the reinforcement requirements in the *y* direction for $f'_c = 3000$ lb/in^2 and $f_y = 60,000$ lb/in^2.

Solution

Part a: Determination of Thickness of Mat
For the critical perimeter column, as shown in Figure 6.16 (ACI 318-95; Section 9.2.1),

$$U = 1.4(DL) + 1.7(LL) = (1.4)(190) + (1.7)(130) = 487 \text{ kip}$$

and

$$b_o = 2(36 + d/2) + (24 + d) = 96 + 2d \text{ (in.)}$$

From ACI 318-95,

$$\phi V_c \geq V_u$$

where V_c = nominal shear strength of concrete
V_u = factored shear strength

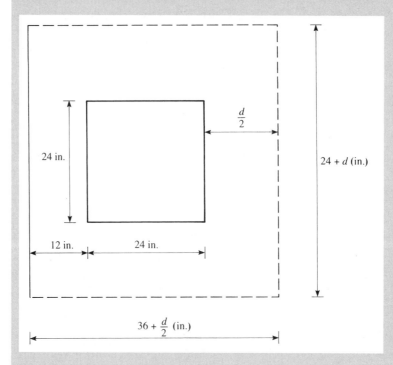

Figure 6.16 Critical perimeter column

we have

$$\phi V_c = \phi(4)\sqrt{f'_c}b_o d = (0.85)(4)(\sqrt{3000})(96 + 2d)d$$

so

$$\frac{(0.85)(4)(\sqrt{3000})(96 + 2d)d}{1000} \geq 487$$

$$(96 + 2d)d \geq 2615.1$$

$$d \approx 19.4 \text{ in.}$$

For the critical internal column shown in Figure 6.17,

$$b_o = 4(24 + d) = 96 + 4d \text{ (in.)}$$

$$U = (1.4)(440) + (1.7)(300) = 1126 \text{ kip}$$

and

$$\frac{(0.85)(4)(\sqrt{3000})(96 + 4d)d}{1000} \geq 1126$$

$$(96 + 4d)d \geq 6046.4$$

$$d \approx 28.7 \text{ in.}$$

Accordingly, use $d = 29$ in.

With a minimum cover of 3 in. over the steel reinforcement and 1-in.-diameter steel bars, the total slab thickness is

$$h = 29 + 3 + 1 = \textbf{33 in.}$$

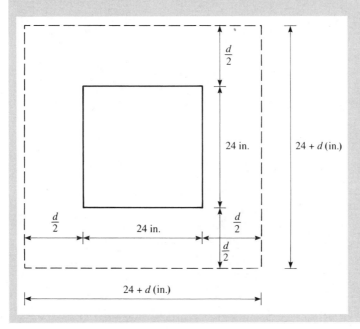

Figure 6.17 Critical internal column

Part b: Average Soil Reaction
In Figure 6.15, for strip $ABMN$ (width $= 14$ ft),

$$q_1 = \frac{q_{(atA)} + q_{(atB)}}{2} = \frac{1.082 + 1.106}{2} = \textbf{1.094 kip/ft}^2$$

and

$$q_2 = \frac{q_{(atM)} + q_{(atN)}}{2} = \frac{1.212 + 1.188}{2} = \textbf{1.20 kip/ft}^2$$

For strip $BCDKLM$ (width $= 24$ ft),

$$q_1 = \frac{1.106 + 1.127 + 1.147}{3} = \textbf{1.127 kip/ft}^2$$

and

$$q_2 = \frac{1.253 + 1.233 + 1.212}{3} = \textbf{1.233 kip/ft}^2$$

For strip $DEFIJK$ (width $= 24$ ft),

$$q_1 = \frac{1.147 + 1.167 + 1.188}{3} = \textbf{1.167 kip/ft}^2$$

and

$$q_2 = \frac{1.294 + 1.273 + 1.253}{3} = \textbf{1.273 kip/ft}^2$$

For strip $FGHI$ (width $= 14$ ft),

$$q_1 = \frac{1.188 + 1.212}{2} = \textbf{1.20 kip/ft}^2$$

and

$$q_2 = \frac{1.318 + 1.294}{2} = \textbf{1.306 kip/ft}^2$$

Check for $\Sigma F_V = 0$:

Soil reaction for strip $ABMN = \frac{1}{2}(1.094 + 1.20)(14)(96) = 1541.6$ kip

Soil reaction for strip $BCDKLM = \frac{1}{2}(1.127 + 1.233)(24)(96) = 2718.7$ kip

Soil reaction for strip $DEFIJK = \frac{1}{2}(1.167 + 1.273)(24)(96) = 2810.9$ kip

Soil reaction for strip $FGHJ = \frac{1}{2}(1.20 + 1.306)(14)(96) = 1684.0$ kip

$$\sum 8755.2 \text{ kip} \approx \sum \text{Column load} = 8761 \text{ kip} - \text{OK}$$

Part c: Reinforcement Requirements

Figure 6.18 gives the design of strip $BCDKLM$ and shows the load diagram, in which

$$Q_1 = (1.4)(180) + (1.7)(120) = 456 \text{ kip}$$

$$Q_2 = (1.4)(360) + (1.7)(200) = 844 \text{ kip}$$

$$Q_3 = (1.4)(400) + (1.7)(240) = 968 \text{ kip}$$

and

$$Q_4 = (1.4)(180) + (1.7)(120) = 456 \text{ kip}$$

The shear and moment diagrams are shown in Figures 6.18b and c, respectively. From Figure 6.18c, the maximum positive moment at the bottom of the foundation is 2281.1/24 = 95.05 kip-ft/ft.

Note that Figure 6.19 shows the design concepts of a rectangular section in bending.

$$\sum \text{Compressive force, } C = 0.85 f'_c ab$$

$$\sum \text{Tensile force, } T = A_s f_y$$

and

$$C = T$$

For this case, $b = 1$ ft = 12 in., so

$$(0.85)(3)(12)a = A_s(60)$$

and

$$A_s = 0.51a$$

From Eq. (6.36),

$$M_u = \phi A_s f_y \left(d - \frac{a}{2} \right)$$

and we have

$$(95.05)(12) = (0.9)(0.51a)(60)\left(29 - \frac{a}{2} \right)$$

or

$$a = 1.47 \text{ in.}$$

Thus,

$$A_s = (0.51)(1.47) = 0.75 \text{ in}^2$$

- Minimum reinforcement s_{min} (ACI 318-95, Section 10.5) = $200/f_y$ = $200/60{,}000 = 0.00333$

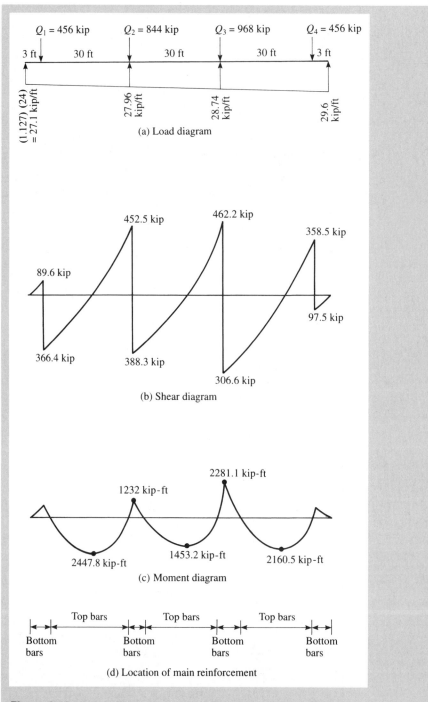

Figure 6.18 Load, shear, and moment diagrams for strip *BCDKLM*

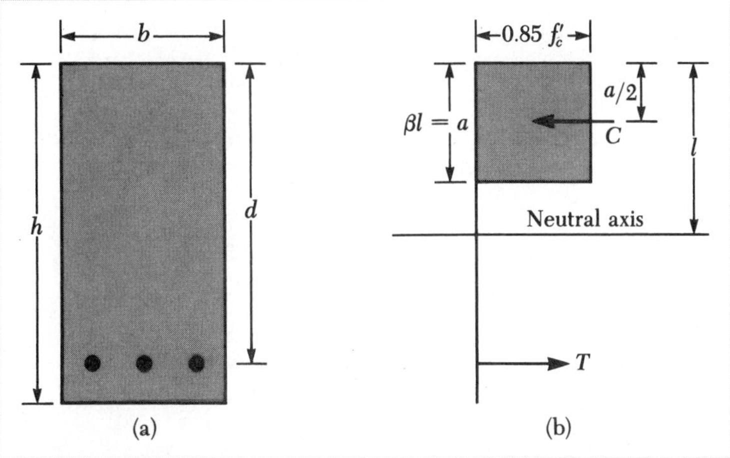

Figure 6.19 Rectangular section in bending; (a) section, (b) assumed stress distribution across the section

- Mainimum $A_s = (0.00333)(12)(29) = 1.16$ in^2/ft. Hence, use minimum reinforcement with $A_s = 1.16$ in^2/ft.
- **Use No. 9 bars at 10 in. center to center $(A_s = 1.2$ in^2/ft) at the bottom of the foundation.**

From Figure 6.18c, the maximum negative moment is 2447.8 kip-ft/24 = 102 kip-ft/ft. By observation, $A_s \leq A_{s(min)}$.

- **Use No. 9 bars at 10 in. center to center at the top of the foundation.**   ■

Problems

6.1 Determine the net ultimate bearing capacity of a mat foundation measuring 45 ft × 30 ft on a saturated clay with $c_u = 1950$ lb/ft^2, $\phi = 0$, and $D_f = 6.5$ ft. Use Eq. (6.10).

6.2 Repeat Problem 6.1, but now let $c_u = 120$ kN/m^2, $\phi = 0$, $B = 8$ m, $L = 18$ m, and $D_f = 3$ m.

6.3 What will be the net allowable bearing capacity of a mat foundation with dimensions of 15 m × 10 m constructed over a sand deposit? Let $D_f = 2$ m, allowable settlement = 30 mm, and corrected average penetration number $(N_1)_{60} = 10$. Use Eq. (6.12).

6.4 Repeat Problem 6.3 for an allowable settlement of 50 mm.

6.5 Consider a mat foundation with dimensions of 20 m × 13 m. The dead and live load on the mat is 42 MN. The mat is to be placed on a clay with $c_u = 40$ kN/m^2. The unit weight of the clay is 17.5 kN/m^3. Find the depth D_f of the mat for a fully compensated foundation.

6.6 What will be the depth D_f of the mat considered in Problem 6.5 for FS = 3 against bearing capacity failure?

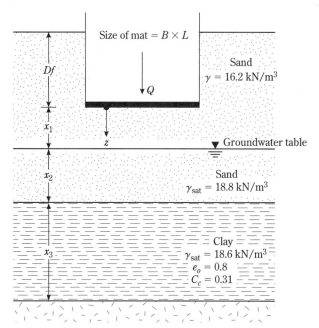

Figure P6.7

6.7 Consider the mat foundation shown in Figure P6.7. Let
$Q = 25$ MN, $D_f = 1.5$ m, $x_1 = 2$ m, $x_2 = 3$ m, and $x_3 = 4$ m. The clay is normally consolidated. Estimate the consolidation settlement under the center
of the mat.

6.8 Estimate the consolidation settlement under the corner of the mat foundation
described in Problem 6.7.

6.9 Redo Problem 6.7, assuming that the preconsolidation pressure of the clay is
120 kN/m^2 and the swelling index is about $1/4C_c$.

6.10 For the mat shown in Figure P6.10, $Q_1 = Q_3 = 40$ tons, $Q_4 = Q_5 = Q_6 =$
60 tons, $Q_2 = Q_9 = 45$ tons, and $Q_7 = Q_8 = 50$ tons. All columns are 20 in.
× 20 in. in cross section. Use the procedure outlined in Section 6.8 to
determine the pressure on the soil at points A, B, C, D, E, F, G, and H.

6.11 The plan of a mat foundation with column loads is shown in Figure P6.11. Calculate the soil pressure at points A, B, C, D, E, and F. (*Note:* All column sections are planned to be 0.5 m × 0.5 m.)

6.12 Divide the mat shown in Figure P6.11 into three strips, such as *AGHF*
($B_1 = 4.25$ m), *GIJH* ($B_1 = 8$ m), and *ICDJ* ($B_1 = 4.25$ m). Use the results
of Problem 6.11, and determine the reinforcement requirements in the y direction. Here, $f'_c = 20.7$ MN/m^2, $f_y = 413.7$ MN/m^2, and the load factor is 1.7.

6.13 From the plate load test on a plate of dimensions 1 ft × 1 ft in the field, the
coefficient of subgrade reaction of a sandy soil is determined to be 80 lb/in^3.
What will be the value of the coefficient of subgrade reaction on the same soil
for a foundation with dimensions of 30 ft × 30 ft?

6.14 The subgrade reaction of a sandy soil obtained from the plate load test on a
plate of dimensions 1 m × 0.7 m is 18 kN/m^3. What will be the value of k on
the same soil for a foundation measuring 5 m × 3.5 m?

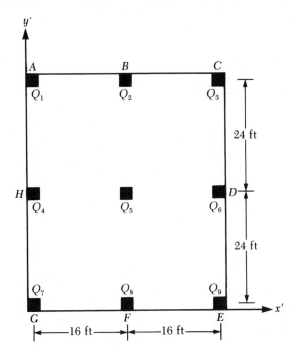

Figure P6.10

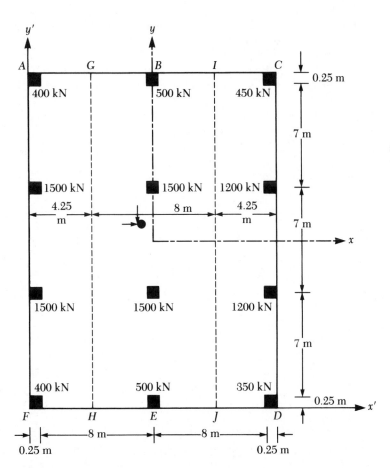

Figure P6.11

References

American Concrete Institute (1995). *ACI Standard Building Code Requirements for Reinforced Concrete,* ACI 318–95, Farmington Hills, MI.

American Concrete Institute Committee 336 (1988). "Suggested Design Procedures for Combined Footings and Mats," *Journal of the American Concrete Institute,* Vol. 63, No. 10, pp. 1041–1077.

Hetenyi, M. (1946). *Beams of Elastic Foundations,* University of Michigan Press, Ann Arbor, MI.

Meyerhof, G. G. (1965). "Shallow Foundations," *Journal of the Soil Mechanics and Foundations Division,* American Society of Civil Engineers, Vol. 91, No. SM2, pp. 21–31.

Rios, L., and Silva, F. P. (1948). "Foundations in Downtown São Paulo (Brazil)," *Proceedings, Second International Conference on Soil Mechanics and Foundation Engineering,* Rotterdam, Vol. 4, p. 69.

Schultze, E. (1962). "Probleme bei der Auswertung von Setzungsmessungen," *Proceedings, Baugrundtagung,* Essen, Germany, p. 343.

Stuart, J. G., and Graham, J. (1975). "Settlement Performance of a Raft Foundation on Sand," in *Settlement of Structures,* Halsted Press, New York, pp. 62–67.

Terzaghi, K. (1955). "Evaluation of the Coefficient of Subgrade Reactions," *Geotechnique,* Institute of Engineers, London, Vol. 5, No. 4, pp. 197–226.

Vargas, M. (1948). "Building Settlement Observations in São Paulo," *Proceedings Second International Conference on Soil Mechanics and Foundation Engineering,* Rotterdam, Vol. 4, p. 13.

Vargas, M. (1961). "Foundations of Tall Buildings on Sand in São Paulo (Brazil)," *Proceedings, Fifth International Conference on Soil Mechanics and Foundation Engineering,* Paris, Vol. 1, p. 841.

Vesic, A. S. (1961). "Bending of Beams Resting on Isotropic Solid," *Journal of the Engineering Mechanics Division,* American Society of Civil Engineers, Vol. 87, No. EM2, pp. 35–53.

7

Lateral Earth Pressure

7.1   *Introduction*

Vertical or near-vertical slopes of soil are supported by retaining walls, cantilever sheet-pile walls, sheet-pile bulkheads, braced cuts, and other, similar structures. The proper design of those structures requires an estimation of lateral earth pressure, which is a function of several factors, such as (a) the type and amount of wall movement, (b) the shear strength parameters of the soil, (c) the unit weight of the soil, and (d) the drainage conditions in the backfill. Figure 7.1 shows a retaining wall of height H. For similar types of backfill,

a. The wall may be restrained from moving (Figure 7.1a). The lateral earth pressure on the wall at any depth is called the *at-rest earth pressure*.

b. The wall may tilt away from the soil that is retained (Figure 7.1b). With sufficient wall tilt, a triangular soil wedge behind the wall will fail. The lateral pressure for this condition is referred to as *active earth pressure*.

c. The wall may be pushed into the soil that is retained (Figure 7.1c). With sufficient wall movement, a soil wedge will fail. The lateral pressure for this condition is referred to as *passive earth pressure*.

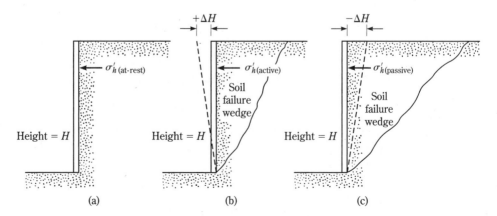

Figure 7.1   Nature of lateral earth pressure on a retaining wall

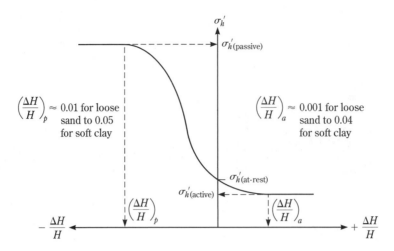

Figure 7.2 Nature of variation of lateral earth pressure at a certain depth

Figure 7.2 shows the nature of variation of the lateral pressure, σ'_h, at a certain depth of the wall with the magnitude of wall movement.

In the sections that follow, we will discuss various relationships to determine the at-rest, active, and passive pressures on a retaining wall. It is assumed that the reader has studied lateral earth pressure in the past, so this chapter will serve as a review.

7.2   *Lateral Earth Pressure at Rest*

Consider a vertical wall of height H, as shown in Figure 7.3, retaining a soil having a unit weight of γ. A uniformly distributed load, q/unit area, is also applied at the ground surface. The shear strength of the soil is

$$s = c' + \sigma' \tan \phi'$$

where c' = cohesion
         ϕ' = effective angle of friction
         σ' = effective normal stress

At any depth z below the ground surface, the vertical subsurface stress is

$$\sigma'_o = q + \gamma z \tag{7.1}$$

If the *wall is at rest and is not allowed to move at all,* either away from the soil mass or into the soil mass (i.e., there is zero horizontal strain), the lateral pressure at a depth z is

$$\sigma_h = K_o \sigma'_o + u \tag{7.2}$$

where u = pore water pressure
         K_o = coefficient of at-rest earth pressure

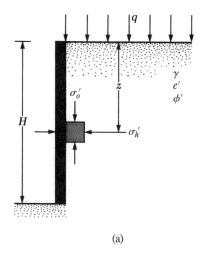
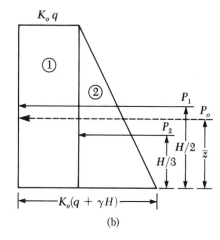

(a)   (b)

Figure 7.3   At-rest earth pressure

For normally consolidated soil, the relation for K_o (Jaky, 1944) is

$$K_o \approx 1 - \sin \phi' \tag{7.3}$$

Equation (7.3) is an empirical approximation.

For normally consolidated clays, the coefficient of earth pressure at rest can be approximated (Brooker and Ireland, 1965) as

$$K_o \approx 0.95 - \sin \phi' \tag{7.4}$$

where ϕ' = drained peak friction angle

On the basis of Brooker and Ireland's experimental results, the value of K_o for normally consolidated clays may be approximately correlated with the plasticity index (PI) via the relationships

$$K_o = 0.4 + 0.007 \,(\text{PI}) \quad (\text{for PI between 0 and 40}) \tag{7.5}$$

and

$$K_o = 0.64 + 0.001 \,(\text{PI}) \quad (\text{for PI between 40 and 80}) \tag{7.6}$$

For overconsolidated clays,

$$K_{o(\text{overconsolidated})} \approx K_{o(\text{normally consolidated})}\sqrt{\text{OCR}} \tag{7.7}$$

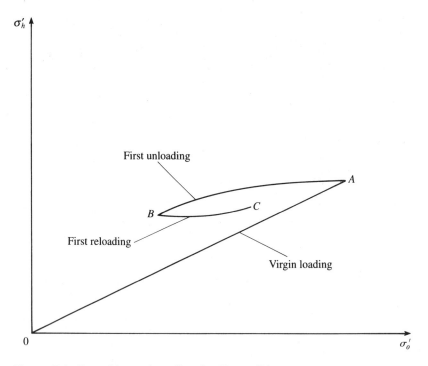

Figure 7.4 Stress history for soil under K_o condition

where   OCR = overconsolidation ratio

Mayne and Kulhawy (1982) analyzed the results of 171 different laboratory-tested soils and proposed the following general empirical relationship to estimate the magnitude of K_o for sand and clay:

$$K_o = (1 - \sin \phi') \left[\frac{\text{OCR}}{\text{OCR}_{\text{max}}^{(1 - \sin \phi)}} + \frac{3}{4} \left(1 - \frac{\text{OCR}}{\text{OCR}_{\text{max}}} \right) \right] \qquad (7.8)$$

In this equation,   OCR = present overconsolidation ratio
OCR_{max} = maximum overconsolidation ratio

In Figure 7.4, OCR_{max} is the value of OCR at point B.

With a properly selected value of the at-rest earth pressure coefficient, Eq. (7.2) can be used to determine the variation of lateral earth pressure with depth z. Figure 7.3b shows the variation of σ_h' with depth for the wall depicted in Figure 7.3a. Note that if the surcharge $q = 0$ and the pore water pressure $u = 0$, the pressure diagram will be a triangle. The total force, P_o, *per unit length* of the wall given in Figure 7.3a can now be obtained from the area of the pressure diagram given in Figure 7.3b and is

$$P_o = P_1 + P_2 = qK_oH + \tfrac{1}{2}\gamma H^2 K_o \qquad (7.9)$$

where P_1 = area of rectangle 1
P_2 = area of triangle 2

The location of the line of action of the resultant force, P_o, can be obtained by taking the moment about the bottom of the wall. Thus,

$$\bar{z} = \frac{P_1\left(\dfrac{H}{2}\right) + P_2\left(\dfrac{H}{3}\right)}{P_o} \qquad (7.10)$$

If the water table is located at a depth $z < H$, the at-rest pressure diagram shown in Figure 7.3b will have to be somewhat modified, as shown in Figure 7.5. If the effective unit weight of soil below the water table equals γ' (i.e., $\gamma_{sat} - \gamma_w$), then

$$\text{at } z = 0, \quad \sigma_h' = K_o\sigma_o' = K_oq$$

$$\text{at } z = H_1, \quad \sigma_h' = K_o\sigma_o' = K_o(q + \gamma H_1)$$

and

$$\text{at } z = H_2, \quad \sigma_h' = K_o\sigma_o' = K_o(q + \gamma H_1 + \gamma'H_2)$$

Note that in the preceding equations, σ_o' and σ_h' are effective vertical and horizontal pressures, respectively. Determining the total pressure distribution on the wall requires adding the hydrostatic pressure u, which is zero from $z = 0$ to $z = H_1$ and is $H_2\gamma_w$ at $z = H_2$. The variation of σ_h' and u with depth is shown in Figure 7.5b. Hence, the total force per unit length of the wall can be determined from the area of the pressure diagram. Specifically,

$$P_o = A_1 + A_2 + A_3 + A_4 + A_5$$

where A = area of the pressure diagram

So

$$P_o = K_oqH_1 + \tfrac{1}{2}K_o\gamma H_1^2 + K_o(q + \gamma H_1)H_2 + \tfrac{1}{2}K_o\gamma'H_2^2 + \tfrac{1}{2}\gamma_wH_2^2 \qquad (7.11)$$

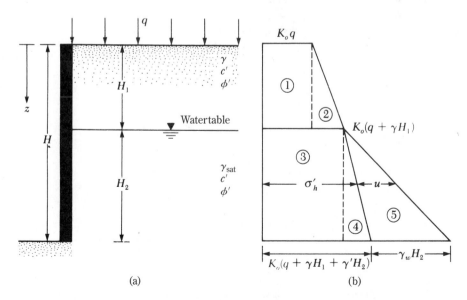

(a)   (b)

Figure 7.5 At-rest earth pressure with water table located at a depth $z < H$

Active Pressure

7.3   *Rankine Active Earth Pressure*

The lateral earth pressure described in Section 7.2 involves walls that do not yield at all. However, if a wall tends to move away from the soil a distance Δx, as shown in Figure 7.6a, the soil pressure on the wall at any depth will decrease. For a wall that is *frictionless*, the horizontal stress, σ'_h, at depth z will equal $K_o\sigma'_o(=K_o\gamma z)$ when Δx is zero. However, with $\Delta x > 0$, σ'_h will be less than $K_o\sigma'_o$.

The Mohr's circles corresponding to wall displacements of $\Delta x = 0$ and $\Delta x > 0$ are shown as circles a and b, respectively, in Figure 7.6b. If the displacement of the wall,

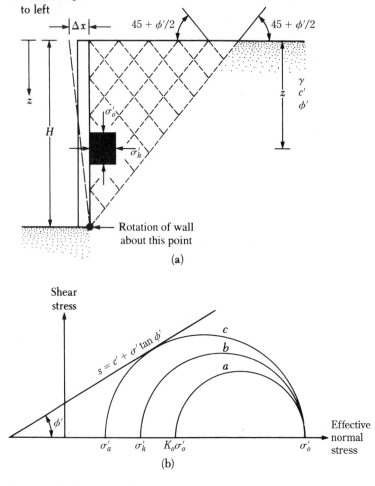

Figure 7.6   Rankine active pressure

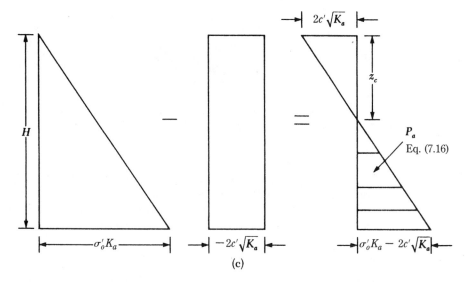

Figure 7.6 (continued)

Δx, continues to increase, the corresponding Mohr's circle eventually will just touch the Mohr–Coulomb failure envelope defined by the equation

$$s = c' + \sigma' \tan \phi'$$

This circle, marked c in the figure, represents the failure condition in the soil mass; the horizontal stress then equals σ'_a, referred to as the *Rankine active pressure*. The *slip lines* (failure planes) in the soil mass will then make angles of $\pm(45 + \phi'/2)$ with the horizontal, as shown in Figure 7.6a.

Equation (1.74) relates the principal stresses for a Mohr's circle that touches the Mohr–Coulomb failure envelope:

$$\sigma'_1 = \sigma'_3 \tan^2\left(45 + \frac{\phi'}{2}\right) + 2c' \tan\left(45 + \frac{\phi'}{2}\right)$$

For the Mohr's circle c in Figure 7.6b,

$$\text{Major principal stress}, \sigma'_1 = \sigma'_o$$

and

$$\text{Minor principal stress}, \sigma'_3 = \sigma'_a$$

Thus,

$$\sigma'_o = \sigma'_a \tan^2\left(45 + \frac{\phi'}{2}\right) + 2c' \tan\left(45 + \frac{\phi'}{2}\right)$$

$$\sigma'_a = \frac{\sigma'_o}{\tan^2\left(45 + \dfrac{\phi'}{2}\right)} - \frac{2c'}{\tan\left(45 + \dfrac{\phi'}{2}\right)}$$

or

$$\sigma'_a = \sigma'_o \tan^2\left(45 - \frac{\phi'}{2}\right) - 2c' \tan\left(45 - \frac{\phi'}{2}\right)$$
$$= \sigma'_o K_a - 2c'\sqrt{K_a} \qquad (7.12)$$

where $K_a = \tan^2(45 - \phi'/2)$ = Rankine active pressure coefficient

The variation of the active pressure with depth for the wall shown in Figure 7.6a is given in Figure 7.6c. Note that $\sigma'_o = 0$ at $z = 0$ and $\sigma'_o = \gamma H$ at $z = H$. The pressure distribution shows that at $z = 0$ the active pressure equals $-2c'\sqrt{K_a}$, indicating a tensile stress that decreases with depth and becomes zero at a depth $z = z_c$, or

$$\gamma z_c K_a - 2c'\sqrt{K_a} = 0$$

and

$$z_c = \frac{2c'}{\gamma\sqrt{K_a}} \qquad (7.13)$$

The depth z_c is usually referred to as the *depth of tensile crack,* because the tensile stress in the soil will eventually cause a crack along the soil–wall interface. Thus, the total Rankine active force per unit length of the wall before the tensile crack occurs is

$$P_a = \int_0^H \sigma'_a \, dz = \int_0^H \gamma z K_a \, dz - \int_0^H 2c'\sqrt{K_a} \, dz$$
$$= \tfrac{1}{2}\gamma H^2 K_a - 2c' H \sqrt{K_a} \qquad (7.14)$$

After the tensile crack appears, the force per unit length on the wall will be caused only by the pressure distribution between depths $z = z_c$ and $z = H$, as shown by the hatched area in Figure 7.6c. This force may be expressed as

$$P_a = \tfrac{1}{2}(H - z_c)(\gamma H K_a - 2c'\sqrt{K_a}) \qquad (7.15)$$

or

$$P_a = \frac{1}{2}\left(H - \frac{2c'}{\gamma\sqrt{K_a}}\right)\left(\gamma H K_a - 2c'\sqrt{K_a}\right) \qquad (7.16)$$

However, it is important to realize that the active earth pressure condition will be reached only if the wall is allowed to "yield" sufficiently. The necessary amount of outward displacement of the wall is about $0.001H$ to $0.004H$ for granular soil backfills and about $0.01H$ to $0.04H$ for cohesive soil backfills.

Note further that if the *total stress* shear strength parameters (c, ϕ) were used, an equation similar to Eq. (7.12) could have been derived, namely,

$$\sigma_a = \sigma_o \tan^2\left(45 - \frac{\phi}{2}\right) - 2c \tan\left(45 - \frac{\phi}{2}\right)$$

Example 7.1

A 6-m-high retaining wall is to support a soil with unit weight $\gamma = 17.4$ kN/m^3, soil friction angle $\phi' = 26°$, and cohesion $c' = 14.36$ kN/m^2. Determine the Rankine active force per unit length of the wall both before and after the tensile crack occurs, and determine the line of action of the resultant in both cases.

Solution
For $\phi' = 26°$,

$$K_a = \tan^2\left(45 - \frac{\phi'}{2}\right) = \tan^2(45 - 13) = 0.39$$

$$\sqrt{K_a} = 0.625$$

$$\sigma_a' = \gamma H K_a - 2c'\sqrt{K_a}$$

From Figure 7.6c,

$$\text{at } z = 0, \sigma_a' = -2c'\sqrt{K_a} = -2(14.36)(0.625) = -17.95 \text{ kN/m}^2$$

and

$$\text{at } z = 6 \text{ m}, \sigma_a' = (17.4)(6)(0.39) - 2(14.36)(0.625)$$

$$= 40.72 - 17.95 = 22.77 \text{ kN/m}^2$$

Active Force before the Tensile Crack Appeared: Eq. (7.14)

$$P_a = \tfrac{1}{2}\gamma H^2 K_a - 2c'H\sqrt{K_a}$$

$$= \tfrac{1}{2}(6)(40.72) - (6)(17.95) = 122.16 - 107.7 = 14.46 \text{ kN/m}$$

The line of action of the resultant can be determined by taking the moment of the area of the pressure diagrams about the bottom of the wall, or

$$P_a \bar{z} = (122.16)\left(\tfrac{6}{3}\right) - (107.7)\left(\tfrac{6}{2}\right)$$

Thus,

$$\bar{z} = \frac{244.32 - 323.1}{14.46} = -5.45 \text{ m}$$

Active Force after the Tensile Crack Appeared: Eq. (7.13)

$$z_c = \frac{2c'}{\gamma\sqrt{K_a}} = \frac{2(14.36)}{(17.4)(0.625)} = 2.64 \text{ m}$$

Using Eq. (7.15) gives

$$P_a = \tfrac{1}{2}(H - z_c)(\gamma H K_a - 2c'\sqrt{K_a}) = \tfrac{1}{2}(6 - 2.64)(22.77) = 38.25 \text{ kN/m}$$

Figure 7.6c indicates that the force $P_a = 38.25$ kN/m is the area of the hatched triangle. Hence, the line of action of the resultant will be located at a height $\bar{z} = (H - z_c)/3$ above the bottom of the wall, or

$$\bar{z} = \frac{6 - 2.64}{3} = \textbf{1.12 m} \qquad \blacksquare$$

Example 7.2

Assume that the retaining wall shown in Figure 7.7a can yield sufficiently to develop an active state. Determine the Rankine active force per unit length of the wall and the location of the resultant line of action.

Solution
If the cohesion, c', is zero, then

$$\sigma_a' = \sigma_o' K_a$$

For the top layer of soil, $\phi_1' = 30°$, so

$$K_{a(1)} = \tan^2\left(45 - \frac{\phi_1'}{2}\right) = \tan^2(45 - 15) = \frac{1}{3}$$

Similarly, for the bottom layer of soil, $\phi_2' = 36°$, and it follows that

$$K_{a(2)} = \tan^2\left(45 - \frac{36}{2}\right) = 0.26$$

The following table shows the calculation of σ_a' and u at various depths below the ground surface.

Depth, z (ft)	σ_o' (lb/ft^2)	K_a	$\sigma_a' = K_a \sigma_o'$ (lb/ft^2)	u (lb/ft^2)
0	0	1/3	0	0
10$^-$	$(102)(10) = 1020$	1/3	340	0
10$^+$	1020	0.26	265.2	0
20	$(102)(10) + (121 - 62.4)(10) = 1606$	0.26	417.6	$(62.4)(10) = 624$

The pressure distribution diagram is plotted in Figure 7.7b. The force per unit length is

$$P_a = \text{area 1} + \text{area 2} + \text{area 3} + \text{area 4}$$

$$= \frac{1}{2}(10)(340) + (265.2)(10) + \frac{1}{2}(417.6 - 265.2)(10) + \frac{1}{2}(624)(10)$$

$$= 1700 + 2652 + 762 + 3120 = \textbf{8234 lb/ft}$$

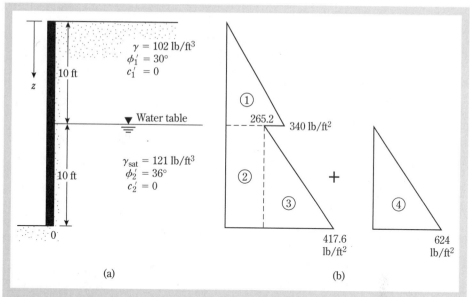

Figure 7.7 Rankine active force behind a retaining wall

The distance of the line of action of the resultant force from the bottom of the wall can be determined by taking the moments about the bottom of the wall (point O in Figure 7.7a) and is

$$\bar{z} = \frac{(1700)\left(10 + \dfrac{10}{3}\right) + (2652)\left(\dfrac{10}{2}\right) + (762 + 3120)\left(\dfrac{10}{3}\right)}{8234} = \textbf{5.93 ft} \quad \blacksquare$$

7.4 *Rankine Active Earth Pressure for Inclined Backfill*

Granular Soil

If the backfill of a frictionless retaining wall is a *granular soil* ($c' = 0$) and rises at an angle α with respect to the horizontal (see Figure 7.8), the *active earth pressure coefficient* may be expressed in the form

$$K_a = \cos \alpha \frac{\cos \alpha - \sqrt{\cos^2 \alpha - \cos^2 \phi'}}{\cos \alpha + \sqrt{\cos^2 \alpha - \cos^2 \phi'}} \tag{7.17}$$

where ϕ' = angle of friction of soil

At any depth z, the *Rankine active pressure* may be expressed as

$$\sigma'_a = \gamma z K_a \tag{7.18}$$

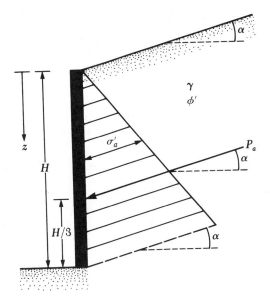

Figure 7.8 Notations for active pressure—Eqs. (7.17), (7.18), (7.19)

Also, the total force per unit length of the wall is

$$P_a = \tfrac{1}{2}\gamma H^2 K_a \tag{7.19}$$

Note that, in this case, the direction of the resultant force P_a is *inclined at an angle α with the horizontal* and intersects the wall at a distance $H/3$ from the base of the wall. Table 7.1 presents the values of K_a (active earth pressure) for varios values of α and ϕ'.

$c' - \phi'$ Soil

The preceding analysis can be extended to the case of an inclined backfill with a $c'-\phi'$ soil. The details of the mathematical derivation are given by Mazindrani and Ganjali (1997). As in Eq. (7.18), for this case

$$\sigma_a' = \gamma z K_a = \gamma z K_a' \cos \alpha \tag{7.20}$$

where

$$K_a' = \frac{1}{\cos^2 \phi'} \left\{ \frac{2\cos^2 \alpha + 2\left(\dfrac{c'}{\gamma z}\right)\cos \phi' \sin \phi'}{-\sqrt{\left[4\cos^2\alpha(\cos^2 \alpha - \cos^2 \phi') + 4\left(\dfrac{c'}{\gamma z}\right)^2 \cos^2 \phi' + 8\left(\dfrac{c'}{\gamma z}\right)\cos^2 \alpha \sin \phi' \cos \phi'\right]}} \right\} - 1 \tag{7.21}$$

Some values of K_a' are given in Table 7.2. For a problem of this type, the depth of tensile crack is given as

$$z_c = \frac{2c'}{\gamma} \sqrt{\frac{1 + \sin \phi'}{1 - \sin \phi'}} \tag{7.22}$$

Table 7.1 Active Earth Pressure Coefficient K_a [from Eq. (7.17)]

				ϕ'(deg) →			
↓α(deg)	28	30	32	34	36	38	40
0	0.361	0.333	0.307	0.283	0.260	0.238	0.217
5	0.366	0.337	0.311	0.286	0.262	0.240	0.219
10	0.380	0.350	0.321	0.294	0.270	0.246	0.225
15	0.409	0.373	0.341	0.311	0.283	0.258	0.235
20	0.461	0.414	0.374	0.338	0.306	0.277	0.250
25	0.573	0.494	0.434	0.385	0.343	0.307	0.275

Table 7.2 Values of K_a'

				$\dfrac{c'}{\gamma z}$	
ϕ' (deg)	α (deg)	0.025	0.05	0.1	0.5
15	0	0.550	0.512	0.435	−0.179
	5	0.566	0.525	0.445	−0.184
	10	0.621	0.571	0.477	−0.186
	15	0.776	0.683	0.546	−0.196
20	0	0.455	0.420	0.350	−0.210
	5	0.465	0.429	0.357	−0.212
	10	0.497	0.456	0.377	−0.218
	15	0.567	0.514	0.417	−0.229
25	0	0.374	0.342	0.278	−0.231
	5	0.381	0.348	0.283	−0.233
	10	0.402	0.366	0.296	−0.239
	15	0.443	0.401	0.321	−0.250
30	0	0.305	0.276	0.218	−0.244
	5	0.309	0.280	0.221	−0.246
	10	0.323	0.292	0.230	−0.252
	15	0.350	0.315	0.246	−0.263

Example 7.3

For the retaining wall shown in Figure 7.8, $H = 7.5$ m, $\gamma = 18$ kN/m^3, $\phi' = 20°$, $c' = 13.5$ kN/m^2, and $\alpha = 10°$. Calculate the Rankine active force, P_a, per unit length of the wall and the location of the resultant force after the occurrence of the tensile crack.

Solution
From Eq. (7.22),

$$z_c = \frac{2c'}{\gamma} \sqrt{\frac{1 + \sin \phi'}{1 - \sin \phi'}} = \frac{(2)(13.5)}{18} \sqrt{\frac{1 + \sin 20}{1 - \sin 20}} = 2.14 \text{ m}$$

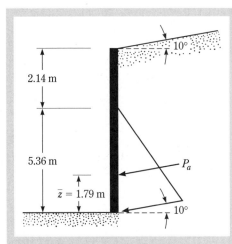

Figure 7.9 Calculation of Rankine active force, c'–ϕ' soil

At $z = 7.5$ m,

$$\frac{c'}{\gamma z} = \frac{13.5}{(18)(7.5)} = 0.1$$

From Table 7.2, for $\phi' = 20°$, $c'/\gamma z = 0.1$, and $\alpha = 10°$, the value of K_a' is 0.377, so at $z = 7.5$ m,

$$\sigma_a' = \gamma z K_a' \cos \alpha = (18)(7.5)(0.377)(\cos 10) = 50.1 \text{ kN/m}^2$$

After the occurrence of the tensile crack, the pressure distribution on the wall will be as shown in Figure 7.9, so

$$P_a = \left(\frac{1}{2}\right)(50.1)(7.5 - 2.14) = \textbf{134.3 kN/m}$$

and

$$\bar{z} = \frac{7.5 - 2.14}{3} = \textbf{1.79 m}$$ ∎

7.5   *Coulomb's Active Earth Pressure*

The Rankine active earth pressure calculations discussed in the preceding sections were based on the assumption that the wall is frictionless. In 1776, Coulomb proposed a theory for calculating the lateral earth pressure on a retaining wall with granular soil backfill. This theory takes wall friction into consideration.

To apply Coulomb's active earth pressure theory, let us consider a retaining wall with its back face inclined at an angle β with the horizontal, as shown in Figure 7.10a. The backfill is a granular soil that slopes at an angle α with the horizontal. Also, let δ be the angle of friction between the soil and the wall (i.e., the angle of wall friction).

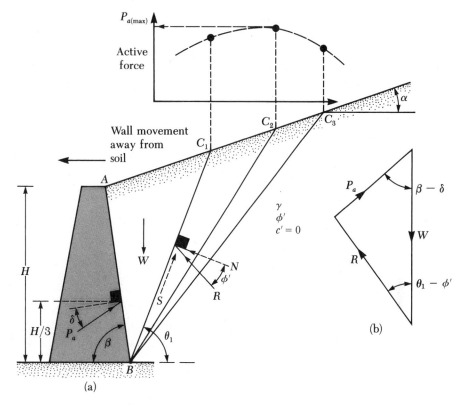

Figure 7.10 Coulomb's active pressure

Under active pressure, the wall will move away from the soil mass (to the left in the figure). Coulomb assumed that, in such a case, the failure surface in the soil mass would be a plane (e.g., $BC_1, BC_2, \dots$). So, to find the active force, consider a possible soil failure wedge ABC_1. The forces acting on this wedge (per unit length at right angles to the cross section shown) are as follows:

1. The weight of the wedge, W.
2. The resultant, R, of the normal and resisting shear forces along the surface, BC_1. The force R will be inclined at an angle ϕ' to the normal drawn to BC_1.
3. The active force per unit length of the wall, P_a, which will be inclined at an angle δ to the normal drawn to the back face of the wall.

For equilibrium purposes, a force triangle can be drawn, as shown in Figure 7.10b. Note that θ_1 is the angle that BC_1 makes with the horizontal. Because the magnitude of W, as well as the directions of all three forces, are known, the value of P_a can now be determined. Similarly, the active forces of other trial wedges, such as $ABC_2, ABC_3, \dots$, can be determined. The maximum value of P_a thus determined is Coulomb's active force (see top part of Figure 7.10), which may be expressed as

$$P_a = \tfrac{1}{2}K_a\gamma H^2 \tag{7.23}$$

Table 7.3   Values of K_a [Eq. (7.24)] for $\beta = 90°$ and $\alpha = 0°$

ϕ' (deg)	δ (deg)					
	0	5	10	15	20	25
28	0.3610	0.3448	0.3330	0.3251	0.3203	0.3186
30	0.3333	0.3189	0.3085	0.3014	0.2973	0.2956
32	0.3073	0.2945	0.2853	0.2791	0.2755	0.2745
34	0.2827	0.2714	0.2633	0.2579	0.2549	0.2542
36	0.2596	0.2497	0.2426	0.2379	0.2354	0.2350
38	0.2379	0.2292	0.2230	0.2190	0.2169	0.2167
40	0.2174	0.2098	0.2045	0.2011	0.1994	0.1995
42	0.1982	0.1916	0.1870	0.1841	0.1828	0.1831

where

$$K_a = \text{Coulomb's active earth pressure coefficient}$$

$$= \frac{\sin^2(\beta + \phi')}{\sin^2 \beta \sin(\beta-\delta)\left[1 + \sqrt{\dfrac{\sin(\phi' + \delta)\sin(\phi'-\alpha)}{\sin(\beta-\delta)\sin(\alpha + \beta)}}\right]^2} \qquad (7.24)$$

and $H = $ height of the wall

The values of the active earth pressure coefficient, K_a, for a vertical retaining wall ($\beta = 90°$) with horizontal backfill ($\alpha = 0°$) are given in Table 7.3. Note that the line of action of the resultant force (P_a) will act at a distance $H/3$ above the base of the wall and will be inclined at an angle δ to the normal drawn to the back of the wall.

In the actual design of retaining walls, the value of the wall friction angle δ is assumed to be between $\phi'/2$ and $\frac{2}{3}\phi'$. The active earth pressure coefficients for various values of ϕ', α, and β with $\delta = \frac{1}{2}\phi'$ and $\frac{2}{3}\phi'$ are respectively given in Tables 7.4 and 7.5. These coefficients are very useful design considerations.

7.6   *Active Pressure for Wall Rotation about the Top: Braced Cut*

In the preceding sections, we have seen that a retaining wall rotates about its bottom. (See Figure 7.11a.) With sufficient yielding of the wall, the lateral earth pressure is approximately equal to that obtained by Rankine's theory or Coulomb's theory. In contrast to retaining walls, braced cuts show a different type of wall yielding. (See Figure 7.11b.) In this case, deformation of the wall gradually increases with the depth of excavation. The variation of the amount of deformation depends on several factors, such as the type of soil, the depth of excavation, and the workmanship involved. However, with very little wall yielding at the top of the cut, the lateral earth pressure will be close to the at-rest pressure. At the bottom of the wall, with a much larger degree of

Table 7.4 Values of K_a [from Eq. (7.24)] for $\delta = \frac{2}{3}\phi'$

α (deg)	ϕ' (deg)	β (deg)					
		90	85	80	75	70	65
0	28	0.3213	0.3588	0.4007	0.4481	0.5026	0.5662
	29	0.3091	0.3467	0.3886	0.4362	0.4908	0.5547
	30	0.2973	0.3349	0.3769	0.4245	0.4794	0.5435
	31	0.2860	0.3235	0.3655	0.4133	0.4682	0.5326
	32	0.2750	0.3125	0.3545	0.4023	0.4574	0.5220
	33	0.2645	0.3019	0.3439	0.3917	0.4469	0.5117
	34	0.2543	0.2916	0.3335	0.3813	0.4367	0.5017
	35	0.2444	0.2816	0.3235	0.3713	0.4267	0.4919
	36	0.2349	0.2719	0.3137	0.3615	0.4170	0.4824
	37	0.2257	0.2626	0.3042	0.3520	0.4075	0.4732
	38	0.2168	0.2535	0.2950	0.3427	0.3983	0.4641
	39	0.2082	0.2447	0.2861	0.3337	0.3894	0.4553
	40	0.1998	0.2361	0.2774	0.3249	0.3806	0.4468
	41	0.1918	0.2278	0.2689	0.3164	0.3721	0.4384
	42	0.1840	0.2197	0.2606	0.3080	0.3637	0.4302
5	28	0.3431	0.3845	0.4311	0.4843	0.5461	0.6190
	29	0.3295	0.3709	0.4175	0.4707	0.5325	0.6056
	30	0.3165	0.3578	0.4043	0.4575	0.5194	0.5926
	31	0.3039	0.3451	0.3916	0.4447	0.5067	0.5800
	32	0.2919	0.3329	0.3792	0.4324	0.4943	0.5677
	33	0.2803	0.3211	0.3673	0.4204	0.4823	0.5558
	34	0.2691	0.3097	0.3558	0.4088	0.4707	0.5443
	35	0.2583	0.2987	0.3446	0.3975	0.4594	0.5330
	36	0.2479	0.2881	0.3338	0.3866	0.4484	0.5221
	37	0.2379	0.2778	0.3233	0.3759	0.4377	0.5115
	38	0.2282	0.2679	0.3131	0.3656	0.4273	0.5012
	39	0.2188	0.2582	0.3033	0.3556	0.4172	0.4911
	40	0.2098	0.2489	0.2937	0.3458	0.4074	0.4813
	41	0.2011	0.2398	0.2844	0.3363	0.3978	0.4718
	42	0.1927	0.2311	0.2753	0.3271	0.3884	0.4625
10	28	0.3702	0.4164	0.4686	0.5287	0.5992	0.6834
	29	0.3548	0.4007	0.4528	0.5128	0.5831	0.6672
	30	0.3400	0.3857	0.4376	0.4974	0.5676	0.6516
	31	0.3259	0.3713	0.4230	0.4826	0.5526	0.6365
	32	0.3123	0.3575	0.4089	0.4683	0.5382	0.6219
	33	0.2993	0.3442	0.3953	0.4545	0.5242	0.6078
	34	0.2868	0.3314	0.3822	0.4412	0.5107	0.5942
	35	0.2748	0.3190	0.3696	0.4283	0.4976	0.5810
	36	0.2633	0.3072	0.3574	0.4158	0.4849	0.5682
	37	0.2522	0.2957	0.3456	0.4037	0.4726	0.5558
	38	0.2415	0.2846	0.3342	0.3920	0.4607	0.5437
	39	0.2313	0.2740	0.3231	0.3807	0.4491	0.5321
	40	0.2214	0.2636	0.3125	0.3697	0.4379	0.5207
	41	0.2119	0.2537	0.3021	0.3590	0.4270	0.5097
	42	0.2027	0.2441	0.2921	0.3487	0.4164	0.4990
15	28	0.4065	0.4585	0.5179	0.5868	0.6685	0.7670

(Continued)

Table 7.4 (Continued)

α (deg)	ϕ' (deg)	β (deg) 90	85	80	75	70	65
	29	0.3881	0.4397	0.4987	0.5672	0.6483	0.7463
	30	0.3707	0.4219	0.4804	0.5484	0.6291	0.7265
	31	0.3541	0.4049	0.4629	0.5305	0.6106	0.7076
	32	0.3384	0.3887	0.4462	0.5133	0.5930	0.6895
	33	0.3234	0.3732	0.4303	0.4969	0.5761	0.6721
	34	0.3091	0.3583	0.4150	0.4811	0.5598	0.6554
	35	0.2954	0.3442	0.4003	0.4659	0.5442	0.6393
	36	0.2823	0.3306	0.3862	0.4513	0.5291	0.6238
	37	0.2698	0.3175	0.3726	0.4373	0.5146	0.6089
	38	0.2578	0.3050	0.3595	0.4237	0.5006	0.5945
	39	0.2463	0.2929	0.3470	0.4106	0.4871	0.5805
	40	0.2353	0.2813	0.3348	0.3980	0.4740	0.5671
	41	0.2247	0.2702	0.3231	0.3858	0.4613	0.5541
	42	0.2146	0.2594	0.3118	0.3740	0.4491	0.5415
20	28	0.4602	0.5205	0.5900	0.6714	0.7689	0.8880
	29	0.4364	0.4958	0.5642	0.6445	0.7406	0.8581
	30	0.4142	0.4728	0.5403	0.6195	0.7144	0.8303
	31	0.3935	0.4513	0.5179	0.5961	0.6898	0.8043
	32	0.3742	0.4311	0.4968	0.5741	0.6666	0.7799
	33	0.3559	0.4121	0.4769	0.5532	0.6448	0.7569
	34	0.3388	0.3941	0.4581	0.5335	0.6241	0.7351
	35	0.3225	0.3771	0.4402	0.5148	0.6044	0.7144
	36	0.3071	0.3609	0.4233	0.4969	0.5856	0.6947
	37	0.2925	0.3455	0.4071	0.4799	0.5677	0.6759
	38	0.2787	0.3308	0.3916	0.4636	0.5506	0.6579
	39	0.2654	0.3168	0.3768	0.4480	0.5342	0.6407
	40	0.2529	0.3034	0.3626	0.4331	0.5185	0.6242
	41	0.2408	0.2906	0.3490	0.4187	0.5033	0.6083
	42	0.2294	0.2784	0.3360	0.4049	0.4888	0.5930

yielding, the lateral earth pressure will be substantially lower than the Rankine active earth pressure. As a result, the distribution of lateral earth pressure will vary substantially in comparison to the linear distribution assumed in the case of retaining walls.

The total lateral force per unit length of the wall, P_a, imposed on a wall may be evaluated theoretically by using Terzaghi's (1943) general wedge theory. (See Figure 7.12.) The failure surface is assumed to be the arc of a logarithmic spiral, defined as

$$r = r_o e^{\theta \tan \phi'} \tag{7.25}$$

where ϕ' = effective angle of friction of soil

In the figure, H is the height of the cut, and the unit weight, angle of friction, and cohesion of the soil are equal to γ, ϕ', and c', respectively. Following are the forces per unit length of the cut acting on the trial failure wedge:

1. Weight of the wedge, W
2. Resultant of the normal and shear forces along ab, R

Table 7.5 Values of K_a [from Eq. (7.24)] for $\delta = \phi'/2$

α (deg)	ϕ' (deg)	β (deg)					
		90	**85**	**80**	**75**	**70**	**65**
0	28	0.3264	0.3629	0.4034	0.4490	0.5011	0.5616
	29	0.3137	0.3502	0.3907	0.4363	0.4886	0.5492
	30	0.3014	0.3379	0.3784	0.4241	0.4764	0.5371
	31	0.2896	0.3260	0.3665	0.4121	0.4645	0.5253
	32	0.2782	0.3145	0.3549	0.4005	0.4529	0.5137
	33	0.2671	0.3033	0.3436	0.3892	0.4415	0.5025
	34	0.2564	0.2925	0.3327	0.3782	0.4305	0.4915
	35	0.2461	0.2820	0.3221	0.3675	0.4197	0.4807
	36	0.2362	0.2718	0.3118	0.3571	0.4092	0.4702
	37	0.2265	0.2620	0.3017	0.3469	0.3990	0.4599
	38	0.2172	0.2524	0.2920	0.3370	0.3890	0.4498
	39	0.2081	0.2431	0.2825	0.3273	0.3792	0.4400
	40	0.1994	0.2341	0.2732	0.3179	0.3696	0.4304
	41	0.1909	0.2253	0.2642	0.3087	0.3602	0.4209
	42	0.1828	0.2168	0.2554	0.2997	0.3511	0.4177
5	28	0.3477	0.3879	0.4327	0.4837	0.5425	0.6115
	29	0.3337	0.3737	0.4185	0.4694	0.5282	0.5972
	30	0.3202	0.3601	0.4048	0.4556	0.5144	0.5833
	31	0.3072	0.3470	0.3915	0.4422	0.5009	0.5698
	32	0.2946	0.3342	0.3787	0.4292	0.4878	0.5566
	33	0.2825	0.3219	0.3662	0.4166	0.4750	0.5437
	34	0.2709	0.3101	0.3541	0.4043	0.4626	0.5312
	35	0.2596	0.2986	0.3424	0.3924	0.4505	0.5190
	36	0.2488	0.2874	0.3310	0.3808	0.4387	0.5070
	37	0.2383	0.2767	0.3199	0.3695	0.4272	0.4954
	38	0.2282	0.2662	0.3092	0.3585	0.4160	0.4840
	39	0.2185	0.2561	0.2988	0.3478	0.4050	0.4729
	40	0.2090	0.2463	0.2887	0.3374	0.3944	0.4620
	41	0.1999	0.2368	0.2788	0.3273	0.3840	0.4514
	42	0.1911	0.2276	0.2693	0.3174	0.3738	0.4410
10	28	0.3743	0.4187	0.4688	0.5261	0.5928	0.6719
	29	0.3584	0.4026	0.4525	0.5096	0.5761	0.6549
	30	0.3432	0.3872	0.4368	0.4936	0.5599	0.6385
	31	0.3286	0.3723	0.4217	0.4782	0.5442	0.6225
	32	0.3145	0.3580	0.4071	0.4633	0.5290	0.6071
	33	0.3011	0.3442	0.3930	0.4489	0.5143	0.5920
	34	0.2881	0.3309	0.3793	0.4350	0.5000	0.5775
	35	0.2757	0.3181	0.3662	0.4215	0.4862	0.5633
	36	0.2637	0.3058	0.3534	0.4084	0.4727	0.5495
	37	0.2522	0.2938	0.3411	0.3957	0.4597	0.5361
	38	0.2412	0.2823	0.3292	0.3833	0.4470	0.5230
	39	0.2305	0.2712	0.3176	0.3714	0.4346	0.5103
	40	0.2202	0.2604	0.3064	0.3597	0.4226	0.4979
	41	0.2103	0.2500	0.2956	0.3484	0.4109	0.4858
	42	0.2007	0.2400	0.2850	0.3375	0.3995	0.4740
15	28	0.4095	0.4594	0.5159	0.5812	0.6579	0.7498

(Continued)

Table 7.5 (Continued)

α (deg)	φ' (deg)	β (deg) 90	85	80	75	70	65
	29	0.3908	0.4402	0.4964	0.5611	0.6373	0.7284
	30	0.3730	0.4220	0.4777	0.5419	0.6175	0.7080
	31	0.3560	0.4046	0.4598	0.5235	0.5985	0.6884
	32	0.3398	0.3880	0.4427	0.5059	0.5803	0.6695
	33	0.3244	0.3721	0.4262	0.4889	0.5627	0.6513
	34	0.3097	0.3568	0.4105	0.4726	0.5458	0.6338
	35	0.2956	0.3422	0.3953	0.4569	0.5295	0.6168
	36	0.2821	0.3282	0.3807	0.4417	0.5138	0.6004
	37	0.2692	0.3147	0.3667	0.4271	0.4985	0.5846
	38	0.2569	0.3017	0.3531	0.4130	0.4838	0.5692
	39	0.2450	0.2893	0.3401	0.3993	0.4695	0.5543
	40	0.2336	0.2773	0.3275	0.3861	0.4557	0.5399
	41	0.2227	0.2657	0.3153	0.3733	0.4423	0.5258
	42	0.2122	0.2546	0.3035	0.3609	0.4293	0.5122
20	28	0.4614	0.5188	0.5844	0.6608	0.7514	0.8613
	29	0.4374	0.4940	0.5586	0.6339	0.7232	0.8313
	30	0.4150	0.4708	0.5345	0.6087	0.6968	0.8034
	31	0.3941	0.4491	0.5119	0.5851	0.6720	0.7772
	32	0.3744	0.4286	0.4906	0.5628	0.6486	0.7524
	33	0.3559	0.4093	0.4704	0.5417	0.6264	0.7289
	34	0.3384	0.3910	0.4513	0.5216	0.6052	0.7066
	35	0.3218	0.3736	0.4331	0.5025	0.5851	0.6853
	36	0.3061	0.3571	0.4157	0.4842	0.5658	0.6649
	37	0.2911	0.3413	0.3991	0.4668	0.5474	0.6453
	38	0.2769	0.3263	0.3833	0.4500	0.5297	0.6266
	39	0.2633	0.3120	0.3681	0.4340	0.5127	0.6085
	40	0.2504	0.2982	0.3535	0.4185	0.4963	0.5912
	41	0.2381	0.2851	0.3395	0.4037	0.4805	0.5744
	42	0.2263	0.2725	0.3261	0.3894	0.4653	0.5582

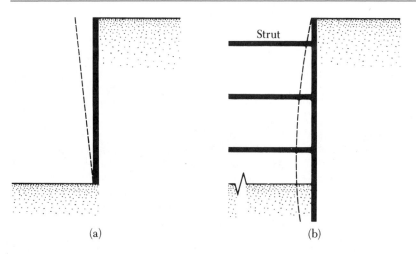

Figure 7.11 Nature of yielding of walls: (a) retaining wall; (b) braced cut

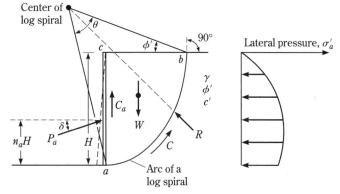

Figure 7.12 Braced cut analysis by general wedge theory: wall rotation about top

3. Cohesive force along *ab*, *C*
4. Adhesive force along *ac*, C_a
5. P_a, which is the force acting a distance $n_a H$ from the bottom of the wall and is inclined at an angle δ to the horizontal

The adhesive force is

$$C_a = c_a' H \qquad (7.26)$$

where c_a' = unit adhesion

A detailed outline for the evaluation of P_a is beyond the scope of this text; those interested should check a soil mechanics text for more information (e.g., Das, 1998). Kim and Preber (1969) provided tabulated values of $P_a/\frac{1}{2}\gamma H^2$ determined by using the principles of general wedge theory, and these values are given in Table 7.6. In developing the theoretical values for that table, it was assumed that

$$\frac{c_a'}{c'} = \frac{\tan \delta}{\tan \phi'} \qquad (7.27)$$

Passive Pressure

7.7 *Rankine Passive Earth Pressure*

Figure 7.13a shows a vertical frictionless retaining wall with a horizontal backfill. At depth z, the effective vertical pressure on a soil element is $\sigma_o' = \gamma z$. Initially, if the wall does not yield at all, the lateral stress at that depth will be $\sigma_h' = K_o \sigma_o'$. This state of stress is illustrated by the Mohr's circle *a* in Figure 7.13b. Now, if the wall is pushed into the soil mass by an amount Δx, as shown in Figure 7.13a, the vertical stress at depth z will stay the same; however, the horizontal stress will increase. Thus, σ_h will be greater than $K_o \sigma_o'$. The state of stress can now be represented by the Mohr's circle *b* in Figure 7.13b. If the wall moves farther inward (i.e., Δx is increased still more), the stresses at depth z will ultimately reach the state represented by

Table 7.6 $P_a/[(\frac{1}{2})\gamma H^2]$ versus ϕ', δ, n_a, and $c'/\gamma H$*

ϕ', in degrees (1)	δ', in degrees (2)	$n_a = 0.3$ $c'/\gamma H$ 0 (3)	0.1 (4)	0.2 (5)	$n_a = 0.4$ $c'/\gamma H$ 0 (6)	0.1 (7)	0.2 (8)	$n_a = 0.5$ $c'/\gamma H$ 0 (9)	0.1 (10)	0.2 (11)	$n_a = 0.6$ $c'/\gamma H$ 0 (12)	0.1 (13)	0.2 (14)
0	0	0.952	0.558	0.164	—	0.652	0.192	—	0.782	0.230	—	0.978	0.288
5	0	0.787	0.431	0.076	0.899	0.495	0.092	1.050	0.580	0.110	1.261	0.697	0.134
	5	0.756	0.345	−0.066	0.863	0.399	−0.064	1.006	0.474	−0.058	1.209	0.573	−0.063
10	0	0.653	0.334	0.015	0.734	0.378	0.021	0.840	0.434	0.027	0.983	0.507	0.032
	5	0.623	0.274	−0.074	0.700	0.312	−0.077	0.799	0.358	−0.082	0.933	0.420	−0.093
	10	0.610	0.242	−0.125	0.685	0.277	−0.131	0.783	0.324	−0.135	0.916	0.380	−0.156
15	0	0.542	0.254	−0.033	0.602	0.285	−0.033	0.679	0.322	−0.034	0.778	0.370	−0.039
	5	0.518	0.214	−0.089	0.575	0.240	−0.094	0.646	0.270	−0.106	0.739	0.310	−0.118
	10	0.505	0.187	−0.131	0.559	0.210	−0.140	0.629	0.238	−0.153	0.719	0.273	−0.174
	15	0.499	0.169	−0.161	0.554	0.191	−0.171	0.623	0.218	−0.187	0.714	0.251	−0.212
20	0	0.499	0.191	−0.067	0.495	0.210	−0.074	0.551	0.236	−0.080	0.622	0.266	−0.090
	5	0.430	0.160	−0.110	0.473	0.179	−0.116	0.526	0.200	−0.126	0.593	0.225	−0.142
	10	0.419	0.140	−0.139	0.460	0.156	−0.149	0.511	0.173	−0.165	0.575	0.196	−0.184
	15	0.413	0.122	−0.169	0.454	0.137	−0.179	0.504	0.154	−0.195	0.568	0.174	−0.219
	20	0.413	0.113	−0.188	0.454	0.124	−0.206	0.504	0.140	−0.223	0.569	0.160	−0.250
25	0	0.371	0.138	−0.095	0.405	0.150	−0.104	0.447	0.167	−0.112	0.499	0.187	−0.125
	5	0.356	0.116	−0.125	0.389	0.128	−0.132	0.428	0.141	−0.146	0.477	0.158	−0.162
	10	0.347	0.099	−0.149	0.378	0.110	−0.158	0.416	0.122	−0.173	0.464	0.136	−0.192
	15	0.342	0.085	−0.172	0.373	0.095	−0.182	0.410	0.106	−0.198	0.457	0.118	−0.221
	20	0.341	0.074	−0.193	0.372	0.083	−0.205	0.409	0.093	−0.222	0.456	0.104	−0.248
	25	0.344	0.065	−0.215	0.375	0.074	−0.228	0.413	0.083	−0.247	0.461	0.093	−0.275
30	0	0.304	0.093	−0.117	0.330	0.103	−0.124	0.361	0.113	−0.136	0.400	0.125	−0.150
	5	0.293	0.078	−0.137	0.318	0.086	−0.145	0.347	0.094	−0.159	0.384	0.105	−0.175
	10	0.286	0.066	−0.154	0.310	0.073	−0.164	0.339	0.080	−0.179	0.374	0.088	−0.198
	15	0.282	0.056	−0.171	0.306	0.060	−0.185	0.334	0.067	−0.199	0.368	0.074	−0.220
	20	0.281	0.047	−0.188	0.305	0.051	−0.204	0.332	0.056	−0.220	0.367	0.062	−0.242
	25	0.284	0.036	−0.211	0.307	0.042	−0.223	0.335	0.047	−0.241	0.370	0.051	−0.267
	30	0.289	0.029	−0.230	0.313	0.033	−0.246	0.341	0.038	−0.265	0.377	0.042	−0.294
35	0	0.247	0.059	−0.129	0.267	0.064	−0.139	0.290	0.069	−0.151	0.318	0.076	−0.165
	5	0.239	0.047	−0.145	0.258	0.052	−0.154	0.280	0.057	−0.167	0.307	0.062	−0.183
	10	0.234	0.038	−0.157	0.252	0.041	−0.170	0.273	0.046	−0.182	0.300	0.050	−0.200
	15	0.231	0.030	−0.170	0.249	0.033	−0.183	0.270	0.035	−0.199	0.296	0.039	−0.218
	20	0.231	0.022	−0.187	0.248	0.025	−0.198	0.269	0.027	−0.215	0.295	0.030	−0.235
	25	0.232	0.015	−0.202	0.250	0.016	−0.218	0.271	0.019	−0.234	0.297	0.020	−0.256
	30	0.236	0.006	−0.224	0.254	0.008	−0.238	0.276	0.011	−0.255	0.302	0.011	−0.281
	35	0.243	0	−0.243	0.262	0.001	−0.260	0.284	0.002	−0.279	0.312	0.002	−0.307
40	0	0.198	0.030	−0.138	0.213	0.032	−0.148	0.230	0.036	−0.159	0.252	0.038	−0.175
	5	0.192	0.021	−0.150	0.206	0.024	−0.158	0.223	0.026	−0.171	0.244	0.029	−0.186
	10	0.189	0.015	−0.158	0.202	0.016	−0.170	0.219	0.018	−0.182	0.238	0.020	−0.199
	15	0.187	0.008	−0.171	0.200	0.010	−0.180	0.216	0.011	−0.195	0.236	0.012	−0.212
	20	0.187	0.003	−0.181	0.200	0.003	−0.195	0.216	0.004	−0.208	0.235	0.004	−0.227
	25	0.188	−0.005		0.202	−0.003		0.218	−0.003		0.237	−0.003	
	30	0.192	−0.010		0.205	−0.010		0.222	−0.011		0.241	−0.012	
	35	0.197	−0.018		0.211	−0.018		0.228	−0.018		0.248	−0.020	

Table 7.6 (Continued)

ϕ', in degrees (1)	δ', in degrees (2)	$n_a = 0.3$ $c'/\gamma H$ 0 (3)	0.1 (4)	0.2 (5)	$n_a = 0.4$ $c'/\gamma H$ 0 (6)	0.1 (7)	0.2 (8)	$n_a = 0.5$ $c'/\gamma H$ 0 (9)	0.1 (10)	0.2 (11)	$n_a = 0.6$ $c'/\gamma H$ 0 (12)	0.1 (13)	0.2 (14)
	40	0.205	−0.025		0.220	−0.025		0.237	−0.027		0.259	−0.030	
45	0	0.156	0.007	−0.142	0.167	0.008	−0.150	0.180	0.009	−0.162	0.196	0.010	−0.177
	5	0.152	0.002	−0.148	0.163	0.002	−0.158	0.175	0.002	−0.170	0.190	0.003	−0.185
	10	0.150	−0.003		0.160	−0.004		0.172	−0.003		0.187	−0.004	
	15	0.148	−0.009		0.159	−0.008		0.171	−0.009		0.185	−0.010	
	20	0.149	−0.013		0.159	−0.014		0.171	−0.014		0.185	−0.016	
	25	0.150	−0.018		0.160	−0.020		0.173	−0.020		0.187	−0.022	
	30	0.153	−0.025		0.164	−0.026		0.176	−0.026		0.190	−0.029	
	35	0.158	−0.030		0.168	−0.031		0.181	−0.034		0.196	−0.037	
	40	0.164	−0.038		0.175	−0.040		0.188	−0.042		0.204	−0.045	
	45	0.173	−0.046		0.184	−0.048		0.198	−0.052		0.215	−0.057	

*After Kim and Preber (1969)

Mohr's circle c. Note that this Mohr's circle touches the Mohr–Coulomb failure envelope, which implies that the soil behind the wall will fail by being pushed upward. The horizontal stress, σ'_o, at this point is referred to as the *Rankine passive pressure,* or $\sigma'_o = \sigma'_p$.

For Mohr's circle c in Figure 7.13b, the major principal stress is σ'_p, and the minor principal stress is σ'_o. Substituting these quantities into Eq. (1.74) yields

$$\sigma'_p = \sigma'_o \tan^2\left(45 + \frac{\phi'}{2}\right) + 2c'\tan\left(45 + \frac{\phi'}{2}\right) \tag{7.28}$$

Now, let

$$K_p = \text{Rankine passive earth pressure coefficient}$$

$$= \tan^2\left(45 + \frac{\phi'}{2}\right) \tag{7.29}$$

Then, from Eq. (7.28), we have

$$\sigma'_p = \sigma'_o K_p + 2c'\sqrt{K_p} \tag{7.30}$$

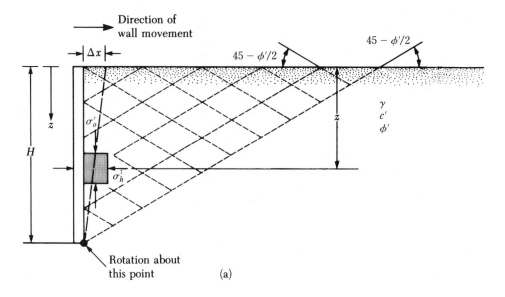

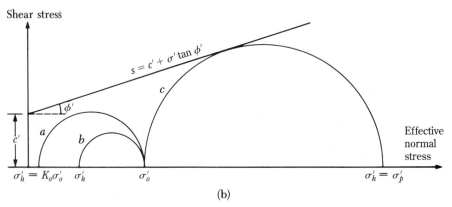

Figure 7.13   Rankine passive pressure

Equation (7.30) produces Figure 7.13c, the passive pressure diagram for the wall shown in Figure 7.13a. Note that at $z = 0$,

$$\sigma'_o = 0 \quad \text{and} \quad \sigma'_p = 2c'\sqrt{K_p}$$

and at $z = H$,

$$\sigma'_o = \gamma H \quad \text{and} \quad \sigma'_p = \gamma H K_p + 2c'\sqrt{K_p}$$

The passive force per unit length of the wall can be determined from the area of the pressure diagram, or

$$P_p = \tfrac{1}{2}\gamma H^2 K_p + 2c'H\sqrt{K_p} \tag{7.31}$$

The approximate magnitudes of the wall movements, Δx, required to develop failure under passive conditions are as follows:

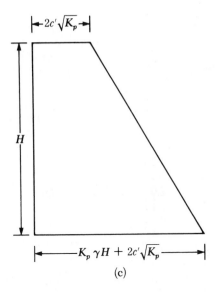

(c)

Figure 7.13 *(continued)*

Soil type	Wall movement for passive condition, Δx
Dense sand	$0.005H$
Loose sand	$0.01H$
Stiff clay	$0.01H$
Soft clay	$0.05H$

If the backfill behind the wall is a granular soil (i.e., $c' = 0$), then, from Eq. (7.31), the passive force per unit length of the wall will be

$$P_p = \frac{1}{2}\gamma H^2 K_p \tag{7.32}$$

The passive force on a *frictionless inclined* retaining wall (see Figure 7.14) with a horizontal granular backfill ($c' = 0$) can also be expressed by Eq. (7.32). The variation

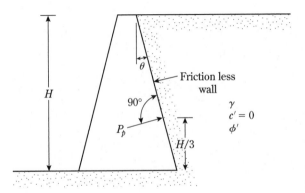

Figure 7.14 Passive force on a frictionless inclined retaining wall

Table 7.7   Variation of K_p [see Eq. (7.32) and Figure 7.14]*

ϕ' (deg)	θ (deg)						
	30	**25**	**20**	**15**	**10**	**5**	**0**
20	1.70	1.69	1.72	1.77	1.83	1.92	2.04
21	1.74	1.73	1.76	1.81	1.89	1.99	2.12
22	1.77	1.77	1.80	1.87	1.95	2.06	2.20
23	1.81	1.81	1.85	1.92	2.01	2.13	2.28
24	1.84	1.85	1.90	1.97	2.07	2.21	2.37
25	1.88	1.89	1.95	2.03	2.14	2.28	2.46
26	1.91	1.93	1.99	2.09	2.21	2.36	2.56
27	1.95	1.98	2.05	2.15	2.28	2.45	2.66
28	1.99	2.02	2.10	2.21	2.35	2.54	2.77
29	2.03	2.07	2.15	2.27	2.43	2.63	2.88
30	2.07	2.11	2.21	2.34	2.51	2.73	3.00
31	2.11	2.16	2.27	2.41	2.60	2.83	3.12
32	2.15	2.21	2.33	2.48	2.68	2.93	3.25
33	2.20	2.26	2.39	2.56	2.77	3.04	3.39
34	2.24	2.32	2.45	2.64	2.87	3.16	3.53
35	2.29	2.37	2.52	2.72	2.97	3.28	3.68
36	2.33	2.43	2.59	2.80	3.07	3.41	3.84
37	2.38	2.49	2.66	2.89	3.18	3.55	4.01
38	2.43	2.55	2.73	2.98	3.29	3.69	4.19
39	2.48	2.61	2.81	3.07	3.41	3.84	4.38
40	2.53	2.67	2.89	3.17	3.53	4.00	4.59
41	2.59	2.74	2.97	3.27	3.66	4.16	4.80
42	2.64	2.80	3.05	3.38	3.80	4.34	5.03
43	2.70	2.88	3.14	3.49	3.94	4.52	5.27
44	2.76	2.94	3.23	3.61	4.09	4.72	5.53
45	2.82	3.02	3.32	3.73	4.25	4.92	5.80

*Based on Zhu and Qian, 2000

of K_p for this case, with wall inclination θ and effective soil friction angle ϕ', is given in Table 7.7 (Zhu and Qian, 2000).

Example 7.4

A 12-ft-high wall is shown in Figure 7.15a. Determine the Rankine passive force per unit length of the wall.

Solution
For the top layer of soil,

$$K_{p(1)} = \tan^2\left(45 + \frac{\phi_1'}{2}\right) = \tan^2(45 + 15) = 3$$

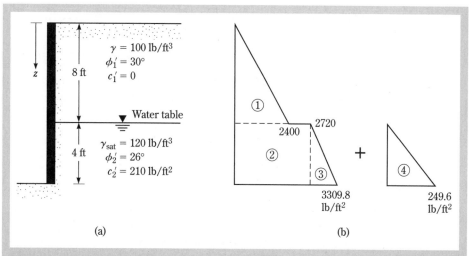

Figure 7.15 Rankine passive pressure on a retaining wall

From the bottom layer,

$$K_{p(2)} = \tan^2\left(45 + \frac{\phi_2'}{2}\right) = \tan^2(45 + 13) = 2.56$$

Now the following table can be prepared:

Depth, z (ft)	σ_o' (lb/ft²)	K_p	$\sigma_o' K_p$ (lb/ft²)	c' (lb/ft²)	$2c'\sqrt{K_p}$ (lb/ft²)	σ_p' [a] (lb/ft²)	u (lb/ft²)
0	0	3	0	0	0	0	0
8⁻	(100)(8) = 800	3	2400	0	0	2400	0
8⁺	800	2.56	2048	210	672	2720	0
12	800 + (120 − 62.4)(4) = 1030.4	2.56	2637.8	210	672	3309.8	(4)(62.4) = 249.6

[a] $\sigma_p' = \sigma_o' K_p + 2c'\sqrt{K_p}$, where σ_o' = effective vertical stress

The passive pressure diagram is plotted in Figure 7.15b. The passive force per unit length of the wall can be determined from the area of the pressure diagram as follows:

Area no.	Area
1	$(\frac{1}{2})(8)(2400) = 9,600$
2	$(2720)(4) = 10,880$
3	$(\frac{1}{2})(4)(3309.8 - 2720) = 1,179.6$
4	$(\frac{1}{2})(4)(249.6) = \underline{499.2}$
	$\approx \mathbf{22,159\ lb/ft}$

■

7.8	*Rankine Passive Earth Pressure: Inclined Backfill*

Granular Soil

For a frictionless vertical retaining wall (Figure 7.8) with a *granular backfill* ($c' = 0$), the Rankine passive pressure at any depth can be determined in a manner similar to that done in the case of active pressure in Section 7.4. The pressure is

$$\sigma_p' = \gamma z K_p \tag{7.33}$$

and the passive force is

$$P_p = \tfrac{1}{2}\gamma H^2 K_p \tag{7.34}$$

$$\text{where}\quad K_p = \cos\alpha\frac{\cos\alpha + \sqrt{\cos^2\alpha - \cos^2\phi'}}{\cos\alpha - \sqrt{\cos^2\alpha - \cos^2\phi'}} \tag{7.35}$$

As in the case of the active force, the resultant force, P_p, is inclined at an angle α with the horizontal and intersects the wall at a distance $H/3$ from the bottom of the wall. The values of K_p (the passive earth pressure coefficient) for various values of α and ϕ' are given in Table 7.8.

c'−ϕ' Soil

If the backfill of the frictionless vertical retaining wall is a $c'-\phi'$ soil (see Figure 7.8), then (Mazindrani and Ganjali, 1997)

$$\sigma_a' = \gamma z K_p = \gamma z K_p' \cos\alpha \tag{7.36}$$

where

$$K_p' = \frac{1}{\cos^2\phi'}\left\{ \frac{2\cos^2\alpha + 2\left(\dfrac{c'}{\gamma z}\right)\cos\phi'\sin\phi'}{+\sqrt{4\cos^2\alpha(\cos^2\alpha - \cos^2\phi') + 4\left(\dfrac{c'}{\gamma z}\right)^2\cos^2\phi' + 8\left(\dfrac{c'}{\gamma z}\right)\cos^2\alpha\sin\phi'\cos\phi'}} \right\} - 1 \tag{7.37}$$

The variation of K_p' with ϕ', α, and $c'/\gamma z$ is given in Table 7.9 (Mazindrani and Ganjali, 1997).

Table 7.8 Passive Earth Pressure Coefficient, K_p [from Eq. 7.35)]

	ϕ' (deg) →						
↓α (deg)	28	30	32	34	36	38	40
0	2.770	3.000	3.255	3.537	3.852	4.204	4.599
5	2.715	2.943	3.196	3.476	3.788	4.136	4.527
10	2.551	2.775	3.022	3.295	3.598	3.937	4.316
15	2.284	2.502	2.740	3.003	3.293	3.615	3.977
20	1.918	2.132	2.362	2.612	2.886	3.189	3.526
25	1.434	1.664	1.894	2.135	2.394	2.676	2.987

Table 7.9 Values of K_p'

ϕ' (deg)	α (deg)	$c'/\gamma z$			
		0.025	0.050	0.100	0.500
15	0	1.764	1.829	1.959	3.002
	5	1.716	1.783	1.917	2.971
	10	1.564	1.641	1.788	2.880
	15	1.251	1.370	1.561	2.732
20	0	2.111	2.182	2.325	3.468
	5	2.067	2.140	2.285	3.435
	10	1.932	2.010	2.162	3.339
	15	1.696	1.786	1.956	3.183
25	0	2.542	2.621	2.778	4.034
	5	2.499	2.578	2.737	3.999
	10	2.368	2.450	2.614	3.895
	15	2.147	2.236	2.409	3.726
30	0	3.087	3.173	3.346	4.732
	5	3.042	3.129	3.303	4.674
	10	2.907	2.996	3.174	4.579
	15	2.684	2.777	2.961	4.394

7.9 *Coulomb's Passive Earth Pressure*

Coulomb (1776) also presented an analysis for determining the passive earth pressure (i.e., when the wall moves *into* the soil mass) for walls possessing friction (δ = angle of wall friction) and retaining a granular backfill material similar to that discussed in Section 7.5.

To understand the determination of Coulomb's passive force, P_p, consider the wall shown in Figure 7.16a. As in the case of active pressure, Coulomb assumed that the potential failure surface in soil is a plane. For a trial failure wedge of soil, such as ABC_1, the forces per unit length of the wall acting on the wedge are

1. The weight of the wedge, W
2. The resultant, R, of the normal and shear forces on the plane BC_1, and
3. The passive force, P_p

Figure 7.16b shows the force triangle at equilibrium for the trial wedge ABC_1. From this force triangle, the value of P_p can be determined, because the direction of all three forces and the magnitude of one force are known.

Similar force triangles for several trial wedges, such as ABC_1, ABC_2, ABC_3, ..., can be constructed, and the corresponding values of P_p can be determined. The top part of Figure 7.16a shows the nature of variation of the P_p values for different wedges. The *minimum value of P_p in this diagram is Coulombs passive force,* mathematically expressed as

$$P_p = \tfrac{1}{2}\gamma H^2 K_p \qquad (7.38)$$

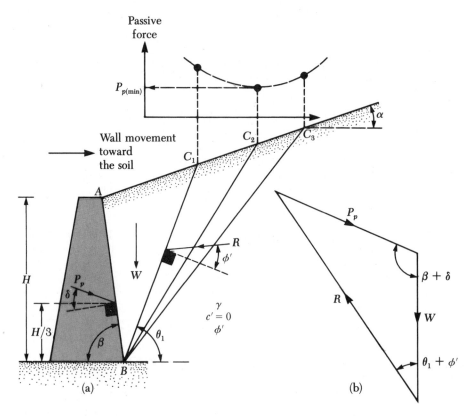

Figure 7.16 Coulomb's passive pressure

where

$$K_p = \text{Coulomb's passive pressure coefficient}$$

$$= \frac{\sin^2(\beta - \phi')}{\sin^2\beta \sin(\beta + \delta)\left[1 - \sqrt{\dfrac{\sin(\phi' + \delta)\sin(\phi' + \alpha)}{\sin(\beta + \delta)\sin(\beta + \alpha)}}\right]^2} \qquad (7.39)$$

The values of the passive pressure coefficient, K_p, for various values of ϕ' and δ are given in Table 7.10 ($\beta = 90°, \alpha = 0°$).

Note that the resultant passive force, P_p, will act at a distance $H/3$ from the bottom of the wall and will be inclined at an angle δ to the normal drawn to the back face of the wall.

Table 7.10 Values of K_p [from Eq. (7.39)] for $\beta = 90°$ and $\alpha = 0°$

ϕ' (deg)	δ (deg)				
	0	**5**	**10**	**15**	**20**
15	1.698	1.900	2.130	2.405	2.735
20	2.040	2.313	2.636	3.030	3.525
25	2.464	2.830	3.286	3.855	4.597
30	3.000	3.506	4.143	4.977	6.105
35	3.690	4.390	5.310	6.854	8.324
40	4.600	5.590	6.946	8.870	11.772

7.10 Comments on the Failure Surface Assumption for Coulomb's Pressure Calculations

Coulomb's pressure calculation methods for active and passive pressure have been discussed in Sections 7.5 and 7.9. The fundamental assumption in these analyses is the acceptance of *plane failure surface.* However, for walls with friction, this assumption does not hold in practice. The nature of *actual* failure surface in the soil mass for active and passive pressure is shown in Figure 7.17a and b, respectively (for a vertical wall with a horizontal backfill). Note that the failure surface *BC* is curved and that the failure surface *CD* is a plane.

Although the actual failure surface in soil for the case of active pressure is somewhat different from that assumed in the calculation of the Coulomb pressure, the results are not greatly different. However, in the case of passive pressure, as the value of δ increases, Coulomb's method of calculation gives increasingly erroneous values of P_p. This factor of error could lead to an unsafe condition because the values of P_p would become higher than the soil resistance.

Several studies have been conducted to determine the passive force P_p, assuming that the curved portion *BC* in Figure 7.17b is an arc of a circle, an ellipse, or a logarithmic spiral. The results of three of these studies are described briefly next.

Analysis of Caquot and Kerisel

Caquot and Kerisel (1948) developed the chart shown in Figure 7.18 for estimating the value of the passive pressure coefficient K_p with a curved failure surface in granular soil ($c' = 0$), such as that shown in Figure 7.17b. In their solution, the portion *BC* of the failure surface was assumed to be an arc of a logarithmic spiral. In using Figure 7.18, the following points should be kept in mind:

1. The curves are for $\phi' = \delta$.
2. If δ/ϕ' is less than unity, then

$$K_{p(\delta)} = RK_{p(\delta = \phi')}$$

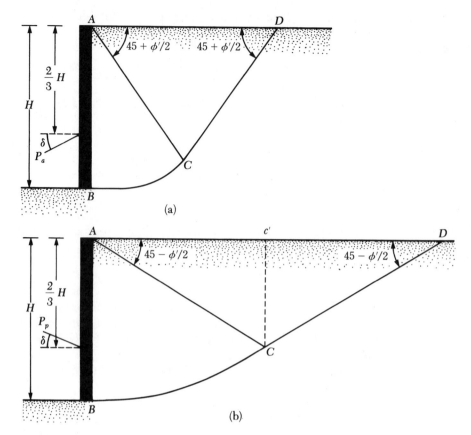

Figure 7.17 Nature of failure surface in soil with wall friction: (a) active pressure; (b) passive pressure

where R = reduction factor

The reduction factor R is given in Table 7.11.

3. The passive pressure is

$$P_p = \tfrac{1}{2}\gamma H^2 K_{p(\delta)}$$

Analysis of Shields and Tolunay

Shields and Tolunay (1973) analyzed the problem of passive pressure for a *vertical* wall with a *horizontal granular soil* backfill ($c' = 0$). This analysis was done by considering the stability of the wedge $ABCC'$ (see Figure 7.17b), using the *method of slices*. From Figure 7.17b, the passive force per unit length of the wall can be expressed as

$$P_p = \frac{1}{2} K_p \gamma H^2$$

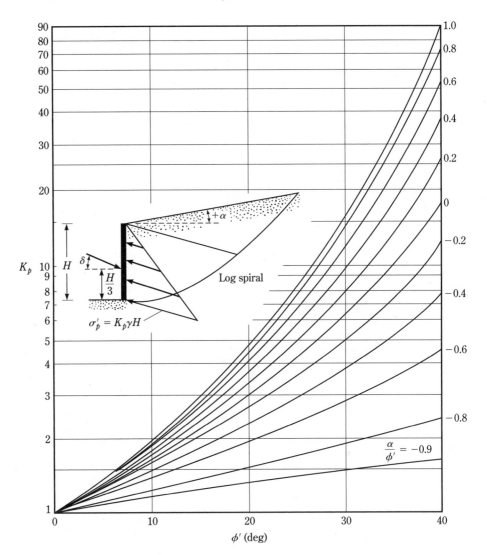

Figure 7.18   Caquot and Kerisel's (1948) passive pressure coefficient, K_p, for granular soil

The values of the passive earth pressure coefficient, K_p, obtained by Shields and Tolunay are given in Table 7.12.

Analysis of Zhu and Qian

Zhu and Qian (2000) determined the passive earth pressure coefficient K_p by using triangular slices within the framework of the limit equilibrium method as shown in Figure 7.19 for a *vertical retaining wall* with a *granular soil backfill*. According to this analysis,

$$P_p = \frac{1}{2}K_p \gamma H^2$$

Table 7.11 Reduction Factor, R, for Use in Conjunction with Figure 7.18

ϕ' (deg)	δ/ϕ							
	0.7	0.6	0.5	0.4	0.3	0.2	0.1	0.0
10	0.978	0.962	0.946	0.929	0.912	0.898	0.880	0.864
15	0.961	0.934	0.907	0.881	0.854	0.830	0.803	0.775
20	0.939	0.901	0.862	0.824	0.787	0.752	0.716	0.678
25	0.912	0.860	0.808	0.759	0.711	0.666	0.620	0.574
30	0.878	0.811	0.746	0.686	0.627	0.574	0.520	0.467
35	0.836	0.752	0.674	0.603	0.536	0.475	0.417	0.362
40	0.783	0.682	0.592	0.512	0.439	0.375	0.316	0.262
45	0.718	0.600	0.500	0.414	0.339	0.276	0.221	0.174

Table 7.12 Values of K_p from the Method of Slices

ϕ' (deg)	δ (deg)									
	0	5	10	15	20	25	30	35	40	45
20	2.04	2.26	2.43	2.55	2.70					
25	2.46	2.77	3.03	3.23	3.39	3.63				
30	3.00	3.43	3.80	4.13	4.40	4.64	5.03			
35	3.69	4.29	4.84	5.34	5.80	6.21	6.59	7.25		
40	4.69	5.44	6.26	7.05	7.80	8.51	9.18	9.83	11.03	
45	5.83	7.06	8.30	9.55	10.80	12.04	13.26	14.46	15.60	18.01

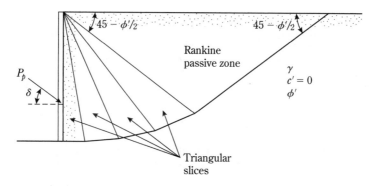

Figure 7.19 Passive earth pressure by method of triangular slices

where $\quad K_p = \eta \tan^2\left(45 + \dfrac{\phi'}{2}\right)$ (7.40)

The variation of η with ϕ' and δ/ϕ' is shown in Figure 7.20.

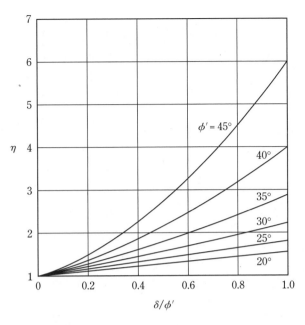

Figure 7.20 Variation of η with δ/ϕ and ϕ'

Example 7.5

A vertical retaining wall with a horizontal granular backfill has a height of 4 m. The unit weight of the backfill, γ, equals 16.5 kN/m^3, $\phi' = 35°$, and $\delta = 20°$. Determine the passive force per unit length of the wall, using

  a. Coulomb's theory,
  b. the analysis of Caquot and Kerisel,
  c. the analysis of Shields and Tolunay, and
  d. the analysis of Zhu and Qian.

Solution
Part a
From Eq. (7.38),

$$P_p = \frac{1}{2}\gamma H^2 K_p$$

For $\phi' = 35°$ and $\delta = 20°$, the value of K_p is 8.324. (See Table 7.10.) So

$$P_p = (1/2)(16.5)(4)^2(8.324) = \mathbf{1098.8 kN/m}$$

Part b
From Figure 7.18, for $\phi' = 35°$ and $\alpha = 0°$, the value of $K_{p(\delta=\phi')}$ is about 10. Hence,

$$\frac{\delta}{\phi'} = \frac{20}{35} = 0.57$$

From Table 7.11 for $\delta/\phi' = 0.57$ and $\phi' = 35°$, the reduction factor $R \approx 0.73$. Thus,

$$P_p = \frac{1}{2}\gamma H^2 K_{p(\delta)} = \frac{1}{2}(16.5)(4)^2(10 \times 0.73) = \textbf{963.5 kN/m}$$

Part c
From Table 7.12, for $\phi' = 35°$ and $\delta = 20°$, the magnitude of $K_p = 5.8$. Consequently,

$$P_p = \frac{1}{2}\gamma H^2 K_p = \frac{1}{2}(16.5)(4)^2(5.8) = \textbf{765.6 kN/m}$$

Part d
From Part a,

$$\frac{\delta}{\phi'} = \frac{20}{35} = 0.57$$

From Figure 7.20, for $\delta/\phi' = 0.57$ and $\phi' = 35°$, the magnitude of η is about 1.9. Thus, using Eq. (7.40), we obtain

$$P_p = \frac{1}{2}\eta \tan^2\left(45 + \frac{\phi'}{2}\right)\gamma H^2 = \frac{1}{2}(1.9)\tan^2\left(45 + \frac{20}{2}\right)(16.5)(4)^2 = \textbf{511.5 kN/m}$$

∎

Problems

7.1 In Figure 7.5a, let $H_1 = 3$ m, $H_2 = 3$ m, $q = 0$, $\gamma = 16.5$ kN/m³, $\gamma_{sat} = 19$ kN/m³, $c' = 0$, and $\phi' = 30°$. Determine the at-rest lateral earth force per meter length of the wall. Also, find the location of the resultant force. Use Eq. (7.3).

7.2 In Figure 7.5a, let $H_1 = 5$ m, $H_2 = 0$, $q = 0$, and $\gamma = 18$ kN/m³. The backfill is an overconsolidated clay with a plasticity index of 26. If the overconsolidation ratio is 2.0, determine the at-rest lateral earth force per meter length of the wall. Also, find the location of the resultant. Use Eqs. (7.5) and (7.7).

7.3 Redo Problem 7.2, assuming that the surcharge $q = 50$ kN/m².

7.4 Use Eq. (7.3), Figure 7.5, and the following values to determine the at-rest lateral earth force per unit length of the wall: $H = 10$ ft, $H_1 = 4$ ft, $H_2 = 6$ ft, $\gamma = 105$ lb/ft³, $\gamma_{sat} = 122$ lb/ft³, $\phi' = 30°$, $c' = 0$, and $q = 300$ lb/ft². Find the location of the resultant force.

7.5 A vertical retaining wall (Figure 7.6a) is 5.5 m high with a horizontal backfill. For the backfill, $\gamma = 18.7$ kN/m³, $\phi' = 24°$, and $c' = 10$ kN/m². Determine the Rankine active force per unit length of the wall after a tensile crack appears.

7.6 In Figure 7.6a, let the height of the retaining wall be $H = 21$ ft. The backfill is a saturated clay with $\phi = 0°$, $c = 630$ lb/ft², and $\gamma_{sat} = 113$ lb/ft³.
   a. Determine the Rankine active pressure distribution diagram behind the wall.
   b. Determine the depth of the tensile crack, z_c.
   c. Estimate the Rankine active force per foot length of the wall after the tensile crack appears.

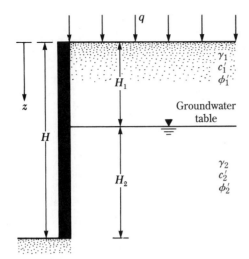

Figure P7.7

7.7 In Figure P7.7, let $H_1 = 8.2$ ft, $H_2 = 14.8$ ft, $\gamma_1 = 107$ lb/ft³, $q = 0$, $\phi_1' = 34°$, $c_1' = 0$, $\gamma_2 = 140$ lb/ft³, $\phi_2' = 25°$, and $c_2' = 209$ lb/ft². Determine the Rankine active force per unit length of the wall.

7.8 For the retaining wall of Figure 7.8, $H = 6$ m, $\phi' = 34°$, $\alpha = 10°$, $\gamma = 17$ kN/m³, and $c' = 0$.
   a. Determine the intensity of the Rankine active force at $z = 2, 4$, and 6 m.
   b. Determine the Rankine active force per meter length of the wall and also the location and direction of the resultant force.

7.9 In Figure 7.8, $H = 22$ ft, $\gamma = 115$ lb/ft³, $\phi' = 25°$, $c' = 250$ lb/ft², and $\alpha = 10°$. Calculate the Rankine active force per unit length of the wall after the tensile crack appears.

7.10 In Figure 7.10a, let $H = 6$ m, $\gamma = 16.5$ kN/m³, $\phi' = 30°$, $\delta = 20°$, $c' = 0$, $\alpha = 10°$, and $\beta = 80°$. Determine the Coulomb's active force per meter length of the wall and the location and direction of the resultant force.

7.11 In Figure 7.10a, let $H = 12$ ft, $\gamma = 105$ lb/ft³, $\phi' = 30°$, $c' = 0$, and $\beta = 85°$. Determine the Coulomb's active force per unit length of the wall and the location and direction of the resultant force for the following cases:
   a. $\alpha = 10°$ and $\delta = 20°$
   b. $\alpha = 20°$ and $\delta = 15°$

7.12 A retaining wall is shown in Figure P7.12. If the wall rotates about its top, determine the magnitude of the active force per unit length of the wall for $n_a = 0.3, 0.4$, and 0.5. Assume unit adhesion, $c_a' = c'(\tan \delta / \tan \phi')$.

7.13 With regard to Problem 7.6,
   a. Draw the Rankine passive pressure distribution diagram behind the wall.
   b. Estimate the Rankine passive force per foot length of the wall and also the location of the resultant force.

7.14 Repeat Problem 7.5 for the Rankine passive case.

7.15 Repeat Problem 7.7 for the Rankine passive case.

7.16 For the retaining wall described in Problem 7.10, determine the Coulomb's passive force per meter length of the wall and the location and direction of the resultant force.

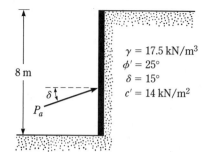

8 m

$\gamma = 17.5 \text{ kN/m}^3$
$\phi' = 25°$
$\delta = 15°$
$c' = 14 \text{ kN/m}^2$

δ

P_a

Figure P7.12

7.17 For the retaining wall with a granular backfill shown in Figure 7.18, let $H = 6 \text{ m}, \alpha = +10°, \phi' = 36°, \gamma = 15 \text{ kN/m}^3$, and $\delta/\phi' = 0.4$. Calculate the passive force per unit length of the wall, assuming a curved failure surface in the soil.

7.18 In Figure 7.17b, which shows a vertical retaining wall with a horizontal backfill, let $H = 13.5 \text{ ft}, \gamma = 105 \text{ lb/ft}^3, \phi' = 35°$, and $\delta = 10°$. Based on Shields and Tolunay's work (see Table 7.12), what would be the passive force per meter length of the wall?

7.19 Solve Problem 7.18, using Eq. (7.40) and Figure 7.20.

References

Brooker, E. W., and Ireland, H. O. (1965). "Earth Pressure at Rest Related to Stress History *Canadian Geotechnical Journal,* Vol. 2, No. 1, pp. 1–15.

Caquot, A., and Kerisel, J. (1948). *Tables for Calculation of Passive Pressure, Active Pressure and Bearing Capacity of Foundations,* Gauthier-Villars, Paris, France.

Coulomb, C. A. (1776). *Essai sur une Application des Règles de Maximis et Minimum à quelques Problemes de Statique Relatifs à l'Architecture,* Mem. Acad. Roy. des Sciences, Paris, Vol. 3, p. 38.

Das, B. M. (1998). *Principles of Geotechnical Engineering,* 4th ed., PWS Publishing Company, Boston.

Jaky, J. (1944). "The Coefficient of Earth Pressure at Rest," *Journal for the Society of Hungarian Architects and Engineers,* October, pp. 355–358.

Kim, J. S., and Preber, T. (1969). "Earth Pressure against Braced Excavations," *Journal of the Soil Mechanics and Foundations Division,* ASCE, Vol. 96, No. 6, pp. 1581–1584.

Mayne, P. W., and Kulhawy, F. H. (1982). "K_o–OCR Relationships in Soil," *Journal of the Geotechnical Engineering Division,* ASCE, Vol. 108, No. GT6, pp. 851–872.

Mazindrani, Z. H., and Ganjali, M. H. (1997). "Lateral Earth Pressure Problem of Cohesive Backfill with Inclined Surface," *Journal of Geotechnical and Geoenvironmental Engineering,* ASCE, Vol. 123, No. 2, pp. 110–112.

Shields, D. H., and Tolunay, A. Z. (1973). "Passive Pressure Coefficients by Method of Slices," *Journal of the Soil Mechanics and Foundations Division,* ASCE, Vol. 99, No. SM12, 1043–1053.

Terzaghi, K. (1943). *Theoretical Soil Mechanics,* Wiley, New York.

Zhu, D. Y., and Qian, Q. (2000). "Determination of Passive Earth Pressure Coefficient by the Method of Triangular Slices," *Canadian Geotechnical Journal,* Vol. 37, No. 2, pp. 485–491.

8

Retaining Walls

Introduction

In Chapter 7, you were introduced to various theories of lateral earth pressure. Those theories will be used in this chapter to design various types of retaining walls. In general, retaining walls can be divided into two major categories: (a) conventional retaining walls and (b) mechanically stabilized earth walls.

Conventional retaining walls can generally be classified into four varieties:

1. Gravity retaining walls
2. Semigravity retaining walls
3. Cantilever retaining walls
4. Counterfort retaining walls

Gravity retaining walls (Figure 8.1a) are constructed with plain concrete or stone masonry. They depend for stability on their own weight and any soil resting on the masonry. This type of construction is not economical for high walls.

In many cases, a small amount of steel may be used for the construction of gravity walls, thereby minimizing the size of wall sections. Such walls are generally referred to as *semigravity walls* (Figure 8.1b).

Cantilever retaining walls (Figure 8.1c) are made of reinforced concrete that consists of a thin stem and a base slab. This type of wall is economical to a height of about 8 m (25 ft).

Counterfort retaining walls (Figure 8.1d) are similar to cantilever walls. At regular intervals, however, they have thin vertical concrete slabs known as *counterforts* that tie the wall and the base slab together. The purpose of the counterforts is to reduce the shear and the bending moments.

To design retaining walls properly, an engineer must know the basic parameters—the *unit weight, angle of friction,* and *cohesion*—of the soil retained behind the wall and the soil below the base slab. Knowing the properties of the soil behind the wall enables the engineer to determine the lateral pressure distribution that has to be designed for.

There are two phases in the design of a conventional retaining wall. First, with the lateral earth pressure known, the structure as a whole is checked for *stability:* The

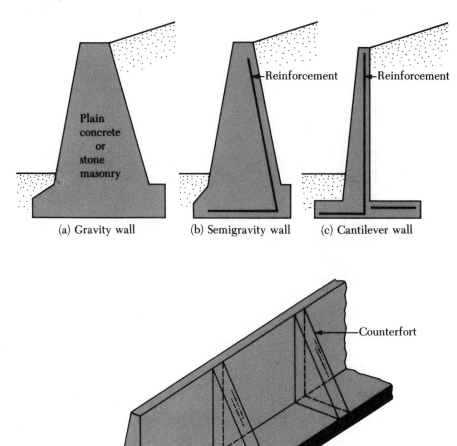

Figure 8.1   Types of retaining wall

structure is examined for possible *overturning, sliding,* and *bearing capacity* failures. Second, each component of the structure is checked for *strength,* and the *steel reinforcement* of each component is determined.

This chapter presents the procedures for determining the stability of the retaining wall. Checks for strength can be found in any textbook on reinforced concrete.

Some retaining walls have their backfills stabilized mechanically by including reinforcing elements such as metal strips, bars, welded wire mats, geotextiles, and geogrids. These walls are relatively flexible and can sustain large horizontal and vertical displacements without much damage.

Gravity and Cantilever Walls

8.2   *Proportioning Retaining Walls*

In designing retaining walls, an engineer must assume some of their dimensions. Called *proportioning,* such assumptions allow the engineer to check trial sections of the walls for stability. If the stability checks yield undesirable results, the sections can be changed and rechecked. Figure 8.2 shows the general proportions of various retaining-wall components that can be used for initial checks.

Note that the top of the stem of any retaining wall should not be less than about 0.3 m. ($\approx$12 in.) for proper placement of concrete. The depth, D, to the bottom of the base slab should be a minimum of 0.6 m ($\approx$2 ft). However, the bottom of the base slab should be positioned below the seasonal frost line.

For counterfort retaining walls, the general proportion of the stem and the base slab is the same as for cantilever walls. However, the counterfort slabs may be about 0.3 m ($\approx$12 in.) thick and spaced at center-to-center distances of $0.3H$ to $0.7H$.

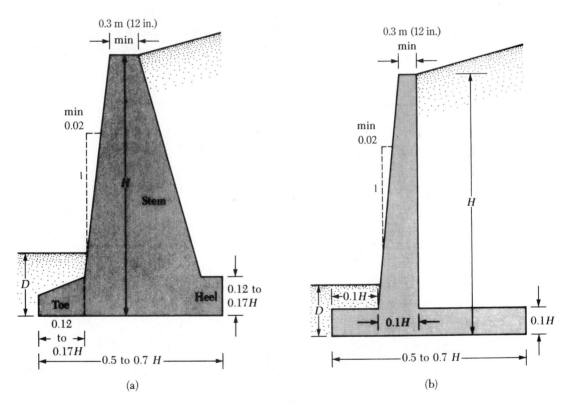

Figure 8.2 Approximate dimensions for various components of retaining wall for initial stability checks: (a) gravity wall; (b) cantilever wall

Application of Lateral Earth Pressure Theories to Design

The fundamental theories for calculating lateral earth pressure were presented in Chapter 7. To use these theories in design, an engineer must make several simple assumptions. In the case of cantilever walls, the use of the Rankine earth pressure theory for stability checks involves drawing a vertical line AB through point A, located at the edge of the heel of the base slab in Figure 8.3a. The Rankine active condition is assumed to exist along the vertical plane AB. Rankine active earth pressure equations may then be used to calculate the lateral pressure on the face AB of the wall. In the analysis of the wall's stability, the force $P_{a(\text{Rankine})}$, the weight of soil above the heel, and the weight W_c of the concrete all should be taken into consideration. The assumption for the development of Rankine active pressure along the soil face AB is theoretically correct if the shear zone bounded by the line AC is not obstructed by the stem of the wall. The angle, η, that the line AC makes with the vertical is

$$\eta = 45 + \frac{\alpha}{2} - \frac{\phi'}{2} - \sin^{-1}\left(\frac{\sin\alpha}{\sin\phi'}\right)$$

A similar type of analysis may be used for gravity walls, as shown in Figure 8.3b. However, *Coulomb's active earth pressure theory* also may be used, as shown in Figure 8.3c. If it is used, the only forces to be considered are $P_{a(\text{Coulomb})}$ and the weight of the wall, W_c.

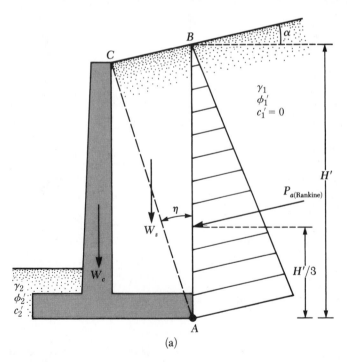

(a)

Figure 8.3 Assumption for the determination of lateral earth pressure: (a) cantilever wall; (b) and (c) gravity wall

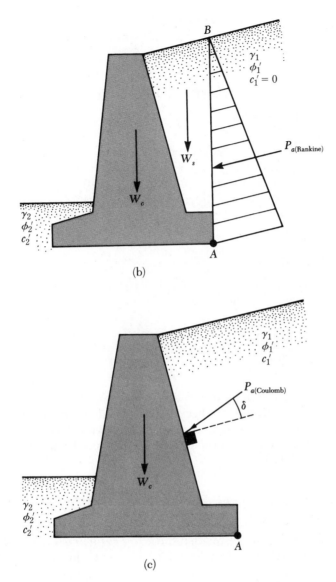

(b)

(c)

Figure 8.3 (Continued)

If Coulomb's theory is used, it will be necessary to know the range of the wall friction angle δ with various types of backfill material. Following are some ranges of wall friction angle for masonry or mass concrete walls:

Backfill material	Range of δ (deg)
Gravel	27–30
Coarse sand	20–28
Fine sand	15–25
Stiff clay	15–20
Silty clay	12–16

In the case of ordinary retaining walls, water table problems and hence hydrostatic pressure are not encountered. Facilities for drainage from the soils that are retained are always provided.

8.4   Equivalent Fluid Method for Determination of Earth Pressure

The *equivalent fluid method* for determining earth pressure during the design of retaining walls was described by Terzaghi and Peck (1967). This method assumes that the retaining wall is backfilled with an "equivalent fluid." Figures 8.4 and 8.5 show semiempirical charts given by Terzaghi and Peck. According to these charts, the horizontal force P_h and the vertical force P_v per unit length of the wall on the plane AB can be expressed as

$$P_h = \frac{1}{2}K_h H'^2$$

and   (8.1)

$$P_v = \frac{1}{2}K_v H'^2$$

Note that K_h and K_v will have a unit of kN/m^3 (or lb/ft^3) and will be a function of the type of soil usually used for backfill. Table 8.1 describes the types of soil referred to in Figures 8.4 and 8.5. Figure 8.4 is for backfills with plane surfaces, and Figure 8.5 is for backfills that slope upward for a limited distance from the crest of the wall and then become horizontal.

8.5   General Comments on Lateral Earth Pressure

In Section 8.3, it was suggested that the *active earth pressure coefficient* be used to estimate the lateral force on a retaining wall due to the backfill. It is important to recognize the fact that the active state of the backfill can be established only if the wall yields sufficiently, which does not happen in all cases. The degree to which the wall yields depends on its height and the section modulus. Furthermore, the lateral force of the backfill depends on several factors identified by Casagrande (1973):

   a. Effect of temperature
   b. Groundwater fluctuation
   c. Readjustment of the soil particles due to creep and prolonged rainfall
   d. Tidal changes
   e. Heavy wave action
   f. Traffic vibration
   g. Earthquakes

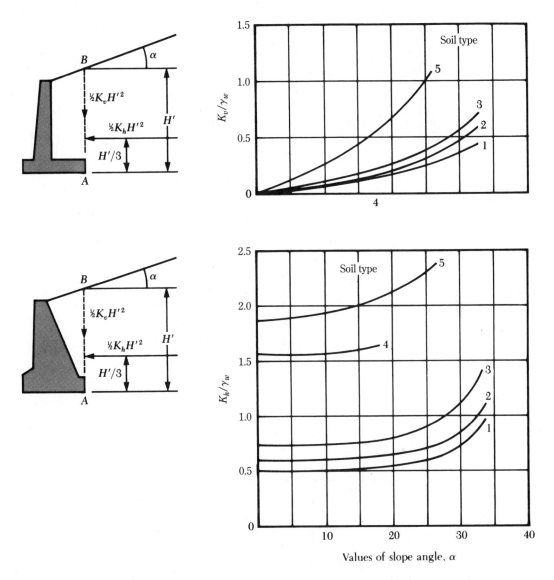

Figure 8.4 Chart for estimating active force of backfill against retaining walls supporting backfills with plane surfaces (after *Soil Mechanics in Engineering Practice,* 2d ed., by K. Terzaghi and R. B. Peck. Copyright 1967 by John Wiley and Sons. Reprinted with permission) (*Note:* For soil type, refer to Table 8.1; γ_w = unit weight of water)

Insufficient wall yielding combined with other unforeseen factors may generate a larger lateral force on the retaining structure, compared with that obtained from the active earth pressure theory. Casagrande (1973) investigated the distribution of lateral earth pressure behind a bridge abutment in Germany with a slag backfill, as shown in Figure 8.6. Laboratory tests on the slag backfill gave effective angles of friction between 37° and 45°, depending on the degree of compaction. For

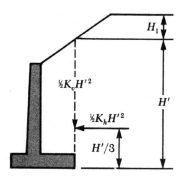

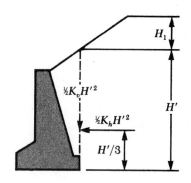

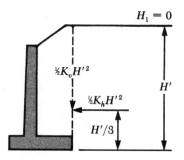

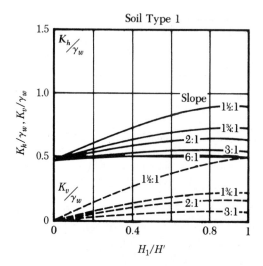

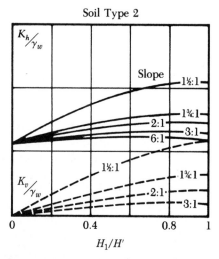

Figure 8.5 Chart for estimating pressure of backfill against retaining walls supporting backfills with surface that slopes upward from crest of wall for limited distance and then becomes horizontal (after *Soil Mechanics in Engineering Practice*, 2d ed., by K. Terzaghi and R. B. Peck. Copyright 1967 by John Wiley and Sons. Reprinted with permission) (*Note:* γ_w = unit weight of water)

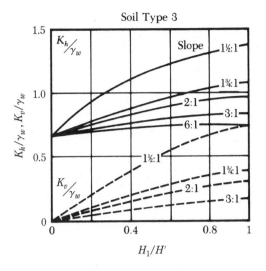

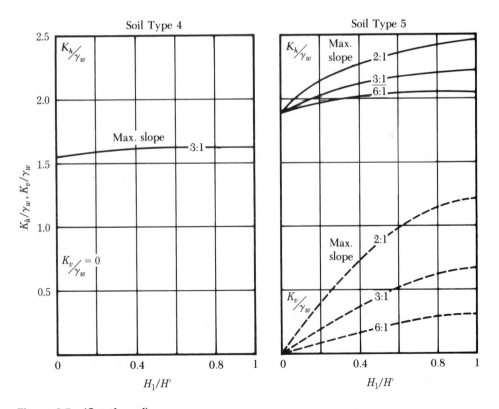

Figure 8.5   (Continued)

Table 8.1   Types of Backfill for Retaining Walls[a]

1. Coarse-grained soil without admixture of fine soil particles, very permeable (clean sand or gravel).
2. Coarse-grained soil of low permeability due to admixture of particles of silt size.
3. Residual soil with stones, fine silty sand, and granular materials with conspicuous clay content.
4. Very soft or soft clay, organic silts, or silty clays.
5. Medium or stiff clay, deposited in chunks and protected in such a way that a negligible amount of water enters the spaces between the chunks during floods or heavy rains. If this condition of protection cannot be satisfied, the clay should not be used as backfill material. With increasing stiffness of the clay, danger to the wall due to infiltration of water increases rapidly.

[a] From *Soil Mechanics in Engineering Practice,* 2d ed., by K. Terzaghi and R. B. Peck. Copyright 1967 by John Wiley and Sons. Reprinted with permission.

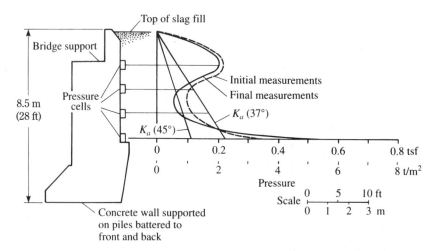

Figure 8.6   Bridge abutment on piles backfilled with granulated slag (after Casagrande, 1973)

purposes of comparison, the variation of the Rankine active earth pressure with $\phi' = 37°$ and $\phi' = 45°$ is also shown in the figure. Comparing the actual and theoretical pressure distribution diagrams indicates that

a. The actual lateral earth pressure distribution may not be triangular.
b. The lateral earth pressure distribution may change with time.
c. The actual active force is greater than the minimum theoretical active force.

The primary reason that many retaining walls designed with theoretical active earth pressure perform satisfactorily is the use of a large factor of safety. In 1993, Goh analyzed the behavior of a retaining wall by means of the finite-element method and proposed the simplified earth pressure distribution shown in Figure 8.7.

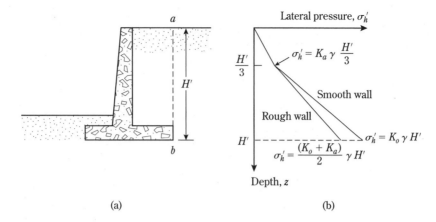

Figure 8.7 Simplified lateral pressure (σ_h') profile: (a) retaining wall; (b) pressure distribution behind virtual wall *ab*. (*Note:* K_o = coefficient of earth pressure at rest, K_a = Rankine active earth pressure coefficient)

8.6 *Stability of Retaining Walls*

A retaining wall may fail in any of the following ways:

- It may *overturn* about its toe. (See Figure 8.8a.)
- It may *slide* along its base. (See Figure 8.8b.)
- It may fail due to the loss of *bearing capacity* of the soil supporting the base. (See Figure 8.8c.)
- It may undergo deep-seated shear failure. (See Figure 8.8d.)
- It may go through excessive settlement.

The checks for stability against overturning, sliding, and bearing capacity failure will be described in Sections 8.7, 8.8, and 8.9. The principles used to estimate settlement were covered in Chapter 5 and will not be discussed further. When a weak soil layer is located at a shallow depth—that is, within a depth of 1.5 times the width of the base slab of the retaining wall—the possibility of excessive settlement should be considered. In some cases, the use of lightweight backfill material behind the retaining wall may solve the problem.

 Deep shear failure can occur along a cylindrical surface, such as *abc* shown in Figure 8.9, as a result of the existence of a weak layer of soil underneath the wall at a depth of about 1.5 times the width of the base slab of the retaining wall. In such cases, the critical cylindrical failure surface *abc* has to be determined by trial and error, using various centers such as *O*. The failure surface along which the minimum factor of safety is obtained is the *critical surface of sliding*. For the backfill slope with α less than about 10°, the critical failure circle apparently passes through the edge of the heel slab (such as *def* in the figure). In this situation, the minimum factor of safety also has to be determined by trial and error by changing the center of the trial circle.

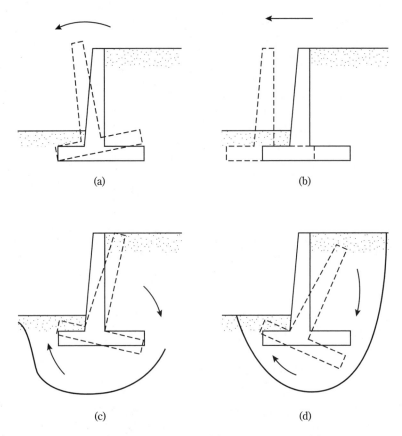

Figure 8.8 Failure of retaining wall: (a) by overturning; (b) by sliding; (c) by bearing capacity failure; (d) by deep-seated shear failure

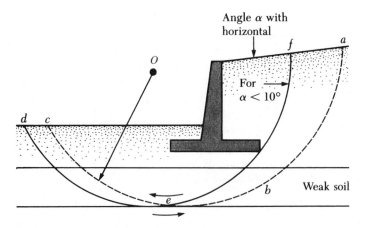

Figure 8.9 Deep-seated shear failure

8.7 *Check for Overturning*

Figure 8.10 shows the forces acting on a cantilever and a gravity retaining wall, based on the assumption that the Rankine active pressure is acting along a vertical plane AB drawn through the heel of the structure. P_p is the Rankine passive pressure; recall that its magnitude is

$$P_p = \tfrac{1}{2}K_p\gamma_2 D^2 + 2c_2'\sqrt{K_p}D \tag{7.31}$$

where $\quad\gamma_2$ = unit weight of soil in front of the heel and under the base slab
$\quad\quad K_p$ = Rankine passive earth pressure coefficient = $\tan^2(45 + \phi_2'/2)$
$\quad c_2', \phi_2'$ = cohesion and effective soil friction angle, respectively

The factor of safety against overturning about the toe—that is, about point C in Figure 8.10—may be expressed as

$$\text{FS}_{(\text{overturning})} = \frac{\Sigma M_R}{\Sigma M_O} \tag{8.2}$$

where $\quad\Sigma M_O$ = sum of the moments of forces tending to overturn about point C
$\quad\quad\Sigma M_R$ = sum of the moments of forces tending to resist overturning about point C

The overturning moment is

$$\Sigma M_O = P_h\left(\frac{H'}{3}\right) \tag{8.3}$$

where $\quad P_h = P_a\cos\alpha$

To calculate the resisting moment, ΣM_R (neglecting P_p), a table such as Table 8.2 can be prepared. The weight of the soil above the heel and the weight of the concrete (or masonry) are both forces that contribute to the resisting moment. Note that the force P_v also contributes to the resisting moment. P_v is the vertical component of the active force P_a, or

$$P_v = P_a\sin\alpha$$

The moment of the force P_v about C is

$$M_v = P_v B = P_a\sin\alpha B \tag{8.4}$$

where $\quad B$ = width of the base slab

Once ΣM_R is known, the factor of safety can be calculated as

$$\text{FS}_{(\text{overturning})} = \frac{M_1 + M_2 + M_3 + M_4 + M_5 + M_6 + M_v}{P_a\cos\alpha(H'/3)} \tag{8.5}$$

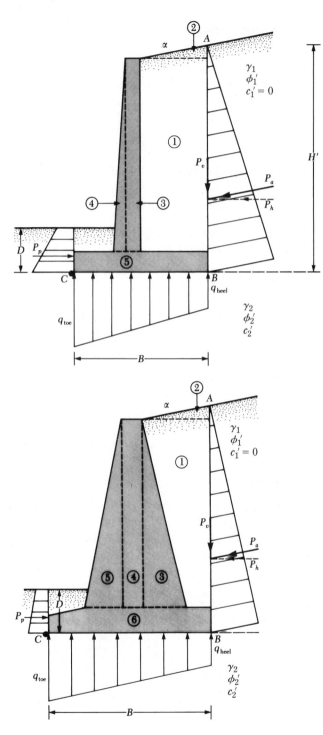

Figure 8.10 Check for overturning, assuming that the Rankine pressure is valid

Table 8.2 Procedure for Calculating ΣM_R

Section (1)	Area (2)	Weight/unit length of wall (3)	Moment arm measured from C (4)	Moment about C (5)
1	A_1	$W_1 = \gamma_1 \times A_1$	X_1	M_1
2	A_2	$W_2 = \gamma_2 \times A_2$	X_2	M_2
3	A_3	$W_3 = \gamma_c \times A_3$	X_3	M_3
4	A_4	$W_4 = \gamma_c \times A_4$	X_4	M_4
5	A_5	$W_5 = \gamma_c \times A_5$	X_5	M_5
6	A_6	$W_6 = \gamma_c \times A_6$	X_6	M_6
		P_v	B	M_v
		ΣV		ΣM_R

Note: γ_l = unit weight of backfill
γ_c = unit weight of concrete

The usual minimum desirable value of the factor of safety with respect to overturning is 2 to 3.

Some designers prefer to determine the factor of safety against overturning with the formula

$$\text{FS}_{\text{(overturning)}} = \frac{M_1 + M_2 + M_3 + M_4 + M_5 + M_6}{P_a \cos\alpha (H'/3) - M_v} \tag{8.6}$$

8.8 *Check for Sliding along the Base*

The factor of safety against sliding may be expressed by the equation

$$\text{FS}_{\text{(sliding)}} = \frac{\Sigma F_{R'}}{\Sigma F_d} \tag{8.7}$$

where $\Sigma F_{R'}$ = sum of the horizontal resisting forces
ΣF_d = sum of the horizontal driving forces

Figure 8.11 indicates that the shear strength of the soil immediately below the base slab may be represented as

$$s = \sigma' \tan\delta + c_a'$$

where δ = angle of friction between the soil and the base slab
c_a' = adhesion between the soil and the base slab

Thus, the maximum resisting force that can be derived from the soil per unit length of the wall along the bottom of the base slab is

$$R' = s(\text{area of cross section}) = s(B \times 1) = B\sigma' \tan\delta + Bc_a'$$

However,

$$B\sigma' = \text{sum of the vertical force} = \Sigma V \text{ (see Table 8.2)}$$

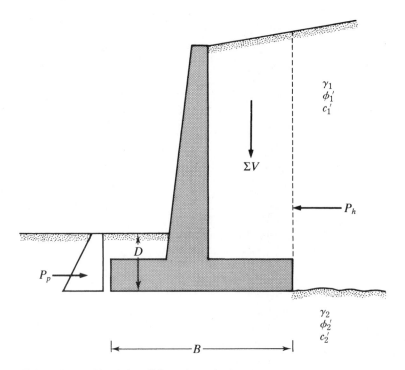

Figure 8.11 Check for sliding along the base

so

$$R' = (\Sigma V)\tan\delta + Bc_a'$$

Figure 8.11 shows that the passive force P_p is also a horizontal resisting force. Hence,

$$\Sigma F_{R'} = (\Sigma V)\tan\delta + Bc_a' + P_p \tag{8.8}$$

The only horizontal force that will tend to cause the wall to slide (a *driving force*) is the horizontal component of the active force P_a, so

$$\Sigma F_d = P_a\cos\alpha \tag{8.9}$$

Combining Eqs. (8.7), (8.8), and (8.9) yields

$$FS_{(sliding)} = \frac{(\Sigma V)\tan\delta + Bc_a + P_p}{P_a\cos\alpha} \tag{8.10}$$

A minimum factor of safety of 1.5 against sliding is generally required.

In many cases, the passive force P_p is ignored in calculating the factor of safety with respect to sliding. In general, we can write $\delta = k_1\phi_2'$ and $c_a' = k_2c_2'$. In most cases, k_1 and k_2 are in the range from $\frac{1}{2}$ to $\frac{2}{3}$. Thus,

$$FS_{(sliding)} = \frac{(\Sigma V)\tan(k_1\phi_2') + Bk_2c_2' + P_p}{P_a\cos\alpha} \tag{8.11}$$

If the desired value of $FS_{(sliding)}$ is not achieved, several alternatives may be investigated (see Figure 8.12):

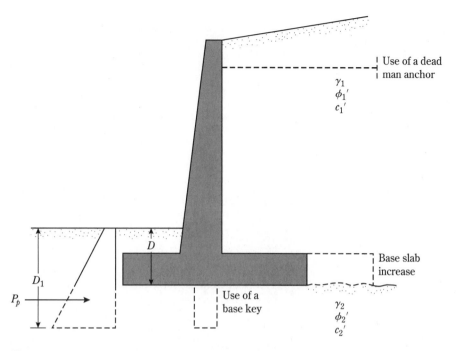

Figure 8.12 Alternatives for increasing the factor of safety with respect to sliding

- Increase the width of the base slab (i.e., the heel of the footing).
- Use a key to the base slab. If a key is included, the passive force per unit length of the wall becomes

$$P_p = \frac{1}{2}\gamma_2 D_1^2 K_p + 2c_2' D_1 \sqrt{K_p}$$

where $K_p = \tan^2\left(45 + \dfrac{\phi_2'}{2}\right)$

- Use a *deadman anchor* at the stem of the retaining wall.

8.9 *Check for Bearing Capacity Failure*

The vertical pressure transmitted to the soil by the base slab of the retaining wall should be checked against the ultimate bearing capacity of the soil. The nature of variation of the vertical pressure transmitted by the base slab into the soil is shown in Figure 8.13. Note that q_{toe} and q_{heel} are the *maximum* and the *minimum* pressures occurring at the ends of the toe and heel sections, respectively. The magnitudes of q_{toe} and q_{heel} can be determined in the following manner:

The sum of the vertical forces acting on the base slab is ΣV (see column 3 of Table 8.2), and the horizontal force $\mathbf{P}_h$ is $P_a \cos\alpha$. Let

$$\mathbf{R} = \Sigma \mathbf{V} + \mathbf{P}_h \tag{8.12}$$

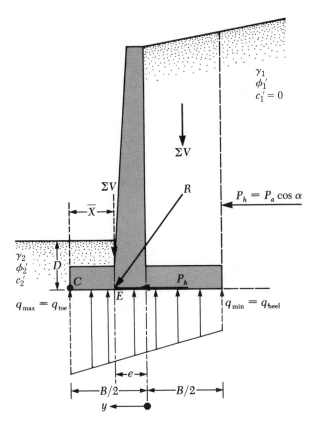

Figure 8.13 Check for bearing capacity failure

be the resultant force. The net moment of these forces about point C in Figure 8.13 is

$$M_{net} = \Sigma M_R - \Sigma M_o \qquad (8.13)$$

Note that the values of ΣM_R and ΣM_o were previously determined. [See column 5 of Table 8.2 and Eq. (8.3)]. Let the line of action of the resultant R intersect the base slab at E. Then the distance

$$\overline{CE} = \overline{X} = \frac{M_{net}}{\Sigma V} \qquad (8.14)$$

Hence, the eccentricity of the resultant R may be expressed as

$$e = \frac{B}{2} - \overline{CE} \qquad (8.15)$$

The pressure distribution under the base slab may be determined by using simple principles from the mechanics of materials. First, we have

$$q = \frac{\Sigma V}{A} \pm \frac{M_{net}y}{I} \qquad (8.16)$$

where M_{net} = moment = $(\Sigma V)e$

I = moment of inertia per unit length of the base section

$= \frac{1}{12}(1)(B^2)$

For maximum and minimum pressures, the value of y in Eq. (8.16) equals $B/2$. Substituting into Eq. (8.16) gives

$$q_{max} = q_{toe} = \frac{\Sigma V}{(B)(1)} + \frac{e(\Sigma V)\dfrac{B}{2}}{\left(\dfrac{1}{12}\right)(B^3)} = \frac{\Sigma V}{B}\left(1 + \frac{6e}{B}\right) \tag{8.17}$$

Similarly,

$$q_{min} = q_{heel} = \frac{\Sigma V}{B}\left(1 - \frac{6e}{B}\right) \tag{8.18}$$

Note that ΣV includes the weight of the soil, as shown in Table 8.2, and that when the value of the eccentricity e becomes greater than $B/6$, q_{min} [Eq. (8.18)] becomes negative. Thus, there will be some tensile stress at the end of the heel section. This stress is not desirable, because the tensile strength of soil is very small. If the analysis of a design shows that $e > B/6$, the design should be reproportioned and calculations redone.

The relationships pertaining to the ultimate bearing capacity of a shallow foundation were discussed in Chapter 3. Recall that

$$q_u = c_2' N_c F_{cd} F_{ci} + q N_q F_{qd} F_{qi} + \tfrac{1}{2}\gamma_2 B' N_\gamma F_{\gamma d} F_{\gamma i} \tag{8.19}$$

where $q = \gamma_2 D$

$B' = B - 2e$

$F_{cd} = 1 + 0.4\dfrac{D}{B'}$

$F_{qd} = 1 + 2\tan\phi_2'(1 - \sin\phi_2')^2\dfrac{D}{B'}$

$F_{\gamma d} = 1$

$F_{ci} = F_{qi} = \left(1 - \dfrac{\psi^\circ}{90^\circ}\right)^2$

$F_{\gamma i} = \left(1 - \dfrac{\psi^\circ}{\phi_2'^\circ}\right)^2$

$\psi^\circ = \tan^{-1}\left(\dfrac{P_a \cos\alpha}{\Sigma V}\right)$

Note that the shape factors F_{cs}, F_{qs}, and $F_{\gamma s}$ given in Chapter 3 are all equal to unity, because they can be treated as a continuous foundation. For this reason, the shape factors are not shown in Eq. (8.19).

Once the ultimate bearing capacity of the soil has been calculated by using Eq. (8.19), the factor of safety against bearing capacity failure can be determined:

$$FS_{(bearing\ capacity)} = \frac{q_u}{q_{max}} \tag{8.20}$$

Generally, a factor of safety of 3 is required. In Chapter 3, we noted that the ultimate bearing capacity of shallow foundations occurs at a settlement of about 10% of the foundation width. In the case of retaining walls, the width B is large. Hence, the ultimate load q_u will occur at a fairly large foundation settlement. A factor of safety of 3 against bearing capacity failure may not ensure that settlement of the structure will be within the tolerable limit in all cases. Thus, this situation needs further investigation.

Example 8.1

The cross section of a cantilever retaining wall is shown in Figure 8.14. Calculate the factors of safety with respect to overturning, sliding, and bearing capacity.

Solution
From the figure,

$$H' = H_1 + H_2 + H_3 = 2.6 \tan 10° + 6 + 0.7$$

$$= 0.458 + 6 + 0.7 = 7.158 \text{ m}$$

The Rankine active force per unit length of wall $= P_p = \frac{1}{2}\gamma_1 H'^2 K_a$. For $\phi_1' = 30°$ and $\alpha = 10°$, K_a is equal to 0.350. (See Table 7.1.) Thus,

$$P_a = \frac{1}{2}(18)(7.158)^2(0.35) = 161.4 \text{ kN/m}$$

$$P_v = P_a \sin 10° = 161.4(\sin 10°) = 28.03 \text{ kN/m}$$

and

$$P_h = P_a \cos 10° = 161.4(\cos 10°) = 158.95 \text{ kN/m}$$

Factor of Safety against Overturning
The following table can now be prepared for determining the resisting moment:

Section no.[a]	Area (m²)	Weight/unit length (kN/m)	Moment arm from point C (m)	Moment (kN-m/m)
1	$6 \times 0.5 = 3$	70.74	1.15	81.35
2	$\frac{1}{2}(0.2)6 = 0.6$	14.15	0.833	11.79
3	$4 \times 0.7 = 2.8$	66.02	2.0	132.04
4	$6 \times 2.6 = 15.6$	280.80	2.7	758.16
5	$\frac{1}{2}(2.6)(0.458) = 0.595$	10.71	3.13	33.52
		$P_v = 28.03$	4.0	112.12
		$\Sigma V = 470.45$		$\Sigma 1128.98 = \Sigma M_R$

[a] For section numbers, refer to Figure 8.14
$\gamma_{\text{concrete}} = 23.58 \text{ kN/m}^3$

The overturning moment

$$M_o = P_h\left(\frac{H'}{3}\right) = 158.95\left(\frac{7.158}{3}\right) = 379.25 \text{ kN-m/m}$$

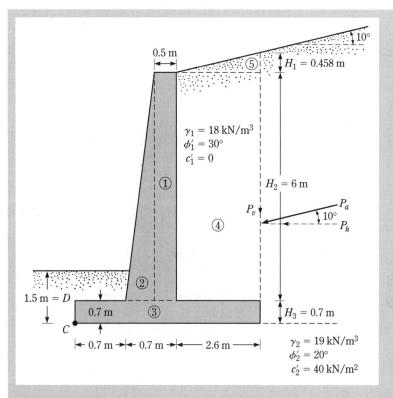

Figure 8.14 Calculation of stability of a retaining wall

and

$$FS_{(overturning)} = \frac{\Sigma M_R}{M_o} = \frac{1128.98}{379.25} = \textbf{2.98} > \textbf{2, OK}$$

Factor of Safety against Sliding
From Eq. (8.11),

$$FS_{(sliding)} = \frac{(\Sigma V)\tan(k_1\phi_2') + Bk_2c_2' + P_p}{P_a\cos\alpha}$$

Let $k_1 = k_2 = \frac{2}{3}$. Also,

$$P_p = \tfrac{1}{2}K_p\gamma_2 D^2 + 2c_2'\sqrt{K_p}D$$

$$K_p = \tan^2\left(45 + \frac{\phi_2'}{2}\right) = \tan^2(45 + 10) = 2.04$$

and

$$D = 1.5 \text{ m}$$

So

$$P_p = \tfrac{1}{2}(2.04)(19)(1.5)^2 + 2(40)(\sqrt{2.04})(1.5)$$

$$= 43.61 + 171.39 = 215 \text{ kN/m}$$

Hence,

$$FS_{(sliding)} = \frac{(470.45)\tan\left(\dfrac{2 \times 20}{3}\right) + (4)\left(\dfrac{2}{3}\right)(40) + 215}{158.95}$$

$$= \frac{111.5 + 106.67 + 215}{158.95} = \mathbf{2.73 > 1.5,\ OK}$$

Note: For some designs, the depth D in a passive pressure calculation may be taken to be *equal to the thickness of the base slab.*

Factor of Safety against Bearing Capacity Failure
Combining Eqs. (8.13), (8.14), and (8.15) yields

$$e = \frac{B}{2} - \frac{\Sigma M_R - \Sigma M_O}{\Sigma V} = \frac{4}{2} - \frac{1128.98 - 379.25}{470.45}$$

$$= 0.406 \text{ m} < \frac{B}{6} = \frac{4}{6} = 0.666 \text{ m}$$

Again, from Eqs. (8.17) and (8.18),

$$q_{heel}^{toe} = \frac{\Sigma V}{B}\left(1 \pm \frac{6e}{B}\right) = \frac{470.45}{4}\left(1 \pm \frac{6 \times 0.406}{4}\right) = 189.2 \text{ kN/m}^2 \text{ (toe)}$$

$$= 45.99 \text{ kN/m}^2 \text{ (heel)}$$

The ultimate bearing capacity of the soil can be determined from Eq. (8.19):

$$q_u = c_2' N_c F_{cd} F_{ci} + q N_q F_{qd} F_{qi} + \tfrac{1}{2}\gamma_2 B' N_\gamma F_{\gamma d} F_{\gamma i}$$

For $\phi_2' = 20°$ (see Table 3.4), $N_c = 14.83$, $N_q = 6.4$, and $N_\gamma = 5.39$. Also,

$$q = \gamma_2 D = (19)(1.5) = 28.5 \text{ kN/m}^2$$

$$B' = B - 2e = 4 - 2(0.406) = 3.188 \text{ m}$$

$$F_{cd} = 1 + 0.4\left(\frac{D}{B'}\right) = 1 + 0.4\left(\frac{1.5}{3.188}\right) = 1.188$$

$$F_{qd} = 1 + 2\tan\phi_2'(1 - \sin\phi_2')^2\left(\frac{D}{B'}\right) = 1 + 0.315\left(\frac{1.5}{3.188}\right) = 1.148$$

$$F_{\gamma d} = 1$$

$$F_{ci} = F_{qi} = \left(1 - \frac{\psi°}{90°}\right)^2$$

and

$$\psi = \tan^{-1}\left(\frac{P_a \cos\alpha}{\Sigma V}\right) = \tan^{-1}\left(\frac{158.95}{470.45}\right) = 18.67°$$

So

$$F_{ci} = F_{qi} = \left(1 - \frac{18.67}{90}\right)^2 = 0.628$$

and

$$F_{\gamma i} = \left(1 - \frac{\psi}{\phi_2'}\right)^2 = \left(1 - \frac{18.67}{20}\right)^2 \approx 0$$

Hence,

$$q_u = (40)(14.83)(1.188)(0.628) + (28.5)(6.4)(1.148)(0.628)$$
$$+ \tfrac{1}{2}(19)(5.93)(3.188)(1)(0)$$
$$= 442.57 + 131.50 + 0 = 574.07 \text{ kN/m}^2$$

and

$$FS_{(\text{bearing capacity})} = \frac{q_u}{q_{\text{toe}}} = \frac{574.07}{189.2} = \textbf{3.03} > \textbf{3, OK} \qquad \blacksquare$$

Example 8.2

A gravity retaining wall is shown in Figure 8.15. Use $\delta = 2/3\phi_1'$ and Coulomb's active earth pressure theory. Determine

a. the factor of safety against overturning
b. the factor of safety against sliding
c. the pressure on the soil at the toe and heel

Solution
The height

$$H' = 5 + 1.5 = 6.5 \text{ m}$$

Coulomb's active force is

$$P_a = \tfrac{1}{2}\gamma_1 H'^2 K_a$$

With $\alpha = 0°$, $\beta = 75°$, $\delta = 2/3\phi_1'$, and $\phi_1 = 32°$, $K_a = 0.4023$. (See Table 7.4.) So

$$P_a = \tfrac{1}{2}(18.5)(6.5)^2(0.4023) = 157.22 \text{ kN/m}$$

$$P_h = P_a \cos\left(15 + \tfrac{2}{3}\phi_1'\right) = 157.22 \cos 36.33 = 126.65 \text{ kN/m}$$

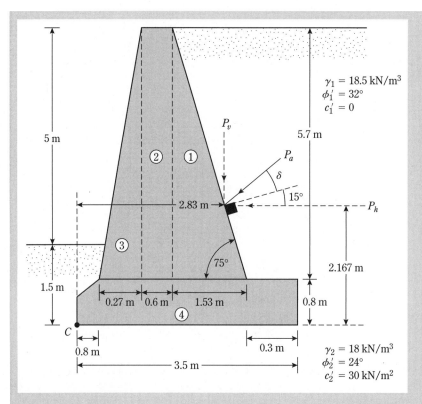

Figure 8.15 Gravity retaining wall (Not to scale)

and

$$P_v = P_a \sin \left(15 + \tfrac{2}{3}\phi_1'\right) = 157.22 \sin 36.33 = 93.14 \text{ kN/m}$$

Part a: Factor of Safety against Overturning
From Figure 8.15, one can prepare the following table:

Area No.	Area (m²)	Weight* (kN/m)	Moment arm from C (m)	Moment (kN-m/m)
1	$\tfrac{1}{2}(5.7)(1.53) = 4.36$	102.81	2.18	224.13
2	$(0.6)(5.7) = 3.42$	80.64	1.37	110.48
3	$\tfrac{1}{2}(0.27)(5.7) = 0.77$	18.16	0.98	17.80
4	$\approx (3.5)(0.8) = 2.8$	66.02	1.75	115.54
		$P_v = 93.14$	2.83	263.59
		$\Sigma V = 360.77 \text{ kN/m}$		$\Sigma M_R = 731.54 \text{ kN-m/m}$

*$\gamma_{\text{concrete}} = 23.58 \text{ kN/m}^3$

Note that the weight of the soil above the back face of the wall is not taken into account in the preceding table. We have

Overturning moment $= M_O = P_h\left(\dfrac{H'}{3}\right) = 126.65(2.167) = 274.45$ kN-m/m

Hence,

$$\text{FS}_{(\text{overturning})} = \frac{\Sigma M_R}{\Sigma M_O} = \frac{731.54}{274.45} = \mathbf{2.67 > 2, OK}$$

Part b: Factor of Safety against Sliding
We have

$$\text{FS}_{(\text{sliding})} = \frac{(\Sigma V)\tan\left(\dfrac{2}{3}\phi_2'\right) + \dfrac{2}{3}c_2'B + P_p}{P_h}$$

$$P_p = \tfrac{1}{2}K_p\gamma_2 D^2 + 2c_2'\sqrt{K_p}D$$

and

$$K_p = \tan^2\left(45 + \frac{24}{2}\right) = 2.37$$

Hence,

$$P_p = \tfrac{1}{2}(2.37)\,(18)\,(1.5)^2 + 2(30)\,(1.54)\,(1.5) = 186.59 \text{ kN/m}$$

So

$$\text{FS}_{(\text{sliding})} = \frac{360.77\tan\left(\dfrac{2}{3} \times 24\right) + \dfrac{2}{3}(30)\,(3.5) + 186.59}{126.65}$$

$$= \frac{103.45 + 70 + 186.59}{126.65} = \mathbf{2.84}$$

If P_p is ignored, the factor of safety is **1.37.**

Part c: Pressure on Soil at Toe and Heel
From Eqs. (8.13), (8.14), and (8.15),

$$e = \frac{B}{2} - \frac{\Sigma M_R - \Sigma M_O}{\Sigma V} = \frac{3.5}{2} - \frac{731.54 - 274.45}{360.77} = 0.483 < \frac{B}{6} = 0.583$$

$$q_{\text{toe}} = \frac{\Sigma V}{B}\left[1 + \frac{6e}{B}\right] = \frac{360.77}{3.5}\left[1 + \frac{(6)\,(0.483)}{3.5}\right] = \mathbf{188.43 \text{ kN/m}^2}$$

and

$$q_{\text{heel}} = \frac{V}{B}\left[1 - \frac{6e}{B}\right] = \frac{360.77}{3.5}\left[1 - \frac{(6)\,(0.483)}{3.5}\right] = \mathbf{17.73 \text{ kN/m}^2}$$

∎

8.10 *Construction Joints and Drainage from Backfill*

Construction Joints

A retaining wall may be constructed with one or more of the following joints:

1. *Construction joints* (see Figure 8.16a) are vertical and horizontal joints that are placed between two successive pours of concrete. To increase the shear at the joints, keys may be used. If keys are not used, the surface of the first pour is cleaned and roughened before the next pour of concrete.
2. *Contraction joints* (Figure 8.16b) are vertical joints (grooves) placed in the face of a wall (from the top of the base slab to the top of the wall) that allow the concrete to shrink without noticeable harm. The grooves may be about 6 to 8 mm ($\approx$0.25 to 0.3 in.) wide and 12 to 16 mm ($\approx$0.5 to 0.6 in.) deep.
3. *Expansion joints* (Figure 8.16c) allow for the expansion of concrete caused by temperature changes; vertical expansion joints from the base to the top of the wall may also be used. These joints may be filled with flexible joint fillers. In most cases, horizontal reinforcing steel bars running across the stem are continuous through all joints. The steel is greased to allow the concrete to expand.

Drainage from the Backfill

As the result of rainfall or other wet conditions, the backfill material for a retaining wall may become saturated, thereby increasing the pressure on the wall and perhaps creating an unstable condition. For this reason, adequate drainage must be provided by means of *weep holes* or *perforated drainage pipes*. (See Figure 8.17.)

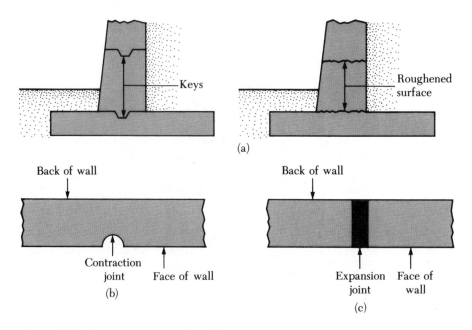

Figure 8.16 (a) Construction joints; (b) contraction joint; (c) expansion joint

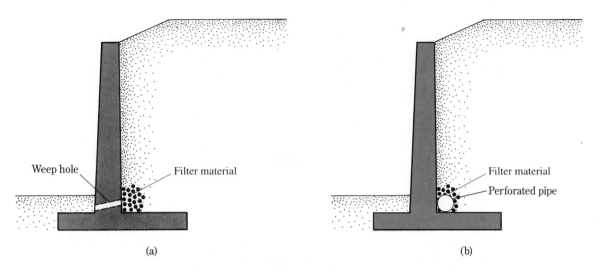

Figure 8.17 Drainage provisions for the backfill of a retaining wall: (a) by weep holes; (b) by a perforated drainage pipe

When provided, weep holes should have a minimum diameter of about 0.1 m (4 in.) and be adequately spaced. Note that there is always a possibility that backfill material may be washed into weep holes or drainage pipes and ultimately clog them. Thus, a filter material needs to be placed behind the weep holes or around the drainage pipes, as the case may be; geotextiles now serve that purpose.

Two main factors influence the choice of filter material: The grain-size distribution of the materials should be such that (a) the soil to be protected is not washed into the filter and (b) excessive hydrostatic pressure head is not created in the soil with a lower coefficient of permeability (in this case, the backfill material). The preceding conditions can be satisfied if the following requirements are met (Terzaghi and Peck, 1967):

$$\frac{D_{15(F)}}{D_{85(B)}} < 5 \qquad [\text{to satisfy condition(a)}] \qquad (8.21)$$

$$\frac{D_{15(F)}}{D_{15(B)}} > 4 \qquad [\text{to satisfy condition(b)}] \qquad (8.22)$$

In these relations, the subscripts F and B refer to the *filter* and the *base* material (i.e., the backfill soil), respectively. Also, D_{15} and D_{85} refer to the diameters through which 15% and 85% of the soil (filter or base, as the case may be) will pass. Example 8.3 gives the procedure for designing a filter.

Example 8.3

Figure 8.18 shows the grain-size distribution of a backfill material. Using the conditions outlined in Section 8.10, determine the range of the grain-size distribution for the filter material.

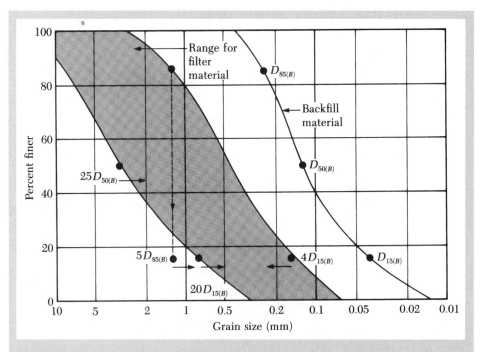

Figure 8.18   Determination of grain-size distribution of filter material

Solution

From the grain-size distribution curve given in the figure, the following values can be determined:

$$D_{15(B)} = 0.04 \text{ mm}$$

$$D_{85(B)} = 0.25 \text{ mm}$$

$$D_{50(B)} = 0.13 \text{ mm}$$

Conditions of Filter

1. $D_{15(F)}$ should be less than $5D_{85(F)}$; that is, $5 \times 0.25 = 1.25$ mm.
2. $D_{15(F)}$ should be greater than $4D_{15(B)}$; that is, $4 \times 0.04 = 0.16$ mm.
3. $D_{50(F)}$ should be less than $25D_{50(B)}$; that is, $25 \times 0.13 = 3.25$ mm.
4. $D_{15(F)}$ should be less than $20D_{15(B)}$; that is, $20 \times 0.04 = 0.8$ mm.

These limiting points are plotted in Figure 8.18. Through them, two curves can be drawn that are similar in nature to the grain-size distribution curve of the backfill material. These curves define the range of the filter material to be used.   ■

Mechanically Stabilized Retaining Walls

More recently, soil reinforcement has been used in the construction and design of foundations, retaining walls, embankment slopes, and other structures. Depending on the type of construction, the reinforcements may be galvanized metal strips, geotextiles, geogrids, or geocomposites. Sections 8.11 and 8.12 provide a general overview of soil reinforcement and various reinforcement materials.

Reinforcement materials such as metallic strips, geotextiles, and geogrids are now being used to reinforce the backfill of retaining walls, which are generally referred to as *mechanically stabilized retaining walls*. The general principles for designing these walls are given in Sections 8.13 through 8.18.

8.11 Soil Reinforcement

The use of reinforced earth is a recent development in the design and construction of foundations and earth-retaining structures. *Reinforced earth* is a construction material made from soil that has been strengthened by tensile elements such as metal rods or strips, nonbiodegradable fabrics (geotextiles), geogrids, and the like. The fundamental idea of reinforcing soil is not new; in fact, it goes back several centuries. However, the present concept of systematic analysis and design was developed by a French engineer, H. Vidal (1966). The French Road Research Laboratory has done extensive research on the applicability and the beneficial effects of the use of reinforced earth as a construction material. This research has been documented in detail by Darbin (1970), Schlosser and Long (1974), and Schlosser and Vidal (1969). The tests that were conducted involved the use of metallic strips as reinforcing material.

Retaining walls with reinforced earth have been constructed around the world since Vidal began his work. The first reinforced-earth retaining wall with metal strips as reinforcement in the United States was constructed in 1972 in southern California.

The beneficial effects of soil reinforcement derive from (a) the soil's increased tensile strength and (b) the shear resistance developed from the friction at the soil-reinforcement interfaces. Such reinforcement is comparable to that of concrete structures. Currently, most reinforced-earth design is done with *free-draining granular soil only*. Thus, the effect of pore water development in cohesive soils, which, in turn, reduces the shear strength of the soil, is avoided.

8.12 Considerations in Soil Reinforcement

Metal Strips

In most instances, galvanized steel strips are used as reinforcement in soil. However, galvanized steel is subject to corrosion. The rate of corrosion depends on several environmental factors. Binquet and Lee (1975) suggested that the average rate of corrosion of galvanized steel strips varies between 0.025 and 0.050 mm/yr. So, in the actual design of reinforcement, allowance must be made for the rate of corrosion. Thus,

$$t_c = t_{\text{design}} + r \ (\text{life span of structure})$$

where t_c = actual thickness of reinforcing strips to be used in construction
t_{design} = thickness of strips determined from design calculations
r = rate of corrosion

Further research needs to be done on corrosion-resistant materials such as fiberglass before they can be used as reinforcing strips.

Nonbiodegradable Fabrics

Nonbiodegradable fabrics are generally referred to as *geotextiles*. Since 1970, the use of geotextiles in construction has increased greatly around the world. The fabrics are usually made from petroleum products—polyester, polyethylene, and polypropylene. They may also be made from fiberglass. Geotextiles are not prepared from natural fabrics, because they decay too quickly. Geotextiles may be woven, knitted, or nonwoven.

Woven geotextiles are made of two sets of parallel filaments or strands of yarn systematically interlaced to form a planar structure. *Knitted geotextiles* are formed by interlocking a series of loops of one or more filaments or strands of yarn to form a planar structure. *Nonwoven geotextiles* are formed from filaments or short fibers arranged in an oriented or random pattern in a planar structure. These filaments or short fibers are arranged into a loose web in the beginning and then are bonded by one or a combination of the following processes:

1. *Chemical bonding*—by glue, rubber, latex, a cellulose derivative, or the like
2. *Thermal bonding*—by heat for partial melting of filaments
3. *Mechanical bonding*—by needle punching

Needle-punched nonwoven geotextiles are thick and have high in-plane permeability.

Geotextiles have four primary uses in foundation engineering:

1. *Drainage:* The fabrics can rapidly channel water from soil to various outlets, thereby providing a higher soil shear strength and hence stability.
2. *Filtration:* When placed between two soil layers, one coarse grained and the other fine grained, the fabric allows free seepage of water from one layer to the other. However, it protects the fine-grained soil from being washed into the coarse-grained soil.
3. *Separation:* Geotextiles help keep various soil layers separate after construction and during the projected service period of the structure. For example, in the construction of highways, a clayey subgrade can be kept separate from a granular base course.
4. *Reinforcement:* The tensile strength of geofabrics increases the load-bearing capacity of the soil.

Geogrids

Geogrids are high-modulus polymer materials, such as polypropylene and polyethylene, and are prepared by tensile drawing. Netlon, Ltd., of the United Kingdom was the first producer of geogrids. In 1982, the Tensar Corporation, presently Tensar Earth Technologies, Inc., introduced geogrids into the United States.

The major function of geogrids is *reinforcement*. Geogrids are relatively stiff netlike materials with openings called *apertures* that are large enough to allow interlocking with the surrounding soil or rock to perform the function of reinforcement or segregation (or both).

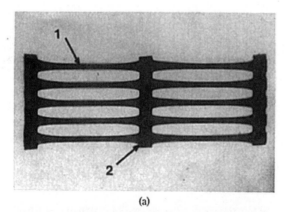

(a)

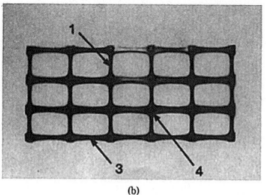

(b)

Figure 8.19 Geogrids:
(a) uniaxial; (b) biaxial
(*Note:* 1—longitudinal rib;
2—transverse bar; 3—transverse
rib; 4—junction)

Geogrids generally are of two types: (a) uniaxial and (b) biaxial. Figures 8.19a and 8.19b show these two types of geogrids, which are produced by Tensar Earth Technologies, Inc. Uniaxial TENSAR grids are manufactured by stretching a punched sheet of extruded high-density polyethylene in one direction under carefully controlled conditions. The process aligns the polymer's long-chain molecules in the direction of draw and results in a product with high one-directional tensile strength and a high modulus. Biaxial TENSAR grids are manufactured by stretching the punched sheet of polypropylene in two orthogonal directions. This process results in a product with high tensile strength and a high modulus in two perpendicular directions. The resulting grid apertures are either square or rectangular.

The commercial geogrids currently available for soil reinforcement have nominal rib thicknesses of about 0.5–1.5 mm (0.02–0.06 in.) and junctions of about 2.5–5 mm (0.1–0.2 in.). The grids used for soil reinforcement usually have apertures that are rectangular or elliptical. The dimensions of the apertures vary from about 25–150 mm (1–6 in.). Geogrids are manufactured so that the open areas of the grids are greater than 50% of the total area. They develop reinforcing strength at low strain levels, such as 2% (Carroll, 1988). Table 8.3 gives some properties of the TENSAR biaxial geogrids that are currently available commercially.

Table 8.3 Properties of TENSAR Biaxial Geogrids

Property	Geogrid		
	BX1000	**BX1100**	**BX1200**
Aperture size			
Machine	25 mm (1 in.)	25 mm (1 in.)	25 mm (1 in.)
direction	(nominal)	(nominal)	(nominal)
Cross-machine	33 mm (1.3 in.)	33 mm (1.3 in.)	33 mm (1.3 in.)
direction	(nominal)	(nominal)	(nominal)
Open area	70% (minimum)	74% (nominal)	77% (nominal)
Junction			
Thickness	2.3 mm (0.09 in.)	2.8 mm (0.11 in.)	4.1 mm (0.16 in.)
	(nominal)	(nominal)	(nominal)
Tensile modulus			
Machine	18.2 kN/m (12,500 lb/ft)	204 kN/m (14,000 lb/ft)	270 kN/m (18,500 lb/ft)
direction	(minimum)	(minimum)	(minimum)
Cross-machine	18.2 kN/m (12,500 lb/ft)	292 kN/m (20,000 lb/ft)	438 kN/m (30,000 lb/ft)
direction	(minimum)	(minimum)	(minimum)
Material			
Polypropylene	97% (minimum)	99% (nominal)	99% (nominal)
Carbon black	2% (minimum)	1% (nominal)	1% (nominal)

8.13 General Design Considerations

The general design procedure of any mechanically stabilized retaining wall can be divided into two parts:

1. Satisfying *internal stability* requirements
2. Checking the *external stability* of the wall

The internal stability checks involve determining tension and pullout resistance in the reinforcing elements and ascertaining the integrity of facing elements. The external stability checks include checks for overturning, sliding, and bearing capacity failure (Figure 8.20). The sections that follow will discuss the retaining-wall design procedures for use with metallic strips, geotextiles, and geogrids.

8.14 Retaining Walls with Metallic Strip Reinforcement

Reinforced-earth walls are flexible walls. Their main components are

1. *backfill,* which is granular soil
2. *reinforcing strips,* which are thin, wide strips placed at regular intervals, and
3. *a cover* or *skin,* on the front face of the wall

Figure 8.21 is a diagram of a reinforced-earth retaining wall. Note that, at any depth, the reinforcing strips or ties are placed with a horizontal spacing of S_H center to center; the vertical spacing of the strips or ties is S_V center to center. The skin can be constructed with sections of relatively flexible thin material. Lee et al. (1973)

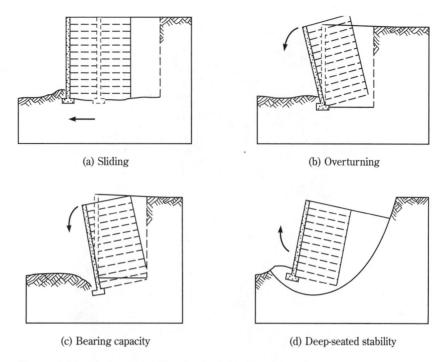

(a) Sliding

(b) Overturning

(c) Bearing capacity

(d) Deep-seated stability

Figure 8.20 External stability checks (after Transportation Research Board, 1995)

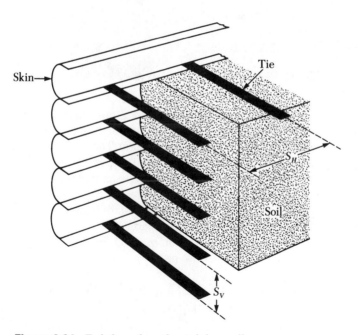

Figure 8.21 Reinforced-earth retaining wall

Figure 8.22   Reinforced-earth retaining wall (with metallic strip) under construction

showed that, with a conservative design, a 5 mm-thick ($\approx$0.2 in.) galvanized steel skin would be enough to hold a wall about 14–15 m (45–50 ft) high. In most cases, precast concrete slabs can also be used as skin. The slabs are grooved to fit into each other so that soil cannot flow out between the joints. When metal skins are used, they are bolted together, and reinforcing strips are placed between the skins.

Figures 8.22 and 8.23 show a reinforced-earth retaining wall under construction; its skin (facing) is a precast concrete slab. Figure 8.24 shows a metallic reinforcement tie attached to the concrete slab.

The simplest and most common method for the design of ties is the *Rankine method*. We discuss this procedure next.

Calculation of Active Horizontal and Vertical Pressure

Figure 8.25 (p. 366) shows a retaining wall with a granular backfill having a unit weight of γ_1 and a friction angle of ϕ_1'. Below the base of the retaining wall, the *in situ* soil has been excavated and recompacted, with granular soil used as backfill. Below the backfill, the *in situ* soil has a unit weight of γ_2, friction angle of ϕ_2', and cohesion of c_2'. A surcharge having an intensity of q per unit area lies atop the retaining wall, which has reinforcement ties at depths $z = 0, S_V, 2S_V, \ldots, NS_V$. The height of the wall is $NS_V = H$.

Figure 8.23 Another view of the retaining wall shown in Figure 8.22

Figure 8.24 Metallic strip attachment to the precast concrete slab used as the skin

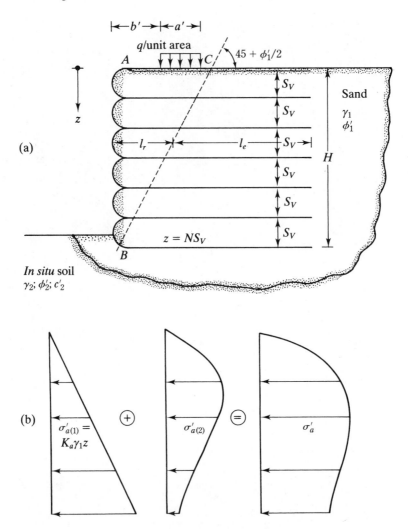

Figure 8.25   Analysis of a reinforced-earth retaining wall

According to the Rankine active pressure theory (Section 7.3)

$$\sigma'_a = \sigma'_o K_a - 2c'\sqrt{K_a}$$

where σ'_a = Rankine active pressure at any depth z

For dry granular soils with no surcharge at the top, $c' = 0$, $\sigma'_o = \gamma_1 z$, and $K_a = \tan^2(45 - \phi'_1/2)$. Thus,

$$\sigma'_{a(1)} = \gamma_1 z K_a \tag{8.23}$$

When a surcharge is added at the top, as shown in Figure 8.25,

$$\sigma'_o = \sigma'_{o(1)} \quad + \sigma'_{o(2)} \tag{8.24}$$

$$\uparrow \qquad\qquad \uparrow$$

$$= \gamma_1 z \quad \text{Due to the}$$

Due to   surcharge

soil only

The magnitude of $\sigma'_{o(2)}$ can be calculated by using the 2:1 method of stress distribution described in Eq. (5.14) and Figure 5.5. The 2:1 method of stress distribution is shown in Figure 8.26a. According to Laba and Kennedy (1986),

$$\sigma'_{o(2)} = \frac{qa'}{a' + z} \qquad (\text{for } z \leqslant 2b') \tag{8.25}$$

and

$$\sigma'_{o(2)} = \frac{qa'}{a' + \dfrac{z}{2} + b'} \qquad (\text{for } z > 2b') \tag{8.26}$$

Also, when a surcharge is added at the top, the lateral pressure at any depth is

$$\sigma'_a = \sigma'_{a(1)} \quad + \sigma'_{a(2)} \tag{8.27}$$

$$\uparrow \qquad\qquad \uparrow$$

$$= K_a \gamma_1 z \quad \text{Due to the}$$

Due to   surcharge

soil only

According to Laba and Kennedy (1986), $\sigma'_{a(2)}$ may be expressed (see Figure 8.26b) as

$$\sigma'_{a(2)} = M\left[\frac{2q}{\pi}(\beta - \sin\beta \cos 2\alpha) \right] \tag{8.28}$$

$$\uparrow$$

$$(\text{in radians})$$

where

$$M = 1.4 - \frac{0.4b'}{0.14H} \geqslant 1 \tag{8.29}$$

The net active (lateral) pressure distribution on the retaining wall calculated by using Eqs. (8.27), (8.28), and (8.29) is shown in Figure 8.25b.

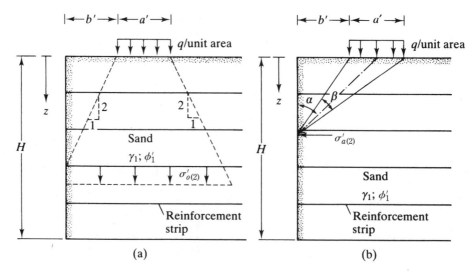

Figure 8.26 (a) Notation for the relationship of $\sigma'_{o(2)}$ in Eqs. (8.25) and (8.26); (b) notation for the relationship of $\sigma'_{o(2)}$ in Eqs. (8.28) and (8.29)

Tie Force

The tie force *per unit length of the wall* developed at any depth z (see Figure 8.25) is

$\quad$ T = active earth pressure at depth z

$\quad\quad$ × area of the wall to be supported by the tie

$$= (\sigma'_a)\,(S_V S_H) \qquad\qquad (8.30)$$

Factor of Safety against Tie Failure

The reinforcement ties at each level, and thus the walls, could fail by either (a) tie breaking or (b) tie pullout.

$\quad$ The factor of safety against *tie breaking* may be determined as

$$
FS_{(B)} = \frac{\text{yield or breaking strength of each tie}}{\text{maximum force in any tie}}
$$

$$
= \frac{wtf_y}{\sigma'_a S_V S_H} \qquad\qquad (8.31)
$$

where w = width of each tie
$\quad\quad\quad$ t = thickness of each tie
$\quad\quad\quad$ f_y = yield or breaking strength of the tie material

A factor of safety of about 2.5–3 is generally recommended for ties at all levels.

$\quad$ Reinforcing ties at any depth z will fail by pullout if the frictional resistance developed along the surfaces of the ties is less than the force to which the ties are being subjected. The *effective length* of the ties along which frictional resistance is developed may be conservatively taken as the length that extends *beyond the limits of the Rankine active failure zone,* which is the zone *ABC* in Figure 8.25. Line *BC* makes an angle of $45 + \phi'_1/2$ with the horizontal. Now, the maximum friction force that can be realized for a tie at depth z is

$$F_R = 2l_e w \sigma'_o \tan \phi'_\mu \tag{8.32}$$

where l_e = effective length
σ'_o = effective vertical pressure at a depth z
ϕ'_μ = soil– tie friction angle

Thus, the factor of safety against *tie pullout* at any depth z is

$$\mathrm{FS}_{(P)} = \frac{F_R}{T} \tag{8.33}$$

Substituting Eqs. (8.30) and (8.32) into Eq. (8.33) yields

$$\mathrm{FS}_{(P)} = \frac{2l_e w \sigma'_o \tan \phi'_\mu}{\sigma'_a S_V S_H} \tag{8.34}$$

Total Length of Tie

The total length of ties at any depth is

$$L = l_r + l_e \tag{8.35}$$

where l_r = length within the Rankine failure zone
l_e = effective length

For a given $\mathrm{FS}_{(P)}$ from Eq. (8.34),

$$l_e = \frac{\mathrm{FS}_{(P)} \sigma'_a S_V S_H}{2w \sigma'_o \tan \phi'_\mu} \tag{8.36}$$

Again, at any depth z,

$$l_r = \frac{(H - z)}{\tan\left(45 + \dfrac{\phi'_1}{2}\right)} \tag{8.37}$$

So, combining Eqs. (8.35), (8.36), and (8.37) gives

$$L = \frac{(H - z)}{\tan\left(45 + \dfrac{\phi'_1}{2}\right)} + \frac{\mathrm{FS}_{(P)} \sigma'_a S_V S_H}{2w \sigma'_o \tan \phi'_\mu} \tag{8.38}$$

Step-by-Step-Design Procedure Using Metallic Strip Reinforcement

Following is a step-by-step procedure for the design of reinforced-earth retaining walls.

General
1. Determine the height of the wall, H, and the properties of the granular backfill material, such as the unit weight (γ_1) and the angle of friction (ϕ'_1).
2. Obtain the soil–tie friction angle, ϕ'_μ, and the required value of $FS_{(B)}$ and $FS_{(P)}$.

Internal Stability
3. Assume values for horizontal and vertical tie spacing. Also, assume the width of reinforcing strip, w, to be used.
4. Calculate σ'_a from Eqs. (8.27), (8.28), and (8.29).
5. Calculate the tie forces at various levels from Eq. (8.30).
6. For the known values of $FS_{(B)}$, calculate the thickness of ties, t, required to resist the tie breakout:

$$T = \sigma'_a S_V S_H = \frac{wtf_y}{FS_{(B)}}$$

or

$$t = \frac{(\sigma'_a S_V S_H)[FS_{(B)}]}{wf_y} \tag{8.39}$$

The convention is to keep the magnitude of t the same at all levels, so σ'_a in Eq. (8.19) should equal $\sigma'_{a(max)}$.
7. For the known values of ϕ'_μ and $FS_{(P)}$, determine the length L of the ties at various levels from Eq. (8.38).
8. The magnitudes of S_V, S_H, t, w, and L may be changed to obtain the most economical design.

External Stability
9. Check for *overturning*, using Figure 8.27 as a guide. Taking the moment about B yields the overturning moment for the unit length of the wall:

$$M_o = P_a z' \tag{8.40}$$

Here, P_a = active force = $\displaystyle\int_0^H \sigma'_a dz$

The resisting moment per unit length of the wall is

$$M_R = W_1 x_1 + W_2 x_2 + \cdots + qa'\left(b' + \frac{a'}{2}\right) \tag{8.41}$$

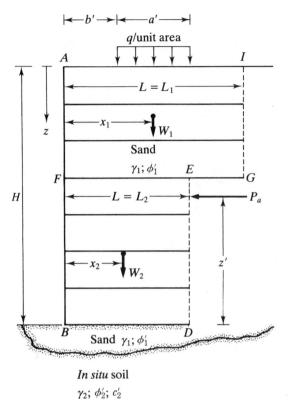

Figure 8.27 Stability check for the retaining wall

where $W_1 = (\text{area } AFEGI)\,(1)\,(\gamma_1)$
$W_2 = (\text{area } FBDE)\,(1)\,(\gamma_1)$
$\vdots$

So

$$\text{FS}_{(\text{overturning})} = \frac{M_R}{M_o}$$

$$= \frac{W_1 x_1 + W_2 x_2 + \cdots + qa'\left(b' + \dfrac{a'}{2}\right)}{\left(\displaystyle\int_0^H \sigma_a'\,dz\right)z'} \tag{8.42}$$

10. The check for *sliding* can be done by using Eq. (8.11), or

$$\text{FS}_{(\text{sliding})} = \frac{(W_1 + W_2 + \cdots + qa')\left[\tan\left(k\phi_1'\right)\right]}{P_a} \tag{8.43}$$

where $k \approx \frac{2}{3}$.

11. Check for ultimate bearing capacity failure, which can be given as

$$q_u = c_2' N_c + \frac{1}{2}\gamma_2 L_2' N_\gamma \qquad (8.44a)$$

The bearing capacity factors N_c and N_γ correspond to the soil friction angle ϕ_2'. (See Table 3.4.) In Eq. (8.44a), L_2' is the effective length; that is,

$$L_2' = L_2 - 2e \qquad (8.44b)$$

where e = eccentricity, given by

$$e = \frac{L_2}{2} - \frac{M_R - M_O}{\Sigma V} \qquad (8.44c)$$

in which $\Sigma V = W_1 + W_2 \ldots + qa'$

From Eq. 8.24, the vertical stress at $z = H$ is

$$\sigma_{o(H)}' = \gamma_1 H + \sigma_{o(2)}' \qquad (8.45)$$

So the factor of safety against bearing capacity failure is

$$FS_{(bearing\ capacity)} = \frac{q_{ult}}{\sigma_{o(H)}'} \qquad (8.46)$$

Generally, minimum values of $FS_{(overturning)} = 3$, $FS_{(sliding)} = 3$, and $FS_{(bearing\ capacity\ failure)} = 3$ to 5 are recommended.

Example 8.4

A 8-m-high retaining wall with galvanized steel-strip reinforcement in a granular backfill has to be constructed. In Figure 8.25, let the following values be given:

Granular backfill: ϕ_1' $= 30°$

                   γ_1 $= 16.6 \text{ kN/m}^3$

Foundation soil: ϕ_2' $= 28°$

                   γ_2 $= 18 \text{ kN/m}^3$

                   c_2' $= 52 \text{ kN/m}^2$

Galvanized steel reinforcement:

   Width of strip, w $= 75 \text{ mm}$

                   S_V $= 0.5 \text{ m center to center}$

$$S_H = 1 \text{ m center to center}$$

$$f_y = 2.4 \times 10^5 \text{kN/m}^2$$

$$\phi'_\mu = 20°$$

$$\text{Required} \quad FS_{(B)} = 3$$

$$\text{Required} \quad FS_{(P)} = 3$$

Check for the external and internal stability of the wall. Assume the corrosion rate of the galvanized steel to be 0.025 mm/year and the life span of the structure to be 50 years.

Solution

Internal Stability Check

a. Tie thickness: The maximum tie force $T_{\max} = \sigma'_{a(\max)} S_V S_H$,

where $\quad \sigma'_{a(\max)} = \gamma_1 H K_a = \gamma H \tan^2\left(45 - \dfrac{\phi'_1}{2} \right)$

So

$$T_{\max} = \gamma_1 H \tan^2\left(45 - \frac{\phi'_1}{2} \right) S_V S_H$$

From Eq. (8.39), for *tie breakout*,

$$t = \frac{(\sigma'_a S_V S_H)[\text{FS}_{(B)}]}{w f_y} = \frac{\left[\gamma_1 H \tan^2\left(45 - \dfrac{\phi'_1}{2} \right) S_V S_H \right]\left[\text{FS}_{(B)} \right]}{w f_y}$$

or

$$t = \frac{\left[(16.6)\,(8) \tan^2\left(45 - \dfrac{30}{2} \right)(0.5)\,(1\text{m}) \right](3)}{(0.075 \text{ m})\,(2.4 \times 10^5 \text{ kN/m}^2)} = 0.00369 \text{ m} = 3.69 \text{ mm}$$

If the rate of corrosion is 0.025 mm/yr and the life span of the structure is 50 yr, then the actual thickness of the ties will be

$$t = 3.69 + (0.025)\,(50) = 4.94 \text{ mm}$$

So a **tie thickness of 5 mm** would be enough.

b. Tie length: We use Eq. (8.38). For this case, $\sigma'_a = \gamma_1 z K_a$ and $\sigma'_o = \gamma_1 z$, so

$$L = \frac{(H - z)}{\tan\left(45 + \dfrac{\phi'_1}{2} \right)} + \frac{\text{FS}_{(P)} \gamma_1 z K_a S_V S_H}{2 w \gamma_1 z \tan \phi'_\mu}$$

Now the following table can be prepared (note: $\text{FS}_{(P)} = 3$, $H = 8$ m, $w = 0.075$ m, and $\phi'_\mu = 20°$):

z (m)	$\dfrac{(H-z)}{\tan\left(45+\dfrac{\phi'_1}{2}\right)}$ (m)	$\dfrac{FS_{(P)}\gamma_1 z K_a S_V S_H}{2w\gamma_1 z \tan \phi'_\mu}$ (m)	L (m)
1	4.04	9.16	13.20
2	3.46	9.16	12.62
3	2.89	9.16	12.05
4	2.31	9.16	11.47
5	1.73	9.16	10.89
6	1.15	9.16	10.31
7	0.58	9.16	9.74

So use a **tie length of L = 13 m**.

External Stability Check

a. Check for overturning: Here, we use Figure 8.28. For this case, from Eq. (8.42), we have

$$\text{FS}_{(\text{overturning})} = \frac{W_1 x_1}{\left[\displaystyle\int_0^H \sigma'_a dz\right] z'}$$

$$W_1 = \gamma_1 HL = (16.6)(8)(13) = 1726.4 \text{ kN}$$

$$x_1 = 6.5 \text{ m}$$

$$P_a = \int_0^H \sigma'_a dz = \tfrac{1}{2}\gamma_1 K_a H^2 = \left(\tfrac{1}{2}\right)(16.6)(0.33)(8)^2 = 175.3 \text{ kN/m}^2$$

$$z' = \frac{8}{3} = 2.67 \text{ m}$$

$$\text{FS}_{(\text{overturning})} = \frac{(1726.4)(6.5)}{(175.3)(2.67)} = \textbf{23.98} > 3, \textbf{ OK}$$

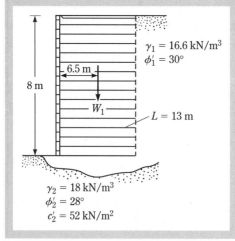

Figure 8.28   Retaining wall with galvanized steel-strip reinforcement in the backfill

b. Check for sliding: From Eq. (8.43),

$$\text{FS}_{(\text{sliding})} = \frac{W_1 \tan(k\phi_1')}{P_a} = \frac{1726.4 \tan\left[\left(\frac{2}{3}\right)(30)\right]}{175.3} = 3.58 > 3, \text{ OK}$$

c. Check for bearing capacity: For $\phi_2' = 28°$, $N_c = 25.8$ and $N_\gamma = 16.72$ (see Table 3.4). From Eq. (8.44a),

$$q_{\text{ult}} = c_2' N_c + \tfrac{1}{2} \gamma_2 L' N_\gamma$$

$$e = \frac{L}{2} - \frac{M_R - M_O}{\Sigma V} = \frac{13}{2} - \left[\frac{(1726.4 \times 6.5) - (175.3 \times 8/3)}{1726.4}\right] = 0.27 \text{ m}$$

$$L' = 13 - (2 \times 0.27) = 12.46 \text{ m}$$

So

$$q_{\text{ult}} = (52)(25.8) + \left(\tfrac{1}{2}\right)(18)(12.46)(16.72) = 3216.6 \text{ kN/m}^2$$

From Eq. (8.45),

$$\sigma_{o(\text{H})}' = \gamma_1 H = (16.6)(8) = 132.8 \text{ kN/m}^2$$

$$\text{FS}_{(\text{bearing capacity})} = \frac{q_{\text{ult}}}{\sigma_{o(\text{H})}'} = \frac{3216.6}{132.8} = 24.2 > 5, \text{ OK} \qquad ■$$

8.16 Retaining Walls with Geotextile Reinforcement

Figure 8.29 shows a retaining wall in which layers of geotextile have been used as reinforcement. As in Figure 8.27, the backfill is a granular soil. In this type of retaining wall, the facing of the wall is formed by lapping the sheets as shown with a lap length of l_l. When construction is finished, the exposed face of the wall must be covered; otherwise, the geotextile will deteriorate from exposure to ultraviolet light. *Bitumen emulsion* or *Gunite* is sprayed on the wall face. A wire mesh anchored to the geotextile facing may be necessary to keep the coating on.

The design of this type of retaining wall is similar to that presented in Section 8.14. Following is a step-by-step procedure for design based on the recommendations of Bell et al. (1975) and Koerner (1990):

Internal Stability

1. Determine the active pressure distribution on the wall from the formula

$$\sigma_a' = K_a \sigma_o' = K_a \gamma_1 z \qquad (8.47)$$

where K_a = Rankine active pressure coefficient = $\tan^2(45 - \phi_1'/2)$
           γ_1 = unit weight of the granular backfill
           ϕ_1' = friction angle of the granular backfill

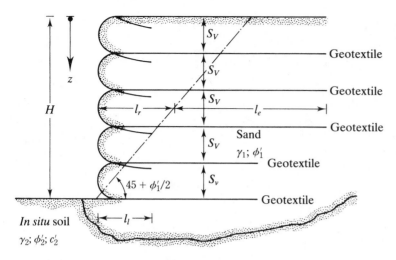

Figure 8.29 Retaining wall with geotextile reinforcement

2. Select a geotextile fabric with an allowable strength of σ_G (lb/ft or kN/m).
3. Determine the vertical spacing of the layers at any depth z from the formula

$$S_V = \frac{\sigma_G}{\sigma_a' FS_{(B)}} = \frac{\sigma_G}{(\gamma_1 z K_a)[FS_{(B)}]} \tag{8.48}$$

Note that Eq. (8.48) is similar to Eq. (8.31). The magnitude of $FS_{(B)}$ is generally 1.3–1.5.
4. Determine the length of each layer of geotextile from the formula

$$L = l_r + l_e \tag{8.49}$$

where

$$l_r = \frac{H - z}{\tan\left(45 + \dfrac{\phi_1'}{2}\right)} \tag{8.50}$$

and

$$l_e = \frac{S_V \sigma_a'[FS_{(P)}]}{2\sigma_o' \tan \phi_F'} \tag{8.51}$$

in which

$$\sigma_a' = \gamma_1 z K_a$$

$$\sigma_o' = \gamma_1 z$$

$$FS_{(P)} = 1.3 \text{ to } 1.5$$

$$\phi_F' = \text{friction angle at geotextile–soil interface}$$

$$\approx \tfrac{2}{3}\phi_1'$$

Note that Eqs. (8.49), (8.50), and (8.51) are similar to Eqs. (8.35), (8.37), and (8.36), respectively.

Based on the published results, the assumption of $\phi'_F/\phi'_1 \approx \frac{2}{3}$ is reasonable and appears to be conservative.

5. Determine the lap length, l_l, from

$$l_l = \frac{S_V \sigma'_a FS_{(P)}}{4\sigma'_o \tan \phi'_F}$$ (8.52)

The minimum lap length should be 1 m (3 ft).

External Stability

6. Check the factors of safety against overturning, sliding, and bearing capacity failure as described in Section 8.15 (Steps 9, 10, and 11).

Example 8.5

A geotextile-reinforced retaining wall 16 ft high is shown in Figure 8.30. For the granular backfill, $\gamma_1 = 110 \text{ lb/ft}^3$ and $\phi'_1 = 36°$. For the geotextile, $\sigma_G = 80 \text{ lb/in}$. For the design of the wall, determine S_V, L, and l_l.

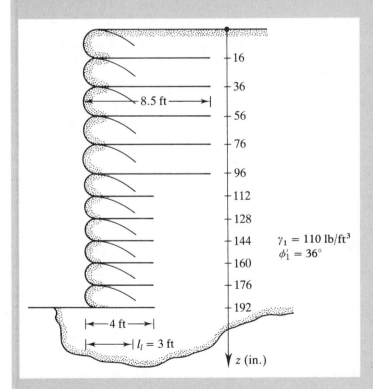

Figure 8.30 Geotextile-reinforced retaining wall

Solution

We have

$$K_a = \tan^2\left(45 - \frac{\phi_1'}{2}\right) = 0.26$$

Determination of S_V

To find S_V, we make a few trials. From Eq. (8.48),

$$S_V = \frac{\sigma_G}{(\gamma_1 z K_a)[FS_{(B)}]}$$

With $FS_{(B)} = 1.5$ at $z = 8$ ft,

$$S_V = \frac{(80 \times 12 \text{ lb/ft})}{(110)(8)(0.26)(1.5)} = 2.8 \text{ ft} \approx 33.6 \text{ in.}$$

At $z = 12$ ft,

$$S_V = \frac{(80 \times 12 \text{ lb/ft})}{(110)(12)(0.26)(1.5)} = 1.87 \text{ ft} \approx 22 \text{ in.}$$

At $z = 16$ ft,

$$S_V = \frac{(80 \times 12 \text{ lb/ft})}{(110)(16)(0.26)(1.5)} = 1.4 \text{ ft} \approx 16.8 \text{ in.}$$

So, **use $S_V = 20$ in. for $z = 0$ to $z = 8$ ft and $S_V = 16$ in. for $z > 8$ ft.** (See Figure 8.29.)

Determination of L

From Eqs. (8.49), (8.50), and (8.51),

$$L = \frac{(H - z)}{\tan\left(45 + \frac{\phi_1'}{2}\right)} + \frac{S_V K_a [FS_{(P)}]}{2 \tan \phi_F'}$$

For $FS_{(P)} = 1.5$, $\tan \phi_F' = \tan\left[\left(\frac{2}{3}\right)(36)\right] = 0.445$, and it follows that

$$L = (0.51)(H - z) + 0.438 S_V$$

Now the following table can be prepared:

z		S_V (ft)	$(0.51)(H-z)$ (ft)	$0.438 S_V$ (ft)	L (ft)
(in.)	(ft)				
16	1.33	1.67	7.48	0.731	8.21
56	4.67	1.67	5.78	0.731	6.51
76	6.34	1.67	4.93	0.731	5.66
96	8.0	1.67	4.08	0.731	4.81
112	9.34	1.33	3.40	0.582	3.982
144	12.0	1.33	2.04	0.582	2.662
176	14.67	1.33	0.68	0.582	1.262

On the basis of the preceding calculations, **use $L = 8.5$ ft for $z \le 8$ ft and $L = 4$ ft for $z > 8$ ft.**

Determination of l_l
From Eq. (8.52),

$$l_l = \frac{S_V \sigma_a' [\mathrm{FS}_{(P)}]}{4\sigma_o' \tan \phi_F'}$$

With $\sigma_a' = \gamma_1 z K_a$, $\mathrm{FS}_{(P)} = 1.5$; with $\sigma_o' = \gamma_1 z$, $\phi_F' = \frac{2}{3}\phi_1'$. So

$$l_l = \frac{S_V K_a [\mathrm{FS}_{(P)}]}{4 \tan \phi_F'} = \frac{S_V (0.26)(1.5)}{4 \tan\left[\left(\frac{2}{3}\right)(36)\right]} = 0.219 S_V$$

At $z = 16$ in.,

$$l_l = 0.219 S_V = (0.219)\left(\frac{20}{12}\right) = 0.365 \text{ ft} \le 3 \text{ ft}$$

So, use $l_l = 3$ ft. ∎

8.17 *Retaining Walls with Geogrid Reinforcement*

Geogrids can also be used as reinforcement in granular backfill for the construction of retaining walls. Figure 8.31 shows typical schematic diagrams of retaining walls with geogrid reinforcement.

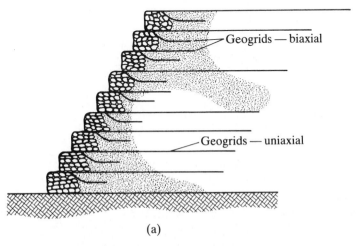

(a)

Figure 8.31 Typical schematic diagrams of retaining walls with geogrid reinforcement: (a) geogrid wraparound wall; (b) wall with gabion facing; (c) concrete panel–faced wall (after The Tensar Corporation, 1986)

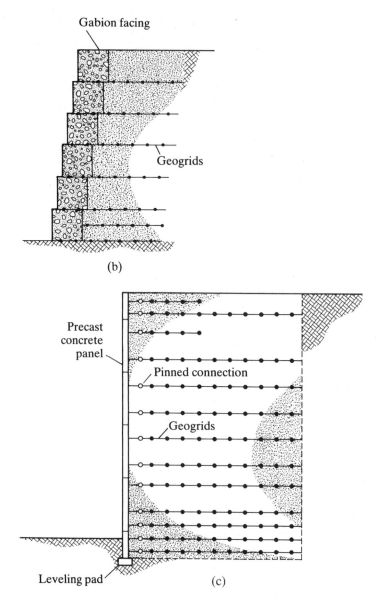

Figure 8.31 (Continued)

Relatively few field measurements are available for lateral earth pressure on retaining walls constructed with geogrid reinforcement. Figure 8.32 shows a comparison of measured and design lateral pressures (Berg et al., 1986) for two retaining walls constructed with precast panel facing. The figure indicates that the measured earth pressures were substantially smaller than those calculated for the Rankine active case.

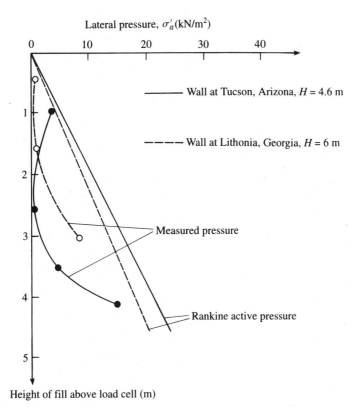

Figure 8.32 Comparison of theoretical and measured lateral pressures in geogrid reinforced retaining walls (based on Berg et al., 1986)

8.18 General Comments

Great progress is being made in the development of rational design procedures for mechanically stabilized earth (MSE) retaining walls. Readers are directed to Transportation Research Circular No. 444 (1995) and Federal Highway Administration Publication No. FHWA-SA-96-071 (1996) for further information. However, following is a summary of a few recent developments:

1. In this chapter, we have used Rankine's active pressure in the design of MSE retaining walls. The appropriate value of the earth pressure coefficient depends, however, on the degree of restraint that the reinforcing elements impose on the soil. If the wall can yield substantially, the Rankine active earth pressure may be appropriate, which is not the case for all types of MSE walls. Figure 8.33 shows the recommended design values for the lateral earth pressure coefficient K. Note that

$$\sigma'_h = K\sigma'_o = K\gamma_1 z$$

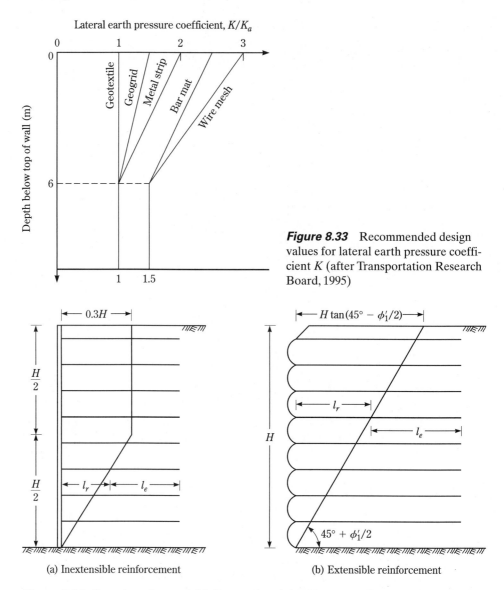

Figure 8.33 Recommended design values for lateral earth pressure coefficient K (after Transportation Research Board, 1995)

(a) Inextensible reinforcement

(b) Extensible reinforcement

Figure 8.34 Location of potential failure surface (after Transportation Research Board, 1995)

where $\sigma_h' =$ effective lateral earth pressure
$\sigma_o' =$ effective vertical stress
$\gamma_1 =$ unit weight of granular backfill

In the figure, $K_a = \tan^2(45 - \phi_1'/2)$, where ϕ_1' is the effective angle of friction of the backfill.

2. In Sections 8.14 and 8.16, the effective length l_e against tie pullout was calculated behind the Rankine failure surface (e.g., see Figure 8.25a). Recent field measure-

ments and theoretical analysis show that the potential failure surface may depend on the type of reinforcement. Figure 8.34a shows the potential failure plane locations for walls with inextensible reinforcement in the granular backfill, and Figure 8.34b shows the locations for such walls with extensible reinforcement.

New developments in the design of MSE walls will be incorporated into future editions of this text.

Problems

In Problems 8.1 through 8.7, use $\gamma_{\text{concrete}} = 23.58 \text{ kN/m}^3 (150 \text{ lb/ft}^3)$. Also, in Eq. (8.11), use $k_1 = k_2 = 2/3$ and $P_p = 0$.

8.1 For the cantilever retaining wall shown in Figure P8.1, let the following data be given:

Wall dimensions: $H = 8$ m, $x_1 = 0.4$ m, $x_2 = 0.6$ m,
$x_3 = 1.5$ m, $x_4 = 3.5$ m, $x_5 = 0.96$ m,
$D = 1.75$ m, $\alpha = 10°$

Soil properties: $\gamma_1 = 16.8 \text{ kN/m}^3$, $\phi_1' = 32°$, $\gamma_2 = 17.6 \text{ kN/m}^3$,
$\phi_2' = 28°$, $c_2' = 30 \text{ kN/m}^2$

Calculate the factor of safety with respect to overturning, sliding, and bearing capacity.

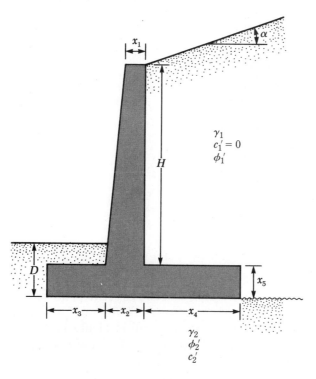

Figure P8.1

8.2 Repeat Problem 8.1 with the following data:

Wall dimensions: $H = 20$ ft, $x_1 = 12$ in., $x_2 = 27$ in., $x_3 = 4.5$ ft, $x_4 = 7.5$ ft, $x_5 = 2.75$ ft, $D = 4$ ft, $\alpha = 5°$

Soil properties: $\gamma_1 = 117$ lb/ft³, $\phi_1' = 34°$, $\gamma_2 = 107$ lb/ft³, $\phi_2' = 18°$, $c_2' = 1050$ lb/ft²

8.3 Repeat Problem 8.1 with the following data:

Wall dimensions: $H = 5.49$ m, $x_1 = 0.46$ m, $x_2 = 0.58$ m, $x_3 = 0.92$ m, $x_4 = 1.55$ m, $x_5 = 0.61$ m, $D = 1.22$ m, $\alpha = 0°$

Soil properties: $\gamma_1 = 18.08$ kN/m³, $\phi_1' = 36°$, $\gamma_2 = 19.65$ kN/m³, $\phi_2' = 15°$, $c_2' = 44$ kN/m²

8.4 A gravity retaining wall is shown in Figure P8.4. Calculate the factor of safety with respect to overturning and sliding, given the following data:

Wall dimensions: $H = 6$ m, $x_1 = 0.6$ m, $x_2 = 2$ m, $x_3 = 2$ m, $x_4 = 0.5$ m, $x_5 = 0.75$ m, $x_6 = 0.8$ m, $D = 1.5$ m

Soil properties: $\gamma_1 = 16.5$ kN/m³, $\phi_1' = 32°$, $\gamma_2 = 18$ kN/m³, $\phi_2' = 22°$, $c_2' = 40$ kN/m²

Use the Rankine active earth pressure in your calculation.

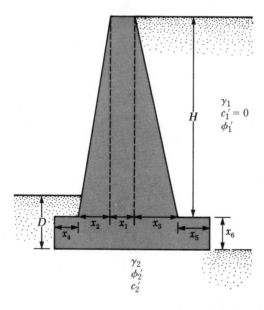

Figure P8.4

8.5 Repeat Problem 8.4, using Coulomb's active pressure in your calculation and letting $\delta = 2/3\phi_1'$.

8.6 Consider the gravity retaining wall in Figure P8.4. Calculate the factor of safety with respect to overturning, sliding, and bearing capacity, given the following data:

Wall dimensions: $H = 15$ ft, $x_1 = 1.5$ ft, $x_2 = 0.8$ ft, $x_3 = 5.25$ ft,
$x_4 = 1.25$ ft, $x_5 = 1.5$ ft, $x_6 = 2.5$ ft, $D = 4$ ft

Soil properties: $\gamma_1 = 121$ lb/ft^3, $\phi_1' = 30°$, $\gamma_2 = 121$ lb/ft^3, $\phi_2' = 20°$,
$c_2' = 1000$ lb/ft^2

Use the Rankine active pressure in your calculation.

8.7 Repeat Problem 8.6, using Coulomb's active pressure in your calculation and letting $\delta = 2/3\phi_1'$.

8.8 In Figure 8.25a, use the following parameters:

Wall: $H = 8$ m
Soil: $\gamma_1 = 17$ kN/m^3 and $\phi_1' = 35°$
Reinforcement: $S_V = 1$ m and $S_H = 1.5$m
Surcharge: $q = 70$ kN/m^2, $a' = 1.5$ m and $b' = 2$ m

Calculate the vertical stress σ_o' [Eqs. (8.24), (8.25), and (8.26)] at $z = 2$ m, 4 m, 6 m, and 8 m.

8.9 For the data given in Problem 8.8, calculate the lateral pressure σ_a' at $z = 2$ m, 4 m, 6 m, and 8 m. Use Eqs. (8.27), (8.28) and (8.29).

8.10 A reinforced-earth retaining wall (Figure 8.25) is to be 10 m high. The following data are given:

Backfill: unit weight, $\gamma_1 = 16$ kN/m^3; soil friction angle, $\phi_1' = 34°$
Reinforcement: vertical spacing, $S_V = 1$ m; horizontal spacing, $S_H = 1.25$ m;
width of reinforcement = 120 mm; $f_y = 260$ MN/m^2; $\phi_\mu' = 25°$;
factor of safety against tie pullout = 3;
factor of safety against tie breaking = 3

Determine:
a. The required thickness of ties
b. The required maximum length of ties

8.11 In Problem 8.10, assume that the ties at all depths are the length determined in Part (b). For the *in situ* soil, $\phi_2' = 25°$, $\gamma_2 = 15.5$ kN/m^3, and $c_2' = 30$ kN/m^2. Calculate the factor of safety against overturning, sliding, and bearing capacity failure.

8.12 A retaining wall with geotextile reinforcement is 6 m high. For the granular backfill, $\gamma_1 = 15.9$ kN/m^3 and $\phi_1' = 30°$. For the geotextile, $\sigma_G = 16$ kN/m. For the design of the wall, determine S_V, L, and l_l. Use FS$_{(B)} = $ FS$_{(P)} = 1.5$.

8.13 For S_V, L, and l_l determined in Problem 8.12, check the overall stability (i.e., factor of safety against overturning, sliding, and bearing capacity failure) of the wall. For the *in situ* soil, $\gamma_2 = 16.8$ kN/m^3, $\phi_2' = 20°$, and $c_2' = 55$ kN/m^2.

References

Bell, J. R., Stilley, A. N., and Vandre, B. (1975). "Fabric Retaining Earth Walls," *Proceedings, Thirteenth Engineering Geology and Soils Engineering Symposium,* Moscow, ID.

Berg, R. R., Bonaparte, R., Anderson, R. P., and Chouery, V. E. (1986). "Design Construction and Performance of Two Tensar Geogrid Reinforced Walls," *Proceedings, Third International Conference on Geotextiles,* Vienna, pp. 401–406.

Binquet, J., and Lee, K. L. (1975). "Bearing Capacity Analysis of Reinforced Earth Slabs," *Journal of the Geotechnical Engineering Division,* American Society of Civil Engineers, Vol. 101, No. GT12, pp. 1257–1276.

Carroll, R., Jr. (1988). "Specifying Geogrids," *Geotechnical Fabric Report,* Industrial Fabric Association International, St. Paul, March/April.

Casagrande, L. (1973). "Comments on Conventional Design of Retaining Structure," *Journal of the Soil Mechanics and Foundations Division,* ASCE, Vol. 99, No. SM2, pp. 181–198.

Darbin, M. (1970). "Reinforced Earth for Construction of Freeways" (in French), *Revue Générale des Routes et Aerodromes,* No. 457, September.

Federal Highway Administration (1996). *Mechanically Stabilized Earth Walls and Reinforced Soil Slopes Design and Construction Guidelines,* Publication No. FHWA-SA-96-071, Washington, DC.

Goh, A. T. C. (1993). "Behavior of Cantilever Retaining Walls," *Journal of Geotechnical Engineering,* ASCE, Vol. 119, No. 11, pp. 1751–1770.

Koerner, R. B. (1990). *Design with Geosynthetics,* 2d ed., Prentice Hall, Englewood Cliffs, NJ.

Laba, J. T., and Kennedy, J. B. (1986). "Reinforced Earth Retaining Wall Analysis and Design," *Canadian Geotechnical Journal,* Vol. 23, No. 3, pp. 317–326.

Lee, K. L., Adams, B. D., and Vagneron, J. J. (1973). "Reinforced Earth Retaining Walls," *Journal of the Soil Mechanics and Foundations Division,* American Society of Civil Engineers, Vol. 99, No. SM10, pp. 745–763.

Schlosser, F., and Long, N. (1974). "Recent Results in French Research on Reinforced Earth," *Journal of the Construction Division,* American Society of Civil Engineers, Vol. 100, No. CO3, pp. 113–237.

Schlosser, F., and Vidal, H. (1969). "Reinforced Earth" (in French), *Bulletin de Liaison des Laboratoires Routier,* Ponts et Chassées, Paris, France, November, pp. 101–144.

Tensar Corporation (1986). Tensar Technical Note. No. TTN:RW1, August.

Terzaghi, K., and Peck, R. B. (1967). *Soil Mechanics in Engineering Practice,* Wiley, New York.

Transportation Research Board (1995). Transportation Research Circular No. 444, National Research Council, Washington, DC.

Vidal, H. (1966). "La terre Armée," *Annales de l'Institut Technique du Bâtiment et des Travaux Publiques,* France, July–August, pp. 888–938.

9

Sheet Pile Walls

9.1 Introduction

Connected or semiconnected sheet piles are often used to build continuous walls for waterfront structures that range from small waterfront pleasure boat launching facilities to large dock facilities. (See Figure 9.1.) In contrast to the construction of other types of retaining wall, the building of sheet pile walls does not usually require dewatering of the site. Sheet piles are also used for some temporary structures, such as braced cuts. (See Chapter 10.) The principles of sheet pile wall design are discussed in the current chapter.

Several types of sheet pile are commonly used in construction: (a) wooden sheet piles, (b) precast concrete sheet piles, and (c) steel sheet piles. Aluminum sheet piles are also marketed.

Wooden sheet piles are used only for temporary, light structures that are above the water table. The most common types are ordinary wooden planks and *Wakefield piles*. The wooden planks are about 50 mm × 300 mm (2 in. × 12 in.) in cross section and are driven edge to edge (Figure 9.2a). Wakefield piles are made by nailing three planks together, with the middle plank offset by 50–75 mm (2–3 in.) (Figure 9.2b). Wooden planks can also be milled to form *tongue-and-groove piles,* as shown in Figure 9.2c. Figure 9.2d shows another type of wooden sheet pile that has precut

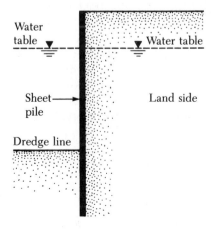

Figure 9.1 Example of waterfront sheet pile wall

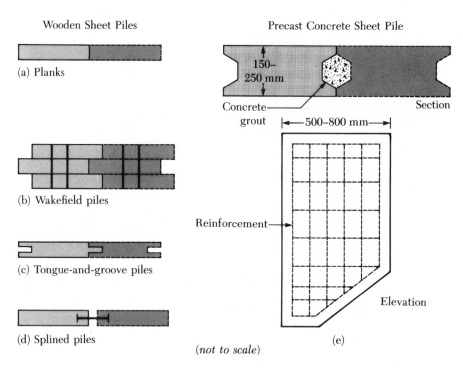

Figure 9.2 Various types of wooden and concrete sheet pile

grooves. Metal *splines* are driven into the grooves of the adjacent sheetings to hold them together after they are sunk into the ground.

Precast concrete sheet piles are heavy and are designed with reinforcements to withstand the permanent stresses to which the structure will be subjected after construction and also to handle the stresses produced during construction. In cross section, these piles are about 500–800 mm (20–32 in.) wide and 150–250 mm (6–10 in.) thick. Figure 9.2e is a schematic diagram of the elevation and the cross section of a reinforced concrete sheet pile.

Steel sheet piles in the United States are about 10–13 mm (0.4–0.5 in.) thick. European sections may be thinner and wider. Sheet pile sections may be *Z, deep arch, low arch,* or *straight web* sections. The interlocks of the sheet pile sections are shaped like a *thumb-and-finger* or *ball-and-socket* joint for watertight connections. Figure 9.3a is a schematic diagram of the thumb-and-finger type of interlocking for straight web sections. The ball-and-socket type of interlocking for *Z* section piles is shown in Figure 9.3b. Figure 9.3c shows a sheet pile wall. Table 9.1 lists the properties of the steel sheet pile sections produced by the Bethlehem Steel Corporation. The allowable design flexural stress for the steel sheet piles is as follows:

Type of steel	Allowable stress	
ASTM A-328	170 MN/m^2	(25,000 lb/in^2)
ASTM A-572	210 MN/m^2	(30,000 lb/in^2)
ASTM A-690	210 MN/m^2	(30,000 lb/in^2)

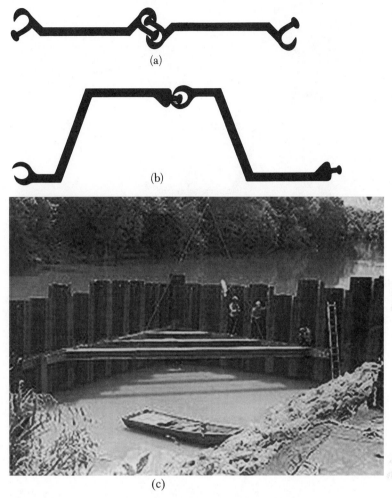

Figure 9.3 (a) Thumb-and-finger type sheet pile connection; (b) ball-and-socket type sheet pile connection; (c) steel sheet pile wall

Table 9.1 Properties of Some Sheet Pile Sections Produced by Bethlehem Steel Corporation

Section designation	Sketch of section	Section modulus		Moment of inertia	
		m^3/m of wall	in^3/ft of wall	m^4/m of wall	in^4/ft of wall
PZ-40		326.4×10^{-5}	60.7	670.5×10^{-6}	490.8

409 mm (16.1 in.)

12.7 mm (0.5 in.)

15.2 mm (0.6 in.)

← Driving distance = 500 mm (19.69 in.) →

Table 9.1 (Continued)

Section designation	Sketch of section	Section modulus		Moment of inertia	
		m^3/m of wall	in^3/ft of wall	m^4/m of wall	in^4/ft of wall
PZ-35	 379 mm (14.0 in.) 12.7 mm (0.5 in.) 15.2 mm (0.6 in.) Driving distance = 575 mm (22.64 in.)	260.5×10^{-5}	48.5	493.4×10^{-6}	361.2
PZ-27	 304.8 mm (12 in.) 9.53 mm ($\frac{3}{8}$ in.) 9.53 mm ($\frac{3}{8}$ in.) Driving distance = 457.2 mm (18 in.)	162.3×10^{-5}	30.2	251.5×10^{-6}	184.2
PZ-22	 228.6 mm (9 in.) 9.53 mm ($\frac{3}{8}$ in.) 9.53 mm ($\frac{3}{8}$ in.) Driving distance = 558.8 mm (22 in.)	97×10^{-5}	18.1	115.2×10^{-6}	84.4
PSA-31	 12.7 mm ($\frac{1}{2}$ in.) Driving distance = 500 mm (19.7 in.)	10.8×10^{-5}	2.01	4.41×10^{-6}	3.23
PSA-23	 9.53 mm ($\frac{3}{8}$ in.) Driving distance = 406.4 mm (16 in.)	12.8×10^{-5}	2.4	5.63×10^{-6}	4.13

Steel sheet piles are convenient to use because of their resistance to the high driving stress that is developed when they are being driven into hard soils. Steel sheet piles are also lightweight and reusable.

9.2 *Construction Methods*

Sheet pile walls may be divided into two basic categories: (a) cantilever and (b) anchored.

In the construction of sheet pile walls, the sheet pile may be driven into the ground and then the backfill placed on the land side, or the sheet pile may first be driven into the ground and the soil in front of the sheet pile dredged. In either case, the soil used for backfill behind the sheet pile wall is usually granular. The soil below the dredge line may be sandy or clayey. The surface of soil on the water side is referred to as the *mud line* or *dredge line*.

Thus, construction methods generally can be divided into two categories (Tsinker, 1983):

1. Backfilled structure
2. Dredged structure

The sequence of construction for a *backfilled structure* is as follows (see Figure 9.4):

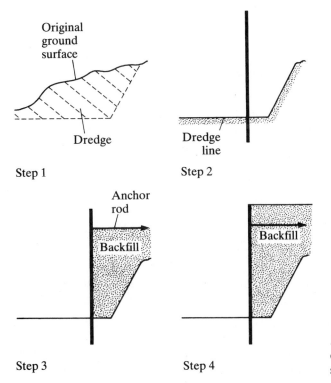

Figure 9.4 Sequence of construction for a backfilled structure

  Step 1. Dredge the *in situ* soil in front and back of the proposed structure.
  Step 2. Drive the sheet piles.
  Step 3. Backfill up to the level of the anchor, and place the anchor system.
  Step 4. Backfill up to the top of the wall.

For a cantilever type of wall, only Steps 1, 2, and 4 apply. The sequence of construction for a *dredged structure* is as follows (see Figure 9.5):

  Step 1. Drive the sheet piles.
  Step 2. Backfill up to the anchor level, and place the anchor system.
  Step 3. Backfill up to the top of the wall.
  Step 4. Dredge the front side of the wall.

With cantilever sheet pile walls, Step 2 is not required.

9.3   *Cantilever Sheet Pile Walls*

Cantilever sheet pile walls are usually recommended for walls of moderate height—about 6 m ($\approx$20 ft) or less, measured above the dredge line. In such walls, the sheet piles act as a wide cantilever beam above the dredge line. The basic principles for estimating net lateral pressure distribution on a cantilever sheet pile wall can be explained with the aid of Figure 9.6. The figure shows the nature of lateral yielding of a cantilever wall penetrating a sand layer below the dredge line. The wall rotates about point O. Because the hydrostatic pressures at any depth from both sides of the wall will cancel each other, we consider only the effective lateral soil pressures. In zone A, the lateral pres-

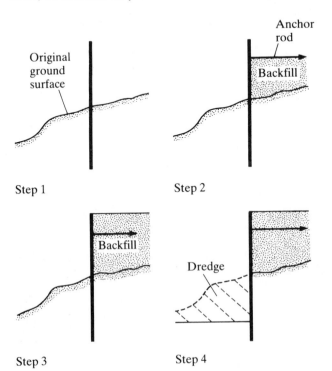

Figure 9.5 Sequence of construction for a dredged structure

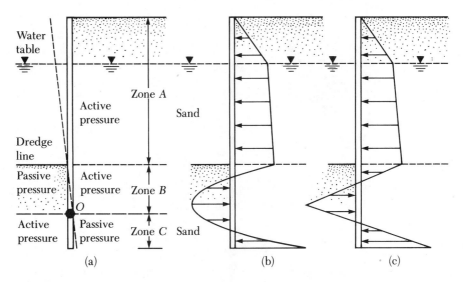

Figure 9.6 Cantilever sheet pile penetrating sand

sure is just the active pressure from the land side. In zone *B,* because of the nature of yielding of the wall, there will be active pressure from the land side and passive pressure from the water side. The condition is reversed in zone *C*—that is, below the point of rotation, *O.* The net actual pressure distribution on the wall is like that shown in Figure 9.6b. However, for design purposes, Figure 9.6c shows a simplified version.

Sections 9.4–9.7 present the mathematical formulation of the analysis of cantilever sheet pile walls. Note that, in some waterfront structures, the water level may fluctuate as the result of tidal effects. Care should be taken in determining the water level that will affect the net pressure diagram.

9.4 *Cantilever Sheet Piling Penetrating Sandy Soils*

To develop the relationships for the proper depth of embedment of sheet piles driven into a granular soil, examine Figure 9.7a. The soil retained by the sheet piling above the dredge line also is sand. The water table is at a depth L_1 below the top of the wall. Let the effective angle of friction of the sand be ϕ'. The intensity of the active pressure at a depth $z = L_1$ is

$$\sigma_1' = \gamma L_1 K_a \tag{9.1}$$

where K_a = Rankine active pressure coefficient = $\tan^2(45 - \phi'/2)$
γ = unit weight of soil above the water table

Similarly, the active pressure at a depth $z = L_1 + L_2$ (i.e., at the level of the dredge line) is

$$\sigma_2' = (\gamma L_1 + \gamma' L_2) K_a \tag{9.2}$$

where γ' = effective unit weight of soil = $\gamma_{sat} - \gamma_w$

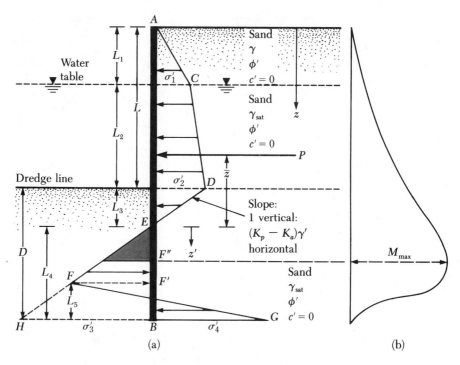

Figure 9.7 Cantilever sheet pile penetrating sand: (a) variation of net pressure diagram; (b) variation of moment

Note that, at the level of the dredge line, the hydrostatic pressures from both sides of the wall are the same magnitude and cancel each other.

To determine the net lateral pressure below the dredge line up to the point of rotation, O, as shown in Figure 9.6a, an engineer has to consider the passive pressure acting from the left side (the water side) toward the right side (the land side) of the wall and also the active pressure acting from the right side toward the left side of the wall. For such cases, ignoring the hydrostatic pressure from both sides of the wall, the active pressure at depth z is

$$\sigma'_a = [\gamma L_1 + \gamma' L_2 + \gamma'(z - L_1 - L_2)]K_a \qquad (9.3)$$

Also, the passive pressure at depth z is

$$\sigma'_p = \gamma'(z - L_1 - L_2)K_p \qquad (9.4)$$

where K_p = Rankine passive pressure coefficient = $\tan^2(45 + \phi'/2)$

Combining Eqs. (9.3) and (9.4) yields the net lateral pressure, namely,

$$\sigma' = \sigma'_a - \sigma'_p = (\gamma L_1 + \gamma' L_2)K_a - \gamma'(z - L_1 - L_2)(K_p - K_a)$$
$$= \sigma'_2 - \gamma'(z - L)(K_p - K_a) \qquad (9.5)$$

where $L = L_1 + L_2$

The net pressure, σ' equals zero at a depth L_3 below the dredge line, so

$$\sigma'_2 - \gamma'(z - L)(K_p - K_a) = 0$$

or

$$(z - L) = L_3 = \frac{\sigma_2'}{\gamma'(K_p - K_a)} \tag{9.6}$$

Equation (9.6) indicates that the slope of the net pressure distribution line *DEF* is 1 vertical to $(K_p - K_a)\gamma'$ horizontal, so, in the pressure diagram,

$$\overline{HB} = \sigma_3' = L_4(K_p - K_a)\gamma' \tag{9.7}$$

At the bottom of the sheet pile, passive pressure, σ_p', acts from the right toward the left side, and active pressure acts from the left toward the right side of the sheet pile, so, at $z = L + D$,

$$\sigma_p' = (\gamma L_1 + \gamma' L_2 + \gamma' D)K_p \tag{9.8}$$

At the same depth,

$$\sigma_a' = \gamma' D K_a \tag{9.9}$$

Hence, the net lateral pressure at the bottom of the sheet pile is

$$\begin{aligned} \sigma_p' - \sigma_a' = \sigma_4' &= (\gamma L_1 + \gamma' L_2)K_p + \gamma' D(K_p - K_a) \\ &= (\gamma L_1 + \gamma' L_2)K_p + \gamma' L_3(K_p - K_a) + \gamma' L_4(K_p - K_a) \\ &= \sigma_5' + \gamma' L_4(K_p - K_a) \end{aligned} \tag{9.10}$$

where
$$\sigma_5' = (\gamma L_1 + \gamma' L_2)K_p + \gamma' L_3(K_p - K_a) \tag{9.11}$$
$$D = L_3 + L_4 \tag{9.12}$$

For the stability of the wall, the principles of statics can now be applied:

$$\Sigma \text{ horizontal forces per unit length of wall} = 0$$

and

$$\Sigma \text{ moment of the forces per unit length of wall about point } B = 0$$

For the summation of the horizontal forces, we have

Area of the pressure diagram $ACDE$ − area of $EFHB$ + area of $FHBG = 0$

or

$$P - \tfrac{1}{2}\sigma_3' L_4 + \tfrac{1}{2}L_5(\sigma_3' + \sigma_4') = 0 \tag{9.13}$$

where P = area of the pressure diagram $ACDE$

Summing the moment of all the forces about point B yields

$$P(L_4 + \bar{z}) - \left(\frac{1}{2}L_4\sigma_3'\right)\left(\frac{L_4}{3}\right) + \frac{1}{2}L_5(\sigma_3' + \sigma_4')\left(\frac{L_5}{3}\right) = 0 \tag{9.14}$$

From Eq. (9.13),

$$L_5 = \frac{\sigma_3' L_4 - 2P}{\sigma_3' + \sigma_4'} \tag{9.15}$$

Combining Eqs. (9.7), (9.10), (9.14), and (9.15) and simplifying them further, we obtain the following fourth-degree equation in terms of L_4:

$$L_4^4 + A_1 L_4^3 - A_2 L_4^2 - A_3 L_4 - A_4 = 0 \tag{9.16}$$

In this equation,

$$A_1 = \frac{\sigma_5'}{\gamma'(K_p - K_a)} \tag{9.17}$$

$$A_2 = \frac{8P}{\gamma'(K_p - K_a)} \tag{9.18}$$

$$A_3 = \frac{6P[2\bar{z}\gamma'(K_p - K_a) + \sigma_5']}{\gamma'^2(K_p - K_a)^2} \tag{9.19}$$

$$A_4 = \frac{P(6\bar{z}\sigma_5' + 4P)}{\gamma'^2(K_p - K_a)^2} \tag{9.20}$$

Step-by-Step Procedure for Obtaining the Pressure Diagram

Based on the preceding theory, a step-by-step procedure for obtaining the pressure diagram for a cantilever sheet pile wall penetrating a granular soil is as follows:

1. Calculate K_a and K_p.
2. Calculate σ_1' [Eq. (9.1)] and σ_2' [Eq. (9.2)]. (*Note:* L_1 and L_2 will be given.)
3. Calculate L_3 [Eq. (9.6)].
4. Calculate P.
5. Calculate $\bar{z}$ (i.e., the center of pressure for the area $ACDE$) by taking the moment about E.
6. Calculate σ_5' [Eq. (9.11)].
7. Calculate A_1, A_2, A_3, and A_4 [Eqs. (9.17) through (9.20)].
8. Solve Eq. (9.16) by trial and error to determine L_4.
9. Calculate σ_4' [Eq. (9.10)].
10. Calculate σ_3' [Eq. (9.7)].
11. Obtain L_5 from Eq. (9.15).
12. Draw a pressure distribution diagram like the one shown in Figure 9.7a.
13. Obtain the theoretical depth [see Eq. (9.12)] of penetration as $L_3 + L_4$. The actual depth of penetration is increased by about 20–30%.

Note that some designers prefer to use a factor of safety on the passive earth pressure coefficient at the beginning. In that case, in Step 1,

$$K_{p(\text{design})} = \frac{K_p}{\text{FS}}$$

where   FS = factor of safety (usually between 1.5 and 2)

For this type of analysis, follow Steps 1–12 with the value of $K_a = \tan^2(45 - \phi'/2)$ and $K_{p(\text{design})}$ (instead of K_p). The actual depth of penetration can now be determined by adding L_3, obtained from Step 3, and L_4, obtained from Step 8.

Calculation of Maximum Bending Moment

The nature of the variation of the moment diagram for a cantilever sheet pile wall is shown in Figure 9.7b. The maximum moment will occur between points E and F'. Obtaining the maximum moment (M_{max}) per unit length of the wall requires determining the point of zero shear. For a new axis z' (with origin at point E) for zero shear,

$$P = \tfrac{1}{2}(z')^2(K_p - K_a)\gamma'$$

or

$$z' = \sqrt{\frac{2P}{(K_p - K_a)\gamma'}} \tag{9.21}$$

Once the point of zero shear force is determined (point F'' in Figure 9.7a), the magnitude of the maximum moment can be obtained as

$$M_{\text{max}} = P(\bar{z} + z') - \left[\tfrac{1}{2}\gamma'z'^2(K_p - K_a)\right]\left(\tfrac{1}{3}\right)z' \tag{9.22}$$

The necessary profile of the sheet piling is then sized according to the allowable flexural stress of the sheet pile material, or

$$S = \frac{M_{\text{max}}}{\sigma_{\text{all}}} \tag{9.23}$$

where S = section modulus of the sheet pile required per unit length of the structure
   σ_{all} = allowable flexural stress of the sheet pile

Example 9.1

Figure 9.8 shows a cantilever sheet pile wall penetrating a granular soil. Here, $L_1 = 2$ m, $L_2 = 3$ m, $\gamma = 15.9$ kN/m³, $\gamma_{\text{sat}} = 19.33$ kN/m³, and $\phi' = 32°$.

   a. What is the theoretical depth of embedment, D?
   b. For a 30% increase in D, what should be the total length of the sheet piles?
   c. What should be the minimum section modulus of the sheet piles? Use $\sigma_{\text{all}} = 172$ MN/m².

Solution

Part a
Using Figure 9.7a for the pressure distribution diagram, one can now prepare the following table for a step-by-step calculation:

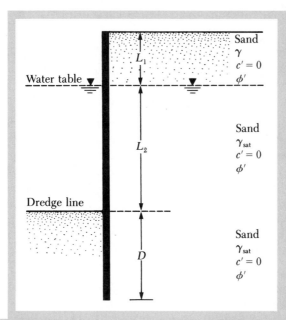

Figure 9.8   Cantilever sheet pile wall

Quantity required	Eq. no.	Equation and calculation
K_a	—	$\tan^2\left(45 - \dfrac{\phi'}{2}\right) = \tan^2\left(45 - \dfrac{32}{2}\right) = 0.307$
K_p	—	$\tan^2\left(45 + \dfrac{\phi'}{2}\right) = \tan^2\left(45 + \dfrac{32}{2}\right) = 3.25$
σ_1'	9.1	$\gamma L_1 K_a = (15.9)(2)(0.307) = 9.763 \text{ kN/m}^2$
σ_2'	9.2	$(\gamma L_1 + \gamma' L_2)K_a = [(15.9)(2) + (19.33 - 9.81)(3)](0.307) = 18.53 \text{ kN/m}^2$
L_3	9.6	$\dfrac{\sigma_2'}{\gamma'(K_p - K_a)} = \dfrac{18.53}{(19.33 - 9.81)(3.25 - 0.307)} = 0.66 \text{ m}$
P	—	$\frac{1}{2}\sigma_1' L_1 + \sigma_1' L_2 + \frac{1}{2}(\sigma_2' - \sigma_1')L_2 + \frac{1}{2}\sigma_2' L_3$ $= \left(\frac{1}{2}\right)(9.763)(2) + (9.763)(3) + \left(\frac{1}{2}\right)(18.53 - 9.763)(3) + \left(\frac{1}{2}\right)(18.53)(0.66)$ $= 9.763 + 29.289 + 13.151 + 6.115 = 58.32 \text{ kN/m}$
$\bar{z}$	—	$\dfrac{\Sigma M_E}{P} = \dfrac{1}{58.32}\left[\begin{array}{l} 9.763(0.66 + 3 + \frac{2}{3}) + 29.289\,(0.66 + \frac{3}{2}) \\ + 13.151(0.66 + \frac{3}{2}) + 6.115(0.66 \times \frac{2}{3}) \end{array}\right] = 2.23 \text{ m}$
σ_5'	9.11	$(\gamma L_1 + \gamma' L_2)K_p + \gamma' L_3(K_p - K_a) = [(15.9)(2) + (19.33 - 9.81)(3)](3.25)$ $+ (19.33 - 9.81)(0.66)(3.25 - 0.307) = 214.66 \text{ kN/m}^2$
A_1	9.17	$\dfrac{\sigma_5'}{\gamma'(K_p - K_a)} = \dfrac{214.66}{(19.33 - 9.81)(3.25 - 0.307)} = 7.66$
A_2	9.18	$\dfrac{8P}{\gamma'(K_p - K_a)} = \dfrac{(8)(58.32)}{(19.33 - 9.81)(3.25 - 0.307)} = 16.65$
A_3	9.19	$\dfrac{6P[2\bar{z}\gamma'(K_p - K_a) + \sigma_5']}{\gamma'^2(K_p - K_a)^2}$ $= \dfrac{(6)(58.32)[(2)(2.23)(19.33 - 9.81)(3.25 - 0.307) + 214.66]}{(19.33 - 9.81)^2(3.25 - 0.307)^2} = 151.93$

A_4	9.20	$\dfrac{P(6\bar{z}\sigma_5' + 4P)}{\gamma'^2(K_p - K_a)^2} = \dfrac{58.32[(6)(2.23)(214.66) + (4)(58.32)]}{(19.33 - 9.81)^2(3.25 - 0.307)^2} = 230.72$
L_4	9.16	$L_4^4 + A_1 L_4^3 - A_2 L_4^2 - A_3 L_4 - A_4 = 0$
		$L_4^4 + 7.66 L_4^3 - 16.65 L_4^2 - 151.93 L_4 - 230.72 = 0; \ L_4 \approx 4.8 \text{ m}$

Thus,

$$D_{\text{theory}} = L_3 + L_4 = 0.66 + 4.8 = \mathbf{5.46 \text{ m}}$$

Part b
The total length of the sheet piles is

$$L_1 + L_2 + 1.3(L_3 + L_4) = 2 + 3 + 1.3(5.46) = \mathbf{12.1 \text{ m}}$$

Part c
Finally, we have the following table:

Quantity required	Eq. no.	Equation and calculation
z'	9.21	$\sqrt{\dfrac{2P}{(K_p - K_a)\gamma'}} = \sqrt{\dfrac{(2)(58.32)}{(3.25 - 0.307)(19.33 - 9.81)}} = 2.04 \text{ m}$
$M_{\max}$	9.22	$P(\bar{z} + z') - \left[\dfrac{1}{2}\gamma' z'^2 (K_p - K_a)\right]\dfrac{z'}{3} = (58.32)(2.23 + 2.04)$ $-\left[\left(\dfrac{1}{2}\right)(19.33 - 9.81)(2.04)^2(3.25 - 0.307)\right]\dfrac{2.04}{3} = 209.39 \text{ kN·m/m}$
S	9.29	$\dfrac{M_{\max}}{\sigma_{\text{all}}} = \dfrac{209.39 \text{ kN·m}}{172 \times 10^3 \text{ kN/m}^2} = \mathbf{1.217 \times 10^{-3} \text{ m}^3/\text{m of wall}}$

■

9.5 *Special Cases for Cantilever Walls Penetrating a Sandy Soil*

Following are two special cases of the mathematical formulation shown in Section 9.4.

Case 1. *Sheet Pile Wall in the Absence of the Water Table*

In the absence of the water table, the net pressure diagram on the cantilever sheet pile wall will be as shown in Figure 9.9, which is a modified version of Figure 9.7. In this case,

$$\sigma_2' = \gamma L K_a \tag{9.24}$$

$$\sigma_3' = L_4 (K_p - K_a)\gamma \tag{9.25}$$

$$\sigma_4' = \sigma_5' + \gamma L_4 (K_p - K_a) \tag{9.26}$$

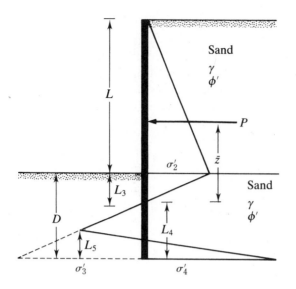

Figure 9.9 Sheet piling penetrating a sandy soil in the absence of the water table

$$\sigma_5' = \gamma L K_p + \gamma L_3 (K_p - K_a) \tag{9.27}$$

$$L_3 = \frac{\sigma_2'}{\gamma(K_p - K_a)} = \frac{L K_a}{(K_p - K_a)} \tag{9.28}$$

$$P = \tfrac{1}{2}\sigma_2' L + \tfrac{1}{2}\sigma_2' L_3 \tag{9.29}$$

$$\bar{z} = L_3 + \frac{L}{3} = \frac{L K_a}{K_p - K_a} + \frac{L}{3} = \frac{L(2K_a + K_p)}{3(K_p - K_a)} \tag{9.30}$$

and Eq. (9.16) transforms to

$$L_4^4 + A_1' L_4^3 - A_2' L_4^2 - A_3' L_4 - A_4' = 0 \tag{9.31}$$

where

$$A_1' = \frac{\sigma_5'}{\gamma(K_p - K_a)} \tag{9.32}$$

$$A_2' = \frac{8P}{\gamma(K_p - K_a)} \tag{9.33}$$

$$A_3' = \frac{6P[2\bar{z}\gamma(K_p - K_a) + \sigma_5']}{\gamma^2(K_p - K_a)^2} \tag{9.34}$$

$$A_4' = \frac{P(6\bar{z}\sigma_5' + 4P)}{\gamma^2(K_p - K_a)^2} \tag{9.35}$$

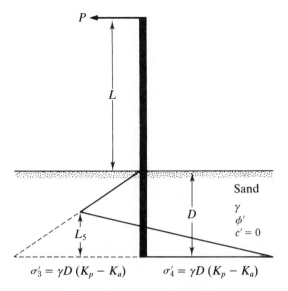

$$\sigma'_3 = \gamma D \,(K_p - K_a) \qquad\qquad \sigma'_4 = \gamma D \,(K_p - K_a)$$

Figure 9.10   Free cantilever sheet piling penetrating a layer of sand

Case 2. Free Cantilever Sheet Piling

Figure 9.10 shows a free cantilever sheet pile wall penetrating a sandy soil and subjected to a line load of P per unit length of the wall. For this case,

$$D^4 - \left[\frac{8P}{\gamma(K_p - K_a)}\right]D^2 - \left[\frac{12PL}{\gamma(K_p - K_a)}\right]D - \left[\frac{2P}{\gamma(K_p - K_a)}\right]^2 = 0 \qquad (9.36)$$

$$L_5 = \frac{\gamma(K_p - K_a)D^2 - 2P}{2D(K_p - K_a)\gamma} \qquad (9.37)$$

$$M_{\max} = P(L + z') - \frac{\gamma z'^3 (K_p - K_a)}{6} \qquad (9.38)$$

and

$$z' = \sqrt{\frac{2P}{\gamma'(K_p - K_a)}} \qquad (9.39)$$

9.6   *Cantilever Sheet Piling Penetrating Clay*

At times, cantilever sheet piles must be driven into a clay layer possessing an undrained cohesion $c(\phi = 0)$. The net pressure diagram will be somewhat different from that shown in Figure 9.7a. Figure 9.11 shows a cantilever sheet pile wall

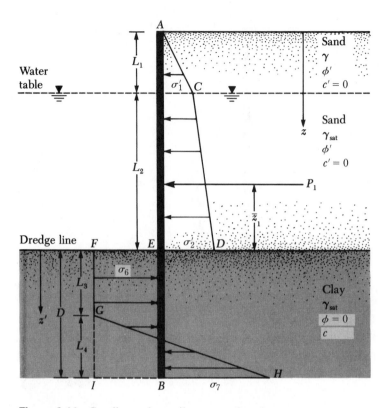

Figure 9.11 Cantilever sheet pile penetrating clay

driven into clay with a backfill of granular soil above the level of the dredge line. The water table is at a depth L_1 below the top of the wall. As before, Eqs. (9.1) and (9.2) give the intensity of the net pressures σ_1' and σ_2', and the diagram for pressure distribution above the level of the dredge line can be drawn. The diagram for net pressure distribution below the dredge line can now be determined as follows.

At any depth greater than $L_1 + L_2$, for $\phi = 0$, the Rankine active earth pressure coefficient $K_a = 1$. Similarly, for $\phi = 0$, the Rankine passive earth pressure coefficient $(K_p) = 1$. Thus, above the point of rotation (point O in Figure 9.6a), the active pressure, from right to left is

$$\sigma_a = [\gamma L_1 + \gamma' L_2 + \gamma_{sat}(z - L_1 - L_2)] - 2c \qquad (9.40)$$

Similarly, the passive pressure from left to right may be expressed as

$$\sigma_p = \gamma_{sat}(z - L_1 - L_2) + 2c \qquad (9.41)$$

Thus, the net pressure is

$$\sigma_6 = \sigma_p - \sigma_a = [\gamma_{sat}(z - L_1 - L_2) + 2c]$$
$$- [\gamma L_1 + \gamma' L_2 + \gamma_{sat}(z - L_1 - L_2)] + 2c$$
$$= 4c - (\gamma L_1 + \gamma' L_2) \qquad (9.42)$$

At the bottom of the sheet pile, the passive pressure from right to left is

$$\sigma_p = (\gamma L_1 + \gamma' L_2 + \gamma_{sat} D) + 2c \qquad (9.43)$$

Similarly, the active pressure from left to right is

$$\sigma_a = \gamma_{sat} D - 2c \qquad (9.44)$$

Hence, the net pressure is

$$\sigma_7 = \sigma_p - \sigma_a = 4c + (\gamma L_1 + \gamma' L_2) \qquad (9.45)$$

For equilibrium analysis, $\Sigma F_H = 0$; that is, the area of the pressure diagram *ACDE* minus the area of *EFIB* plus the area of *GIH* = 0, or

$$P_1 - [4c - (\gamma L_1 + \gamma' L_2)]D + \tfrac{1}{2}L_4[4c - (\gamma L_1 + \gamma' L_2) + 4c + (\gamma L_1 + \gamma' L_2)] = 0$$

where P_1 = area of the pressure diagram *ACDE*

Simplifying the preceding equation produces

$$L_4 = \frac{D[4c - (\gamma L_1 + \gamma' L_2)] - P_1}{4c} \qquad (9.46)$$

Now, taking the moment about point B ($\Sigma M_B = 0$) yields

$$P_1(D + \bar{z}_1) - [4c - (\gamma L_1 + \gamma' L_2)]\frac{D^2}{2} + \frac{1}{2}L_4(8c)\left(\frac{L_4}{3}\right) = 0 \qquad (9.47)$$

where $\bar{z}_1$ = distance of the center of pressure of the pressure diagram *ACDE*, measured from the level of the dredge line

Combining Eqs. (9.46) and (9.47) yields

$$D^2[4c - (\gamma L_1 + \gamma' L_2)] - 2DP_1 - \frac{P_1(P_1 + 12c\bar{z}_1)}{(\gamma L_1 + \gamma' L_2) + 2c} = 0 \qquad (9.48)$$

Equation (9.48) may be solved to obtain D, the theoretical depth of penetration of the clay layer by the sheet pile.

Step-by-Step Procedure for Obtaining the Pressure Diagram

1. Calculate $K_a = \tan^2(45 - \phi'/2)$ for the granular soil (backfill).
2. Obtain σ_1' and σ_2'. [See Eqs. (9.1) and (9.2).]
3. Calculate P_1 and $\bar{z}_1$.
4. Use Eq. (9.48) to obtain the theoretical value of D.
5. Using Eq. (9.46), calculate L_4.
6. Calculate σ_6 and σ_7. [See Eqs. (9.42) and (9.45).]
7. Draw the pressure distribution diagram as shown in Figure 9.11.
8. The actual depth of penetration is

$$D_{actual} = 1.4 \text{ to } 1.6(D_{theoretical})$$

Maximum Bending Moment

According to Figure 9.11, the maximum moment (zero shear) will be between $L_1 + L_2 < z < L_1 + L_2 + L_3$. Using a new coordinate system z' (with $z' = 0$ at the dredge line) for zero shear gives

$$P_1 - \sigma_6 z' = 0$$

or

$$z' = \frac{P_1}{\sigma_6} \tag{9.49}$$

The magnitude of the maximum moment may now be obtained:

$$M_{max} = P_1(z' + \bar{z}_1) - \frac{\sigma_6 z'^2}{2} \tag{9.50}$$

Knowing the maximum bending moment, we determine the section modulus of the sheet pile section from Eq. (9.23).

Example 9.2

In Figure 9.12, for the sheet pile wall, determine

a. the theoretical and actual depth of penetration. Use $D_{actual} = 1.5 D_{theory}$.
b. the minimum size of sheet pile section necessary. Use $\sigma_{all} = 172.5$ MN/m².

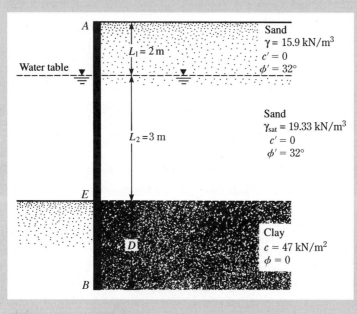

Figure 9.12 Cantilever sheet pile penetrating into saturated clay

Solution

We will follow the step-by-step procedure given in Section 9.6:

Step 1.

$$K_a = \tan^2\left(45 - \frac{\phi'}{2}\right) = \tan^2\left(45 - \frac{32}{2}\right) = 0.307$$

Step 2.

$$\sigma_1' = \gamma L_1 K_a = (15.9)(2)(0.307) = 9.763 \text{ kN/m}^2$$

$$\sigma_2' = (\gamma L_1 + \gamma' L_2)K_a = [(15.9)(2) + (19.33 - 9.81)3]0.307$$

$$= 18.53 \text{ kN/m}^2$$

Step 3. From the net pressure distribution diagram given in Figure 9.11, we have

$$P_1 = \frac{1}{2}\sigma_1' L_1 + \sigma_1' L_2 + \frac{1}{2}(\sigma_2' - \sigma_1')L_2$$

$$= 9.763 + 29.289 + 13.151 = 52.2 \text{ kN/m}$$

and

$$\bar{z}_1 = \frac{1}{52.2}\left[9.763\left(3 + \frac{2}{3}\right) + 29.289\left(\frac{3}{2}\right) + 13.151\left(\frac{3}{3}\right)\right]$$

$$= 1.78 \text{ m}$$

Step 4. From Eq. (9.48),

$$D^2[4c - (\gamma L_1 + \gamma' L_2)] - 2DP_1 - \frac{P_1(P_1 + 12c\bar{z}_1)}{(\gamma L_1 + \gamma' L_2) + 2c} = 0$$

Substituting proper values yields

$$D^2\{(4)(47) - [(2)(15.9) + (19.33 - 9.81)3]\} - 2D(52.2)$$

$$- \frac{52.2[52.2 + (12)(47)(1.78)]}{[(15.9)(2) + (19.33 - 9.81)3] + (2)(47)} = 0$$

or

$$127.64D^2 - 104.4D - 357.15 = 0$$

Solving the preceding equation, we obtain $D = 2.13$ m.

Step 5. From Eq. (9.46),

$$L_4 = \frac{D[4c - (\gamma L_1 + \gamma' L_2)] - P_1}{4c}$$

and

$$4c - (\gamma L_1 + \gamma' L_2) = (4)(47) - [(15.9)(2) + (19.33 - 9.81)3]$$

$$= 127.64 \text{ kN/m}^2$$

So

$$L_4 = \frac{2.13(127.64) - 52.2}{(4)(47)} = 1.17 \text{ m}$$

Step 6.

$$\sigma_6 = 4c - (\gamma L_1 + \gamma' L_2) = 127.64 \text{ kN/m}^2$$

$$\sigma_7 = 4c + (\gamma L_1 + \gamma' L_2) = 248.36 \text{ kN/m}^2$$

Step 7. The net pressure distribution diagram can now be drawn, as shown in Figure 9.11.

Step 8. $D_{\text{actual}} \approx 1.5 D_{\text{theoretical}} = 1.5(2.13) \approx$ **3.2 m**

Maximum-Moment Calculation

From Eq. (9.49),

$$z' = \frac{P_1}{\sigma_6} = \frac{52.2}{127.64} \approx 0.41 \text{ m}$$

Again, from Eq. (9.50),

$$M_{\text{max}} = P_1(z' + \bar{z}_1) - \frac{\sigma_6 z'^2}{2}$$

So

$$M_{\text{max}} = 52.2(0.41 + 1.78) - \frac{127.64(0.41)^2}{2}$$

$$= 114.32 - 10.73 = 103.59 \text{ kN-m/m}$$

The minimum required section modulus (assuming that $\sigma_{\text{all}} = 172.5 \text{ MN/m}^2$) is

$$S = \frac{103.59 \text{ kN-m/m}}{172.5 \times 10^3 \text{ kN/m}^2} = \textbf{0.6} \times \textbf{10}^{-3} \textbf{ m}^3\textbf{/m of the wall} \quad \blacksquare$$

9.7 *Special Cases for Cantilever Walls Penetrating Clay*

As in Section 9.5, relationships for special cases for cantilever walls penetrating clay may also be derived.

Case 1. Sheet Pile Wall in the Absence of the Water Table

Referring to Figure 9.13, we can write

$$\sigma'_2 = \gamma L K_a \tag{9.51}$$

$$\sigma_6 = 4c - \gamma L \tag{9.52}$$

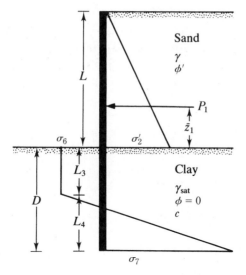

Figure 9.13 Sheet pile wall penetrating clay

$$\sigma_7 = 4c + \gamma L \tag{9.53}$$

$$P_1 = \tfrac{1}{2}L\sigma_2' = \tfrac{1}{2}\gamma L^2 K_a \tag{9.54}$$

and

$$L_4 = \frac{D(4c - \gamma L) - \tfrac{1}{2}\gamma L^2 K_a}{4c} \tag{9.55}$$

The theoretical depth of penetration, D, can be calculated [in a manner similar to the calculation of Eq. (9.48)] as

$$D^2(4c - \gamma L) - 2DP_1 - \frac{P_1(P_1 + 12c\bar{z}_1)}{\gamma L + 2c} = 0 \tag{9.56}$$

where $\quad \bar{z}_1 = \dfrac{L}{3} \tag{9.57}$

The magnitude of the maximum moment in the wall is

$$M_{max} = P_1(z' + \bar{z}_1) - \frac{\sigma_6 z'^2}{2} \tag{9.58}$$

where $\quad z' = \dfrac{P_1}{\sigma_6} = \dfrac{\tfrac{1}{2}\gamma L^2 K_a}{4c - \gamma L} \tag{9.59}$

Case 2. Free Cantilever Sheet Pile Wall Penetrating Clay

Figure 9.14 shows a free cantilever sheet pile wall penetrating a clay layer. The wall is being subjected to a line load of P per unit length. For this case,

$$\sigma_6 = \sigma_7 = 4c \tag{9.60}$$

The depth of penetration, D, may be obtained from the relation

$$4D^2c - 2PD - \frac{P(P + 12cL)}{2c} = 0 \tag{9.61}$$

Also, note that, for a construction of the pressure diagram,

$$L_4 = \frac{4cD - P}{4c} \tag{9.62}$$

The maximum moment in the wall is

$$M_{\max} = P(L + z') - \frac{4cz'^2}{2} \tag{9.63}$$

where $z' = \dfrac{P}{4c}$ (9.64)

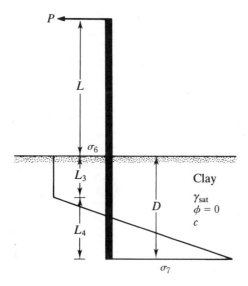

Figure 9.14 Free cantilever sheet piling penetrating clay

Anchored Sheet Pile Walls

When the height of the backfill material behind a cantilever sheet pile wall exceeds about 6 m ($\approx$20 ft), tying the wall near the top to anchor plates, anchor walls, or anchor piles becomes more economical. This type of construction is referred to as *anchored sheet pile wall* or an *anchored bulkhead.* Anchors minimize the depth of penetration required by the sheet piles and also reduce the cross-sectional area and weight of the sheet piles needed for construction. However, the tie rods and anchors must be carefully designed.

The two basic methods of designing anchored sheet pile walls are (a) the *free earth support* method and (b) the *fixed earth support* method. Figure 9.15 shows the assumed nature of deflection of the sheet piles for the two methods.

The free earth support method involves a minimum penetration depth. Below the dredge line, no pivot point exists for the static system. The nature of the variation

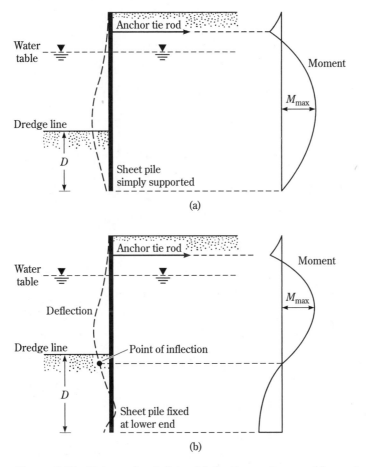

Figure 9.15 Nature of variation of deflection and moment for anchored sheet piles: (a) free earth support method; (b) fixed earth support method

of the bending moment with depth for both methods is also shown in Figure 9.15. Note that

$$D_{\text{free earth}} < D_{\text{fixed earth}}$$

9.9 Free Earth Support Method for Penetration of Sandy Soil

Figure 9.16 shows an anchor sheet pile wall with a granular soil backfill; the wall has been driven into a granular soil. The tie rod connecting the sheet pile and the anchor is located at a depth l_1 below the top of the sheet pile wall.

The diagram of the net pressure distribution above the dredge line is similar to that shown in Figure 9.7. At depth $z = L_1$, $\sigma_1' = \gamma L_1 K_a$, and at $z = L_1 + L_2$, $\sigma_2' = (\gamma L_1 + \gamma' L_2) K_a$. Below the dredge line, the net pressure will be zero at $z = L_1 + L_2 + L_3$. The relation for L_3 is given by Eq. (9.6), or

$$L_3 = \frac{\sigma_2'}{\gamma'(K_p - K_a)}$$

At $z = L_1 + L_2 + L_3 + L_4$, the net pressure is given by

$$\sigma_8' = \gamma'(K_p - K_a)L_4 \tag{9.65}$$

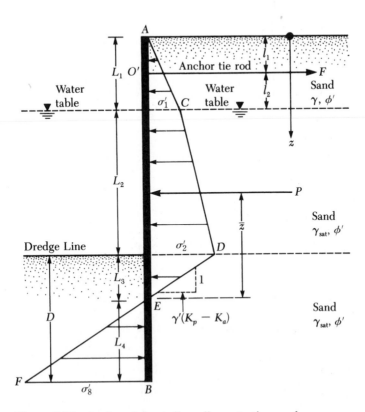

Figure 9.16 Anchored sheet pile wall penetrating sand

Note that the slope of the line DEF is 1 vertical to $\gamma'(K_p - K_a)$ horizontal.

For equilibrium of the sheet pile, Σ horizontal forces = 0, and Σ moment about $O' = 0$. (*Note:* Point O' is located at the level of the tie rod.)

Summing the forces in the horizontal direction (per unit length of the wall) gives

$$\text{Area of the pressure diagram } ACDE - \text{area of } EBF - F = 0$$

where $\quad F$ = tension in the tie rod/unit length of the wall, or

$$P - \tfrac{1}{2}\sigma'_8 L_4 - F = 0$$

or

$$F = P - \tfrac{1}{2}[\gamma'(K_p - K_a)]L_4^2 \qquad (9.66)$$

where $\quad P$ = area of the pressure diagram $ACDE$

Now, taking the moment about point O' gives

$$-P[(L_1 + L_2 + L_3) - (\bar{z} + l_1)] + \tfrac{1}{2}[\gamma'(K_p - K_a)]L_4^2(l_2 + L_2 + L_3 + \tfrac{2}{3}L_4) = 0$$

or

$$L_4^3 + 1.5L_4^2(l_2 + L_2 + L_3) - \frac{3P[(L_1 + L_2 + L_3) - (\bar{z} + l_1)]}{\gamma'(K_p - K_a)} = 0 \qquad (9.67)$$

Equation (9.67) may be solved by trial and error to determine the theoretical depth, L_4:

$$D_{\text{theoretical}} = L_3 + L_4$$

The theoretical depth is increased by about 30–40% for actual construction, or

$$D_{\text{actual}} = 1.3 \text{ to } 1.4 D_{\text{theoretical}} \qquad (9.68)$$

The step-by-step procedure in Section 9.4 indicated that a factor of safety can be applied to K_p at the beginning [i.e., $K_{p(\text{design})} = K_p/\text{FS}$]. If that is done, there is no need to increase the theoretical depth by 30–40%. This approach is often more conservative.

The maximum theoretical moment to which the sheet pile will be subjected occurs at a depth between $z = L_1$ and $z = L_1 + L_2$. The depth z for zero shear and hence maximum moment may be evaluated from

$$\tfrac{1}{2}\sigma'_1 L_1 - F + \sigma'_1(z - L_1) + \tfrac{1}{2}K_a\gamma'(z - L_1)^2 = 0 \qquad (9.69)$$

Once the value of z is determined, the magnitude of the maximum moment is easily obtained.

Sometimes, the dredge line slopes at an angle β with respect to the horizontal, as shown in Figure 9.17a. In that case, the passive pressure coefficient will not be equal to $\tan^2(45 + \phi'/2)$. The variations of K_p (Coulomb's passive earth pressure

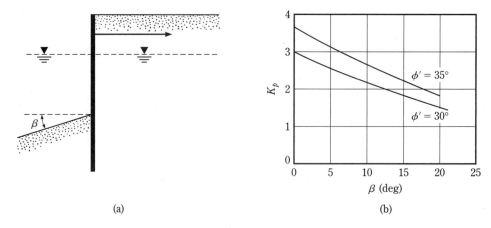

Figure 9.17 (a) Anchored sheet pile wall with sloping dredge line; (b) variation of K_p with β and ϕ'

analysis for a wall friction angle of zero) with β for $\phi' = 30°$ and $35°$ are shown in Figure 9.17b. With these values of K_p, the procedure described in this section may be used to determine the depth of penetration, D.

9.10 Moment Reduction for Anchored Sheet Pile Walls

Sheet piles are flexible, and hence sheet pile walls yield (i.e., become displaced laterally), which redistributes the lateral earth pressure. This change tends to reduce the maximum bending moment, M_{max}, as calculated by the procedure outlined in Section 9.9. For that reason, Rowe (1952, 1957) suggested a procedure for reducing the maximum design moment on the sheet pile walls *obtained from the free earth support method*. This section discusses the procedure of moment reduction for sheet piles *penetrating into sand*.

In Figure 9.18, which is valid for the case of a sheet pile penetrating sand, the following notation is used:

1. H' = total height of pile driven (i.e., $L_1 + L_2 + D_{actual}$)

2. Relative flexibility of pile = $\rho = 10.91 \times 10^{-7} \left(\dfrac{H'^4}{EI} \right)$ (9.70)

where H' is in meters
  E = modulus of elasticity of the pile material (MN/m^2)
  I = moment of inertia of the pile section per meter of the wall (m^4/m of wall)

3. M_d = design moment
4. M_{max} = maximum theoretical moment

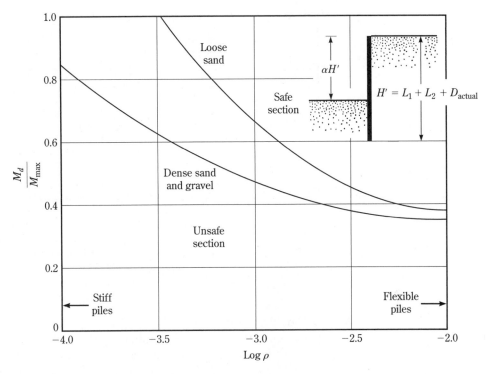

Figure 9.18 Plot of log ρ against $M_d/M_{\max}$ for sheet pile walls penetrating sand (after Rowe, 1952)

In English units, Eq. (9.70) takes the form

$$\rho = \frac{H'^4}{EI} \tag{9.71}$$

where H' is in ft, E is in lb/in², and I is in in⁴/ft of the wall

The procedure for the use of the moment reduction diagram (see Figure 9.18) is as follows:

Step 1. Choose a sheet pile section (e.g., from among those given in Table 9.1).
Step 2. Find the modulus S of the selected section (Step 1) per unit length of the wall.
Step 3. Determine the moment of inertia of the section (Step 1) per unit length of the wall.
Step 4. Obtain H' and calculate ρ [see Eq. (9.70) or Eq. (9.71)].
Step 5. Find log ρ.
Step 6. Find the moment capacity of the pile section chosen in Step 1 as $M_d = \sigma_{\text{all}}S$.
Step 7. Determine $M_d/M_{\max}$. Note that $M_{\max}$ is the maximum theoretical moment determined before.

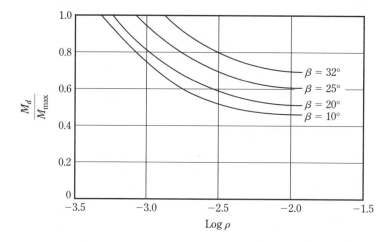

Figure 9.19 Plot of ρ against M_d/M_{max} for sheet pile walls penetrating into sand with a sloping dredge line (after Schroeder and Roumillac, 1983)

Step 8. Plot $\log \rho$ (Step 5) and M_d/M_{max} in Figure 9.18.

Step 9. Repeat Steps 1–8 for several sections. The points that fall above the curve (in loose sand or dense sand, as the case may be) are *safe sections.* The points that fall below the curve are *unsafe sections.* The cheapest section may now be chosen from those points which fall above the proper curve. Note that the section chosen will have an $M_d < M_{max}$.

For anchor sheet pile walls penetrating into sand with a sloping dredge line (see Figure 9.17), a moment reduction procedure similar to that just outlined may be adopted. For this procedure, Figure 9.19 (which was developed by Schroeder and Roumillac, 1983) should be used.

Example 9.3

Let $L_1 = 3.05$ m, $L_2 = 6.1$ m, $l_1 = 1.53$ m, $l_2 = 1.52$ m, $c' = 0$, $\phi' = 30°$, $\gamma = 16$ kN/m³, $\gamma_{sat} = 19.5$ kN/m³, and $E = 207 \times 10^3$ MN/m² in Figure 9.16.

a. Determine the theoretical and actual depths of penetration. (*Note:* $D_{actual} = 1.3D_{theory}$.)
b. Find the anchor force per unit length of the wall.
c. Determine the maximum moment, M_{max}.
d. Use Rowe's moment reduction technique and find a suitable sheet pile section. Take $\sigma_{all} = 172,500$ kN/m².

Solution

Part a
We use the following table:

Quantity required	Eq. no.	Equation and calculation
K_a	—	$\tan^2\left(45 - \dfrac{\phi'}{2}\right) = \tan^2\left(45 - \dfrac{30}{2}\right) = \dfrac{1}{3}$
K_p	—	$\tan^2\left(45 + \dfrac{\phi'}{2}\right) = \tan^2\left(45 + \dfrac{30}{2}\right) = 3$
$K_a - K_p$	—	$3 - 0.333 = 2.667$
γ'	—	$\gamma_{sat} - \gamma_\omega = 19.5 - 9.81 = 9.69 \text{ kN/m}^3$
σ'_1	9.1	$\gamma L_1 K_a = (16)(3.05)\left(\tfrac{1}{3}\right) = 16.27 \text{ kN/m}^2$
σ'_2	9.2	$(\gamma L_1 + \gamma' L_2)K_a = [(16)(3.05) + (9.69)(6.1)]\tfrac{1}{3} = 35.97 \text{ kN/m}^2$
L_3	9.6	$\dfrac{\sigma'_2}{\gamma'(K_p - K_a)} = \dfrac{35.97}{(9.69)(2.667)} = 1.39 \text{ m}$
P	—	$\tfrac{1}{2}\sigma'_1 L_1 + \sigma'_2 L_2 + \tfrac{1}{2}(\sigma'_2 - \sigma'_1)L_2 + \tfrac{1}{2}\sigma'_2 L_3 = \left(\tfrac{1}{2}\right)(16.27)(3.05)$
		$+ (16.27)(6.1) + \left(\tfrac{1}{2}\right)(35.97 - 16.27)(6.1) + \left(\tfrac{1}{2}\right)(35.97)(1.39)$
		$= 24.81 + 99.25 + 60.01 + 25.0 = 209.07 \text{ kN/m}$
$\bar{z}$	—	$\dfrac{\Sigma M_E}{P} = \left[\begin{array}{l} (24.81)\left(1.39 + 6.1 + \dfrac{3.05}{3}\right) + (99.25)\left(1.39 + \dfrac{6.1}{2}\right) \\ + (60.01)\left(1.39 + \dfrac{6.1}{3}\right) + (25.0)\left(\dfrac{2\times1.39}{3}\right) \end{array}\right]\dfrac{1}{209.07} = 4.21 \text{ m}$
L_4	9.67	$L_4^3 + 1.5L_4^2(l_2 + L_2 + L_3) - \dfrac{3P[(L_1 + L_2 + L_3) - (\bar{z} + l_1)]}{\gamma'(K_p - K_a)} = 0$
		$L_4^3 + 1.5L_4^2(1.52 + 6.1 + 1.39)$
		$-\dfrac{(3)(209.07)[(3.05 + 6.1 + 1.39) - (4.21 + 1.53)]}{(9.69)(2.667)} = 0$
		$L_4 = 2.7 \text{ m}$
D_{theory}	—	$L_3 + L_4 = 1.39 + 2.7 = 4.09 \approx \mathbf{4.1 \text{ m}}$
D_{actual}	—	$1.3D_{theory} = (1.3)(4.1) = \mathbf{5.33 \text{ m}}$

Part b

The anchor force per unit length of the wall is

$$F = P - \tfrac{1}{2}\gamma'(K_p - K_a)L_4^2$$

$$= 209.07 - \left(\tfrac{1}{2}\right)(9.69)(2.667)(2.7)^2 = 114.87 \text{ kN/m} \approx \mathbf{115 \text{kN/m}}$$

Part c

From Eq. (9.69), for zero shear,

$$\tfrac{1}{2}\sigma'_1 L_1 - F + \sigma'_1(z - L_1) + \tfrac{1}{2}K_a\gamma'(z - L_1)^2 = 0$$

Let $z - L_1 = x$, so that

$$\tfrac{1}{2}\sigma'_1 l_1 - F + \sigma'_1 x + \tfrac{1}{2}K_a\gamma' x^2 = 0$$

or

$$\left(\tfrac{1}{2}\right)(16.27)(3.05) - 115 + (16.27)(x) + \left(\tfrac{1}{2}\right)\left(\tfrac{1}{3}\right)(9.69)x^2 = 0$$

giving $$x^2 + 10.07x - 55.84 = 0$$

Now, $x = 4$ m and $z = x + L_1 = 4 + 3.05 = 7.05$ m. Taking the moment about the point of zero shear, we obtain

$$M_{\max} = -\frac{1}{2}\sigma_1' L_1 \left(x + \frac{3.05}{3}\right) + F(x + 1.52) - \sigma_1' \frac{x^2}{2} - \frac{1}{2}K_a \gamma' x^2 \left(\frac{x}{3}\right)$$

or

$$M_{\max} = -\left(\frac{1}{2}\right)(16.27)(3.05)\left(4 + \frac{3.05}{3}\right) + (115)(4 + 1.52) - (16.27)\left(\frac{4^2}{2}\right)$$

$$-\left(\frac{1}{2}\right)\left(\frac{1}{3}\right)(9.69)(4)^2\left(\frac{4}{3}\right) = \mathbf{344.9\ kN \cdot m/m}$$

Part d

The total length of sheet pile is

$$H' = L_1 + L_2 + D_{\text{actual}} = 3.05 + 6.1 + 5.33 = 14.48\ \text{m}$$

and we have the following table:

Section	$I(\text{m}^4/\text{m})$	H' (m)	$\times 10^{-7}\left(\dfrac{H'^4}{EI}\right)$	$\log \rho$	$S(\text{m}^3/\text{m})$	$M_d = S\sigma_{\text{all}}$ (kN·m/m)	$\dfrac{M_d}{M_{\max}}$
			$\rho = 10.91$				
PZ-22	115.2×10^{-6}	14.48	20.11×10^{-4}	-2.7	97×10^{-5}	167.33	0.485
PZ-27	251.5×10^{-6}	14.48	9.21×10^{-4}	-3.04	162.3×10^{-5}	284.84	0.826

Figure 9.20 gives a plot of $M_d/M_{\max}$ versus ρ. It can be seen that **PZ-27** will be sufficient.

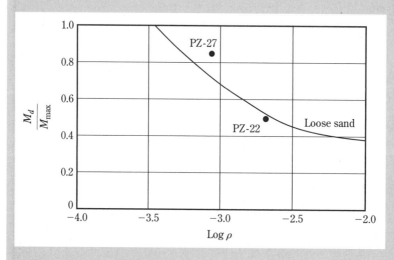

Figure 9.20 Plot of M_d/M_{max} vs. log ρ

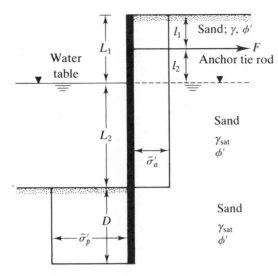

Figure 9.21 Computational pressure diagram method (*Note:* $L_1 + L_2 = L$)

9.11 Computational Pressure Diagram Method for Penetration into Sandy Soil

The computational pressure diagram (CPD) method for sheet pile penetrating a sandy soil is a simplified method of design and an alternative to the free earth method described in Sections 9.9 and 9.10 (Nataraj and Hoadley, 1984). In this method, the net pressure diagram shown in Figure 9.16 is replaced by rectangular pressure diagrams, as in Figure 9.21. Note that $\overline{\sigma}'_a$ is the width of the net active pressure diagram above the dredge line and $\overline{\sigma}'_p$ is the width of the net passive pressure diagram below the dredge line. The magnitudes of $\overline{\sigma}'_a$ and $\overline{\sigma}'_p$ may respectively be expressed as

$$\overline{\sigma}'_a = CK_a\gamma'_{av}L \tag{9.72}$$

and

$$\overline{\sigma}'_p = RCK_a\gamma'_{av}L = R\overline{\sigma}'_a \tag{9.73}$$

where γ'_{av} = average effective unit weight of sand

$$\approx \frac{\gamma L_1 + \gamma' L_2}{L_1 + L_2} \tag{9.74}$$

C = coefficient

$$R = \text{coefficient} = \frac{L(L - 2l_1)}{D(2L + D - 2l_1)} \tag{9.75}$$

The range of values for C and R is given in Table 9.2.

The depth of penetration, D, anchor force per unit length of the wall, F, and maximum moment in the wall, M_{max}, are obtained from the following relationships.

Table 9.2   Range of Values for C and R [from Eqs. (9.72) and (9.73)]

Soil type	C^a	R
Loose sand	0.8–0.85	0.3–0.5
Medium sand	0.7–0.75	0.55–0.65
Dense sand	0.55–0.65	0.60–0.75

[a] Valid for the case in which there is no surcharge above the granular backfill (i.e., on the right side of the wall, as shown in Figure 9.21)

Depth of Penetration

For the depth of penetration, we have

$$D^2 + 2DL\left[1 - \left(\frac{l_1}{L}\right)\right] - \left(\frac{L^2}{R}\right)\left[1 - 2\left(\frac{l_1}{L}\right)\right] = 0 \qquad (9.76)$$

Anchor Force

The anchor force is

$$F = \overline{\sigma}_a'(L - RD) \qquad (9.77)$$

Maximum Moment

The maximum moment is calculated from

$$M_{\max} = 0.5\overline{\sigma}_a'L^2\left[\left(1 - \frac{RD}{L}\right)^2 - \left(\frac{2l_1}{L}\right)\left(1 - \frac{RD}{L}\right)\right] \qquad (9.78)$$

Note the following qualifications:

1. The magnitude of D obtained from Eq. (9.76) is about 1.25 to 1.5 times the value of D_{theory} obtained by the conventional free earth support method (see Section 9.9), so

$$D \approx D_{\text{actual}}$$
$$\uparrow \qquad \uparrow$$
$$\text{Eq. (9.76)} \quad \text{Eq. (9.68)}$$

2. The magnitude of F obtained by using Eq. (9.77) is about 1.2 to 1.6 times the value obtained by using Eq. (9.66). Thus, an additional factor of safety for the actual design of anchors need not be used.

3. The magnitude of $M_{\max}$ obtained from Eq. (9.78) is about 0.6 to 0.75 times the value of $M_{\max}$ obtained by the conventional free earth support method. Hence,

the former value of M_{max} can be used as the actual design value, and Rowe's moment reduction need not be applied.

Example 9.4

In Figure 9.21, let $L_1 = 9$ ft, $L_2 = 26$ ft, $l_1 = 5$ ft, $\gamma = 108.5$ lb/ft³, $\gamma_{sat} = 128.5$ lb/ft³, and $\phi' = 35°$. Using the CPD method, determine the following:

a. D
b. F
c. M_{max}

Solution

Part a
We have

$$\gamma' = \gamma_{sat} - \gamma_w = 128.5 - 62.4 = 66.1 \text{ lb/ft}^3$$

From Eq. (9.74),

$$\gamma'_{av} = \frac{\gamma L_1 + \gamma' L_2}{L_1 + L_2} = \frac{(108.5)(9) + (66.1)(26)}{9 + 26} = 77 \text{ lb/ft}^3$$

Also,

$$K_a = \tan^2\left(45 - \frac{\phi'}{2}\right) = \tan^2\left(45 - \frac{35}{2}\right) = 0.271$$

and

$$\overline{\sigma}'_a = CK_a\gamma'_{av}L = (0.65)(0.271)(77)(35) = 474.7 \text{ lb/ft}^2$$

Note that *medium sand* is assumed and that $C = 0.65$ (from Table 9.2). Now,

$$\overline{\sigma}'_p = R\overline{\sigma}'_a$$

Assume that $R = 0.6$ (using the range from Table 9.2). Then

$$\overline{\sigma}'_p = (0.6)(474.7) = 284.82 \text{ lb/ft}^2$$

From Eq. (9.76),

$$D^2 + 2DL\left[1 - \left(\frac{l_1}{L}\right)\right] - \frac{L^2}{R}\left[1 - 2\left(\frac{l_1}{L}\right)\right] = 0$$

or

$$D^2 + 2(D)(35)\left[1 - \left(\frac{5}{35}\right)\right] - \frac{(35)^2}{0.6}\left[1 - 2\left(\frac{5}{35}\right)\right] = 0$$

giving

$$D^2 + 60D - 1458.3 = 0$$

Solving the latter equation yields

$$D \approx \textbf{18.6 ft}$$

Checking for the assumption of R, we obtain

$$R = \frac{L(L - 2l_1)}{D(2L + D - 2l_1)} = \frac{35[35 - (2)(5)]}{18.6[(2)(35) + 18.6 - (2)(5)]}$$

$$= \textbf{0.599} \approx \textbf{0.6, OK}$$

Part b
From Eq. (9.77), we get

$$F = \overline{\sigma}_a'(L - RD) = 474.7[35 - (0.6)(18.6)]$$

$$= \textbf{11,316.8 lb/ft}$$

Part c
From Eq. (9.78), we have

$$M_{max} = 0.5\overline{\sigma}_a'L^2\left[\left(1 - \frac{RD}{L}\right)^2 - \left(\frac{2l_1}{L}\right)\left(1 - \frac{RD}{L}\right)\right]$$

in which

$$1 - \frac{RD}{L} = 1 - \frac{(0.6)(18.6)}{35} = 0.681$$

So

$$M_{max} = (0.5)(474.7)(35)^2\left[(0.681)^2 - \frac{(2)(5)(0.681)}{35}\right]$$

$$= \textbf{78,260 lb-ft/ft}$$

The following table compares the values of D, F, and M_{max} obtained from the CPD method with those that would be obtained if the problem were solved by the conventional free earth support method (see Section 9.9):

Item	CPD method	Conventional free earth method	
D	18.6 ft	11.23 ft	$\dfrac{D_{(CPD)}}{D_{actual-free\ earth}} = 1.66$
F	11,316.8 lb/ft	8070 lb/ft	$\dfrac{D_{(CPD)}}{D_{actual-free\ earth}} = 1.4$
M_{max}	78,260 lb-ft/ft	92,790 lb-ft/ft	$\dfrac{M_{max(CPD)}}{M_{max-free\ earth}} = 0.84$

■

9.12 *Free Earth Support Method for Penetration of Clay*

Figure 9.22 shows an anchored sheet pile wall penetrating a clay soil and with a granular soil backfill. The diagram of pressure distribution above the dredge line is similar to that shown in Figure 9.11. From Eq. (9.42), the net pressure distribution below the dredge line (from $z = L_1 + L_2$ to $z = L_1 + L_2 + D$) is

$$\sigma_6 = 4c - (\gamma L_1 + \gamma' L_2)$$

For static equilibrium, the sum of the forces in the horizontal direction is

$$P_1 - \sigma_6 D = F \tag{9.79}$$

where P_1 = area of the pressure diagram ACD
F = anchor force per unit length of the sheet pile wall

Again, taking the moment about O' produces

$$P_1(L_1 + L_2 - l_1 - \bar{z}_1) - \sigma_6 D\left(l_2 + L_2 + \frac{D}{2}\right) = 0$$

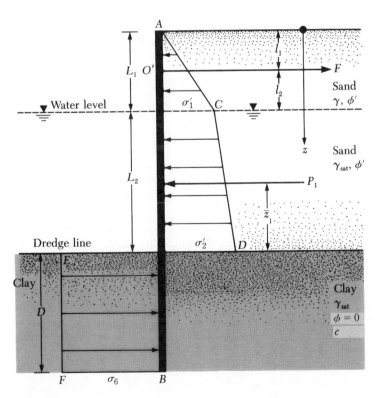

Figure 9.22 Anchored sheet pile wall penetrating clay

Simplification yields

$$\sigma_6 D^2 + 2\sigma_6 D(L_1 + L_2 - l_1) - 2P_1(L_1 + L_2 - l_1 - \bar{z}_1) = 0 \qquad (9.80)$$

Equation (9.80) gives the theoretical depth of penetration, D.

As in Section 9.9, the maximum moment in this case occurs at a depth $L_1 < z < L_1 + L_2$. The depth of zero shear (and thus the maximum moment) may be determined from Eq. (9.69).

A moment reduction technique similar to that in Section 9.10 for anchored sheet piles penetrating into clay has also been developed by Rowe (1952, 1957). This technique is presented in Figure 9.23, in which the following notation is used:

1. The stability number is

$$S_n = 1.25\frac{c}{(\gamma L_1 + \gamma' L_2)} \qquad (9.81)$$

where c = undrained cohesion ($\phi = 0$)

For the definition of γ, γ', L_1, and L_2, see Figure 9.22.

2. The nondimensional wall height is

$$\alpha = \frac{L_1 + L_2}{L_1 + L_2 + D_{\text{actual}}} \qquad (9.82)$$

3. The flexibility number is ρ [see Eq. (9.70) or Eq. (9.71)]
4. M_d = design moment

   M_{max} = maximum theoretical moment

The procedure for moment reduction, using Figure 9.23, is as follows:

   Step 1.   Obtain $H' = L_1 + L_2 + D_{\text{actual}}$.
   Step 2.   Determine $\alpha = (L_1 + L_2)/H'$.
   Step 3.   Determine S_n [from Eq. (9.81)].
   Step 4.   For the magnitudes of α and S_n obtained in Steps 2 and 3, determine M_d/M_{max} for various values of log ρ from Figure 9.23, and plot M_d/M_{max} against log ρ.
   Step 5.   Follow Steps 1–9 as outlined for the case of moment reduction of sheet pile walls penetrating granular soil. (See Section 9.10.)

Example 9.5

In Figure 9.22, let $L_1 = 3$ m, $L_2 = 6$ m, and $l_1 = 1.5$ m. Also, let $\gamma = 17$ kN/m³, $\gamma_{\text{sat}} = 20$ kN/m³, $\phi' = 35°$, and $c = 41$ kN/m².

   a. Determine the theoretical depth of embedment of the sheet pile wall
   b. Calculate the anchor force per unit length of the wall.

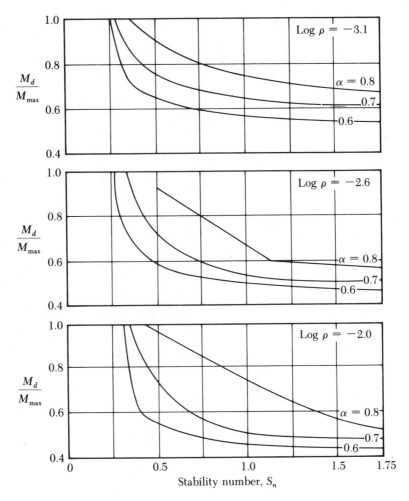

Figure 9.23 Plot of M_d/M'_{max} against stability number for sheet pile wall penetrating clay (after Rowe, 1957)

Solution

Part a
We have

$$K_a = \tan^2\left(45 - \frac{\phi'}{2}\right) = \tan^2\left(45 - \frac{35}{2}\right) = 0.271$$

and

$$K_p = \tan^2\left(45 + \frac{\phi'}{2}\right) = \tan^2\left(45 + \frac{35}{2}\right) = 3.69$$

From the pressure diagram in Figure 9.24,

$\sigma_1' = \gamma L_1 K_a = (17)(3)(0.271) = 13.82 \text{ kN/m}^2$

$\sigma_2' = (\gamma L_1 + \gamma' L_2) K_a = [(17)(3) + (20 - 9.81)(6)](0.271) = 30.39 \text{ kN/m}^2$

$P_1 = \text{areas } 1 + 2 + 3 = 1/2(3)(13.82) + (13.82)(6) + 1/2(30.39 - 13.82)(6)$

$\quad = 20.73 + 82.92 + 49.71 = 153.36 \text{ kN/m}$

and

$$\bar{z}_1 = \frac{(20.73)\left(6 + \dfrac{3}{3}\right) + (82.92)\left(\dfrac{6}{2}\right) + (49.71)\left(\dfrac{6}{3}\right)}{153.36} = 3.2 \text{ m}$$

From Eq. (9.80),

$$\sigma_6 D^2 + 2\sigma_6 D(L_1 + L_2 - l_1) - 2P_1(L_1 + L_2 - l_1 - \bar{z}_1) = 0$$

or

$$\sigma_6 = 4c - (\gamma L_1 + \gamma' L_2) = (4)(41) - [(17)(3)$$
$$+ (20 - 9.81)(6)] = 51.86 \text{ kN/m}^2$$

So

$$(51.86)D^2 + (2)(51.86)(D)(3 + 6 - 1.5)$$
$$- (2)(153.36)(3 + 6 - 1.5 - 3.2) = 0$$

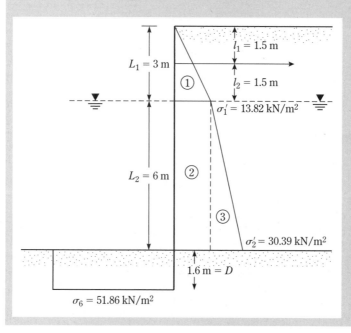

Figure 9.24 Free earth support method, sheet pile penetrating into clay

or

$$D^2 + 15D - 25.43 = 0$$

Hence,

$$D \approx \textbf{1.6 m}$$

Part b
From Eq. (9.79),

$$F = P_1 - \sigma_6 D = 153.36 - (51.86)(1.6) = \textbf{70.38 kN/m} \qquad \blacksquare$$

9.13   *Anchors*

Sections 9.9–9.12 gave an analysis of anchored sheet pile walls and discussed how to obtain the force F per unit length of the sheet pile wall that has to be sustained by the anchors. The current section covers in more detail the various types of anchor generally used and the procedures for evaluating their ultimate holding capacities.

The general types of anchor used in sheet pile walls are as follows:

1. Anchor plates and beams (deadman)
2. Tie backs
3. Vertical anchor piles
4. Anchor beams supported by batter (compression and tension) piles

Anchor plates and beams are generally made of cast concrete blocks. (See Figure 9.25a.) The anchors are attached to the sheet pile by *tie-rods*. A *wale* is placed at the front or back face of a sheet pile for the purpose of conveniently attaching the tie-rod to the wall. To protect the tie rod from corrosion, it is generally coated with paint or asphaltic materials.

In the construction of *tiebacks,* bars or cables are placed in predrilled holes (see Figure 9.25b) with concrete grout (cables are commonly high-strength, pre-stressed steel tendons). Figures 9.25c and 9.25d show a vertical anchor pile and an anchor beam with batter piles.

Placement of Anchors

The resistance offered by anchor plates and beams is derived primarily from the passive force of the soil located in front of them. Figure 9.25a, in which AB is the sheet pile wall, shows the best location for maximum efficiency of an anchor plate. If the anchor is placed inside wedge ABC, which is the Rankine active zone, it would not provide any resistance to failure. Alternatively, the anchor could be placed in zone $CFEH$. Note that line DFG is the slip line for the Rankine passive pressure. If part of the passive wedge is located inside the active wedge ABC, full passive resistance of the anchor cannot be realized upon failure of the sheet pile wall. However, if the anchor is placed in zone ICH, the Rankine passive zone in front of the anchor slab or plate is located completely outside the Rankine active zone ABC. In this case, full passive resistance from the anchor can be realized.

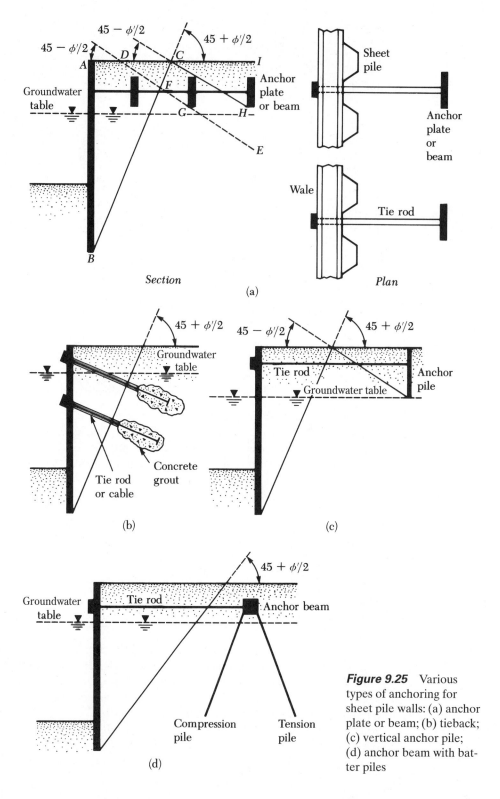

Figure 9.25 Various types of anchoring for sheet pile walls: (a) anchor plate or beam; (b) tieback; (c) vertical anchor pile; (d) anchor beam with batter piles

Figures 9.25b, 9.25c, and 9.25d also show the proper locations for the placement of tiebacks, vertical anchor piles, and anchor beams supported by batter piles.

9.14 Holding Capacity of Anchor Plates in Sand

Ovesen and Stromann (1972) proposed a semi-empirical method for determining the ultimate resistance of anchors in sand. Their calculations, made in three steps, are carried out as follows:

Step 1. **Basic Case.** Determine the depth of embedment, H. Assume that the anchor slab has height H and is continuous (i.e., B = length of anchor slab perpendicular to the cross section = ∞), as shown in Figure 9.26, in which the following notation is used:

P_p = passive force per unit length of anchor

P_a = active force per unit length of anchor

ϕ' = effective soil friction angle

δ = friction angle between anchor slab and soil

P'_u = ultimate resistance per unit length of anchor

W = effective weight per unit length of anchor slab

Also,

$$P'_u = \tfrac{1}{2}\gamma H^2 K_p \cos \delta - P_a \cos \phi' = \tfrac{1}{2}\gamma H^2 K_p \cos \delta - \tfrac{1}{2}\gamma H^2 K_a \cos \phi'$$
$$= \tfrac{1}{2}\gamma H^2 (K_p \cos \delta - K_a \cos \phi') \qquad (9.83)$$

where K_a = active pressure coefficient with $\delta = \phi'$
(see Figure 9.27a)

K_p = passive pressure coefficient

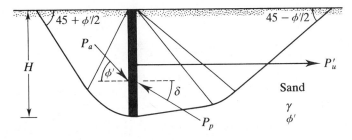

Figure 9.26 Basic case: continuous vertical anchor in granular soil

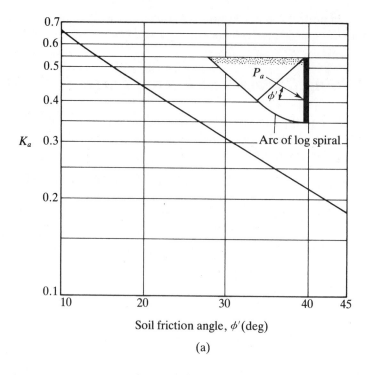

(a)

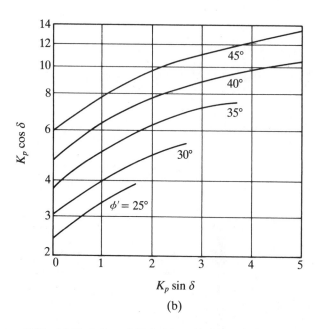

(b)

Figure 9.27 (a) Variation of K_a for $\delta = \phi'$, (b) variation of $K_p \cos \delta$ with $K_p \sin \delta$ (based on Ovesen and Stromann, 1972)

To obtain $K_p \cos \delta$, first calculate

$$K_p \sin \delta = \frac{W + P_a \sin \phi'}{\frac{1}{2}\gamma H^2} = \frac{W + \frac{1}{2}\gamma H^2 K_a \sin \phi'}{\frac{1}{2}\gamma H^2} \qquad (9.84)$$

Then use the magnitude of $K_p \sin \delta$ obtained from Eq. (9.84) to esti-mate the magnitude of $K_p \cos \delta$ from the plots given in Figure 9.27b.

Step 2. **Strip Case.** Determine the actual height h of the anchor to be con-structed. If a continuous anchor (i.e., an anchor for which $B = \infty$) of height h is placed in the soil so that its depth of embedment is H, as shown in Figure 9.28, the ultimate resistance per unit length is

$$P'_{us} = \left[\frac{C_{ov} + 1}{C_{ov} + \left(\dfrac{H}{h}\right)} \right] \underset{\text{Eq. 9.83}}{\overset{\uparrow}{P'_u}} \qquad (9.85)$$

where P'_{us} = ultimate resistance for the *strip case*
    C_{ov} = 19 for dense sand and 14 for loose sand

Step 3. **Actual Case.** In practice, the anchor plates are placed in a row with center-to-center spacing S', as shown in Figure 9.29a. The ultimate resistance of each anchor is

$$P_u = P'_{us} B_e \qquad (9.86)$$

where B_e = equivalent length

The equivalent length is a function of S', B, H, and h. Figure 9.29b shows a plot of $(B_e - B)/(H + h)$ against $(S' - B)/(H + h)$ for the cases of loose and dense sand. With known values of S', B, H, and h, the value of B_e can be calculated and used in Eq. (9.86) to obtain P_u.

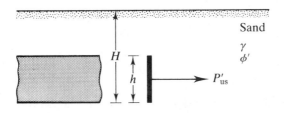

Sand
γ
ϕ'

H

h

P'_{us}

Figure 9.28 Strip case: vertical anchor

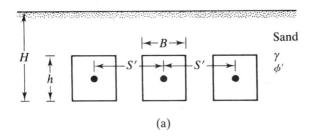

(a)

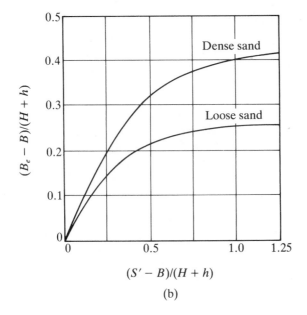

(b)

Figure 9.29 (a) Actual case for row of anchors; (b) variation of $(B_e - B)/(H + h)$ with $(S' - B)/(H + h)$ (based on Ovesen and Stromann, 1972)

Empirical Correlation Based on Model Tests

Ghaly (1997) used the results of 104 laboratory tests, 15 centrifugal model tests, and 9 field tests to propose an empirical correlation for the ultimate resistance of single anchors. The correlation can be written as

$$P_u = \frac{5.4}{\tan \phi'}\left(\frac{H^2}{A}\right)^{0.28} \gamma A H \tag{9.87}$$

where A = area of the anchor = Bh

Ghaly also used the model test results of Das and Seeley (1975) to develop a load–displacement relationship for single anchors. The relationship can be given as

$$\frac{P}{P_u} = 2.2\left(\frac{u}{H}\right)^{0.3} \tag{9.88}$$

where u = horizontal displacement of the anchor at a load level P

Equations (9.87) and (9.88) apply to single anchors (i.e., anchors for which $S'/B = \infty$). For all practical purposes, when $S'/B \approx 2$ the anchors behave as single anchors.

Factor of Safety for Anchor Plates

The allowable resistance per anchor plate may be given as

$$P_{all} = \frac{P_u}{FS}$$

where   FS = factor of safety

Generally, a factor of safety of 2 is suggested when the method of Ovesen and Stromann is used. A factor of safety of 3 is suggested for P_u calculated by Eq. (9.87).

Spacing of Anchor Plates

The center-to-center spacing of anchors, S', may be obtained from

$$S' = \frac{P_{all}}{F}$$

where F = force per unit length of the sheet pile

Example 9.6

A row of vertical anchors embedded in sand is shown in Figure 9.30. The anchor plates are made of 0.15-m-thick concrete. The design parameters are $B = h = 0.4$ m, $S' = 1.25$ m, $H = 1$ m, $\gamma = 16.5$ kN/m^3, $\phi' = 35°$, and the unit weight of concrete = 23.58 kN/m^3.

Determine the allowable resistance of each anchor by using

a. the method of Ovesen and Stromann, together with a safety factor of 2.
b. the empirical correlation method [Eq. (8.87)], together with a safety factor of 3.

Solution

Part a
From Figure 9.27a, for $\phi' = 35°$, the magnitude of K_a is about 0.26. Also,

$$W = Ht\gamma_{concrete} = (1)(0.15)(23.58) = 3.537 \text{ kN/m}$$

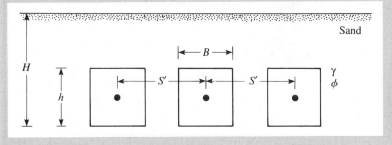

Figure 9.30   A row of vertical anchors

From Eq. (9.84),

$$K_p \sin \delta = \dfrac{W + \dfrac{1}{2}\gamma H^2 K_a \sin \phi'}{\dfrac{1}{2}\gamma H^2}$$

$$= \dfrac{3.537 + \dfrac{1}{2}(16.5)(1)^2(0.26)(\sin 35)}{\dfrac{1}{2}(16.5)(1)^2} = 0.578$$

From Figure 9.27b with $\phi' = 35°$ and $K_p \sin \delta = 0.578$, the magnitude of $K_p \cos \delta$ is about 4.5. Now, from Eq. (9.83),

$$P'_u = \dfrac{1}{2}\gamma H^2(K_p \cos \delta - K_a \cos \phi')$$

$$= \dfrac{1}{2}(16.5)(1)^2[4.5 - (0.26)(\cos 35)] = 35.37 \text{ kN/m}$$

To calculate P'_{us}, we assume the sand to be loose. So, C_{ov} in Eq. (9.85) is 14. Hence,

$$P'_{us} = \left[\dfrac{C_{ov} + 1}{C_{ov} + \left(\dfrac{H}{h}\right)}\right]P'_u = \left[\dfrac{14 + 1}{14 + \left(\dfrac{1}{0.4}\right)}\right](35.37) = 32.15 \text{ kN/m}$$

and

$$\dfrac{S' - B}{H + h} = \dfrac{1.25 - 0.4}{1 + 0.4} = 0.607$$

For $(S' - B)/(H + h) = 0.607$ and loose sand, Figure 9.29b yields

$$\dfrac{B_e - B}{H - h} = 0.22$$

So

$$B_e = (0.22)(H + h) + B = (0.22)(1 + 0.4) + 0.4 = 0.708 \text{ m}$$

Hence, from Eq. (9.86),

$$P_u = P'_{us}B_e = (32.15)(0.708) = \textbf{22.76 kN}$$

and

$$P_{all} = \dfrac{P_u}{\text{FS}} = \dfrac{22.76}{2} = \textbf{11.38 kN}$$

Part b
From Eq. (9.87),

$$P_u = \frac{5.4}{\tan \phi'}\left(\frac{H^2}{A}\right)^{0.28}\gamma A H$$

$$= \frac{5.4}{\tan 35}\left(\frac{1^2}{0.4 \times 0.4}\right)^{0.28}(16.5)(0.4 \times 0.4)(1) = 34\ \text{kN}$$

and

$$P_{\text{all}} = \frac{P_u}{\text{FS}} = \frac{34}{3} \approx \mathbf{11.33\ kN} \qquad \blacksquare$$

9.15   *Ultimate Resistance of Tiebacks*

According to Figure 9.31, the ultimate resistance offered by a tieback in sand is

$$P_u = \pi d l \overline{\sigma}'_o K \tan \phi' \tag{9.89}$$

where ϕ' = effective angle of friction of soil
$\overline{\sigma}'_o$ = average effective vertical stress ($=\gamma z$ in dry sand)
K = earth pressure coefficient

The magnitude of K can be taken to be equal to the earth pressure coefficient at rest (K_o) if the concrete grout is placed under pressure (Littlejohn, 1970). The lower limit of K can be taken to be equal to the Rankine active earth pressure coefficient.

In clays, the ultimate resistance of tiebacks may be approximated as

$$P_u = \pi d l c_a \tag{9.90}$$

where c_a = adhesion

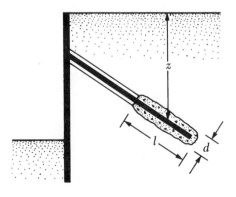

Figure 9.31 Parameters for defining the ultimate resistance of tiebacks

The value of c_a may be approximated as $\frac{2}{3}c_u$ (where c_u = undrained cohesion). A factor of safety of 1.5–2 may be used over the ultimate resistance to obtain the allowable resistance offered by each tieback.

9.16 *Field Observations for Anchored Sheet Pile Walls*

In the preceding sections, large factors of safety were used for the depth of penetration, D. In most cases, designers use smaller magnitudes of the soil friction angle ϕ', thereby ensuring a built-in factor of safety for the active earth pressure. This procedure is followed primarily because of the uncertainties involved in predicting the actual earth pressure a sheet pile wall will be subjected to in the field. In addition, Casagrande (1973) observed that if the soil behind the sheet pile wall has grain sizes that are predominantly smaller than those of coarse sand, the active earth pressure after construction sometimes increases to an at-rest earth pressure condition. Such an increase causes a large increase in the anchor force, F, and also the maximum moment. Casagrande recounted the case history of a bulkhead in Toledo to illustrate his point. We summarize his account next. A typical cross section of the Toledo bulkhead that was completed in 1961 is shown in Figure 9.32.

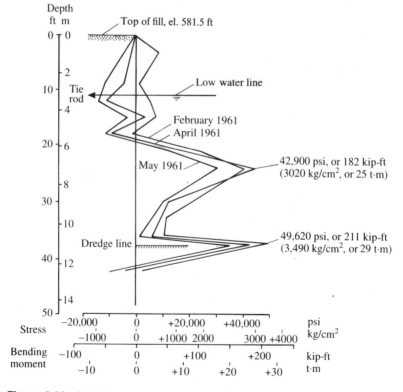

Figure 9.32 Bending moment and sresses from strain gage measurements at test location 3, Toledo bulkhead (after Casagrande, 1973)

The foundation soil was primarily fine to medium sand, but the dredge line did cut into highly overconsolidated clay. The figure also shows the actual measured values of stress and bending moment along the sheet pile wall. Casagrande used the Rankine active earth pressure distribution to calculate the maximum bending moment according to the free earth support method with and without Rowe's moment reduction. His results were as follows:

Design method	Maximum predicted bending moment, M_{max}
Free earth support method	14.9 ton·m (108 kip-ft)
Free earth support method with Rowe's moment reduction	8.0 ton·m (58 kip-ft)

Comparisons of these magnitudes of M_{max} with those actually observed show that the field values are substantially larger. The reason probably is that the backfill was primarily fine sand and the measured active earth pressure distribution was larger than that predicted theoretically.

Problems

9.1   Figure P9.1 shows a cantilever sheet pile wall penetrating a granular soil. Here, $L_1 = 8$ ft, $L_2 = 15$ ft, $\gamma = 100$ lb/ft³, $\gamma_{sat} = 110$ lb/ft³, and $\phi' = 35°$.
   a. What is the theoretical depth of embedment, D?
   b. For a 30% increase in D, what should be the total length of the sheet piles?
   c. Determine the theoretical maximum moment of the sheet pile.

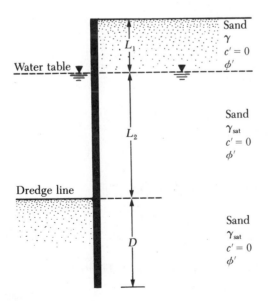

Figure P9.1

9.2 Solve Problem 9.1 with the following data: $L_1 = 3$ m, $L_2 = 8$ m, $\gamma = 16$ kN/m³, $\gamma_{sat} = 18$ kN/m³, and $\phi' = 30°$.

9.3 Redo Problem 9.1 with the following changes: $L_1 = 10$ ft, $L_2 = 15$ ft, $\gamma = 105$ lb/ft³, $\gamma_{sat} = 124$ lb/ft³, and $\phi' = 32°$.

9.4 In Figure 9.9, let $L = 5$ m, $\gamma = 15$ kN/m³, and $\phi' = 36°$. Calculate the theoretical depth of penetration, D, and the maximum moment.

9.5 Repeat Problem 9.4 with the following data: $L = 4$ m, $\gamma = 16$ kN/m³, and $\phi' = 32°$.

9.6 Figure 9.10 shows a free cantilever sheet pile wall penetrating a layer of sand. Determine the theoretical depth of penetration and the maximum moment if $\gamma = 18$ kN/m³, $\phi' = 36°$, $L = 5$ m, and $P = 20$ kN/m.

9.7 In Figure P9.7, let $L_1 = 3$ m, $L_2 = 5$ m, $\gamma = 16$ kN/m³, $\gamma_{sat} = 18$ kN/m³, $c = 35$ kN/m², and $\phi' = 30°$.
   a. Find the theoretical depth of penetration, D.
   b. Increase D by 40%. What length of sheet piles is needed?
   c. Determine the theoretical maximum moment in the sheet pile.

9.8 Solve Problem 9.7 with the following data: $L_1 = 6$ ft, $L_2 = 12$ ft, $\gamma = 110$ lb/ft³, $\gamma_{sat} = 125$ lb/ft³, $\phi' = 32°$, and $c = 800$ lb/ft².

9.9 In Figure 9.14, let $L = 4$ m, $P = 8$ kN/m, $c = 45$ kN/m², and $\gamma_{sat} = 19.2$ kN/m³. Find the theoretical value of D and the maximum moment.

9.10 An anchored sheet pile bulkhead is shown in Figure P9.10. Let $L_1 = 4$ m, $L_2 = 9$ m, $l_1 = 2$ m, $\gamma = 17$ kN/m³, $\gamma_{sat} = 19$ kN/m³, and $\phi' = 34°$.
   a. Calculate the theoretical value of the depth of embedment, D.
   b. Draw the pressure distribution diagram.
   c. Determine the anchor force per unit length of the wall.
   Use the free earth support method.

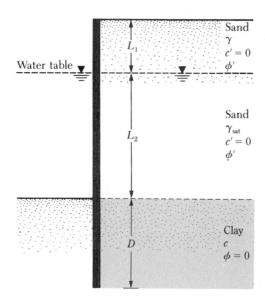

Figure P9.7

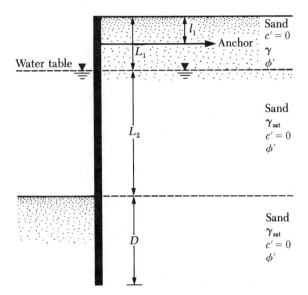

Figure P9.10

9.11 In Problem 9.10, assume that $D_{actual} = 1.3D_{theory}$.
   a. Determine the theoretical maximum moment.
   b. Using Rowe's moment reduction technique, choose a sheet pile section. Take $E = 210 \times 10^3$ MN/m² and $\sigma_{all} = 210{,}000$ kN/m².

9.12 Redo Problem 9.11 with the following data: $L_1 = 9$ ft, $L_2 = 26$ ft, $l_1 = 5$ ft, $\gamma = 108.5$ lb/ft³, $\gamma_{sat} = 128.5$ lb/ft³, and $\phi' = 35°$. Use the free earth support method.

9.13 In Problem 9.12, assume that $D_{actual} = 1.4D_{theory}$. Using Rowe's moment reduction method, choose a sheet pile section. Take $E = 30 \times 10^6$ lb/in.² and $\sigma_{all} = 25$ kip/in.².

9.14 In Figure P9.10, let $L_1 = 4$ m, $L_2 = 8$ m, $l_1 = 1.5$ m, $\gamma = 17$ kN/m³, $\gamma_{sat} = 19$ kN/m³, and $\phi' = 35°$. Use the computational diagram method to determine D, F, and M_{max}. Assume that $C = 0.68$ and $R = 0.6$.

9.15 An anchored sheet pile bulkhead is shown in Figure P9.15. Let $L_1 = 3$ m, $L_2 = 8$ m, $l_1 = 1.5$ m, $\gamma = 17$ kN/m³, $\gamma_{sat} = 19.5$ kN/m³, $\phi' = 36°$, and $c = 40$ kN/m².
   a. Determine the theoretical depth of embedment, D.
   b. Calculate the anchor force per unit length of the sheet pile wall.
   Use the free earth support method.

9.16 Repeat Problem 9.15 with the following data: $L_1 = 8$ ft, $L_2 = 20$ ft, $l_1 = 4$ ft, $c = 1500$ lb/ft², $\gamma = 115$ lb/ft³, $\gamma_{sat} = 128$ lb/ft³, and $\phi' = 40°$.

9.17 In Figure 9.29a, for the anchor slab in sand, $H = 5$ ft, $h = 3$ ft, $B = 4$ ft, $S' = 7$ ft, $\phi' = 30°$, and $\gamma = 110$ lb/ft³. The anchor plates are made of concrete and have a thickness of 3 in. Using Ovesen and Stromann's method, calculate the ultimate holding capacity of each anchor. Take $\gamma_{concrete} = 150$ lb/ft³.

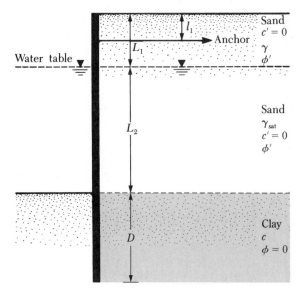

Figure P9.15

References

Casagrande, L. (1973). "Comments on Conventional Design of Retaining Structures," *Journal of the Soil Mechanics and Foundations Division,* ASCE, Vol. 99, No. SM2, pp. 181–198.

Das, B. M., and Seeley, G. R. (1975). "Load–Displacement Relationships for Vertical Anchor Plates," *Journal of the Geotechnical Engineering Division,* American Society of Civil Engineers, Vol. 101, No, GT7, pp. 711–715.

Ghaly, A. M. (1997). "Load–Displacement Prediction for Horizontally Loaded Vertical Plates." *Journal of Geotechnical and Geoenvironmental Engineering,* ASCE, Vol. 123, No. 1, pp. 74–76.

Littlejohn, G. S. (1970). "Soil Anchors," *Proceedings, Conference on Ground Engineering,* Institute of Civil Engineers, London, pp. 33–44.

Nataraj, M. S., and Hoadley, P. G. (1984). "Design of Anchored Bulkheads in Sand," *Journal of Geotechnical Engineering,* American Society of Civil Engineers, Vol. 110, No. GT4, pp. 505–515.

Ovesen, N. K., and Stromann, H. (1972). "Design Methods for Vertical Anchor Slabs in Sand," *Proceedings, Specialty Conference on Performance of Earth and Earth-Supported Structures.* American Society of Civil Engineers, Vol. 2.1, pp. 1481–1500.

Rowe, P. W. (1952). "Anchored Sheet Pile Walls," *Proceedings, Institute of Civil Engineers,* Vol. 1, Part 1, pp. 27–70.

Rowe, P. W. (1957). "Sheet Pile Walls in Clay," *Proceedings, Institute of Civil Engineers,* Vol. 7, pp. 654–692.

Schroeder, W. L., and Roumillac, P. (1983). "Anchored Bulkheads with Sloping Dredge Lines," *Journal of Geotechnical Engineering,* American Society of Civil Engineers, Vol. 109. No. 6, pp. 845–851.

Tsinker, G. P. (1983). "Anchored Street Pile Bulkheads: Design Practice," *Journal of Geotechnical Engineering,* American Society of Civil Engineers, Vol. 109, No. GT8, pp. 1021–1038.

10

Braced Cuts

10.1 Introduction

Sometimes construction work requires ground excavations with vertical or near-vertical faces—for example, basements of buildings in developed areas or underground transportation facilities at shallow depths below the ground surface (a cut-and-cover type of construction). The vertical faces of the cuts need to be protected by temporary bracing systems to avoid failure that may be accompanied by considerable settlement or by bearing capacity failure of nearby foundations.

Figure 10.1 shows two types of braced cut commonly used in construction work. One type uses the *soldier beam* (Figure 10.1a), which is driven into the ground before excavation and is a vertical steel or timber beam. *Laggings,* which are horizontal timber planks, are placed between soldier beams as the excavation proceeds. When the excavation reaches the desired depth, *wales* and *struts* (horizontal steel beams) are installed. The struts are compression members. Figure 10.1b shows another type of braced excavation. In this case, interlocking *sheet piles* are driven into the soil before excavation. Wales and struts are inserted immediately after excavation reaches the appropriate depth.

Figure 10.2 shows the braced-cut construction used for the Chicago subway in 1940. Timber lagging, timber struts, and steel wales were used. Figure 10.3 shows a braced cut made during the construction of the Washington, DC, metro in 1974. In this cut, timber lagging, steel H-soldier piles, steel wales, and pipe struts were used.

To design braced excavations (i.e., to select wales, struts, sheet piles, and soldier beams), an engineer must estimate the lateral earth pressure to which the braced cuts will be subjected. The theoretical aspects of the lateral earth pressure on a braced cut were discussed in Section 7.6. The total active force per unit length of the wall (P_a) was calculated using the general wedge theory. However, that analysis does not provide the relationships required for estimating the variation of lateral pressure with depth, which is a function of several factors, such as the type of soil, the experience of the construction crew, the type of construction equipment used, and so forth. For that reason, empirical pressure envelopes developed from field observations are used for the design of braced cuts. This procedure is discussed in the next section.

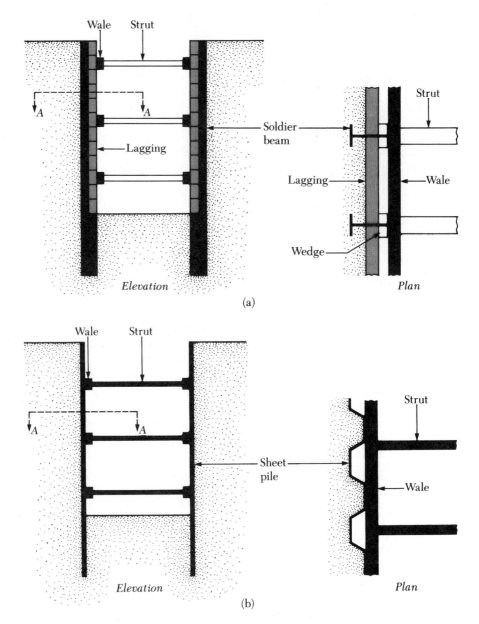

Figure 10.1 Types of braced cut: (a) use of soldier beams; (b) use of sheet piles

10.2 *Pressure Envelope for Braced-Cut Design*

As mentioned in Section 10.1, the lateral earth pressure in a braced cut is dependent on the type of soil, construction method, and type of equipment used. The lateral earth pressure changes from place to place. Each strut should also be designed for the maximum load to which it may be subjected. Therefore, the braced cuts

Figure 10.2   Braced cut in Chicago subway construction, January 1940 (courtesy of Ralph B. Peck)

should be designed using apparent-pressure diagrams that are envelopes of all the pressure diagrams determined from measured strut loads in the field. Figure 10.4 shows the method for obtaining the apparent-pressure diagram at a section from strut loads. In this figure, let $P_1, P_2, P_3, P_4, \ldots$ be the measured strut loads. The apparent horizontal pressure can then be calculated as

$$\sigma_1 = \frac{P_1}{(s)\left(d_1 + \dfrac{d_2}{2}\right)}$$

$$\sigma_2 = \frac{P_2}{(s)\left(\dfrac{d_2}{2} + \dfrac{d_3}{2}\right)}$$

$$\sigma_3 = \frac{P_3}{(s)\left(\dfrac{d_3}{2} + \dfrac{d_4}{2}\right)}$$

$$\sigma_4 = \frac{P_4}{(s)\left(\dfrac{d_4}{2} + \dfrac{d_5}{2}\right)}$$

Figure 10.3 Braced cut in the construction of Washington, DC, metro, May 1974 (courtesy of Ralph B. Peck)

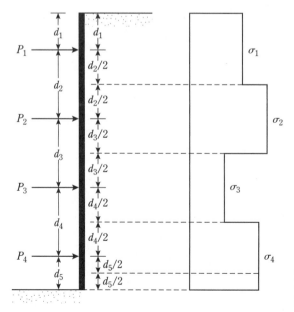

Figure 10.4 Procedure for calculating apparent-pressure diagram from measured strut loads

where $\sigma_1, \sigma_2, \sigma_3, \sigma_4$ = apparent pressures

s = center-to-center spacing of the struts

Using the procedure just described for strut loads observed from the Berlin subway cut, Munich subway cut, and New York subway cut, Peck (1969) provided the envelope of apparent-lateral-pressure diagrams for design of cuts in *sand*. This envelope is illustrated in Figure 10.5, in which

$$\sigma_a = 0.65\gamma H K_a \tag{10.1}$$

where γ = unit weight

H = height of the cut

K_a = Rankine active pressure coefficient = $\tan^2(45 - \phi'/2)$

ϕ' = effective friction angle of sand

Cuts in Clay

In a similar manner, Peck (1969) also provided the envelopes of apparent-lateral-pressure diagrams for cuts in *soft to medium clay* and in *stiff clay*. The pressure envelope for soft to medium clay is shown in Figure 10.6 and is applicable to the condition

$$\frac{\gamma H}{c} > 4$$

where c = undrained cohesion ($\phi = 0$)

The pressure, σ_a, is the larger of

$$\sigma_a = \gamma H \left[1 - \left(\frac{4c}{\gamma H} \right) \right]$$

and

$$\sigma_a = 0.3\gamma H \tag{10.2}$$

where γ = unit weight of clay

The pressure envelope for cuts in stiff clay is shown in Figure 10.7, in which

$$\sigma_a = 0.2\gamma H \text{ to } 0.4\gamma H \qquad (\text{with an average of } 0.3\gamma H) \tag{10.3}$$

is applicable to the condition $\gamma H/c \leqslant 4$.

When using the pressure envelopes just described, keep the following points in mind:

1. They apply to excavations having depths greater than about 6 m ($\approx$20 ft).

2. They are based on the assumption that the water table is below the bottom of the cut.

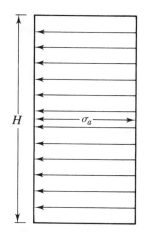

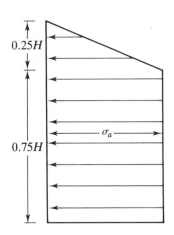

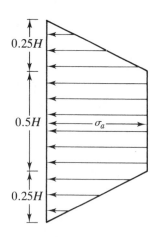

Figure 10.5 Peck's (1969) apparent-pressure envelope for cuts in sand

Figure 10.6 Peck's (1969) apparent-pressure envelope for cuts in soft to medium clay

Figure 10.7 Peck's (1969) apparent-pressure envelope for cuts in stiff clay

3. Sand is assumed to be drained with zero pore water pressure.
4. Clay is assumed to be undrained and pore water pressure is not considered.

10.3 Pressure Envelope for Cuts in Layered Soil

Sometimes, layers of both sand and clay are encountered when a braced cut is being constructed. In this case, Peck (1943) proposed that an equivalent value of cohesion ($\phi = 0$) should be determined according to the formula (see Figure 10.8a).

$$c_{av} = \frac{1}{2H}[\gamma_s K_s H_s^2 \tan \phi_s' + (H - H_s)n'q_u]$$ (10.4)

where H = total height of the cut
      γ_s = unit weight of sand
      H_s = height of the sand layer
      K_s = a lateral earth pressure coefficient for the sand layer (≈ 1)
      ϕ_s' = effective angle of friction of sand
      q_u = unconfined compression strength of clay
      n' = a coefficient of progressive failure (ranging from 0.5 to 1.0; average value 0.75)

The average unit weight of the layers may be expressed as

$$\gamma_a = \frac{1}{H}[\gamma_s H_s + (H - H_s)\gamma_c]$$ (10.5)

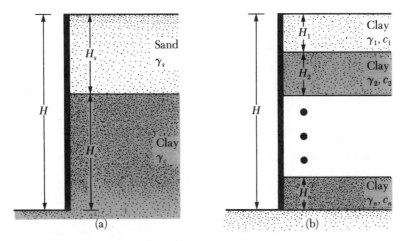

Figure 10.8 Layered soils in braced cuts

where γ_c = saturated unit weight of clay layer

Once the average values of cohesion and unit weight are determined, the pressure envelopes in clay can be used to design the cuts.

   Similarly, when several clay layers are encountered in the cut (Figure 10.8b), the average undrained cohesion becomes

$$c_{\text{av}} = \frac{1}{H}(c_1 H_1 + c_2 H_2 + \cdots + c_n H_n) \tag{10.6}$$

where $c_1, c_2, \ldots, c_n$ = undrained cohesion in layers 1, 2, ..., n
   $H_1, H_2, \ldots, H_n$ = thickness of layers 1, 2, ..., n

The average unit weight is now

$$\gamma_a = \frac{1}{H}(\gamma_1 H_1 + \gamma_2 H_2 + \gamma_3 H_3 + \cdots + \gamma_n H_n) \tag{10.7}$$

10.4 *Design of Various Components of a Braced Cut*

Struts

In construction work, struts should have a minimum vertical spacing of about 2.75 m (9 ft) or more. Struts are horizontal columns subject to bending. The load-carrying capacity of columns depends on their *slenderness ratio*, which can be reduced by providing vertical and horizontal supports at intermediate points. For wide cuts, splicing the struts may be necessary. For braced cuts in clayey soils, the depth of the first strut below the ground surface should be less than the depth of tensile crack, z_c. From Eq. (7.12),

$$\sigma_a' = \gamma z K_a - 2c' \sqrt{K_a}$$

where K_a = coefficient of Rankine active pressure

For determining the depth of tensile crack,

$$\sigma_a' = 0 = \gamma z_c K_a - 2c'\sqrt{K_a}$$

or

$$z_c = \frac{2c'}{\sqrt{K_a}\gamma}$$

With $\phi = 0$, $K_a = \tan^2(45 - \phi/2) = 1$, so

$$z_c = \frac{2c}{\gamma}$$

A simplified conservative procedure may be used to determine the strut loads. Although this procedure will vary, depending on the engineers involved in the project, the following is a step-by-step outline of the general methodology (see Figure 10.9):

1. Draw the pressure envelope for the braced cut. (See Figures 10.5, 10.6, and 10.7.) Also, show the proposed strut levels. Figure 10.9a shows a pressure envelope for a sandy soil; however, it could also be for a clay. The strut levels are marked *A, B, C,* and *D.* The sheet piles (or soldier beams) are assumed to be hinged at the strut levels, except for the top and bottom ones. In Figure 10.9a, the hinges are at the level of struts *B* and *C.* (Many designers also assume the sheet piles or soldier beams to be hinged at all strut levels except for the top.)
2. Determine the reactions for the two simple cantilever beams (top and bottom) and all the simple beams between. In Figure 10.9b, these reactions are A, B_1, B_2, C_1, C_2, and D.
3. The strut loads in the figure may be calculated via the formulas

$$P_A = (A)(s)$$
$$P_B = (B_1 + B_2)(s) \tag{10.8}$$
$$P_C = (C_1 + C_2)(s)$$

and

$$P_D = (D)(s)$$

where $\quad P_A, P_B, P_C, P_D$ = loads to be taken by the individual struts at levels *A, B, C,* and *D,* respectively

A, B_1, B_2, C_1, C_2, D = reactions calculated in Step 2 (note the unit: force/unit length of the braced cut)

s = horizontal spacing of the struts (see plan in Figure 10.9a)

4. Knowing the strut loads at each level and the intermediate bracing conditions allows selection of the proper sections from the steel construction manual.

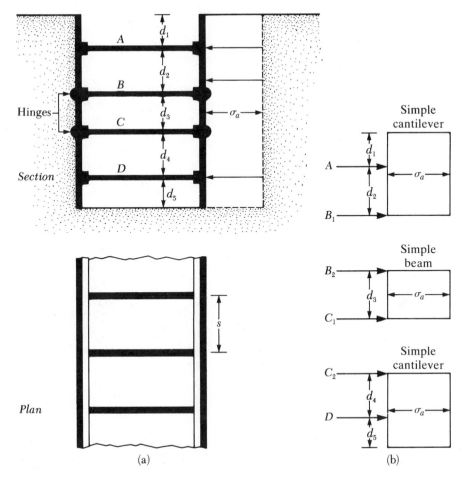

Figure 10.9 Determination of strut loads: (a) section and plan of the cut; (b) method for determining strut loads

Sheet Piles

The following steps are involved in designing the sheet piles:

1. For each of the sections shown in Figure 10.9b, determine the maximum bending moment.
2. Determine the maximum value of the maximum bending moments (M_{max}) obtained in Step 1. Note that the unit of this moment will be, for example, kN · m/m (lb-ft/ft) length of the wall.
3. Obtain the required section modulus of the sheet piles, namely,

$$S = \frac{M_{max}}{\sigma_{all}} \tag{10.9}$$

where σ_{all} = allowable flexural stress of the sheet pile material

4. Choose a sheet pile having a section modulus greater than or equal to the required section modulus from a table such as Table 9.1.

Wales

Wales may be treated as continuous horizontal members if they are spliced properly. Conservatively, they may also be treated as though they are pinned at the struts. For the section shown in Figure 10.9a, the maximum moments for the wales (assuming that they are pinned at the struts) are,

$$\text{at level } A, \quad M_{\text{max}} = \frac{(A)(s^2)}{8}$$

$$\text{at level } B, \quad M_{\text{max}} = \frac{(B_1 + B_2)s^2}{8}$$

$$\text{at level } C, \quad M_{\text{max}} = \frac{(C_1 + C_2)s^2}{8}$$

and

$$\text{at level } D, \quad M_{\text{max}} = \frac{(D)(s^2)}{8}$$

where A, B_1, B_2, C_1, C_2, and D are the reactions under the struts per unit length of the wall (see Step 2 of strut design)

Now determine the section modulus of the wales:

$$S = \frac{M_{\text{max}}}{\sigma_{\text{all}}}$$

The wales are sometimes fastened to the sheet piles at points that satisfy the lateral support requirements.

Example 10.1

The cross section of a long braced cut is shown in Figure 10.10a.

a. Draw the earth pressure envelope.
b. Determine the strut loads at levels A, B, and C.
c. Determine the section modulus of the sheet pile section required.
d. Determine a design section modulus for the wales at level B.

(*Note:* The struts are placed at 3 m, center to center, in the plan. Use

$$\sigma_{\text{all}} = 170 \times 10^3 \text{ kN/m}^2$$

Solution

Part a
We are given that $\gamma = 18 \text{ kN/m}^2$, $c = 35 \text{ kN/m}^2$, and $H = 7 \text{ m}$. So

$$\frac{\gamma H}{c} = \frac{(18)(7)}{35} = 3.6 < 4$$

Thus, the pressure envelope will be like the one in Figure 10.7. The envelope is plotted in Figure 10.10a with maximum pressure intensity, σ_a, equal to $0.3\gamma H = 0.3(18)(7) = \mathbf{37.8\ kN/m^2}$.

Part b

To calculate the strut loads, examine Figure 10.10b. Taking the moment about B_1, we have $\Sigma\ M_{B_1} = 0$, and

$$A(2.5) - \left(\frac{1}{2}\right)(37.8)(1.75)\left(1.75 + \frac{1.75}{3}\right) - (1.75)(37.8)\left(\frac{1.75}{2}\right) = 0$$

or

$$A = 54.02\ \text{kN/m}$$

Also, Σ vertical forces $= 0$. Thus,

$$\tfrac{1}{2}(1.75)(37.8) + (37.8)(1.75) = A + B_1$$

or

$$33.08 + 66.15 - A = B_1$$

So

$$B_1 = 45.2\ \text{kN/m}$$

Due to symmetry,

$$B_2 = 45.2\ \text{kN/m}$$

and

$$C = 54.02\ \text{kN/m}$$

Hence, the strut loads at the levels indicated by the subscripts are

$$P_a = 54.02 \times \text{horizontal spacing},\ s = 54.02 \times 3 = \mathbf{162.06\ kN}$$

$$P_B = (B_1 + B_2)3 = (45.2 + 45.2)3 = \mathbf{271.2\ kN}$$

and

$$P_C = 54.02 \times 3 = \mathbf{162.06\ kN}$$

Part c

At the left side of Figure 10.10b, for the maximum moment, the shear force should be zero. The nature of the variation of the shear force is shown in Figure 10.10c. The location of point E can be given as

$$x = \frac{\text{reaction at } B_1}{37.8} = \frac{45.2}{37.8} = 1.196\ \text{m}$$

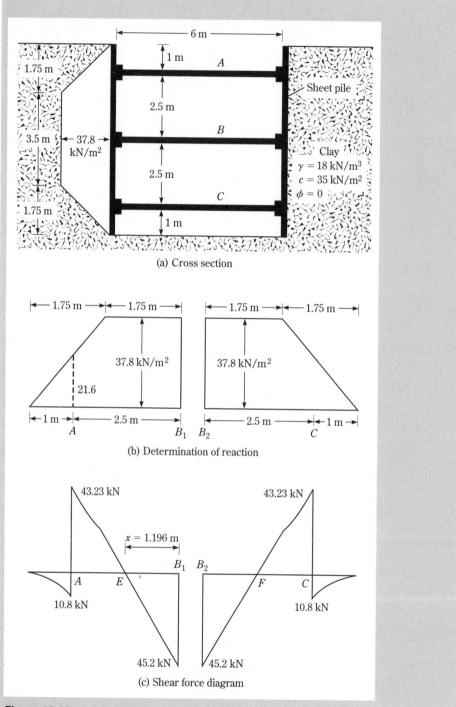

(a) Cross section

(b) Determination of reaction

(c) Shear force diagram

Figure 10.10   Analysis of a braced cut

Also,

$$\text{magnitude of moment at } A = \frac{1}{2}(1)\left(\frac{37.8}{1.75} \times 1\right)\left(\frac{1}{3}\right)$$

$$= 3.6 \text{ kN-m/meter of wall}$$

and

$$\text{magnitude of moment at } E = (45.2 \times 1.196) - (37.8 \times 1.196)\left(\frac{1.196}{2}\right)$$

$$= 54.06 - 27.03 = 27.03 \text{ kN-m/meter of wall}$$

Because the loading on the left and right sections of Figure 10.10b are the same, the magnitudes of the moments at F and C (see Figure 10.10c) will be the same as those at E and A, respectively. Hence, the maximum moment is 27.03 kN-m/meter of wall.

The section modulus of the sheet piles is thus

$$S = \frac{M_{max}}{\sigma_{all}} = \frac{27.03 \text{ kN-m}}{170 \times 10^3 \text{ kN/m}^2} = \mathbf{15.9 \times 10^{-5} m^3/m} \textbf{ of the wall}$$

Part d
The reaction at level B has been calculated in Part b. Hence,

$$M_{max} = \frac{(B_1 + B_2)s^2}{8} = \frac{(45.2 + 45.2)3^2}{8} = 101.7 \text{ kN-m}$$

and

$$\text{Section modulus, } S = \frac{101.7}{\sigma_{all}} = \frac{101.7}{(170 \times 1000)}$$

$$= \mathbf{0.598 \times 10^{-3} m^3} \qquad \blacksquare$$

10.5 Bottom Heave of a Cut in Clay

Braced cuts in clay may become unstable as a result of heaving of the bottom of the excavation. Terzaghi (1943) analyzed the factor of safety of long braced excavations against bottom heave. The failure surface for such a case in a homogeneous soil is shown in Figure 10.11. In the figure, the following notations are used: B = width of the cut, H = depth of the cut, T = thickness of the clay below the base of excavation, and q = uniform surcharge adjacent to the excavation.

The ultimate bearing capacity at the base of a soil column with a width of B' can be given as

$$q_u = cN_c \qquad (10.10)$$

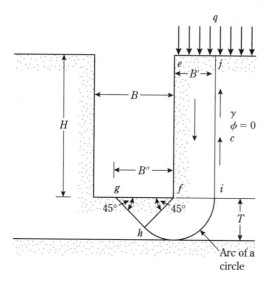

Figure 10.11   Heaving in braced cuts in clay

where $N_c = 5.7$ (for a perfectly rough foundation)

The vertical load per unit area along *fi* is

$$q = \gamma H + q - \frac{cH}{B'} \tag{10.11}$$

Hence, the factor of safety against bottom heave is

$$FS = \frac{q_u}{q} = \frac{cN_c}{\gamma H + q - \dfrac{cH}{B'}} = \frac{cN_c}{\left(\gamma + \dfrac{q}{H} - \dfrac{c}{B'}\right)H} \tag{10.12}$$

For excavations of limited length L, the factor of safety can be modified to

$$FS = \frac{cN_c\left(1 + 0.2\dfrac{B'}{L}\right)}{\left(\gamma + \dfrac{q}{H} - \dfrac{c}{B'}\right)H} \tag{10.13}$$

where $B' = T$ or $B/\sqrt{2}$ (whichever is smaller)

In 2000, Chang suggested a revision of Eq. (10.13) with the following changes:

1. The shearing resistance along *ij* may be considered as an increase in resistance rather than a reduction in loading.
2. In Figure 10.11, *fg* with a width of B'' at the base of the excavation may be treated as a negatively loaded footing.
3. The value of the bearing capacity factor N_c should be 5.14 (not 5.7) for a perfectly smooth footing, because of the restraint-free surface at the base of the excavation.

With the foregoing modifications, Eq. (10.13) takes the form

$$FS = \frac{5.14c\left(1 + \frac{0.2B''}{L}\right) + \frac{cH}{B'}}{\gamma H + q} \tag{10.14}$$

where $B' = T$ if $T \leq B/\sqrt{2}$
$B' = B/\sqrt{2}$ if $T > B/\sqrt{2}$
$B'' = \sqrt{2}B'$

Bjerrum and Eide (1956) compiled a number of case records for the bottom heave of cuts in clay. Chang (2000) used those records to calculate FS by means of Eq. (10.14); his findings are summarized in Table 10.1. It can be seen from this table that the actual field observations agree well with the calculated factors of safety.

Equation (10.14) is recommended for use in this test. In most cases, a factor of safety of about 1.5 is recommended.

In homogeneous clay, if FS becomes less than 1.5, the sheet pile is driven deeper. (See Figure 10.12.) Usually, the depth d is kept less than or equal to $B/2$, in which case the force P per unit length of the buried sheet pile (aa' and bb') may be expressed as (U.S. Department of the Navy, 1971)

$$P = 0.7(\gamma H B - 1.4cH - \pi cB) \qquad \text{for } d > 0.47B \tag{10.15}$$

Table 10.1 Calculated Factors of Safety for Selected Case Records Compiled by Bjerrum and Eide (1956) and Calculated by Chang (2000)

Site	B (m)	B/L	H (m)	H/B	γ (kN/m³)	c (kN/m²)	q (kN/m²)	FS [Eq. (10.14)]	Type of failure
Pumping station, Fornebu, Oslo	5.0	1.0	3.0	0.6	17.5	7.5	0	1.05	Total failure
Storehouse, Drammen	4.8	0	2.4	0.5	19.0	12	15	1.05	Total failure
Sewerage tank, Drammen	5.5	0.69	3.5	0.64	18.0	10	10	0.92	Total failure
Excavation, Grey Wedels Plass, Oslo	5.8	0.72	4.5	0.78	18.0	14	10	1.07	Total failure
Pumping station, Jernbanetorget, Oslo	8.5	0.70	6.3	0.74	19.0	22	0	1.26	Partial failure
Storehouse, Freia, Oslo	5.0	0	5.0	1.00	19.0	16	0	1.10	Partial failure
Subway, Chicago	16	0	11.3	0.70	19.0	35	0	1.00	Near failure

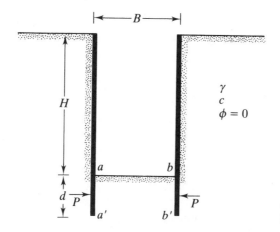

Figure 10.12 Force on the buried length of sheet pile

and

$$P = 1.5d\left(\gamma H - \frac{1.4cH}{B} - \pi c\right) \qquad \text{for } d < 0.47B \qquad (10.16)$$

Example 10.2

In Figure 10.13, for a braced cut in clay, $B = 3$ m, $L = 20$ m, $H = 5.5$ m, $T = 1.5$ m, $\gamma = 17$ kN/m³, $c = 30$ kN/m², and $q = 0$. Calculate the factor of safety against heave. Use Eq. (10.14).

Solution
From Eq. (10.14),

$$FS = \frac{5.14c\left(1 + \frac{0.2B''}{L}\right) + \frac{cH}{B'}}{\gamma H + q}$$

with $T = 1.5$ m,

$$\frac{B}{\sqrt{2}} = \frac{3}{\sqrt{2}} = 2.12 \text{ m}$$

So

$$T \leqslant \frac{B}{\sqrt{2}}$$

Hence, $B' = T = 1.5$ m, and it follows that

$$B'' = \sqrt{2}B' = (\sqrt{2})(1.5) = 2.12 \text{ m}$$

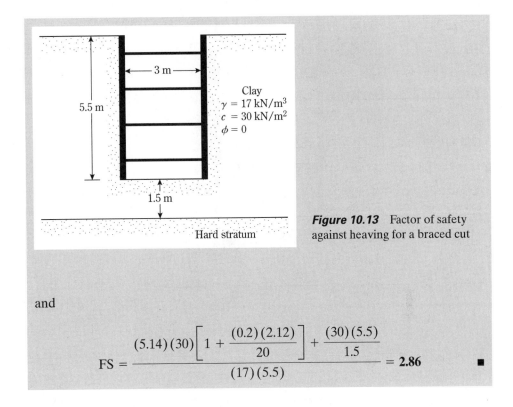

Figure 10.13 Factor of safety against heaving for a braced cut

and

$$FS = \frac{(5.14)(30)\left[1 + \dfrac{(0.2)(2.12)}{20}\right] + \dfrac{(30)(5.5)}{1.5}}{(17)(5.5)} = \mathbf{2.86}$$

■

10.6 *Stability of the Bottom of a Cut in Sand*

The bottom of a cut in sand is generally stable. When the water table is encountered, the bottom of the cut is stable as long as the water level inside the excavation is higher than the groundwater level. In case dewatering is needed (see Figure 10.14), the factor of safety against piping should be checked. [*Piping* is another term for failure by heave, as defined in Section 1.10; see Eq. (1.46).] Piping may occur when a high hydraulic gradient is created by water flowing into the excavation. To check the factor of safety, draw flow nets and determine the maximum exit gradient $[i_{max(exit)}]$ that will occur at points A and B. Figure 10.15 shows such a flow net, for which the maximum exit gradient is

$$i_{max(exit)} = \frac{\dfrac{h}{N_d}}{a} = \frac{h}{N_d a} \tag{10.17}$$

where a = length of the flow element at A (or B)
N_d = number of drops (*Note:* in Figure 10.15, $N_d = 8$; see also Section 1.9)

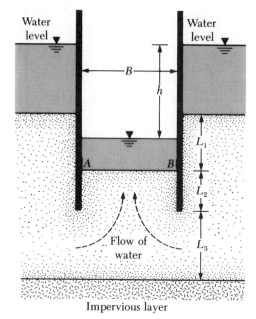

Figure 10.14 Stability of the bottom of a cut in sand

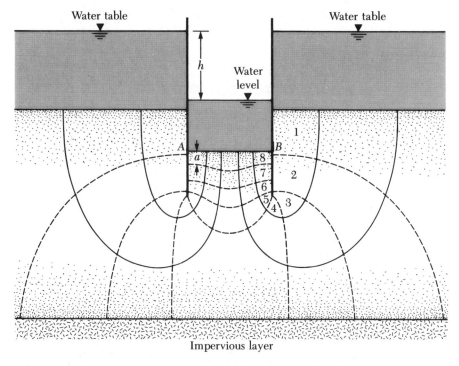

Figure 10.15 Determining the factor of safety against piping by drawing a flow net

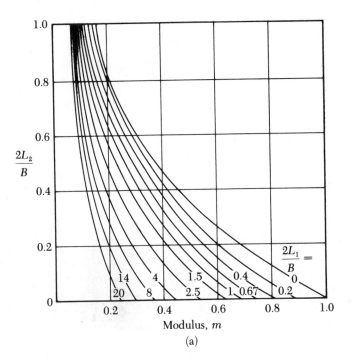

(a)

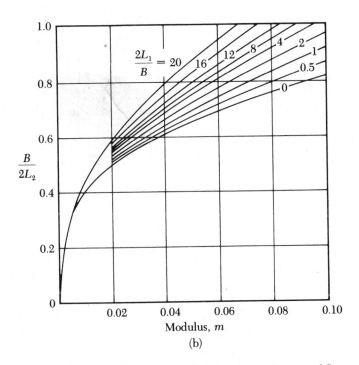

(b)

Figure 10.16 Variation of modulus (from *Groundwater and Seepage,* by M. E. Harr. Copyright © 1962 by McGraw-Hill. Used with permission.)

The factor of safety against piping may be expressed as

$$FS = \frac{i_{cr}}{i_{max(exit)}} \tag{10.18}$$

where i_{cr} = critical hydraulic gradient

The relationship for i_{cr} was given in Chapter 1 as

$$i_{cr} = \frac{G_s - 1}{e + 1}$$

The magnitude of i_{cr} varies between 0.9 and 1.1 in most soils, with an average of about 1. A factor of safety of about 1.5 is desirable.

The maximum exit gradient for sheeted excavations in sands with $L_3 = \infty$ can also be evaluated theoretically (Harr, 1962). (Only the results of these mathematical derivations will be presented here. For further details, see the original work.) To calculate the maximum exit gradient, examine Figures 10.16 and 10.17 and perform the following steps:

1. Determine the modulus, m, from Figure 10.16 by obtaining $2L_2/B$ (or $B/2L_2$) and $2L_1/B$.
2. With the known modulus and $2L_1/B$, examine Figure 10.17 and determine $L_2 i_{exit(max)}/h$. Because L_2 and h will be known, $i_{exit(max)}$ can be calculated.
3. The factor of safety against piping can be evaluated by using Eq. (10.18).

Marsland (1958) presented the results of model tests conducted to study the influence of seepage on the stability of sheeted excavations in sand. The results were summarized by the U.S. Department of the Navy (1971) in NAVFAC DM-7 and are given in Figure 10.18a, b, and c. Note that Figure 10.18b is for the case of determining the sheet pile penetration L_2 needed for the required factor of safety against piping when the sand layer extends to a great depth below the excavation. By contrast, Figure 10.18c represents the case in which an impervious layer lies at a depth $L_2 + L_3$ below the bottom of the excavation.

Example 10.3

In Figure 10.14, let $h = 4.5$ m, $L_1 = 5$ m, $L_2 = 4$ m, $B = 5$ m, and $L_3 = \infty$. Determine the factor of safety against piping.

Solution
We have

$$\frac{2L_1}{B} = \frac{2(5)}{5} = 2$$

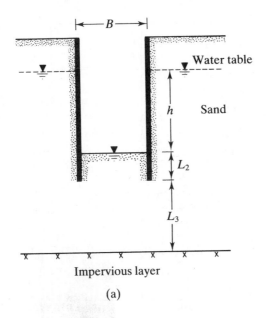

Impervious layer

(a)

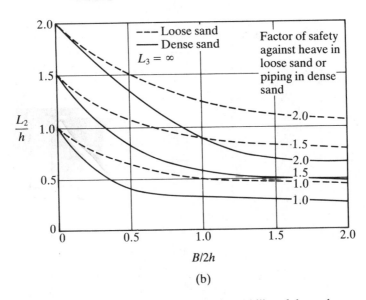

B/2h

(b)

Figure 10.18 Influence of seepage on the stability of sheeted excavation (after U.S. Department of the Navy, 1971)

and

$$\frac{B}{2L_2} = \frac{5}{2(4)} = 0.625$$

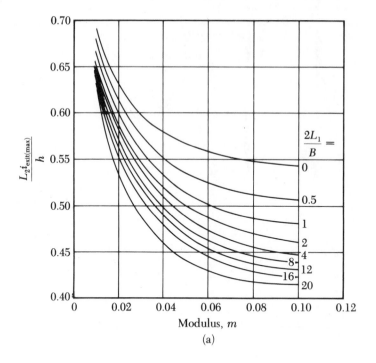

(a)

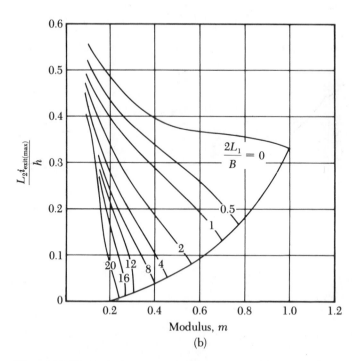

(b)

Figure 10.17   Variation of maximum exit gradient with modulus (from *Groundwater and Seepage,* by M. E. Harr. Copyright © 1962 by McGraw-Hill. Used with permission.)

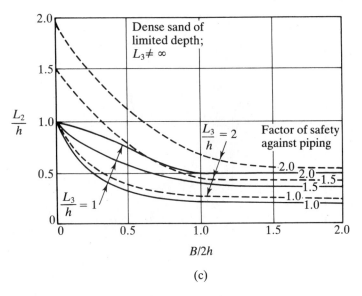

Figure 10.18 (Continued)

According to Figure 10.16b, for $2L_1/B = 2$ and $B/2L_2 = 0.625$, $m \approx 0.033$. From Figure 10.17a, for $m = 0.033$ and $2L_1/B = 2$, $L_2 i_{\text{exit(max)}}/h = 0.54$. Hence,

$$i_{\text{exit(max)}} = \frac{0.54(h)}{L_2} = 0.54(4.5)/4 = 0.608$$

and

$$FS = \frac{i_{\text{cr}}}{i_{\text{max(exit)}}} = \frac{1}{0.608} = 1.645$$ ∎

10.7 *Lateral Yielding of Sheet Piles and Ground Settlement*

In braced cuts, some lateral movement of sheet pile walls may be expected. (See Figure 10.19.) The amount of lateral yield depends on several factors, the most important of which is the elapsed time between excavation and the placement of wales and struts. Mana and Clough (1981) analyzed the field records of several braced cuts in clay from the San Francisco, Oslo (Norway), Boston, Chicago, and Bowline Point (New York) areas and found that, under ordinary construction conditions, the maximum lateral wall yield, $\delta_{H(\text{max})}$, has a definite relationship to the factor of safety against heave, as shown in Figure 10.19. Note that the factor of safety against heave plotted in that figure was calculated by using Eq. (10.13).

As discussed before, in several instances the sheet piles (or the soldier, piles as the case may be) are driven to a certain depth below the bottom of the excavation.

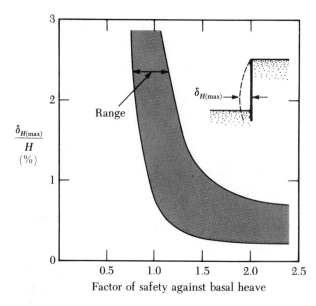

Figure 10.19 Range of variation of $\delta_{H(max)}/H$ with FS against basal heave from field observation (redrawn after Mana and Clough, 1981)

The reason is to reduce the lateral yielding of the walls during the last stages of excavation. Lateral yielding of the walls will cause the ground surface surrounding the cut to settle. The degree of lateral yielding, however, depends mostly on the type of soil below the bottom of the cut. If clay below the cut extends to a great depth and $\gamma H/c$ is less than about 6, extension of the sheet piles or soldier piles below the bottom of the cut will help considerably in reducing the lateral yield of the walls.

However, under similar circumstances, if $\gamma H/c$ is about 8, the extension of sheet piles into the clay below the cut does not help greatly. In such circumstances, we may expect a great degree of wall yielding that could result in the total collapse of the bracing systems. If a hard layer of soil lies below a clay layer at the bottom of the cut, the piles should be embedded in the stiffer layer. This action will greatly reduce lateral yield.

The lateral yielding of walls will generally induce ground settlement, δ_V, around a braced cut. Such settlement is generally referred to as *ground loss*. On the basis of several field observations, Peck (1969) provided curves for predicting ground settlement in various types of soil. (See Figure 10.20.) The magnitude of ground loss varies extensively; however, the figure may be used as a general guide.

On the basis of the field data obtained from various cuts in the areas of San Francisco, Oslo, and Chicago, Mana and Clough (1981) provided a correlation between the maximum lateral yield of sheet piles, $\delta_{H(max)}$, and the maximum ground settlement, $\delta_{V(max)}$. The correlation is shown in Figure 10.21. Note that

$$\delta_{V(max)} \approx 0.5\delta_{H(max)} \quad \text{to} \quad 1.0\delta_{H(max)} \tag{10.19}$$

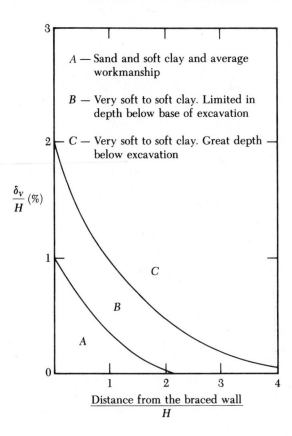

Figure 10.20 Variation of ground settlement with distance (after Peck, 1969)

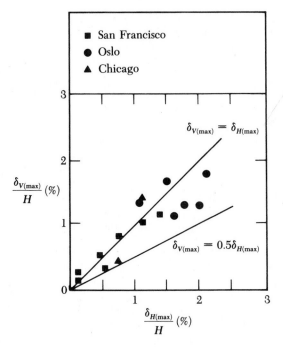

Figure 10.21 Variation of maximum lateral yield with maximum ground settlement (after Mana and Clough, 1981)

10.8   *Subway Extension of the Massachusetts Bay Transportation Authority (MBTA): A Case Study*

The procedure for determining strut loads and the design of sheet piles and wales presented in the preceding sections appears to be fairly straightforward. However, it is made possible only if a proper pressure envelope is chosen for the design—a difficult task to carry out. This section describes a case study of braced cuts made for the excavation of a subway for the Massachusetts Bay Transportation Authority (MBTA). It highlights the difficulties and degree of judgment needed for the successful completion of various projects.

Lambe (1970) provided data on the performance of three excavations for the subway extension of the MBTA in Boston (test sections *A, B,* and *D*), all of which were well instrumented. Figure 10.22 gives the details, including subsoil conditions, of test section *B,* where the cut was 18 m (≈58 ft). The subsoil consisted of gravel, sand, silt, and clay (fill) to a depth of about 8 m (26 ft), followed by a light gray, slightly organic silt to a depth of 14 m (46 ft). A layer of coarse sand and gravel with some clay was present from 14 m to 16.5 m (46 ft to 54 ft) below the ground surface. Rock was encountered below 16.5 m (54 ft). The horizontal spacing of the struts was 3.66 m (12 ft) center to center.

Because the apparent-pressure envelopes that are available (see Section 10.2) are for *sand* and *clay* only, questions may arise about how to treat the fill, silt, and till. Figure 10.23 shows the apparent-pressure envelopes proposed by Peck (1969) to

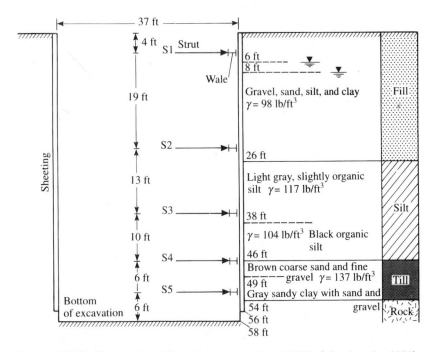

Figure 10.22   Test section *B* for subway extension, MBTA (after Lambe, 1970)

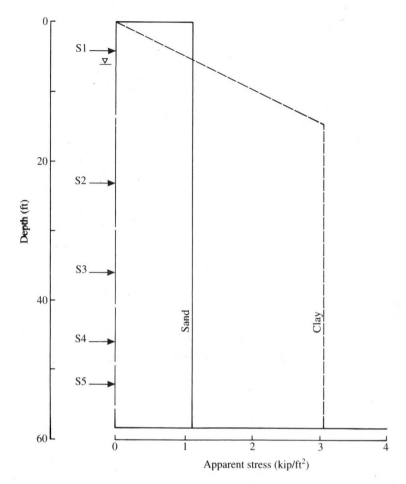

Figure 10.23 Apparent-pressure envelopes for test section *B*, MBTA (after Lambe, 1970)

overcome that problem, considering the soil as *sand* and also as *clay*. For the average soil parameters of the profile, the following values of σ_a were used to develop the pressure envelopes shown in the figure.

Sand

$$\sigma_a = 0.65\gamma H K_a \tag{10.1}$$

For $\gamma = 114$ lb/ft^3, $H = 58$ ft, and $K_a = 0.26$,

$$\sigma_a = (0.65)(114)(58)(0.26) = 1117 \text{ lb/ft}^2 \approx 1.12 \text{ kip/ft}^2$$

Clay

$$\sigma_a = \gamma H \left[1 - \left(\frac{4c}{\gamma H} \right) \right] \tag{10.2}$$

For $c = 890$ lb/ft^2,

$$\sigma_a = (114)(58)\left[1 - \frac{(4)(890)}{(114)(58)}\right] = 3052 \text{ lb/ft}^2 \approx 3.05 \text{ kip/ft}^2$$

Figure 10.24 shows the variations of the strut load, based on the assumed pressure envelopes shown in Figure 10.23. Also shown in Figure 10.24 are the measured strut loads in the field. A comparison of these two variables indicates that

1. In most cases the measured strut loads differed widely from the predicted ones, due primarily to the uncertainties involved in the assumption of the soil parameters.
2. The actual design strut loads were substantially higher than those measured.

Figure 10.25 shows test section A for the subway extension, along with the soil profile. The measured and predicted strut loads for this section (similar to those shown in Figure 10.24) are shown in Figure 10.26. Considering the uncertainties involved, the general agreement between the measured and predicted values is quite good.

Figure 10.27 shows soil movements in the vicinity of sections A and B in nondimensional form (δ_V/H and δ_H/H). They appear to be in general agreement with those presented in Section 10.7.

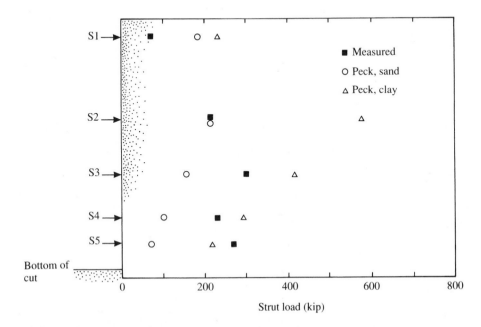

Figure 10.24   Measured and predicted strut loads, test section B, MBTA (after Lambe, 1970)

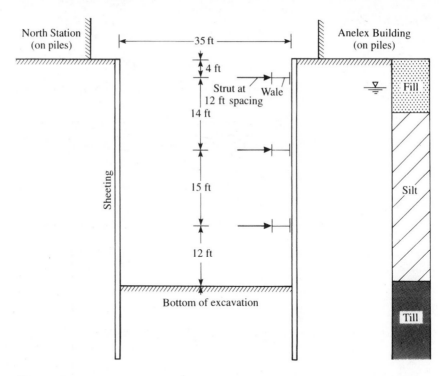

Figure 10.25 Test section *A* for subway extension, MBTA (after Lambe, 1970)

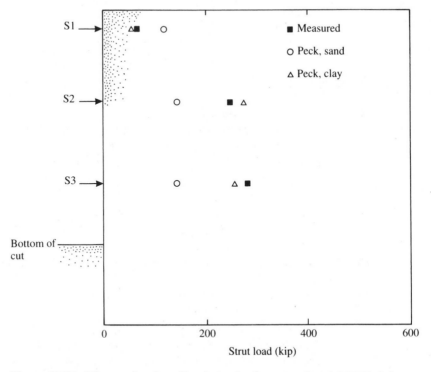

Figure 10.26 Measured and predicted strut loads, test section *A*, MBTA (after Lambe, 1970)

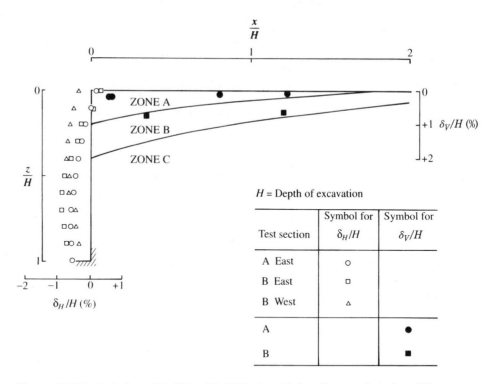

Figure 10.27   Variation of δ_H/H and δ_V/H in the vicinity of test sections A and B, MBTA (after Lambe, 1970)

Problems

10.1 For the braced cut shown in Figure P10.1, $\gamma = 112$ lb/ft^3, $\phi' = 32°$, and $c' = 0$. The struts are located at 12 ft on center in the plan. Draw the earth pressure envelope and determine the strut loads at levels *A, B,* and *C.*

10.2 For the braced cut described in Problem 10.1, assume that $\sigma_{all} = 25{,}000$ lb/in.^{2} and determine
  a. The section modulus of the sheet pile section.
  b. The section modulus of the wales at level *A.*

10.3 Redo Problem 10.1 for $\gamma = 116$ lb/ft^3, $\phi' = 35°$, $c' = 0$, and a planned center-to-center strut spacing of 10 ft.

10.4 Determine the sheet pile section required for the braced cut in Problem 10.3 for $\sigma_{all} = 25{,}000$ lb/in.^{2}.

10.5 For the braced cut shown in Figure 10.8a, $H = 6$ m; $H_s = 2$ m; $\gamma_s = 16.2$ kN/m^3; the angle of friction of sand, $\phi'_s = 34°$; $H_c = 4$ m; $\gamma_c = 17.5$ kN/m^3; and the unconfined compression strength of clay layer, $q_u = 68$ kN/m^2.
  a. Estimate the average cohesion, c_{av}, and average unit weight, γ_{av}, for development of the earth pressure envelope. Use $n' = 0.5$ in Eq. (10.4).
  b. Plot the earth pressure envelope.

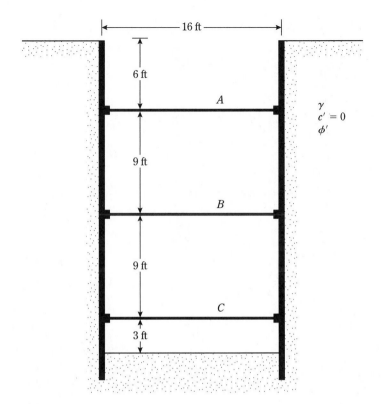

Figure P10.1

10.6 Figure 10.8b shows a braced cut in clay. Let $H = 22$ ft, $H_1 = 6$ ft, $c_1 = 2125$ lb/ft², $\gamma_1 = 111$ lb/ft³, $H_2 = 8$ ft, $c_2 = 1565$ lb/ft², $\gamma_2 = 107$ lb/ft³, $H_3 = 8$ ft, $c_3 = 1670$ lb/ft², and $\gamma_3 = 109$ lb/ft³.

    a. Estimate the average cohesion, c_{av}, and average unit weight, γ_{av}, for development of the earth pressure envelope.

    b. Plot the earth pressure envelope.

10.7 In Figure P10.7, $\gamma = 17.5$ kN/m³, $c = 30$ kN/m², and the center-to-center spacing of the struts is 5 m. Draw the earth pressure envelope and determine the strut loads at levels A, B, and C.

10.8 For the braced cut described in Problem 10.7, determine the section modulus of the sheet pile required. Use $\sigma_{all} = 170$ MN/m².

10.9 Redo Problem 10.7 with $c = 60$ kN/m².

10.10 Determine the factor of safety against bottom heave for the braced cut described in Problem 10.7. Use Eq. (10.14), and assume that the length of the cut, $L = 18$ m.

10.11 Determine the factor of safety against bottom heave for the braced cut described in Problem 10.9. Use Eq. (10.14). The length of the cut is 12.5 m.

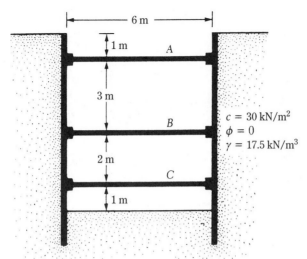

$c = 30 \text{ kN/m}^2$
$\phi = 0$
$\gamma = 17.5 \text{ kN/m}^3$

Figure P10.7

References

Bjerrum, L, and Eide, O. (1956). "Stability of Strutted Excavation in Clay," *Geotechnique,* Vol. 6, No. 1, pp. 32–47.

Chang, M. F. (2000). "Basal Stability Analysis of Braced Cuts in Clay," *Journal of Geotechnical and Geoenvironmental Engineering,* ASCE, Vol. 126, No. 3, pp. 276–279.

Harr, M. E. (1962). *Groundwater and Seepage,* McGraw-Hill, New York.

Lambe, T. W. (1970). "Braced Excavations," *Proceedings of the Specialty Conference on Lateral Stresses in the Ground and Design of Earth-Retaining Structures,* American Society of Civil Engineers, pp. 149–218.

Mana, A. I., and Clough, G. W. (1981). "Prediction of Movements for Braced Cuts in Clay," *Journal of the Geotechnical Engineering Division,* American Society of Civil Engineers, Vol. 107, No. GT8, pp. 759–777.

Marsland, A. (1958). "Model Experiments to Study the Influence of Seepage on the Stability of a Sheeted Excavation in Sand," *Geotechnique,* London, Vol. 3, p. 223.

Peck, R. B. (1943). "Earth Pressure Measurements in Open Cuts, Chicago (Ill.) Subway," *Transactions,* American Society of Civil Engineers, Vol. 108, pp. 1008–1058.

Peck, R. B. (1969). "Deep Excavation and Tunneling in Soft Ground," *Proceedings Seventh International Conference on Soil Mechanics and Foundation Engineering,* Mexico City, State-of-the-Art Volume, pp. 225–290.

Terzaghi, K. (1943). *Theoretical Soil Mechanics,* Wiley, New York.

U.S. Department of the Navy (1971). "Design Manual—Soil Mechanics, Foundations, and Earth Structures," NAVFAC DM-7, Washington, DC.

11

Pile Foundations

Introduction

Piles are structural members that are made of steel, concrete, or timber. They are used to build pile foundations, which are deep and which cost more than shallow foundations. (See Chapters 3, 4, and 5.) Despite the cost, the use of piles often is necessary to ensure structural safety. The following list identifies some of the conditions that require pile foundations (Vesic, 1977):

1. When one or more upper soil layers are highly compressible and too weak to support the load transmitted by the superstructure, piles are used to transmit the load to underlying bedrock or a stronger soil layer, as shown in Figure 11.1a. When bedrock is not encountered at a reasonable depth below the ground surface, piles are used to transmit the structural load to the soil gradually. The resistance to the applied structural load is derived mainly from the frictional resistance developed at the soil–pile interface. (See Figure 11.1b.)

2. When subjected to horizontal forces (see Figure 11.1c), pile foundations resist by bending, while still supporting the vertical load transmitted by the superstructure. This type of situation is generally encountered in the design and construction of earth-retaining structures and foundations of tall structures that are subjected to high wind or to earthquake forces.

3. In many cases, expansive and collapsible soils may be present at the site of a proposed structure. These soils may extend to a great depth below the ground surface. Expansive soils swell and shrink as their moisture content increases and decreases, and the pressure of the swelling can be considerable. If shallow foundations are used in such circumstances, the structure may suffer considerable damage. However, pile foundations may be considered as an alternative when piles are extended beyond the active zone, which is where swelling and shrinking occur. (See Figure 11.1d)

    Soils such as loess are collapsible in nature. When the moisture content of these soils increases, their structures may break down. A sudden decrease in the void ratio of soil induces large settlements of structures supported by shallow foundations. In such cases, pile foundations may be used in which the piles are extended into stable soil layers beyond the zone where moisture will change.

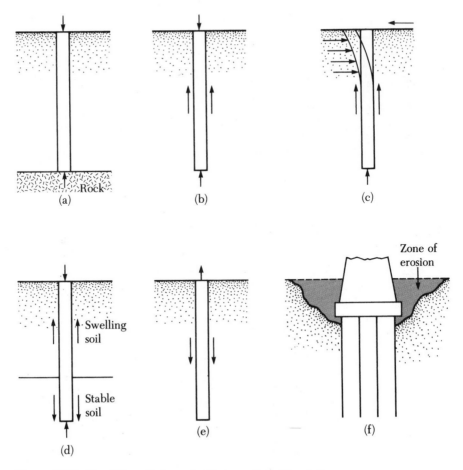

Figure 11.1 Conditions that require the use of pile foundations

4. The foundations of some structures, such as transmission towers, offshore platforms, and basement mats below the water table, are subjected to uplifting forces. Piles are sometimes used for these foundations to resist the uplifting force. (See Figure 11.1e.)

5. Bridge abutments and piers are usually constructed over pile foundations to avoid the loss of bearing capacity that a shallow foundation might suffer because of soil erosion at the ground surface. (See Figure 11.1f.)

Although numerous investigations, both theoretical and experimental, have been conducted in the past to predict the behavior and the load-bearing capacity of piles in granular and cohesive soils, the mechanisms are not yet entirely understood and may never be. The design and analysis of pile foundations may thus be considered somewhat of an art as a result of the uncertainties involved in working with some subsoil conditions. This chapter discusses the present state of the art.

11.2   *Types of Piles and Their Structural Characteristics*

Different types of piles are used in construction work, depending on the type of load to be carried, the subsoil conditions, and the location of the water table. Piles can be divided into the following categories: (a) steel piles, (b) concrete piles, (c) wooden (timber) piles, and (d) composite piles.

Steel Piles

Steel piles generally are either *pipe piles* or *rolled steel* H-*section piles.* Pipe piles can be driven into the ground with their ends open or closed. Wide-flange and I-section steel beams can also be used as piles. However, H-section piles are usually preferred because their web and flange thicknesses are equal. (In wide-flange and I-section beams, the web thicknesses are smaller than the thicknesses of the flange.) Table 11.1 gives the dimensions of some standard H-section steel piles used in the United States. Table 11.2 shows selected pipe sections frequency used for piling purposes. In many cases, the pipe piles are filled with concrete after they have been driven.

The allowable structural capacity for steel piles is

$$Q_{all} = A_s f_s \tag{11.1}$$

where A_s = cross-sectional area of the steel

f_s = allowable stress of steel (≈ 0.33–$0.5 f_y$)

Once the design load for a pile is fixed, one should determine, on the basis of geotechnical considerations, whether $Q_{(design)}$ is within the allowable range as defined by Eq. 11.1.

When necessary, steel piles are spliced by welding or by riveting. Figure 11.2a shows a typical splice by welding for an H-pile. A typical splice by welding for a pipe pile is shown in Figure 11.2b. Figure 11.2c is a diagram of a splice of an H-pile by rivets or bolts.

When hard driving conditions are expected, such as driving through dense gravel, shale, or soft rock, steel piles can be fitted with driving points or shoes. Figures 11.2d and 11.2e are diagrams of two types of shoe used for pipe piles.

Steel piles may be subject to corrosion. For example, swamps, peats, and other organic soils are corrosive. Soils that have a pH greater than 7 are not so corrosive. To offset the effect of corrosion, an additional thickness of steel (over the actual designed cross-sectional area) is generally recommended. In many circumstances factory-applied epoxy coatings on piles work satisfactorily against corrosion. These coatings are not easily damaged by pile driving. Concrete encasement of steel piles in most corrosive zones also protects against corrosion.

Following are some general facts about steel piles:

– Usual length: 15 m–60 m (50 ft–200 ft)
– Usual load: 300 kN–1200 kN (67 kip–265 kip)
– Advantages:

　a. Easy to handle with respect to cutoff and extension to the desired length
　b. Can stand high driving stresses

Table 11.1a   Common H-Pile Sections Used in the United States (SI Units)

Designation, size (mm) × weight (kg/m)	Depth d_1 (mm)	Section area (m^2 × 10^{-3})	Flange and web thickness w (mm)	Flange width d_2 (mm)	Moment of inertia (m^4 × 10^{-6})	
					I_{xx}	I_{yy}
HP 200 × 53	204	6.84	11.3	207	49.4	16.8
HP 250 × 85	254	10.8	14.4	260	123	42
× 62	246	8.0	10.6	256	87.5	24
HP 310 × 125	312	15.9	17.5	312	271	89
× 110	308	14.1	15.49	310	237	77.5
× 93	303	11.9	13.1	308	197	63.7
× 79	299	10.0	11.05	306	164	62.9
HP 330 × 149	334	19.0	19.45	335	370	123
× 129	329	16.5	16.9	333	314	104
× 109	324	13.9	14.5	330	263	86
× 89	319	11.3	11.7	328	210	69
HP 360 × 174	361	22.2	20.45	378	508	184
× 152	356	19.4	17.91	376	437	158
× 132	351	16.8	15.62	373	374	136
× 108	346	13.8	12.82	371	303	109

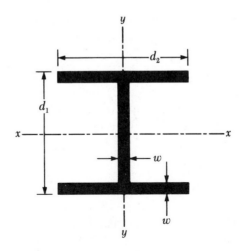

    c. Can penetrate hard layers such as dense gravel and soft rock
    d. High load-carrying capacity

– Disadvantages:

    a. Relatively costly
    b. High level of noise during pile driving
    c. Subject to corrosion
    d. H-piles may be damaged or deflected from the vertical during driving through hard layers or past major obstructions

Table 11.1b Common H-Pile Sections Used in the United States (English Units)

Designation size (in.) × weight (lb/ft)	Depth d_1 (in.)	Section area (in²)	Flange and web thickness w (in.)	Flange width d_2 (in.)	Moment of inertia (in⁴)	
					I_{xx}	I_{yy}
HP 8 × 36	8.02	10.6	0.445	8.155	119	40.3
HP 10 × 57	9.99	16.8	0.565	10.225	294	101
× 42	9.70	12.4	0.420	10.075	210	71.7
HP 12 × 84	12.28	24.6	0.685	12.295	650	213
× 74	12.13	21.8	0.610	12.215	570	186
× 63	11.94	18.4	0.515	12.125	472	153
× 53	11.78	15.5	0.435	12.045	394	127
HP 13 × 100	13.15	29.4	0.766	13.21	886	294
× 87	12.95	25.5	0.665	13.11	755	250
× 73	12.74	21.6	0.565	13.01	630	207
× 60	12.54	17.5	0.460	12.90	503	165
HP 14 × 117	14.21	34.4	0.805	14.89	1220	443
× 102	14.01	30.0	0.705	14.78	1050	380
× 89	13.84	26.1	0.615	14.70	904	326
× 73	13.61	21.4	0.505	14.59	729	262

Table 11.2a Selected Pipe Pile Sections (SI Units)

Outside diameter (mm)	Wall thickness (mm)	Area of steel (cm²)
219	3.17	21.5
	4.78	32.1
	5.56	37.3
	7.92	52.7
254	4.78	37.5
	5.56	43.6
	6.35	49.4
305	4.78	44.9
	5.56	52.3
	6.35	59.7
406	4.78	60.3
	5.56	70.1
	6.35	79.8
457	5.56	80
	6.35	90
	7.92	112
508	5.56	88
	6.35	100
	7.92	125
610	6.35	121
	7.92	150
	9.53	179
	12.70	238

Table 11.2b Selected Pipe Pile Sections (English Units)

Outside diameter (in.)	Wall thickness (in.)	Area of steel (in²)
$8\frac{5}{8}$	0.125	3.34
	0.188	4.98
	0.219	5.78
	0.312	8.17
10	0.188	5.81
	0.219	6.75
	0.250	7.66
12	0.188	6.96
	0.219	8.11
	0.250	9.25
16	0.188	9.34
	0.219	10.86
	0.250	12.37
18	0.219	12.23
	0.250	13.94
	0.312	17.34
20	0.219	13.62
	0.250	15.51
	0.312	19.30
24	0.250	18.7
	0.312	23.2
	0.375	27.8
	0.500	36.9

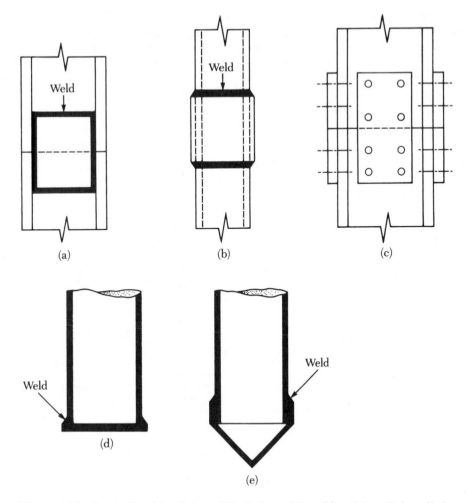

Figure 11.2 Steel piles: (a) splicing of H-pile by welding; (b) splicing of pipe pile by welding; (c) splicing of H-pile by rivets and bolts; (d) flat driving point of pipe pile; (e) conical driving point of pipe pile

Concrete Piles

Concrete piles may be divided into two basic categories: (a) precast piles and (b) cast-in-situ piles. *Precast piles* can be prepared by using ordinary reinforcement, and they can be square or octagonal in cross section. (See Figure 11.3.) Reinforcement is provided to enable the pile to resist the bending moment developed during pickup and transportation, the vertical load, and the bending moment caused by a lateral load. The piles are cast to desired lengths and cured before being transported to the work sites.

Some general facts about concrete piles are as follows:

– Usual length: 10 m–15 m (30 ft–50 ft)
– Usual load: 300 kN–3000 kN (67 kip–675 kip)

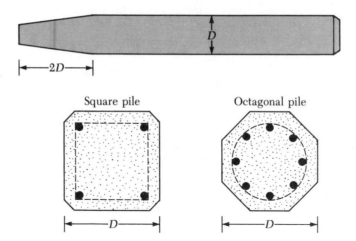

Figure 11.3   Precast piles with ordinary reinforcement

– Advantages:

    a. Can be subjected to hard driving
    b. Corrosion resistant
    c. Can be easily combined with a concrete superstructure

– Disadvantages:

    a. Difficult to achieve proper cutoff
    b. Difficult to transport

    Precast piles can also be prestressed by the use of high-strength steel pre-stressing cables. The ultimate strength of these cables is about 1800 MN/m^2 ($\approx$260 ksi). During casting of the piles, the cables are pretensioned to about $900–1300 \text{ MN/m}^2$ ($\approx$130–190 ksi), and concrete is poured around them. After curing, the cables are cut, producing a compressive force on the pile section. Table 11.3 gives additional information about prestressed concrete piles with square and octagonal cross sections.

    Some general facts about precast prestressed piles are as follows:

– Usual length: 10 m–45 m (30 ft–150 ft)
– Maximum length: 60 m (200 ft)
– Maximum load: 7500 kN–8500 kN (1700 kip–1900 kip)

The advantages and disadvantages are the same as those of precast piles.

    Cast-in-situ, or *cast-in-place, piles* are built by making a hole in the ground and then filling it with concrete. Various types of cast-in-place concrete piles are currently used in construction, and most of them have been patented by their manufacturers. These piles may be divided into two broad categories: (a) cased and (b) uncased. Both types may have a pedestal at the bottom.

    Cased piles are made by driving a steel casing into the ground with the help of a mandrel placed inside the casing. When the pile reaches the proper depth the

Table 11.3a Typical Prestressed Concrete Pile in Use (SI Units)

Pile shape[a]	D (mm)	Area of cross section (cm²)	Perimeter (mm)	Number of strands 12.7-mm diameter	Number of strands 11.1-mm diameter	Minimum effective prestress force (kN)	Section modulus (m³ × 10⁻³)	Design bearing capacity (kN) Strength of concrete (MN/m²) 34.5	Design bearing capacity (kN) Strength of concrete (MN/m²) 41.4
S	254	645	1016	4	4	312	2.737	556	778
O	254	536	838	4	4	258	1.786	462	555
S	305	929	1219	5	6	449	4.719	801	962
O	305	768	1016	4	5	369	3.097	662	795
S	356	1265	1422	6	8	610	7.489	1091	1310
O	356	1045	1168	5	7	503	4.916	901	1082
S	406	1652	1626	8	11	796	11.192	1425	1710
O	406	1368	1346	7	9	658	7.341	1180	1416
S	457	2090	1829	10	13	1010	15.928	1803	2163
O	457	1729	1524	8	11	836	10.455	1491	1790
S	508	2581	2032	12	16	1245	21.844	2226	2672
O	508	2136	1677	10	14	1032	14.355	1842	2239
S	559	3123	2235	15	20	1508	29.087	2694	3232
O	559	2587	1854	12	16	1250	19.107	2231	2678
S	610	3658	2438	18	23	1793	37.756	3155	3786
O	610	3078	2032	15	19	1486	34.794	2655	3186

[a]S = square section; O = octagonal section

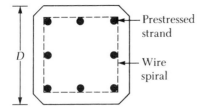
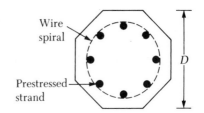

mandrel is withdrawn and the casing is filled with concrete. Figures 11.4a, 11.4b, 11.4c, and 11.4d show some examples of cased piles without a pedestal. Figure 11.4e shows a cased pile with a pedestal. The pedestal is an expanded concrete bulb that is formed by dropping a hammer on fresh concrete.

Some general facts about cased cast-in-place piles are as follows:

- Usual length: 5 m–15 m (15 ft–50 ft)
- Maximum length: 30 m–40 m (100 ft–130 ft)
- Usual load: 200 kN–500 kN (45 kip–115 kip)
- Approximate maximum load: 800 kN (180 kip)
- Advantages:

    a. Relatively cheap
    b. Allow for inspection before pouring concrete
    c. Easy to extend

Table 11.3b Typical Prestressed Concrete Pile in Use (English Units)

Pile shape[a]	D (in.)	Area of cross section (in²)	Perimeter (in.)	Number of strands $\frac{1}{2}$-in diameter	Number of strands $\frac{7}{16}$-in diameter	Minimum effective prestress force (kip)	Section modulus (in³)	Design bearing capacity (kip) Strength of Concrete 5000 psi	Design bearing capacity (kip) Strength of Concrete 6000 psi
S	10	100	40	4	4	70	167	125	175
O	10	83	33	4	4	58	109	104	125
S	12	144	48	5	6	101	288	180	216
O	12	119	40	4	5	83	189	149	178
S	14	196	56	6	8	137	457	245	295
O	14	162	46	5	7	113	300	203	243
S	16	256	64	8	11	179	683	320	385
O	16	212	53	7	9	148	448	265	318
S	18	324	72	10	13	227	972	405	486
O	18	268	60	8	11	188	638	336	402
S	20	400	80	12	16	280	1333	500	600
O	20	331	66	10	14	234	876	414	503
S	22	484	88	15	20	339	1775	605	727
O	22	401	73	12	16	281	1166	502	602
S	24	576	96	18	23	403	2304	710	851
O	24	477	80	15	19	334	2123	596	716

[a]S = square section; O = octagonal section

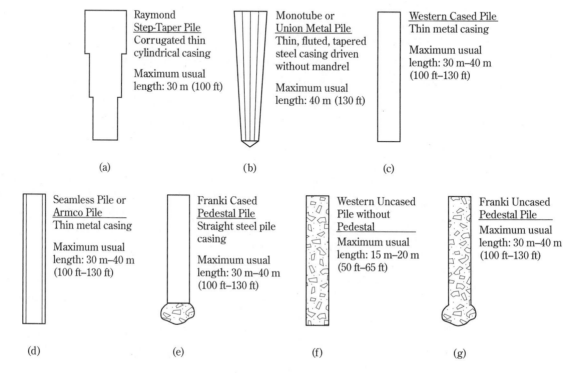

(a) Raymond Step-Taper Pile Corrugated thin cylindrical casing Maximum usual length: 30 m (100 ft)

(b) Monotube or Union Metal Pile Thin, fluted, tapered steel casing driven without mandrel Maximum usual length: 40 m (130 ft)

(c) Western Cased Pile Thin metal casing Maximum usual length: 30 m–40 m (100 ft–130 ft)

(d) Seamless Pile or Armco Pile Thin metal casing Maximum usual length: 30 m–40 m (100 ft–130 ft)

(e) Franki Cased Pedestal Pile Straight steel pile casing Maximum usual length: 30 m–40 m (100 ft–130 ft)

(f) Western Uncased Pile without Pedestal Maximum usual length: 15 m–20 m (50 ft–65 ft)

(g) Franki Uncased Pedestal Pile Maximum usual length: 30 m–40 m (100 ft–130 ft)

Figure 11.4 Cast-in-place concrete piles

– Disadvantages:

    a. Difficult to splice after concreting
    b. Thin casings may be damaged during driving

– Allowable load: $Q_{all} = A_s f_s + A_c f_c$ (11.2)

    where A_s = area of cross section of steel
              A_c = area of cross section of concrete
              f_s = allowable stress of steel
              f_c = allowable stress of concrete

Figures 11.4f and 11.4g are two types of uncased pile, one with a pedestal and the other without. The uncased piles are made by first driving the casing to the desired depth and then filling it with fresh concrete. The casing is then gradually withdrawn.

Following are some general facts about uncased cast-in-place concrete piles:

– Usual length: 5 m–15 m (15 ft–50 ft)
– Maximum length: 30 m–40 m (100 ft–130 ft)
– Usual load: 300 kN–500 kN (67 kip–115 kip)
– Approximate maximum load: 700 kN (160 kip)

– Advantages:

    a. Initially economical
    b. Can be finished at any elevation

– Disadvantages:

    a. Voids may be created if concrete is placed rapidly
    b. Difficult to splice after concreting
    c. In soft soils, the sides of the hole may cave in, squeezing the concrete

– Allowable load: $Q_{all} = A_c f_c$ (11.3)

    where A_c = area of cross section of concrete
              f_c = allowable stress of concrete

Timber Piles

Timber piles are tree trunks that have had their branches and bark carefully trimmed off. The maximum length of most timber piles is 10–20 m (30–65 ft). To qualify for use as a pile, the timber should be straight, sound, and without any defects. The American Society of Civil Engineers' *Manual of Practice,* No. 17 (1959), divided timber piles into three classes:

1. *Class A piles* carry heavy loads. The minimum diameter of the butt should be 356 mm (14 in.).
2. *Class B piles* are used to carry medium loads. The minimum butt diameter should be 305–330 mm (12–13 in.).
3. *Class C piles* are used in temporary construction work. They can be used permanently for structures when the entire pile is below the water table. The minimum butt diameter should be 305 mm (12 in.).

In any case, a pile tip should not have a diameter less than 150 mm (6 in.).

Timber piles cannot withstand hard driving stress; therefore, the pile capacity is generally limited. Steel shoes may be used to avoid damage at the pile tip (bottom). The tops of timber piles may also be damaged during the driving operation. The crushing of the wooden fibers caused by the impact of the hammer is referred to as *brooming*. To avoid damage to the top of the pile, a metal band or a cap may be used.

Splicing of timber piles should be avoided, particularly when they are expected to carry a tensile load or a lateral load. However, if splicing is necessary, it can be done by using *pipe sleeves* (see Figure 11.5a) or *metal straps and bolts* (see Figure 11.5b). The length of the sleeve should be at least five times the diameter of the pile. The butting ends should be cut square so that full contact can be maintained. The spliced portions should be carefully trimmed so that they fit tightly to the inside of the pipe sleeve. In the case of metal straps and bolts, the butting ends should also be cut square. The sides of the spliced portion should be trimmed plane for putting the straps on.

Timber piles can stay undamaged indefinitely if they are surrounded by saturated soil. However, in a marine environment, timber piles are subject to attack by various organisms and can be damaged extensively in a few months. When located above the water table, the piles are subject to attack by insects. The life of the piles may be increased by treating them with preservatives such as creosote.

The allowable load-carrying capacity of wooden piles is

$$Q_{\text{all}} = A_p f_w \tag{11.4}$$

where A_p = average area of cross section of the pile
f_w = allowable stress on the timber

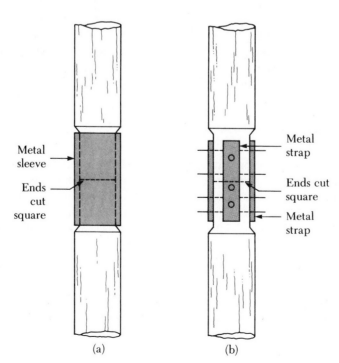

Metal sleeve

Ends cut square

Metal strap

Ends cut square

Metal strap

(a)   (b)

Figure 11.5 Splicing of timber piles: (a) use of pipe sleeves; (b) use of metal straps and bolts

The following allowable stresses are for pressure-treated round timber piles made from Pacific Coast Douglas fir and Southern pine used in hydraulic structures (ASCE, 1993):

Pacific Coast Douglas Fir

- Compression parallel to grain: 6.04 MN/m^2 (875 lb/in.^{2})
- Bending: 11.7 MN/m^2 (1700 lb/in.^{2})
- Horizontal shear: 0.66 MN/m^2 (95 lb/in.^{2})
- Compression perpendicular to grain: 1.31 MN/m^2 (190 lb/in.^{2})

Southern Pine

- Compression parallel to grain: 5.7 MN/m^2 (825 lb/in.^{2})
- Bending: 11.4 MN/m^2 (1650 lb/in.^{2})
- Horizontal shear: 0.62 MN/m^2 (90 lb/in.^{2})
- Compression perpendicular to grain: 1.41 MN/m^2 (205 lb/in.^{2})

The usual length of wooden piles is 5 m to 15 m (15 ft to 50 ft). The maximum length is about 30 m to 40 m (100 ft to 130 ft). The usual load carried by wooden piles is 300 kN to 500 kN (67 kip to 115 kip).

Composite Piles

The upper and lower portions of *composite piles* are made of different materials. For example, composite piles may be made of steel and concrete or timber and concrete. Steel-and-concrete piles consist of a lower portion of steel and an upper portion of cast-in-place concrete. This type of pile is used when the length of the pile required for adequate bearing exceeds the capacity of simple cast-in-place concrete piles. Timber-and-concrete piles usually consist of a lower portion of timber pile below the permanent water table and an upper portion of concrete. In any case, forming proper joints between two dissimilar materials is difficult, and for that reason, composite piles are not widely used.

11.3   *Estimating Pile Length*

Selecting the type of pile to be used and estimating its necessary length are fairly difficult tasks that require good judgment. In addition to being broken down into the classification given in Section 11.2, piles can be divided into three major categories, depending on their lengths and the mechanisms of load transfer to the soil: (a) point bearing piles, (b) friction piles, and (c) compaction piles.

Point Bearing Piles

If soil-boring records establish the presence of bedrock or rocklike material at a site within a reasonable depth, piles can be extended to the rock surface. (See Figure 11.6a.) In this case, the ultimate capacity of the piles depends entirely on the load-bearing capacity of the underlying material; thus, the piles are called *point bearing piles.* In most of these cases, the necessary length of the pile can be fairly well established.

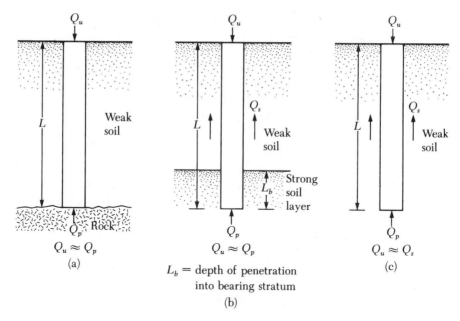

Figure 11.6 (a) and (b) Point bearing piles; (c) friction piles

If, instead of bedrock, a fairly compact and hard stratum of soil is encountered at a reasonable depth, piles can be extended a few meters into the hard stratum. (See Figure 11.6b.) Piles with pedestals can be constructed on the bed of the hard stratum, and the ultimate pile load may be expressed as

$$Q_u = Q_p + Q_s \tag{11.5}$$

where Q_p = load carried at the pile point
   Q_s = load carried by skin friction developed at the side of the pile (caused by shearing resistance between the soil and the pile)

If Q_s is very small,

$$Q_s \approx Q_p \tag{11.6}$$

In this case, the required pile length may be estimated accurately if proper subsoil exploration records are available.

Friction Piles

When no layer of rock or rocklike material is present at a reasonable depth at a site, point bearing piles become very long and uneconomical. In this type of subsoil, piles are driven through the softer material to specified depths. (See Figure 11.6c.) The ultimate load of the piles may be expressed by Eq. (11.5). However, if the value of Q_p is relatively small, then

$$Q_u \approx Q_s \tag{11.7}$$

These piles are called *friction piles,* because most of their resistance is derived from skin friction. However, the term *friction pile,* although used often in the literature, is a misnomer: In clayey soils, the resistance to applied load is also caused by *adhesion*.

The lengths of friction piles depend on the shear strength of the soil, the applied load, and the pile size. To determine the necessary lengths of these piles, an engineer needs a good understanding of soil–pile interaction, good judgment, and experience. Theoretical procedures for calculating the load-bearing capacity of piles are presented later in the chapter.

Compaction Piles

Under certain circumstances, piles are driven in granular soils to achieve proper compaction of soil close to the ground surface. These piles are called *compaction piles*. The lengths of compaction piles depend on factors such as (a) the relative density of the soil before compaction, (b) the desired relative density of the soil after compaction, and (c) the required depth of compaction. These piles are generally short; however, some field tests are necessary to determine a reasonable length.

11.4   *Installation of Piles*

Most piles are driven into the ground by means of *hammers* or *vibratory drivers*. In special circumstances, piles can also be inserted by *jetting* or *partial augering*. The types of hammer used for pile driving include (a) the drop hammer, (b) the single-acting air or steam hammer, (c) the double-acting and differential air or steam hammer, and (d) the diesel hammer. In the driving operation, a cap is attached to the top of the pile. A cushion may be used between the pile and the cap. The cushion has the effect of reducing the impact force and spreading it over a longer time; however, the use of the cushion is optional. A hammer cushion is placed on the pile cap. The hammer drops on the cushion.

Figure 11.7 illustrates various hammers. A drop hammer (see Figure 11.7a) is raised by a winch and allowed to drop from a certain height *H*. It is the oldest type of hammer used for pile driving. The main disadvantage of the drop hammer is its slow rate of blows. The principle of the single-acting air or steam hammer is shown in Figure 11.7b. The striking part, or ram, is raised by air or steam pressure and then drops by gravity. Figure 11.7c shows the operation of the double-acting and differential air or steam hammer. Air or steam is used both to raise the ram and to push it downward, thereby increasing the impact velocity of the ram. The diesel hammer (see Figure 11.7d) consists essentially of a ram, an anvil block, and a fuel-injection system. First the ram is raised and fuel is injected near the anvil. Then the ram is released. When the ram drops, it compresses the air–fuel mixture, which ignites. This action, in effect, pushes the pile downward and raises the ram. Diesel hammers work well under hard driving conditions. In soft soils, the downward movement of the pile is rather large, and the upward movement of the ram is small. This differential may not be sufficient to ignite the air–fuel system, so the ram may have to be lifted manually.

The principles of operation of a vibratory pile driver are shown in Figure 11.7e. This driver consists essentially of two counterrotating weights. The horizontal com-

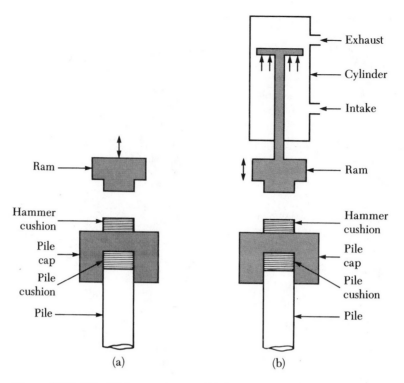

Figure 11.7 Pile-driving equipment: (a) drop hammer; (b) single-acting air or steam hammer. Following page: (c) double-acting and differential air or steam hammer; (d) diesel hammer; (e) vibratory pile driver; (f) photograph of a vibratory pile driver (Courtesy of Michael W. O'Neill, University of Houston)

ponents of the centrifugal force generated as a result of rotating masses cancel each other. As a result, a sinusoidal dynamic vertical force is produced on the pile and helps drive the pile downward.

Figure 11.7f is a photograph of a vibratory pile driver. Figure 11.8 shows a pile-driving operation in the field.

Jetting is a technique that is sometimes used in pile driving when the pile needs to penetrate a thin layer of hard soil (such as sand and gravel) overlying a layer of softer soil. In this technique, water is discharged at the pile point by means of a pipe 50–75 mm (2–3 in.) in diameter to wash and loosen the sand and gravel.

Piles driven at an angle to the vertical, typically 14° to 20°, are referred to as *batter piles*. Batter piles are used in group piles when higher lateral load-bearing capacity is required. Piles also may be advanced by partial augering, with power augers (see Chapter 2) used to predrill holes part of the way. The piles can then be inserted into the holes and driven to the desired depth.

Piles may be divided into two categories based on the nature of their placement: *displacement piles* and *nondisplacement piles*. Driven piles are displacement piles, because they move some soil laterally; hence, there is a tendency for

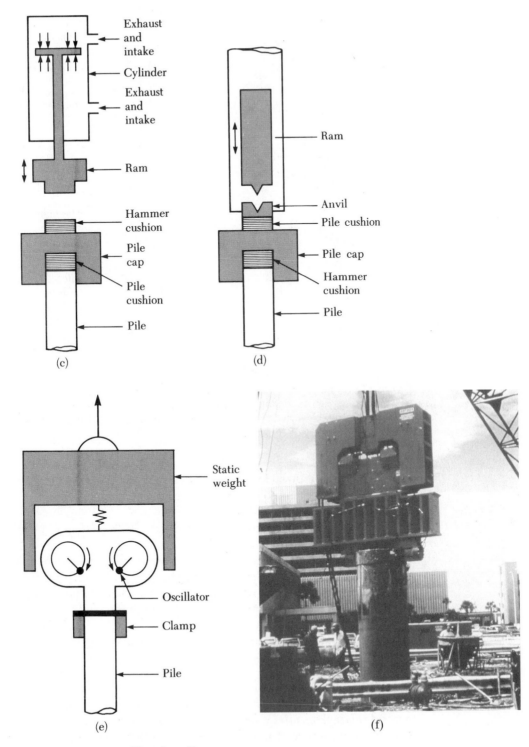

Figure 11.7 (Continued)

Figure 11.8 A pile-driving operation in the field (Courtesy of E. C. Shin, University of Incheon, Korea)

densification of soil surrounding them. Concrete piles and closed-ended pipe piles are high-displacement piles. However, steel H-piles displace less soil laterally during driving, so they are low-displacement piles. In contrast, bored piles are nondisplacement piles because their placement causes very little change in the state of stress in the soil.

11.5   *Load Transfer Mechanism*

The load transfer mechanism from a pile to the soil is complicated. To understand it, consider a pile of length L, as shown in Figure 11.9a. The load on the pile is gradually increased from zero to $Q_{(z=0)}$ at the ground surface. Part of this load will be resisted by the side friction developed along the shaft, Q_1, and part by the soil below the tip of

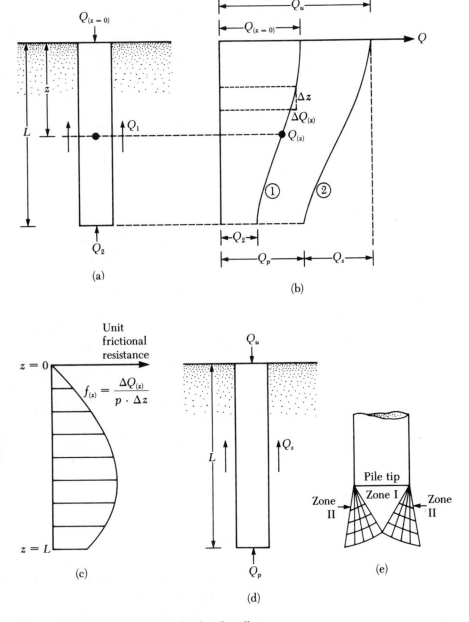

Figure 11.9   Load transfer mechanism for piles

the pile, Q_2. Now, how are Q_1 and Q_2 related to the total load? If measurements are made to obtain the load carried by the pile shaft, $Q_{(z)}$, at any depth z, the nature of the variation found will be like that shown in curve 1 of Figure 11.9b. The *frictional resistance per unit area* at any depth z may be determined as

$$f_{(z)} = \frac{\Delta Q_{(z)}}{(p)(\Delta z)} \tag{11.8}$$

where p = perimeter of the cross section of the pile

Figure 11.9c shows the variation of $f_{(z)}$ with depth.

   If the load Q at the ground surface is gradually increased, maximum frictional resistance along the pile shaft will be fully mobilized when the relative displacement between the soil and the pile is about 5–10 mm (0.2–0.3 in.), irrespective of the pile size and length L. However, the maximum point resistance $Q_2 = Q_p$ will not be mobilized until the tip of the pile has moved about 10–25% of the pile width (or diameter). (The lower limit applies to driven piles and the upper limit to bored piles). At ultimate load (Figure 11.9d and curve 2 in Figure 11.9b), $Q_{(z=0)} = Q_u$. Thus,

$$Q_1 = Q_s$$

and

$$Q_2 = Q_p$$

The preceding explanation indicates that Q_s (or the unit skin friction, f, along the pile shaft) is developed at a *much smaller pile displacement compared with the point resistance, Q_p.*

   At ultimate load, the failure surface in the soil at the pile tip (a bearing capacity failure caused by Q_p) is like that shown in Figure 11.9e. Note that pile foundations are deep foundations and that the soil fails mostly in a *punching mode,* as illustrated previously in Figures 3.1c and 3.3. That is, a *triangular zone,* I, is developed at the pile tip, which is pushed downward without producing any other visible slip surface. In dense sands and stiff clayey soils, a *radial shear zone,* II, may partially develop. Hence, the load displacement curves of piles will resemble those shown in Figure 3.1c.

11.6   *Equations for Estimating Pile Capacity*

The ultimate load-carrying capacity Q_u of a pile is given by the equation

$$Q_u = Q_p + Q_s \tag{11.9}$$

where Q_p = load-carrying capacity of the pile point
$\quad\quad\quad Q_s$ = frictional resistance (skin friction) derived from the soil–pile interface (see Figure 11.10)

Numerous published studies cover the determination of the values of Q_p and Q_s. Excellent reviews of many of these investigations have been provided by Vesic (1977), Meyerhof (1976), and Coyle and Castello (1981). These studies afford an insight into the problem of determining the ultimate pile capacity.

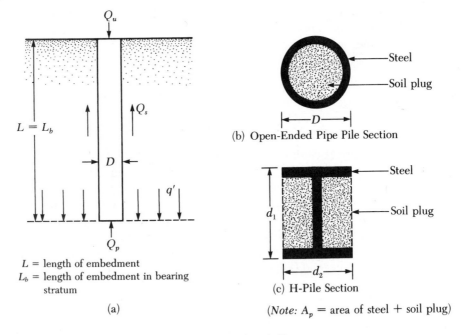

L = length of embedment
L_b = length of embedment in bearing
    stratum

(a)

(b) Open-Ended Pipe Pile Section

(c) H-Pile Section

(*Note:* A_p = area of steel + soil plug)

Figure 11.10 Ultimate load-carrying capacity of pile

Point Bearing Capacity, Q_p

The ultimate bearing capacity of shallow foundations was discussed in Chapter 3. According to Terzaghi's equations,

$$q_u = 1.3c'N_c + qN_q + 0.4\gamma BN_\gamma \qquad \text{(for shallow square foundations)}$$

and

$$q_u = 1.3c'N_c + qN_q + 0.3\gamma BN_\gamma \qquad \text{(for shallow circular foundations)}$$

Similarly, the general bearing capacity equation for shallow foundations was given in Chapter 3 (for vertical loading) as

$$q_u = c'N_cF_{cs}F_{cd} + qN_qF_{qs}F_{qd} + \tfrac{1}{2}\gamma BN_\gamma F_{\gamma s}F_{\gamma d}$$

Hence, in general, the ultimate load-bearing capacity may be expressed as

$$q_u = c'N_c^* + qN_q^* + \gamma BN_\gamma^* \qquad (11.10)$$

where N_c^*, N_q^*, and N_γ^* are the bearing capacity factors that include the necessary shape and depth factors

Pile foundations are deep. However, the ultimate resistance per unit area developed at the pile tip, q_p, may be expressed by an equation similar in form to Eq. (11.10), although the values of N_c^*, N_q^*, and N_γ^* will change. The notation used in this chapter for the width of a pile is D. Hence, substituting D for B in Eq. (11.10) gives

$$q_u = q_p = c'N_c^* + qN_q^* + \gamma DN_\gamma^* \qquad (11.11)$$

Because the width D of a pile is relatively small, the term $\gamma D N_\gamma^*$ may be dropped from the right side of the preceding equation without introducing a serious error; thus, we have

$$q_p = c'N_c^* + q'N_q^* \tag{11.12}$$

Note that the term q has been replaced by q' in Eq. (11.12), to signify effective vertical stress. Thus, the point bearing of piles is

$$Q_p = A_p q_p = A_p(c'N_c^* + q'N_q^*) \tag{11.13}$$

where A_p = area of pile tip
 c' = cohesion of the soil supporting the pile tip
 q_p = unit point resistance
 q' = effective vertical stress at the level of the pile tip
 N_c^*, N_q^* = the bearing capacity factors

Frictional Resistance, Q_s

The frictional, or skin, resistance of a pile may be written as

$$Q_s = \Sigma\, p\, \Delta L f \tag{11.14}$$

where p = perimeter of the pile section
 ΔL = incremental pile length over which p and f are taken to be constant
 f = unit friction resistance at any depth z

The various methods for estimating Q_p and Q_s are discussed in the next several sections. It needs to be reemphasized that, in the field, for full mobilization of the point resistance (Q_p), the pile tip must go through a displacement of 10 to 25% of the pile width (or diameter).

11.7   *Meyerhof's Method for Estimating Q_p*

Sand

The point bearing capacity, q_p, of a pile in sand generally increases with the depth of embedment in the bearing stratum and reaches a maximum value at an embedment ratio of $L_b/D = (L_b/D)_{cr}$. Note that in a homogeneous soil L_b is equal to the actual embedment length of the pile, L. (See Figure 11.10a.) However, where a pile has penetrated into a bearing stratum, $L_b < L$. (See Figure 11.6b.) Beyond the critical embedment ratio, $(L_b/D)_{cr}$, the value of q_p remains constant ($q_p = q_l$). That is, as shown in Figure 11.11 for the case of a homogeneous soil, $L = L_b$.

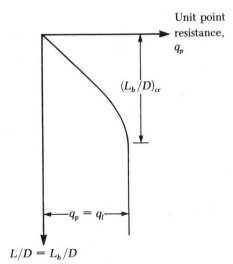

Figure 11.11   Nature of variation of unit point resistance in a homogeneous sand

For piles in sand, $c' = 0$, and Eq. (11.13) simpifies to

$$Q_p = A_p q_p = A_p q' N_q^*$$
(11.15)

The variation of N_q^* with soil friction angle ϕ' is shown in Figure 11.12. However, Q_p should not exceed the limiting value $A_p q_l$; that is,

$$Q_p = A_p q' N_q^* \leq A_p q_l$$
(11.16)

The limiting point resistance is

$$q_l = 0.5\, p_a N_q^* \tan \phi'$$
(11.17)

where p_a = atmospheric pressure ($=100$ kN/m^2 or 2000 lb/ft^2)
  ϕ' = effective soil friction angle of the bearing stratum

On the basis of field observations, Meyerhof (1976) also suggested that the ultimate point resistance q_p in a homogeneous granular soil ($L = L_b$) may be obtained from standard penetration numbers as

$$q_l = 0.4\, p_a (N_1)_{60} \frac{L}{D} \leq 4\, p_a (N_1)_{60}$$
(11.18)

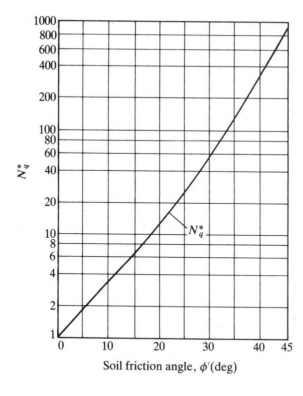

Figure 11.12   Variation of the maximum values of N_q^* with soil friction angle ϕ' (after Meyerhof, 1976)

where $(N_1)_{60}$ = the average corrected value of the standard penetration number near the pile point (about $10D$ above and $4D$ below the pile point)
   p_a = atmospheric pressure (≈ 100 kN/m^2 or 2000 lb/ft^2)

Clay ($\phi = 0$)

For piles in *saturated clays* under undrained conditions ($\phi = 0$),

$$Q_p = N_c^* c_u A_p = 9 c_u A_p \tag{11.19}$$

where c_u = undrained cohesion of the soil below the tip of the pile

11.8   Vesic's Method for Estimating Q_p

Vesic (1977) proposed a method for estimating the pile point bearing capacity based on the theory of *expansion of cavities*. According to this theory, on the basis of effective stress parameters, we may write

$$Q_p = A_p q_p = A_p(c' N_c^* + \overline{\sigma}_o' N_\sigma^*) \tag{11.20}$$

where $\quad\quad \bar{\sigma}_o$ = mean effective normal ground stress at the level of the pile point

$$= \left(\frac{1 + 2K_o}{3}\right)q' \tag{11.21}$$

$\quad\quad K_o$ = earth pressure coefficient at rest = $1 - \sin \phi'$ $\quad\quad$ (11.22)

N_c^*, N_σ^* = bearing capacity factors

Note that Eq. (11.20) is a modification of Eq. (11.13) with

$$N_\sigma^* = \frac{3N_q^*}{(1 + 2K_o)} \tag{11.23}$$

Note also that N_c^* in Eq. (11.20) may be expressed as

$$N_c^* = (N_q^* - 1)\cot \phi' \tag{11.24}$$

According to Vesic's theory,

$$N_\sigma^* = f(I_{rr}) \tag{11.25}$$

where $\quad I_{rr}$ = reduced rigidity index for the soil

However, $$I_{rr} = \frac{I_r}{1 + I_r\Delta} \tag{11.26}$$

where $\quad I_r$ = rigidity index = $\dfrac{E_s}{2(1 + \mu_s)(c' + q' \tan \phi')} = \dfrac{G_s}{c' + q' \tan \phi'}$ $\quad$ (11.27)

$\quad E_s$ = modulus of elasticity of soil

$\quad \mu_s$ = Poisson's ratio of soil

$\quad G_s$ = shear modulus of soil

$\quad \Delta$ = average volumatic strain in the plastic zone below the pile point

When the volume does not change (e.g., for dense sand or saturated clay), $\Delta = 0$, so

$$I_r = I_{rr} \tag{11.28}$$

Table 11.4 gives the values of N_c^* and N_σ^* for various values of the soil friction angle ϕ' and I_{rr}. For $\phi = 0$ (an undrained condition),

$$N_c^* = \frac{4}{3}(\ln I_{rr} + 1) + \frac{\pi}{2} + 1 \tag{11.29}$$

The values of I_r can be estimated from laboratory consolidation and triaxial tests corresponding to the proper stress levels. However, for preliminary use, the following values are recommended:

Type of soil	I_r
Sand	70–150
Silts and clays (drained condition)	50–100
Clays (undrained condition)	100–200

Table 11.4 Bearing Capacity Factors N_c^* and N_σ^* Based on the Theory of Expansion of Cavities

ϕ'	I_{rr}									
	10	20	40	60	80	100	200	300	400	500
0	6.97	7.90	8.82	9.36	9.75	10.04	10.97	11.51	11.89	12.19
	1.00	1.00	1.00	1.00	1.00	1.00	1.00	1.00	1.00	1.00
1	7.34	8.37	9.42	10.04	10.49	10.83	11.92	12.57	13.03	13.39
	1.13	1.15	1.16	1.18	1.18	1.19	1.21	1.22	1.23	1.23
2	7.72	8.87	10.06	10.77	11.28	11.69	12.96	13.73	14.28	14.71
	1.27	1.31	1.35	1.38	1.39	1.41	1.45	1.48	1.50	1.51
3	8.12	9.40	10.74	11.55	12.14	12.61	14.10	15.00	15.66	16.18
	1.43	1.49	1.56	1.61	1.64	1.66	1.74	1.79	1.82	1.85
4	8.54	9.96	11.47	12.40	13.07	13.61	15.34	16.40	17.18	17.80
	1.60	1.70	1.80	1.87	1.91	1.95	2.07	2.15	2.20	2.24
5	8.99	10.56	12.25	13.30	14.07	14.69	16.69	17.94	18.86	19.59
	1.79	1.92	2.07	2.16	2.23	2.28	2.46	2.57	2.65	2.71
6	9.45	11.19	13.08	14.26	15.14	15.85	18.17	19.62	20.70	21.56
	1.99	2.18	2.37	2.50	2.59	2.67	2.91	3.06	3.18	3.27
7	9.94	11.85	13.96	15.30	16.30	17.10	19.77	21.46	22.71	23.73
	2.22	2.46	2.71	2.88	3.00	3.10	3.43	3.63	3.79	3.91
8	10.45	12.55	14.90	16.41	17.54	18.45	21.51	23.46	24.93	26.11
	2.47	2.76	3.09	3.31	3.46	3.59	4.02	4.30	4.50	4.67
9	10.99	13.29	15.91	17.59	18.87	19.90	23.39	25.64	27.35	28.73
	2.74	3.11	3.52	3.79	3.99	4.15	4.70	5.06	5.33	5.55
10	11.55	14.08	16.97	18.86	20.29	21.46	25.43	28.02	29.99	31.59
	3.04	3.48	3.99	4.32	4.58	4.78	5.48	5.94	6.29	6.57
11	12.14	14.90	18.10	20.20	21.81	23.13	27.64	30.61	32.87	34.73
	3.36	3.90	4.52	4.93	5.24	5.50	6.37	6.95	7.39	7.75
12	12.76	15.77	19.30	21.64	23.44	24.92	30.03	33.41	36.02	38.16
	3.71	4.35	5.10	5.60	5.98	6.30	7.38	8.10	8.66	9.11
13	13.41	16.69	20.57	23.17	25.18	26.84	32.60	36.46	39.44	41.89
	4.09	4.85	5.75	6.35	6.81	7.20	8.53	9.42	10.10	10.67
14	14.08	17.65	21.92	24.80	27.04	28.89	35.38	39.75	43.15	45.96
	4.51	5.40	6.47	7.18	7.74	8.20	9.82	10.91	11.76	12.46
15	14.79	18.66	23.35	26.53	29.02	31.08	38.37	43.32	47.18	50.39
	4.96	6.00	7.26	8.11	8.78	9.33	11.28	12.61	13.64	14.50
16	15.53	19.73	24.86	28.37	31.13	33.43	41.58	47.17	51.55	55.20
	5.45	6.66	8.13	9.14	9.93	10.58	12.92	14.53	15.78	16.83

Table 11.4 (Continued)

I_{rr}

ϕ'	10	20	40	60	80	100	200	300	400	500
17	16.30	20.85	26.46	30.33	33.37	35.92	45.04	51.32	56.27	60.42
	5.98	7.37	9.09	10.27	11.20	11.98	14.77	16.99	18.20	19.47
18	17.11	22.03	28.15	32.40	35.76	38.59	48.74	55.80	61.38	66.07
	6.56	8.16	10.15	11.53	12.62	13.54	16.84	19.13	20.94	22.47
19	17.95	23.26	29.93	34.59	38.30	41.42	52.71	60.61	66.89	72.18
	7.18	9.01	11.31	12.91	14.19	15.26	19.15	21.87	24.03	25.85
20	18.83	24.56	31.81	36.92	40.99	44.43	56.97	65.79	72.82	78.78
	7.85	9.94	12.58	14.44	15.92	17.17	21.73	24.94	27.51	29.67
21	19.75	25.92	33.80	39.38	43.85	47.64	61.51	71.34	79.22	85.90
	8.58	10.95	13.97	16.12	17.83	19.29	24.61	28.39	31.41	33.97
22	20.71	27.35	35.89	41.98	46.88	51.04	66.37	77.30	86.09	93.57
	9.37	12.05	15.50	17.96	19.94	21.62	27.82	32.23	35.78	38.81
23	21.71	28.84	38.09	44.73	50.08	54.66	71.56	83.68	93.47	101.83
	10.21	13.24	17.17	19.99	22.26	24.20	31.37	36.52	40.68	44.22
24	22.75	30.41	40.41	47.63	53.48	58.49	77.09	90.51	101.39	110.70
	11.13	14.54	18.99	22.21	24.81	27.04	35.32	41.30	46.14	50.29
25	23.84	32.05	42.85	50.69	57.07	62.54	82.98	97.81	109.88	120.23
	12.12	15.95	20.98	24.64	27.61	30.16	39.70	46.61	52.24	57.06
26	24.98	33.77	45.42	53.93	60.87	66.84	89.25	105.61	118.96	130.44
	13.18	17.47	23.15	27.30	30.69	33.60	44.53	52.51	59.02	64.62
27	26.16	35.57	48.13	57.34	64.88	71.39	95.02	113.92	128.67	141.39
	14.33	19.12	25.52	30.21	34.06	37.37	49.88	59.05	66.56	73.04
28	27.40	37.45	50.96	60.93	69.12	76.20	103.01	122.79	139.04	153.10
	15.57	20.91	28.10	33.40	37.75	41.51	55.77	66.29	74.93	82.40
29	28.69	39.42	53.95	64.71	73.58	81.28	110.54	132.23	150.11	165.61
	16.90	22.85	30.90	36.87	41.79	46.05	62.27	74.30	84.21	92.80
30	30.03	41.49	57.08	68.69	78.30	86.64	118.53	142.27	161.91	178.98
	18.24	24.95	33.95	40.66	46.21	51.02	69.43	83.14	94.48	104.33
31	31.43	43.64	60.37	72.88	83.27	92.31	126.99	152.95	174.49	193.23
	19.88	27.22	37.27	44.79	51.03	56.46	77.31	92.90	105.84	117.11
32	32.89	45.90	63.82	77.29	88.50	98.28	135.96	164.29	187.87	208.43
	21.55	29.68	40.88	49.30	56.30	62.41	85.96	103.66	118.39	131.24
33	34.41	48.26	67.44	81.92	94.01	104.58	145.46	176.33	202.09	224.62
	23.34	32.34	44.80	54.20	62.05	68.92	95.46	115.51	132.24	146.87
34	35.99	50.72	71.24	86.80	99.82	111.22	155.51	189.11	217.21	241.84
	25.28	35.21	49.05	59.54	68.33	76.02	105.90	128.55	147.51	164.12

Table 11.4 (Continued)

φ'	10	20	40	60	80	100	200	300	400	500
						I_{rr}				
35	37.65	53.30	75.22	91.91	105.92	118.22	166.14	202.64	233.27	260.15
	27.36	38.32	53.67	65.36	75.17	83.78	117.33	142.89	164.33	183.16
36	39.37	55.99	79.39	97.29	112.34	125.59	177.38	216.98	250.30	279.60
	29.60	41.68	58.68	71.69	82.62	92.24	129.87	158.65	182.85	204.14
37	41.17	58.81	83.77	102.94	119.10	133.34	189.25	232.17	268.36	300.26
	32.02	45.31	64.13	78.57	90.75	101.48	143.61	175.95	203.23	227.26
38	43.04	61.75	88.36	108.86	126.20	141.50	201.78	248.23	287.50	322.17
	34.63	49.24	70.03	86.05	99.60	111.56	158.65	194.94	225.62	252.71
39	44.99	64.83	93.17	115.09	133.66	150.09	215.01	265.23	307.78	345.41
	37.44	53.50	76.45	94.20	109.24	122.54	175.11	215.78	250.23	280.71
40	47.03	68.04	98.21	121.62	141.51	159.13	228.97	283.19	329.24	370.04
	40.47	58.10	83.40	103.05	119.74	134.52	193.13	238.62	277.26	311.50
41	49.16	71.41	103.49	128.48	149.75	168.63	243.69	302.17	351.95	396.12
	43.74	63.07	90.96	112.68	131.18	147.59	212.84	263.67	306.94	345.34
42	51.38	74.92	109.02	135.68	158.41	178.62	259.22	322.22	375.97	423.74
	47.27	68.46	99.16	123.16	143.64	161.83	234.40	291.13	339.52	382.53
43	53.70	78.60	114.82	143.23	167.51	189.13	275.59	343.40	401.36	452.96
	51.08	74.30	108.08	134.56	157.21	177.36	257.99	321.22	375.28	423.39
44	56.13	82.45	120.91	151.16	177.07	200.17	292.85	365.75	428.21	483.88
	55.20	80.62	117.76	146.97	172.00	194.31	283.80	354.20	414.51	468.28
45	58.66	86.48	127.28	159.48	187.12	211.79	311.04	389.35	456.57	516.58
	59.66	87.48	128.28	160.48	188.12	212.79	312.03	390.35	457.57	517.58
46	61.30	90.70	133.97	168.22	197.67	224.00	330.20	414.26	486.54	551.16
	64.48	94.92	139.73	175.20	205.70	232.96	342.94	429.98	504.82	571.74
47	64.07	95.12	140.99	177.40	208.77	236.85	350.41	440.54	518.20	587.72
	69.71	103.00	152.19	191.24	224.88	254.99	376.77	473.42	556.70	631.25
48	66.97	99.75	148.35	187.04	220.43	250.36	371.70	468.28	551.64	626.36
	75.38	111.78	165.76	208.73	245.81	279.06	413.82	521.08	613.65	696.64
49	70.01	104.60	156.09	197.17	232.70	264.58	394.15	497.56	586.96	667.21
	81.54	121.33	180.56	227.82	268.69	305.37	454.42	573.38	676.22	768.53
50	73.19	109.70	164.21	207.83	245.60	279.55	417.82	528.46	624.28	710.39
	88.23	131.73	196.70	248.68	293.70	334.15	498.94	630.80	744.99	847.61

From "Design of Pile Foundations," by A. S. Vesic, in *NCHRP Synthesis of Highway Practice 42*, Transportation Research Board, 1977. Reprinted by permission.

Note: Upper number N_c^*, lower number N_σ^*.

On the basis of cone penetration tests in the field, Baldi et al. (1981) gave the following correlations for I_r:

$$I_r = \frac{300}{F_r(\%)} \quad \text{(for mechanical cone penetration)} \quad (11.30a)$$

and

$$I_r = \frac{170}{F_r(\%)} \quad \text{(for electric cone penetration)} \quad (11.30b)$$

For the definition of F_r, see Eq. (2.37).

11.9 *Janbu's Method for Estimating* Q_p

Janbu (1976) proposed calculating Q_p as follows:

$$Q_p = A_p(c'N_c^* + q'N_q^*) \quad (11.31)$$

Note that Eq. (11.31) has the same form as Eq. (11.13). The bearing capacity factors N_c^* and N_q^* are calculated by assuming a failure surface in soil at the pile tip similar to that shown in Figure 11.13. The bearing capacity relationships are then

$$N_q^* = (\tan \phi' + \sqrt{1 + \tan^2 \phi'})^2 (e^{2\eta' \tan \phi'}) \quad (11.32a)$$

(the angle η' is defined in the figure) and

$$N_c^* = (N_q^* - 1) \cot \phi' \quad (11.32b)$$
$$\uparrow$$
from Eq. (11.32a)

The angle η' varies from 60° for soft clays to about 105° for dense sandy soils. It is recommended that, for practical use,

$$60° \leqslant \eta' \leqslant 90°$$

Table 11.5 gives the variation of N_c^* and N_q^* for $\eta' = 60°, 75°$, and 90°.

11.10 *Coyle and Castello's Method for Estimating* Q_p *in Sand*

Coyle and Castello (1981) analyzed 24 large-scale field load tests of driven piles in sand. On the basis of the test results, they suggested that, in sand,

$$Q_p = q'N_q^*A_p \quad (11.33)$$

where
$q' = $ effective vertical stress at the pile tip
$N_q^* = $ bearing capacity factor

Figure 11.14 shows the variation of N_q^* with L/D and the soil friction angle ϕ'.

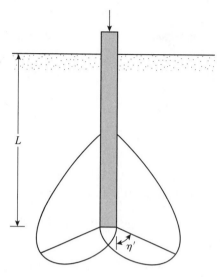

Figure 11.13 Failure surface at the pile tip

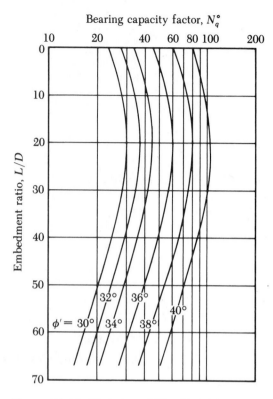

Figure 11.14 Variation of N_q^* with L/D (redrawn after Coyle and Castello, 1981)

Table 11.5 Janbu's Bearing Capacity Factors

ϕ'°	$\eta' = 60°$		$\eta' = 75°$		$\eta' = 90°$	
	N_c^*	N_q^*	N_c^*	N_q^*	N_c^*	N_q^*
0	5.74	1.0	5.74	1.0	5.74	1.0
10	5.95	2.05	7.11	2.25	8.34	2.47
20	9.26	4.37	11.78	5.29	14.83	6.40
30	19.43	10.05	21.82	13.60	30.14	18.40
40	30.58	26.66	48.11	41.37	75.31	64.20
45	46.32	47.32	78.90	79.90	133.87	134.87

11.11 Other Correlations for Calculating Q_p with SPT and CPT Results

There are several correlations in the literature for calculating Q_p on the basis of standard penetration test and cone penetration test results conducted in the field. We summarize some of these correlations in this section. Table 11.6 gives the correlation

Table 11.6 Correlations with Standard Penetration Resistance

Reference	Relationship	Applicability
Briaud et al. (1985)	$q_p = 19.7 p_a (N_{60})^{0.36}$	Sand
Shioi and Fukui (1982)	$q_p = 3 p_a$	Cast in place, sand
	$q_p = 0.1 p_a N_{60}$	Bored pile, sand
	$q_p = 0.15 p_a N_{60}$	Bored pile, gravelly sand
	$q_p = 0.3 p_a N_{60}$	Driven piles, all soils

of q_p with the standard penetration number N_{60}. It is important to note that the N_{60} value is the average condition near the pile tip (i.e., $4D$ below and $10D$ above the pile tip).

There are two major methods for estimating the magnitude of q_p using the cone penetration resistance q_c:

1. The LCPC method, developed by Laboratoire Central des Ponts at Chaussées (Bustamante and Gianeselli, 1982); and
2. The Dutch method (DeRuiter and Beringen, 1979).

LCPC Method

According to the LCPC method,

$$q_p = q_{c(eq)} k_b \tag{11.34}$$

where $q_{c(eq)}$ = equivalent average cone resistance
   k_b = empirical bearing capacity factor

The magnitude of $q_{c(eq)}$ is calculated in the following manner:

1. Consider the cone tip resistance q_c within a range of $1.5D$ below the pile tip to $1.5D$ above the pile tip, as shown in Figure 11.15.
2. Calculate the average value of $q_c [q_{c(av)}]$ within the zone shown in Figure 11.15.
3. Eliminate the q_c values that are higher than $1.3 q_{c(av)}$ and the q_c values that are lower than $0.7 q_{c(av)}$.
4. Calculate $q_{c(eq)}$ by averaging the remaining q_c values.

Briaud and Miran (1991) suggested that

$$k_b = 0.6 \text{ (for clays and silts)}$$

and

$$k_b = 0.375 \text{ (for sands and gravels)}$$

Dutch Method

According to the Dutch method, one considers the variation of q_c in the range of $4D$ below the pile tip to $8D$ above the pile tip, as shown in Figure 11.16. Then one conducts the following operations:

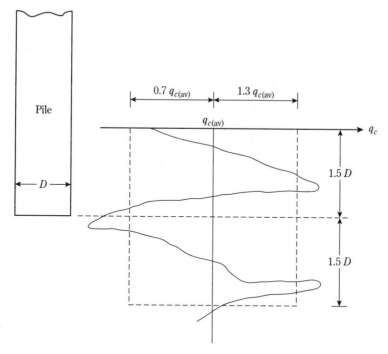

Figure 11.15 LCPC method

1. Average the q_c values over a distance yD below the pile tip. This is path *a–b–c*. Sum q_c values along the downward path *a–b* (i.e., the actual path *a*) and the upward path *b–c* (i.e., the minimum path). Determine the minimum value q_{c1} = average value of q_c for $0.7 < y < 4$.
2. Average the q_c values (q_{c2}) between the pile tip and $8D$ above the pile tip along the path *c–d–e–f–g*, using the minimum path and ignoring minor peak depressions.
3. Calculate

$$q_p = \frac{(q_{c1} + q_{c2})}{2} k_b' \le 150 p_a \tag{11.35}$$

where p_a = atmospheric pressure (≈ 100 kN/m², or 2000 lb/ft²)

DeRuiter and Beringen (1979) recommended the following values for k_b' for sand:

- 1.0 for OCR (overconsolidation ratio) = 1
- 0.67 for OCR = 2 to 4

Nottingham and Schmertmann (1975) and Schmertmann (1978) recommended the following relationship for q_p in clay:

$$q_p = R_1 R_2 \frac{(q_{c1} + q_{c2})}{2} k_b' \le 150 p_a \tag{11.36}$$

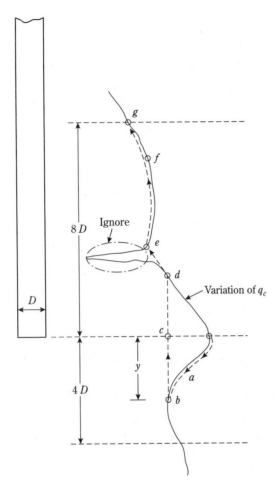

Figure 11.16 Dutch method

In this equation, R_1 = reduction factor, which is a function of the undrained shear strength c_u

R_2 = 1 for electric cone penetrometer; = 0.6 for mechanical cone penetrometer

The interpolated values of R_1 with c_u provided by Schmertmann (1978) are as follows:

$\dfrac{c_u}{p_a}$	R_1
⩾0.5	1
0.75	0.64
1.0	0.53
1.25	0.42
1.5	0.36
1.75	0.33
2.0	0.30

11.12   *Frictional Resistance (Q_s) in Sand*

According to Eq. (11.14), the frictional resistance

$$Q_s = \Sigma p \, \Delta L f$$

The unit frictional resistance, f, is hard to estimate. In making an estimation of f, several important factors must be kept in mind:

1. The nature of the pile installation. For driven piles in sand, the vibration caused during pile driving helps densify the soil around the pile. Figure 11.17 shows the contours of the soil friction angle ϕ' around a driven pile (Meyerhof, 1961). Note that, in this case, the original effective soil friction angle of the sand was 32°. The zone of sand densification is about 2.5 times the pile diameter, in the sand surrounding the pile.

2. It has been observed that the nature of variation of f in the field is approximately as shown in Figure 11.18. The unit skin friction increases with depth more or less

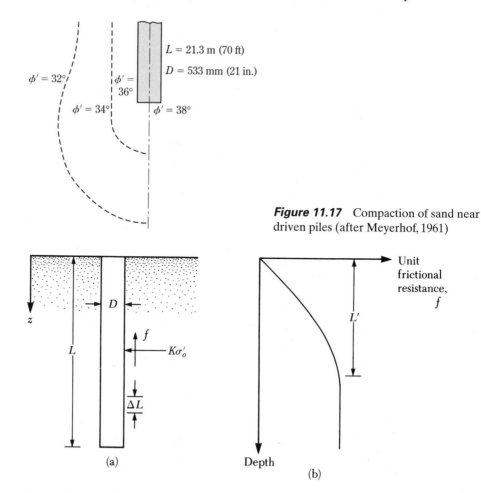

Figure 11.17   Compaction of sand near driven piles (after Meyerhof, 1961)

Figure 11.18   Unit frictional resistance for piles in sand

linearly to a depth of L' and remains constant thereafter. The magnitude of the critical depth L' may be 15 to 20 pile diameters. A conservative estimate would be

$$L' \approx 15D \tag{11.37}$$

3. At similar depths, the unit skin friction in loose sand is higher for a high-displacement pile, compared with a low-displacement pile.
4. At similar depths, bored, or jetted, piles will have a lower unit skin friction, compared with driven piles.

Taking into account the preceding factors, we can give the following approximate relationship for f (see Figure 11.18):

For $z = 0$ to L',

$$f = K\sigma_o' \tan \delta \tag{11.38}$$

and for $z = L'$ to L,

$$f = f_{z=L'} \tag{11.39}$$

In these equations, K = effective earth coefficient
σ_o' = effective vertical stress at the depth under consideration
δ = soil-pile friction angle

In reality, the magnitude of K varies with depth; it is approximately equal to the Rankine passive earth pressure coefficient, K_p, at the top of the pile and may be less than the at-rest pressure coefficient, K_o, at a greater depth. Based on presently available results, the following average values of K are recommended for use in Eq. (11.38):

Pile type	K
Bored or jetted	$\approx K_o = 1 - \sin \phi'$
Low-displacement driven	$\approx K_o = 1 - \sin \phi'$ to $1.4K_o = 1.4(1 - \sin \phi')$
High-displacement driven	$\approx K_o = 1 - \sin \phi'$ to $1.8K_o = 1.8(1 - \sin \phi')$

The values of δ from various investigations appear to be in the range from $0.5\phi'$ to $0.8\phi'$. Judgment must be used in choosing the value of δ. For high-displacement driven piles, Bhusan (1982) recommended

$$K \tan \delta = 0.18 + 0.0065D_r \tag{11.40}$$

and

$$K = 0.5 + 0.008D_r \tag{11.41}$$

where D_r = relative density (%)

Coyle and Castello (1981), in conjunction with the material presented in Section 11.10, proposed that

$$Q_s = f_{av}pL = (K\overline{\sigma}_o' \tan \delta)pL \tag{11.42}$$

where $\overline{\sigma}'_o$ = average effective overburden pressure
δ = soil–pile friction angle = $0.8\phi'$

The lateral earth pressure coefficient K, which was determined from field observations, is shown in Figure 11.19. Thus, if that figure is used,

$$Q_s = K\overline{\sigma}'_o \tan(0.8\phi') pL \tag{11.43}$$

Correlation with Standard Penetration Test Results

Meyerhof (1976) indicated that the average unit frictional resistance, f_{av}, for high-displacement driven piles may be obtained from average corrected standard penetration resistance values as

$$f_{av} = 0.02 p_a (\overline{N}_1)_{60} \tag{11.44}$$

where $(\overline{N}_1)_{60}$ = average corrected value of standard penetration resistance
p_a = atmospheric pressure (≈ 100 kN/m^2 or 2000 lb/ft^2)

For low-displacement driven piles

$$f_{av} = 0.01 p_a (\overline{N}_1)_{60} \tag{11.45}$$

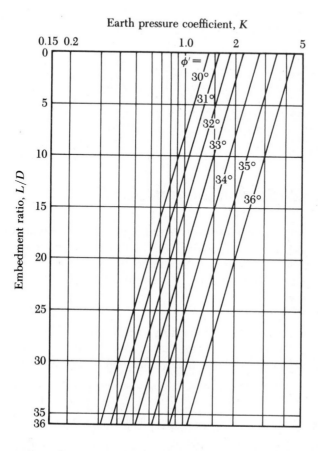

Figure 11.19 Variation of K with L/D (redrawn after Coyle and Castello, 1981)

Thus,

$$Q_s = pLf_{av} \tag{11.46}$$

Briaud et al. (1985) proposed another correlation for unit skin friction with the standard penetration resistance, in the form

$$f = 0.224p_a(N_{60})^{0.29} \tag{11.47}$$

Hence,

$$Q_s = \Sigma p(\Delta L)f = \Sigma 0.224p\, p_a(\Delta L)(N_{60})^{0.29} \tag{11.48}$$

In a fairly homogeneous soil, we can estimate the average value of N_{60}. In that case,

$$Q_s = pLf_{av}$$

where $f_{av} = 0.224p_a(N_{60})_{av}^{0.29}$ $\qquad\qquad$ (11.49)

Correlation with Cone Penetration Test Results

In Section 11.11, the Dutch method for calculating pile tip capacity Q_p using cone penetration test results was described. In conjunction with using that method, Nottingham and Schmertmann (1975) and Schmertmann (1978) provided correlations for estimating Q_s using the frictional resistance (f_c) obtained during cone penetration tests. According to this method

$$f = \alpha' f_c \tag{11.50}$$

The variations of α' with z/D for electric cone and mechanical cone penetrometers are shown in Figures 11.20 and 11.21, respectively. We have

$$Q_s = \Sigma p(\Delta L)f = \Sigma p(\Delta L)\alpha' f_c \tag{11.51}$$

11.13 *Frictional (Skin) Resistance in Clay*

Estimating the frictional (or skin) resistance of piles in clay is almost as difficult a task as estimating that in sand (see Section 11.12), due to the presence of several variables that cannot easily be quantified. Several methods for obtaining the unit frictional resistance of piles are described in the literature. We examine some of them next.

λ Method

This method, proposed by Vijayvergiya and Focht (1972), is based on the assumption that the displacement of soil caused by pile driving results in a passive lateral pressure at any depth and that the average unit skin resistance is

$$f_{av} = \lambda(\overline{\sigma}'_o + 2c_u) \tag{11.52}$$

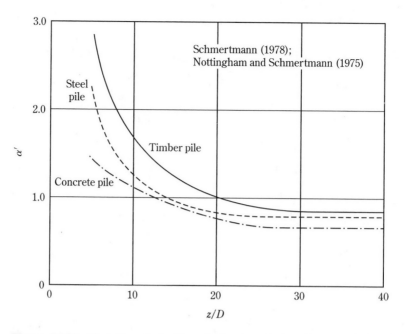

Figure 11.20 Variation of α' with embedment ratio for pile in sand: electric cone penetrometer

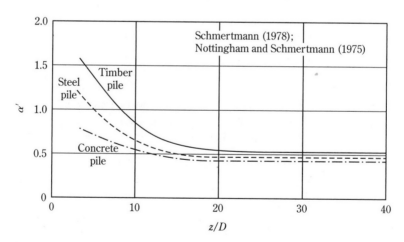

Figure 11.21 Variation of α' with embedment ratio for piles in sand: mechanical cone penetrometer

where $\bar{\sigma}_o'$ = mean effective vertical stress for the entire embedment length
  c_u = mean undrained shear strength ($\phi = 0$)

The value of λ changes with the depth of penetration of the pile. (See Figure 11.22.) Thus, the total frictional resistance may be calculated as

$$Q_s = pLf_{av}$$

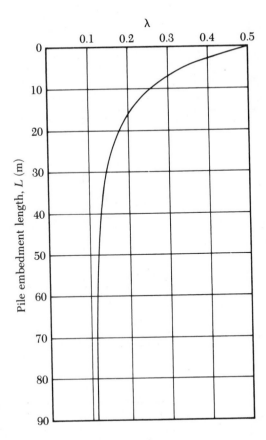

Figure 11.22 Variation of λ with pile embedment length (redrawn after McClelland, 1974)

Care should be taken in obtaining the values of $\overline{\sigma}'_o$ and c_u in layered soil. Figure 11.23 helps explain the reason. Figure 11.23a shows a pile penetrating three layers of clay. According to Figure 11.23b, the mean value of c_u is $(c_{u(1)}L_1 + c_{u(2)}L_2 + \cdots)/L$. Similarly, Figure 11.23c shows the plot of the variation of effective stress with depth. The mean effective stress is

$$\overline{\sigma}'_o = \frac{A_1 + A_2 + A_3 + \cdots}{L} \tag{11.53}$$

where $A_1, A_2, A_3, \ldots$ = areas of the vertical effective stress diagrams

α Method

According to the α method, the unit skin resistance in clayey soils can be represented by the equation

$$f = \alpha c_u \tag{11.54}$$

where α = empirical adhesion factor

The approximate variation of the value of α is shown in Figure 11.24,

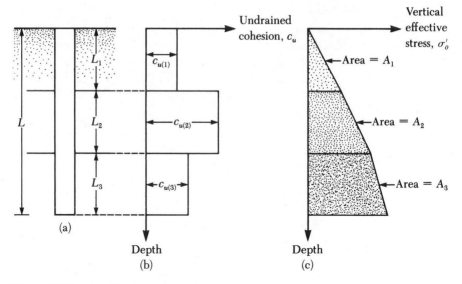

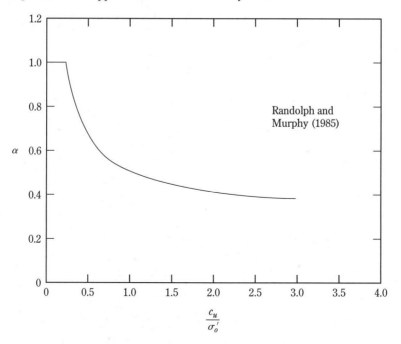

Figure 11.23 Application of λ method in layered soil

Figure 11.24 Variation of α with c_u/σ_o'

where σ_o' is the vertical effective stress. This variation of α with c_u/σ_o' was obtained by Randolph and Murphy (1985). With it, we have

$$Q_s = \Sigma f p \, \Delta L = \Sigma \alpha c_u p \, \Delta L \tag{11.55}$$

β Method

When piles are driven into saturated clays, the pore water pressure in the soil around the piles increases. The excess pore water pressure in normally consolidated clays may be four to six times c_u. However, within a month or so, this pressure gradually dissipates. Hence, the unit frictional resistance for the pile can be determined on the basis of the effective stress parameters of the clay in a remolded state ($c' = 0$). Thus, at any depth,

$$f = \beta\sigma_o' \tag{11.56}$$

where $\sigma_o' =$ vertical effective stress
$$\beta = K \tan \phi_R \tag{11.57}$$
$\phi_R' =$ drained friction angle of remolded clay
$K =$ earth pressure coefficient

Conservatively, the magnitude of K is the earth pressure coefficient at rest, or

$$K = 1 - \sin \phi_R' \qquad \text{(for normally consolidated clays)} \tag{11.58}$$

and

$$K = (1 - \sin \phi_R')\sqrt{\text{OCR}} \qquad \text{(for overconsolidated clays)} \tag{11.59}$$

where   OCR = overconsolidation ratio

Combining Eqs. (11.56), (11.57), (11.58), and (11.59), for normally consolidated clays yields

$$f = (1 - \sin \phi_R') \tan \phi_R \sigma_o' \tag{11.60}$$

and for overconsolidated clays,

$$f = (1 - \sin \phi_R')\tan \phi_R'\sqrt{\text{OCR}}\ \sigma_o' \tag{11.61}$$

With the value of f determined, the total frictional resistance may be evaluated as

$$Q_s = \Sigma f p\ \Delta L$$

Correlation with Cone Penetration Test Results

Nottingham and Schmertmann (1975) and Schmertmann (1978) found the correlation for unit skin friction in clay (with $\phi = 0$) to be

$$f = \alpha' f_c \tag{11.62}$$

The variation of α' with the frictional resistance f_c is shown in Figure 11.25. Thus,

$$Q_s = \Sigma f p(\Delta L) = \Sigma \alpha' f_c p(\Delta L) \tag{11.63}$$

11.14   *General Comments and Allowable Pile Capacity*

Although calculations for estimating the ultimate load-bearing capacity of a pile can be made by using the relationships presented in Sections 11.6 through 11.13, an engineer needs to keep the following points in mind:

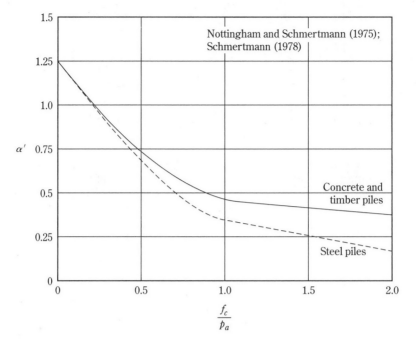

Figure 11.25   Variation of α' with f_c/p_a for piles in clay (p_a = atmosphic pressure $\approx 100 \text{ kN/m}^2$ or 2000 lb/ft^2)

1. In calculating the area of cross section, A_p, and the perimeter, p, of piles with developed profiles, such as H-piles and open-ended pipe piles, the effect of soil plug should be considered. According to Figures 11.10b and 11.10c, for pipe piles,

$$A_p = \left(\frac{\pi}{4}\right)D^2$$

and

$$p = \pi D$$

Similarly, for H-piles,

$$A_p = d_1 d_2$$

and

$$p = 2(d_1 + d_2)$$

Note that for H-piles, because $d_2 > d_1$, $D = d_1$.

2. The ultimate point load relations given in Eqs. (11.13), (11.20), and (11.30) are for the gross ultimate point load; that is, they include the weight of the pile. So the net ultimate point load is approximately

$$Q_{p(\text{net})} = Q_{p(\text{gross})} - q' A_p$$

However, in practice, for soils with $\phi' > 0$, the assumption is made that $Q_{p(\text{net})} = Q_{p(\text{gross})}$. In cohesive soils with $\phi = 0$, $N_q^* = 1$. (See Figure 11.12.) Hence, from Eq. (11.13),

$$Q_{p(\text{gross})} = (c_u N_c^* + q')A_p$$

so

$$Q_{p(\text{net})} = [(c_u N_c^* + q') - q']A_p = c_u N_c^* A_p = 9c_u A_p = Q_p$$

This relation is the one given in Eq. (11.19).

After the total ultimate load-carrying capacity of a pile has been determined by summing the point bearing capacity and the frictional (or skin) resistance, a reasonable factor of safety should be used to obtain the total allowable load for each pile, or

$$Q_{\text{all}} = \frac{Q_u}{\text{FS}} \tag{11.64}$$

where Q_{all} = allowable load-carrying capacity for each pile
  FS = factor of safety

The factor of safety generally used ranges from 2.5 to 4, depending on the uncertainties surrounding the calculation of ultimate load.

11.15   *Point Bearing Capacity of Piles Resting on Rock*

Sometimes piles are driven to an underlying layer of rock. In such cases, the engineer must evaluate the bearing capacity of the rock. The ultimate unit point resistance in rock (Goodman, 1980) is approximately

$$q_p = q_u(N_\phi + 1) \tag{11.65}$$

where $N_\phi = \tan^2 (45 + \phi'/2)$
  q_u = unconfined compression strength of rock
  ϕ' = drained angle of friction

The unconfined compression strength of rock can be determined by laboratory tests on rock specimens collected during field investigation. However, extreme caution should be used in obtaining the proper value of q_u, because laboratory specimens usually are small in diameter. As the diameter of the specimen increases, the unconfined compression strength decreases—a phenomenon referred to as the *scale effect*. For specimens larger than about 1 m (3 ft) in diameter, the value of q_u remains approximately constant. There appears to be a fourfold to fivefold reduction of the magnitude of q_u in this process. The scale effect in rock is caused primarily by randomly distributed large and small fractures and also by progressive ruptures along the slip lines. Hence, we always recommend that

$$q_{u(\text{design})} = \frac{q_{u(\text{lab})}}{5} \tag{11.66}$$

Table 11.7 Typical Unconfined Compressive Strength of Rocks

Type of rock	q_u MN/m²	q_u lb/in²
Sandstone	70–140	10,000–20,000
Limestone	105–210	15,000–30,000
Shale	35–70	5000–10,000
Granite	140–210	20,000–30,000
Marble	60–70	8500–10,000

Table 11.8 Typical Values of Angle of Friction ϕ' of Rocks

Type of rock	Angle of friction, ϕ' (deg)
Sandstone	27–45
Limestone	30–40
Shale	10–20
Granite	40–50
Marble	25–30

Table 11.7 lists some representative values of (laboratory) unconfined compression strengths of rock. Representative values of the rock friction angle ϕ' are given in Table 11.8.

A factor of safety of at least 3 should be used to determine the allowable point bearing capacity of piles. Thus,

$$Q_{p(\text{all})} = \frac{[q_{u(\text{design})}(N_\phi + 1)]A_p}{\text{FS}} \tag{11.67}$$

Example 11.1

A concrete pile is 16 m (L) long and 410 mm $\times$ 410 mm in cross section. The pile is fully embedded in sand for which $\gamma = 17$ kN/m³ and $\phi' = 30°$. Calculate the ultimate point load, Q_p, by

a. Meyerhof's method (Section 11.7).
b. Vesic's method (Section 11.8). Use $I_r = I_{rr} = 50$.
c. Janbu's method (Section 11.9). Use $\eta' = 90°$.

Solution

Part a
From Eq. (11.15),

$$Q_p = A_p q' N_q^* = A_p \gamma L N_q^*$$

For $\phi' = 30°$, $N_q^* \approx 55$ (see Figure 11.12), so

$$Q_p = (0.41 \times 0.41 \text{ m}^2)(16 \times 17)(55) = 2515 \text{ kN}$$

Again, from Eq. (11.17),

$$q_p = (0.5 p_a N_q^* \tan \phi') A_p$$
$$= [(0.5)(100)(55)\tan 30](0.41 \times 0.41) = 267 \text{ kN}$$

Hence,

$$Q_p = \textbf{267 kN}$$

Part b
From Eqs. (11.20), (11.21), and (11.22) with $c' = 0$,

$$Q_p = A_p \sigma_o' N_\sigma^* = A_p \left[\frac{1 + 2(1 - \sin \phi')}{3} \right] q' N_\sigma^*$$

For $\phi' = 30°$ and $I_{rr} = 50$, the value of N_σ^* is about 36. (See Table 11.4.) So

$$Q_p = (0.41 \times 0.41) \left[\frac{1 + 2(1 - \sin 30)}{3} \right] (16 \times 17)(36) = \textbf{1097 kN}$$

Part c
From Eq. (11.31) with $c' = 0$

$$Q_p = A_p q' N_q^*$$

For $\phi' = 30°$ and $\eta' = 90°$, the value of $N_q^* \approx 18.4$. (See Table 11.5.) Therefore,

$$Q_p = (0.41 \times 0.41)(16 \times 17)(18.4) = \textbf{841 kN} \qquad \blacksquare$$

Example 11.2

For the pile described in Example 11.1,

a. Given that $K = 1.3$ and $\delta = 0.8\phi'$, determine the frictional resistance Q_s. Use Eqs. (11.14), (11.38), and (11.39).
b. Using the results of Example 11.1 and Part a of this problem, estimate the allowable load-carrying capacity of the pile. Let FS = 4.

Solution
Part a
From Eq. (11.37),

$$L \approx 15D = 15(0.41 \text{ m}) = 6.15 \text{ m}$$

From Eq. (11.38), at $z = 0$, $\sigma_o' = 0$, so $f = 0$. Again, at $z = L' = 6.15$ m

$$\sigma_o' = \gamma L' = (17)(6.15) = 104.55 \text{ kN/m}^2$$

So

$$f = K\sigma_o' \tan \delta = (1.3)(104.55)[\tan(0.8 \times 30)] = 60.51 \text{ kN/m}^2$$

Thus,

$$Q_s = \left(\frac{f_{z=0} + f_{z=6.15 \text{ m}}}{2} \right) pL' + f_{6.15 \text{ m}} p(L - L')$$

$$= \left(\frac{0 + 60.51}{2} \right)(4 \times 0.41)(6.15) + (60.51)(4 \times 0.41)(16 - 6.15)$$

$$= 305.2 + 977.5 = \textbf{1282.7 kN}$$

Part b
We have $Q_u = Q_p + Q_s$. From Example 11.1, the average value of Q_p is

$$\frac{267 + 1097 + 841}{3} \approx 735 \text{ kN}$$

So

$$Q_{\text{all}} = \frac{Q_u}{\text{FS}} = \frac{1}{4}(735 + 1282.7) = \textbf{504.4 kN} \qquad \blacksquare$$

Example 11.3

For the pile described in Example 11.1, estimate Q_{all} using Coyle and Castello's method. [See Section 11.10 and Eq. (11.43).]

Solution
From Eqs. (11.33) and (11.43),

$$Q_u = Q_p + Q_s = q'N_q^*A_p + K\overline{\sigma}_o'\tan(0.8\phi')pL$$

and

$$\frac{L}{D} = \frac{16}{0.41} = 39$$

For $\phi' = 30°$ and $L/D = 39$, $N_q^* = 25$ (see Figure 11.14) and $K \approx 0.2$ (see Figure 11.19). Thus,

$$Q_u = (17 \times 16)(25)(0.41 \times 0.41)$$
$$+ (0.2)\left(\frac{17 \times 16}{2}\right)\tan(0.8 \times 30)(4 \times 0.41)(16)$$
$$= 1143 + 317.8 = 1460.8 \text{ kN}$$

and

$$Q_{\text{all}} = \frac{Q_u}{\text{FS}} = \frac{1460.8}{4} = \textbf{365.2 kN} \qquad \blacksquare$$

Example 11.4

A driven pipe pile in clay is shown in Figure 11.26a. The pipe has an outside diameter of 406 mm and a wall thickness of 6.35 mm.

a. Calculate the net point bearing capacity. Use Eq. (11.19).

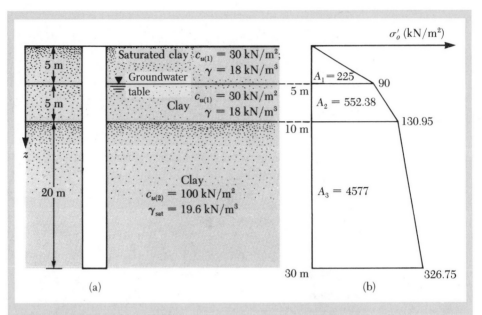

Figure 11.26 Estimation of the load bearing capacity of a driven pipe pile

b. Calculate the skin resistance (1) by using Eqs. (11.54) and (11.55) (α method), (2) by using Eq. (11.52) (λ method), and (3) by using Eq. (11.56) (β method). For all clay layers, $\phi'_R = 30°$. The top 10 m of clay is normally consolidated. The bottom clay layer has an OCR of 2.

c. Estimate the net allowable pile capacity. Use FS = 4.

Solution
The area of cross section of the pile, including the soil inside the pile, is

$$A_p = \frac{\pi}{4}D^2 = \frac{\pi}{4}(0.406)^2 = 0.1295 \text{ m}^2$$

Part a: Calculation of Net Point Bearing Capacity
From Eq. (11.19),

$$Q_p = A_p q_p = A_p N_c^* c_{u(2)} = (0.1295)(9)(100) = \textbf{116.55 kN}$$

Part b: Calculation of Skin Resistance
(1) Using Eqs. (11.54) and (11.55), we have

$$Q_s = \Sigma \alpha c_u p \Delta L$$

The variation of vertical effective stress with depth is shown in Figure 11.26b. Now the following table can be prepared:

Depth (m)	Average depth (m)	Average vertical effective stress, $\overline{\sigma}'_o$ (kN/m^2)	c_u (kN/m^2)	$\dfrac{c_u}{\overline{\sigma}'_o}$ (kN/m^2)	α (Figure 11.24)
0–5	2.5	$\dfrac{0 + 90}{2} = 45$	30	0.67	0.6
5–10	7.5	$\dfrac{90 + 130.95}{2} = 110.5$	30	0.27	0.9
10–30	20	$\dfrac{130.95 + 326.75}{2} = 228.85$	100	0.44	0.725

Thus,

$$Q_s = [\alpha_1 c_{u(1)}L_1 + \alpha_2 c_{u(1)}L_2 + \alpha_3 c_{u(2)}L_3]p$$
$$= [(0.6)(30)(5) + (0.9)(30)(5)$$
$$+ (0.725)(100)(20)](\pi \times 0.406) = \textbf{2136 kN}$$

(2) From Eq. 11.52, $f_{av} = \lambda(\overline{\sigma}'_o + 2c_u)$. Now, the average value of c_u is

$$\frac{c_{u(1)}(10) + c_{u(2)}(20)}{30} = \frac{(30)(10) + (100)(20)}{30} = 76.7 \text{ kN/m}^2$$

To obtain the average value of $\overline{\sigma}'_o$, the diagram for vertical effective stress variation with depth is plotted in Figure 11.26b. From Eq. (11.53),

$$\overline{\sigma}'_o = \frac{A_1 + A_2 + A_3}{L} = \frac{225 + 552.38 + 4577}{30} = 178.48 \text{ kN/m}^2$$

From Figure 11.22, the magnitude of λ is 0.14. So

$$f_{av} = 0.14[178.48 + (2)(76.7)] = 46.46 \text{ kN/m}^2$$

Hence,

$$Q_s = pLf_{av} = \pi(0.406)(30)(46.46) = \textbf{1778 kN}$$

(3) The top layer of clay (10 m) is normally consolidated, and $\phi'_R = 30°$. For $z = 0$–5 m, from Eq. (11.60), we have

$$f_{av(1)} = (1 - \sin \phi'_R) \tan \phi'_R \, \overline{\sigma}'_o$$
$$= (1 - \sin 30°)(\tan 30°)\left(\frac{0 + 90}{2}\right) = 13.0 \text{ kN/m}^2$$

Similarly, for $z = 5$–10 m.

$$f_{av(2)} = (1 - \sin 30°)(\tan 30°)\left(\frac{90 + 130.95}{2}\right) = 31.9 \text{ kN/m}^2$$

For $z = 10{-}30$ m from Eq. (11.61),

$$f_{av} = (1 - \sin \phi_R)\tan \phi_R\sqrt{\text{OCR}}\ \sigma'_o$$

For OCR = 2,

$$f_{av(3)} = (1 - \sin 30°)(\tan 30°)\sqrt{2}\left(\frac{130.95 + 326.75}{2}\right) = 93.43 \text{ kN/m}^2$$

So

$$Q_s = p[f_{av(1)}(5) + f_{av(2)}(5) + f_{av(3)}(20)]$$
$$= (\pi)(0.406)[(13)(5) + (31.9)(5) + (93.43)(20)] = \mathbf{2670\ kN}$$

Part c: Calculation of Net Ultimate Capacity, Q_u
We have

$$Q_{s(\text{average})} = \frac{2136 + 1778 + 2670}{3} \approx 2195 \text{ kN}$$

Thus,

$$Q_u = Q_p + Q_s = 116.55 + 2195 = 2311.55 \text{ kN}$$

and

$$3Q_{\text{all}} = \frac{Q_u}{F_s} = \frac{2311.55}{4} \approx \mathbf{578\ kN} \qquad \blacksquare$$

Example 11.5

A concrete pile 305 mm $\times$ 305 mm in cross section is driven to a depth of 20 m below the ground surface in a saturated clay soil. A summary of the variation of frictional resistance f_c obtained from a cone penetration test is as follows:

Depth (m)	Friction resistance, f_c (kg/cm²)
0–6	0.35
6–12	0.56
12–20	0.72

Estimate the frictional resistance Q_s for the pile.

Solution
We can prepare the following table:

Depth (m)	f_c (kN/m²)	α' (Figure 11.25)	ΔL (m)	$\alpha' f_c p(\Delta L)$ [Eq. (11.63)] (kN)
0–6	34.34	0.84	6	211.5
6–12	54.94	0.71	6	285.5
12–20	70.63	0.63	8	434.2

Note: $p = (4)(0.305) = 1.22$ m

Thus,

$$Q_s = \Sigma \alpha' f_c p(\Delta L) = \mathbf{931\ kN}$$

∎

Example 11.6

An H-pile (size HP 310 × 125) having a length of embedment of 26 m is driven through a soft clay layer to rest on sandstone. The sandstone has a laboratory unconfined compression strength of 76 MN/m² and a friction angle of 28°. Use a factor of safety of 5, and estimate the allowable point bearing capacity.

Solution
From Eqs. (11.66) and (11.67),

$$Q_{p(\text{all})} = \frac{\left\{\left[\dfrac{q_{u(\text{lab})}}{5}\right]\left[\tan^2\left(45 + \dfrac{\phi'}{2}\right) + 1\right]\right\}A_p}{\text{FS}}$$

From Table 11.1a, for HP 310 × 125 piles, $A_p = 15.9 \times 10^{-3}\,\text{m}^2$, so

$$Q_{p(\text{all})} = \frac{\left\{\left[\dfrac{76 \times 10^3\ \text{kN/m}^2}{5}\right]\left[\tan^2\left(45 + \dfrac{28}{2}\right) + 1\right]\right\}(15.9 \times 10^{-3}\,\text{m}^2)}{5}$$

$$= \mathbf{182\ kN}$$

∎

11.16   *Pile Load Tests*

In most large projects, a specific number of load tests must be conducted on piles. The primary reason is the unreliability of prediction methods. The vertical and lateral load-bearing capacity of a pile can be tested in the field. Figure 11.27a shows a schematic diagram of the pile load arrangement for testing *axial compression* in the field. The load is applied to the pile by a hydraulic jack. Step loads are applied to the pile, and

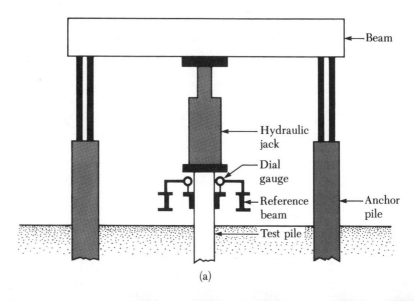

(a)

(b)

Figure 11.27 (a) Schematic diagram of pile load test arrangement; (b) a load test in progress (Courtesy of E. C. Shin, University of Incheon, Korea); (c) plot of load against total settlement; (d) plot of load against net settlement

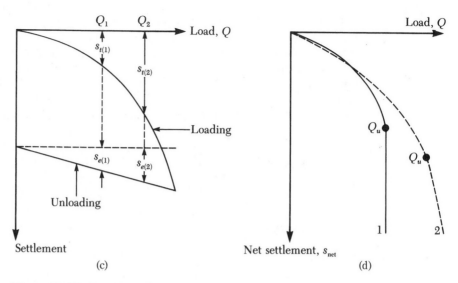

Figure 11.27 (Continued)

sufficient time is allowed to elapse after each load so that a small amount of settlement occurs. The settlement of the pile is measured by dial gauges. The amount of load to be applied for each step will vary, depending on local building codes. Most building codes require that each step load be about one-fourth of the proposed working load. The load test should be carried out to at least a total load of two times the proposed working load. After the desired pile load is reached, the pile is gradually unloaded.

Figure 11.27b shows a pile load test in progress.

Figure 11.27c shows a load settlement diagram obtained from field loading and unloading. For any load Q, the net pile settlement can be calculated as follows: When $Q = Q_1$,

$$\text{Net settlement, } s_{\text{net}(1)} = s_{t(1)} - s_{e(1)}$$

When $Q = Q_2$,

$$\text{Net settlement, } s_{\text{net}(2)} = s_{t(2)} - s_{e(2)}$$

$$\vdots$$

where s_{net} = net settlement
  s_e = elastic settlement of the pile itself
  s_t = total settlement

These values of Q can be plotted in a graph against the corresponding net settlement, s_{net}, as shown in Figure 11.27d. The ultimate load of the pile can then be determined from the graph. Pile settlement may increase with load to a certain point, beyond which the load–settlement curve becomes vertical. The load corresponding to the point where the curve of Q vs. s_{net} becomes vertical is the ultimate load, Q_u, for the pile; it is shown by curve 1 in Figure 11.27d. In many cases, the latter stage of the load–settlement curve is almost linear, showing a large degree of settlement for a small increment of load; this is shown by curve 2 in the figure. The ultimate load, Q_u, for such a case is determined from the point of the curve of Q vs. s_{net} where this steep linear portion starts.

The load test procedure just described requires the application of step loads on the piles and the measurement of settlement and is called a *load-controlled* test. Another technique used for a pile load test is the *constant-rate-of-penetration* test, wherein the load on the pile is continuously increased to maintain a constant rate of penetration, which can vary from 0.25–2.5 mm/min (0.01–0.1 in./min). This test gives a load–settlement plot similar to that obtained from the load-controlled test. Another type of pile load test is *cyclic loading,* in which an incremental load is repeatedly applied and removed.

In order to conduct a load test on piles, it is important to take into account the time lapse after the end of driving (EOD). When piles are driven into soft clay, a certain zone surrounding the clay becomes remolded or compressed, as shown in Figure 11.28a. This results in a reduction of undrained shear strength, c_u (Figure 11.28b). With time, the loss of undrained shear strength is partially or fully regained. The time lapse may range from 30 to 60 days. Figure 11.29, based on results reported by

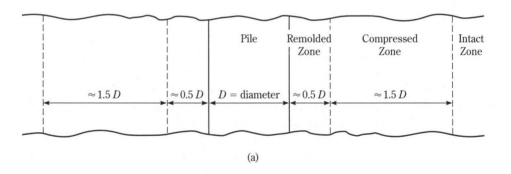

(a)

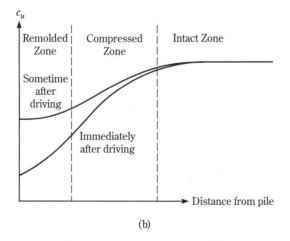

(b)

Figure 11.28 (a) Remolded or compacted zone around a pile driven into soft clay; (b) Nature of variation of undrained shear strength (c_u) with time around a pile driven into soft clay

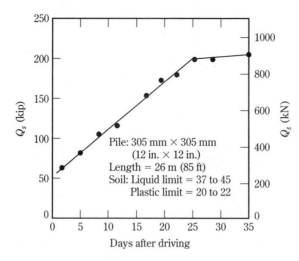

Figure 11.29 Variation of Q_s with time for a pile driven into soft clay (based on load test results of Terzaghi and Peck, 1967)

Terzaghi and Peck (1967), shows the magnitude of the variation of Q_s with time for a pile driven into soft clay. It can be seen that Q_s increased by about 300% during a time of about 25 days.

For piles driven in dilative (dense to very dense) saturated fine sands, relaxation is possible. Negative pore water pressure, if developed during pile driving, will dissipate over time, resulting in a reduction in pile capacity with time after the driving operation is completed. At the same time, excess pore water pressure may be generated in contractive fine sands during pile driving. The excess pore water pressure will dissipate over time, which will result in greater pile capacity.

Several empirical relationships have been developed to predict changes in pile capacity with time.

Skov and Denver (1988)

Skov and Denver proposed the equation

$$Q_t = Q_{\text{OED}}\left[A \log\left(\frac{t}{t_o}\right) + 1 \right] \tag{11.68}$$

where Q_t = pile capacity t days after the end of driving
Q_{OED} = pile capacity at the end of driving
t = time, in days

For sand, $A = 0.2$ and $t_o = 0.5$ days; for clay, $A = 0.6$ and $t_o = 1.0$ days.

Guang-Yu (1988)

According to Guang-Yu,

$$Q_{14} = (0.375S_t + 1)Q_{\text{OED}} \qquad \text{(applicable to clay soil)} \tag{11.69}$$

where Q_{14} = pile capacity 14 days after pile driving
S_t = sensitivity of clay

Svinkin (1996)

Svinkin suggests the relationship

$$Q_t = 1.4Q_{OED}t^{0.1} \qquad \text{(upper limit for sand)} \qquad (11.70)$$

$$Q_t = 1.025Q_{OED}t^{0.1} \qquad \text{(lower limit for sand)} \qquad (11.71)$$

where t = time after driving, in days

11.17   *Comparison of Theory with Field Load Test Results*

Details of many field studies related to the estimation of the ultimate load-carrying capacity of various types of piles are available in the literature. In some cases, the results agree reasonably well with the theoretical predictions, and in others, they vary widely. The discrepancy between theory and field-test results may be attributed to factors such as improper interpretation of subsoil properties, incorrect theoretical assumptions, erroneous acquisition of field-test results, and other factors.

We saw from Example 11.1 that, for similar soil properties, the ultimate point load Q_p can vary over 400% or more, depending on which theory and equation are used. Also, from the calculation of part a of the example, it is easy to see that, in most cases, for long piles embedded in sand, the limiting point resistance q_l [Eq. 11.17] controls the unit point resistance q_p. Meyerhof (1976) provided the results of several field load tests on long piles $(L/D \geqslant 10)$, from which the derived values of q_p were calculated and plotted in Figure 11.30. Also plotted in this figure is the variation of q_l calculated from Eq. 11.17. It can be seen that, for a

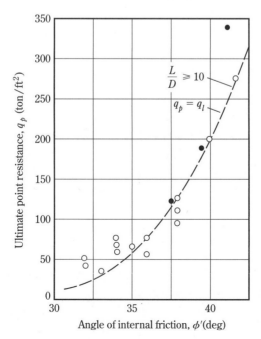

Figure 11.30   Ultimate point resistance of driven piles in sand (after Meyerhof, 1976)

Table 11.9 Pile Load Test Results

Pile no.	Pile type	Batter	Q_p (ton)	Q_s (ton)	Q_u (ton)	Pile length (ft)
1–3A	HP14 × 73	Vertical	152	161	313	54
1–6	HP14 × 73	Vertical	75	353	428	53
1–9	HP14 × 73	1:2.5	85	252	337	58
2–5	HP14 × 73	1:2.5	46	179	225	59

Table 11.10 Summary of Briaud et al.'s Statistical Analysis for H-Piles, Plugged Case

Theoretical method	Q_p			Q_s			Q_u		
	Mean	Standard deviation	Coefficient of variation	Mean	Standard deviation	Coefficient of variation	Mean	Standard deviation	Coefficient of variation
Coyle and Castello (1981)	2.38	1.31	0.55	0.87	0.36	0.41	1.17	0.44	0.38
Briaud and Tucker (1984)	1.79	1.02	0.59	0.81	0.32	0.40	0.97	0.39	0.40
Meyerhof (1976)	4.37	2.76	0.63	0.92	0.43	0.46	1.68	0.76	0.45
API (1984)	1.62	1.00	0.62	0.59	0.25	0.43	0.79	0.34	0.43

given friction angle ϕ', the magnitude of q_p can deviate substantially from that given in the theory.

Briaud et al. (1989) reported the results of 28 axial load tests performed by the U.S. Army Engineering District (St. Louis) on impact-driven H-piles and pipe piles in sand during the construction of the New Lock and Dam No. 26 on the Mississippi River. The results of load tests on four of the H-piles are summarized in Table 11.9.

Briaud and his colleagues made a statistical analysis to determine the ratio of theoretical ultimate load to measured ultimate load. The results of this analysis are summarized in Table 11.10 for the plugged case. (See Figure 11.10.) Note that a perfect prediction would have a mean of 1.0, a standard deviation of 0, and a coefficient of variation of 0. The table indicates that no method gave a perfect prediction; in general, Q_p was overestimated and Q_s was underestimated. Again, this shows the uncertainty in predicting the load-bearing capacity of piles.

Lessons from the preceding case studies and others available in the literature show that previous experience and good practical judgment, along with a knowledge of theoretical developments, are required to design safe pile foundations.

11.18   *Elastic Settlement of Piles*

The total settlement of a pile under a vertical working load Q_w is given by

$$s_e = s_{e(1)} + s_{e(2)} + s_{e(3)} \tag{11.72}$$

where $s_{e(1)}$ = elastic settlement of pile
$s_{e(2)}$ = settlement of pile caused by the load at the pile tip
$s_{e(3)}$ = settlement of pile caused by the load transmitted along the pile shaft

If the pile material is assumed to be elastic, the deformation of the pile shaft can be evaluated, in accordance with the fundamental principles of mechanics of materials, as

$$s_{e(1)} = \frac{(Q_{wp} + \xi Q_{ws})L}{A_p E_p}$$

(11.73)

where Q_{wp} = load carried at the pile point under working load condition
Q_{ws} = load carried by frictional (skin) resistance under working load condition
A_p = area of cross section of pile
L = length of pile
E_p = modulus of elasticity of the pile material

The magnitude of ξ will depend on the nature of the distribution of the unit friction (skin) resistance f along the pile shaft. If the distribution of f is uniform or parabolic, as shown in Figures 11.31a and 11.31b, then $\xi = 0.5$. However, for a triangular distribution of f (Figure 11.31c), the magnitude of ξ is about 0.67 (Vesic, 1977).

The settlement of a pile caused by the load carried at the pile point may be expressed in the form:

$$s_{e(2)} = \frac{q_{wp}D}{E_s}(1 - \mu_s^2)I_{wp}$$

(11.74)

where D = width or diameter of pile
q_{wp} = point load per unit area at the pile point = Q_{wp}/A_p
E_s = modulus of elasticity of soil at or below the pile point
μ_s = Poisson's ratio of soil
I_{wp} = influence factor ≈ 0.85

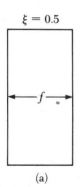
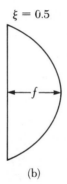
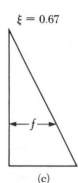

$\xi = 0.5$ $\xi = 0.5$ $\xi = 0.67$

(a)   (b)   (c)

Figure 11.31 Various types of distribution of unit friction (skin) resistance along the pile shaft

Vesic (1977) also proposed a semi-empirical method for obtaining the magnitude of the settlement of $s_{e(2)}$. His equation is

$$s_{e(2)} = \frac{Q_{wp}C_p}{Dq_p}$$

(11.75)

where q_p = ultimate point resistance of the pile
  C_p = an empirical coefficient

Representative values of C_p for various soils are given in Table 11.11.

The settlement of a pile caused by the load carried by the pile shaft is given by a relation similar to Eq. (11.74), namely,

$$s_{e(3)} = \left(\frac{Q_{ws}}{pL}\right)\frac{D}{E_s}(1 - \mu_s^2)I_{ws}$$

(11.76)

where p = perimeter of the pile
  L = embedded length of pile
  I_{ws} = influence factor

Note that the term Q_{ws}/pL in Eq. (11.76) is the average value of f along the pile shaft. The influence factor, I_{ws}, has a simple empirical relation (Vesic, 1977):

$$I_{ws} = 2 + 0.35\sqrt{\frac{L}{D}}$$

(11.77)

Vesic (1977) also proposed a simple empirical relation similar to Eq. (11.75) for obtaining $s_{e(3)}$:

$$s_{e(3)} = \frac{Q_{ws}C_s}{Lq_p}$$

(11.78)

In this equation, C_s = an empirical constant = $(0.93 + 0.16\sqrt{L/D})C_p$ (11.79)

The values of C_p for use in Eq. (11.78) may be estimated from Table 11.11.

Table 11.11   Typical Values of C_p [from Eq. (11.75)]

Type of soil	Driven pile	Bored pile
Sand (dense to loose)	0.02–0.04	0.09–0.18
Clay (stiff to soft)	0.02–0.03	0.03–0.06
Silt (dense to loose)	0.03–0.05	0.09–0.12

From "Design of Pile Foundations," by A. S. Vesic, in *NCHRP Synthesis of Highway Practice 42,* Transportation Research Board, 1977. Reprinted by permission.

Example 11.7

A fully embedded precast, prestressed concrete pile is 12 m long and is driven into a homogeneous layer of sand $(c' = 0)$. The pile is square in cross section, with sides measuring 305 mm. The dry unit weight of sand (γ_d) is 16.8 kN/m³, and the average effective soil friction angle is 35°. The allowable working load is 338 kN. If 240 kN is contributed by the frictional resistance and 98 kN is from the point load, determine the elastic settlement of the pile. Use $E_p = 21 \times 10^6$ kN/m², $E_s = 30,000$ kN/m², and $\mu_s = 0.3$.

Solution

We will use Eq. (11.72):

$$s_e = s_{e(1)} + s_{e(2)} + s_{e(3)}$$

From Eq. (11.73),

$$s_{e(1)} = \frac{(Q_{wp} + \xi Q_{ws})L}{A_p E_p}$$

Let $\xi = 0.6$ and $E_p = 21 \times 10^6$ kN/m². Then

$$s_{e(1)} = \frac{[97 + (0.6)(240)]12}{(0.305)^2(21 \times 10^6)} = 0.00148 \text{ m} = 1.48 \text{ mm}$$

From Eq. (11.74),

$$s_{e(2)} = \frac{q_{wp}D}{E_s}(1 - \mu_s^2)I_{wp}$$

Now, $I_{wp} = 0.85$ and

$$q_{wp} = \frac{Q_{wp}}{A_p} = \frac{97}{(0.305)^2} = 1042.7 \text{ kN/m}^2$$

So

$$s_{e(2)} = \left[\frac{(1042.7)(0.305)}{30,000}\right](1 - 0.3^2)(0.85) = 0.0082 \text{ m} = 8.2 \text{ mm}$$

Again, from Eq. (11.76)

$$s_{e(3)} = \left(\frac{Q_{ws}}{pL}\right)\frac{D}{E_s}(1 - \mu_s^2)I_{ws}$$

and

$$I_{ws} = 2 + 0.35\sqrt{\frac{L}{D}} = 2 + 0.35\sqrt{\frac{12}{0.305}} = 4.2$$

So

$$s_{e(3)} = \frac{240}{(4 \times 0.305)(12)}\left(\frac{0.305}{30{,}000}\right)(1 - 0.3^2)(4.2) = 0.00064 \text{ m} = 0.64 \text{ mm}$$

Hence, the total settlement

$$s_e = 1.48 + 8.2 + 0.64 = \textbf{10.32 mm}$$ ■

11.19 *Laterally Loaded Piles*

A vertical pile resists a lateral load by mobilizing passive pressure in the soil surrounding it. (See Figure 11.1c.) The degree of distribution of the soil's reaction depends on (a) the stiffness of the pile, (b) the stiffness of the soil, and (c) the fixity of the ends of the pile. In general, laterally loaded piles can be divided into two major categories: (1) short or rigid piles and (2) long or elastic piles. Figures 11.32a and 11.32b show the nature of the variation of the pile deflection and the distribution of the moment and shear force along the pile length when the pile is subjected to lateral loading. We next summarize the current solutions for laterally loaded piles.

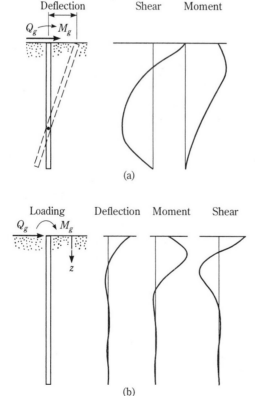

Figure 11.32 Nature of variation of pile deflection, moment, and shear force for (a) a rigid pile and (b) an elastic pile

Elastic Solution

A general method for determining moments and displacements of a vertical pile embedded in a *granular soil* and subjected to lateral load and moment at the ground surface was given by Matlock and Reese (1960). Consider a pile of length L subjected to a lateral force Q_g and a moment M_g at the ground surface ($z = 0$), as shown in Figure 11.33a. Figure 11.33b shows the general deflected shape of the pile and the soil resistance caused by the applied load and the moment.

According to a simpler Winkler's model, an elastic medium (soil in this case) can be replaced by a series of infinitely close independent elastic springs. Based on this assumption,

$$k = \frac{p'(\text{kN/m or lb/ft})}{x(\text{m or ft})} \tag{11.80}$$

where k = modulus of subgrade reaction
  p' = pressure on soil
  x = deflection

The subgrade modulus for *granular soils* at a depth z is defined as

$$k_z = n_h z \tag{11.81}$$

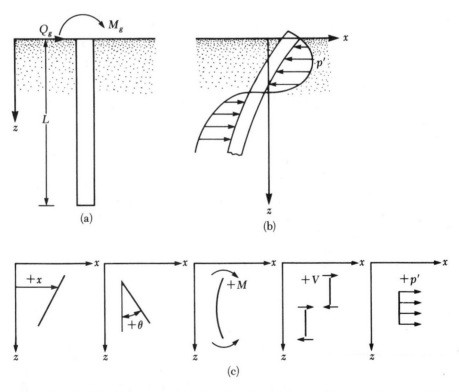

Figure 11.33 (a) Laterally loaded pile; (b) soil resistance on pile caused by lateral load; (c) sign conventions for displacement, slope, moment, shear, and soil reaction

where n_h = constant of modulus of horizontal subgrade reaction

Referring to Figure 11.33b and using the theory of beams on an elastic foundation, we can write

$$E_p I_p \frac{d^4 x}{dz^4} = p' \tag{11.82}$$

where E_p = modulus of elasticity in the pile material
I_p = moment of inertia of the pile section

Based on Winkler's model

$$p' = -kx \tag{11.83}$$

The sign in Eq. (11.83) is negative because the soil reaction is in the direction opposite that of the pile deflection.

Combining Eqs. (11.82) and (11.83) gives

$$E_p I_p \frac{d^4 x}{dz^4} + kx = 0 \tag{11.84}$$

The solution of Eq. (11.84) results in the following expressions:

Pile Deflection at Any Depth [$x_z(z)$]

$$x_z(z) = A_x \frac{Q_g T^3}{E_p I_p} + B_x \frac{M_g T^2}{E_p I_p} \tag{11.85}$$

Slope of Pile at Any Depth [$\theta_z(z)$]

$$\theta_z(z) = A_\theta \frac{Q_g T^2}{E_p I_p} + B_\theta \frac{M_g T}{E_p I_p} \tag{11.86}$$

Moment of Pile at Any Depth [$M_z(z)$]

$$M_z(z) = A_m Q_g T + B_m M_g \tag{11.87}$$

Shear Force on Pile at Any Depth [$V_z(z)$]

$$V_z(z) = A_v Q_g + B_v \frac{M_g}{T} \tag{11.88}$$

Soil Reaction at Any Depth $[p_z'(z)]$

$$p_z'(z) = A_{p'}\frac{Q_g}{T} + B_{p'}\frac{M_g}{T^2} \tag{11.89}$$

where $A_x, B_x, A_\theta, B_\theta, A_m, B_m, A_v, B_v, A_{p'},$ and $B_{p'}$ are coefficients
T = characteristic length of the soil–pile system

$$= \sqrt[5]{\frac{E_p I_p}{n_h}} \tag{11.90}$$

n_h has been defined in Eq. (11.81)

When $L \geqslant 5T$, the pile is considered to be a *long pile*. For $L \leqslant 2T$, the pile is considered to be a *rigid pile*. Table 11.12 gives the values of the coefficients for long piles $(L/T \geqslant 5)$ in Eqs. (11.85) through (11.89). Note that, in the first column of the table,

$$Z = \frac{z}{T} \tag{11.91}$$

is the nondimensional depth.

The positive sign conventions for $x_z(z), \theta_z(z), M_z(z), V_z(z),$ and $p_z'(z)$ assumed in the derivations in Table 11.12 are shown in Figure 11.33c. Figure 11.34 shows the variation of $A_x, B_x, A_m,$ and B_m for various values of $L/T = Z_{max}$. It indicates that, when L/T is greater than about 5, the coefficients do not change, which is true of long piles only.

Table 11.12   Coefficients for Long Piles, $k_z = n_h z$

Z	A_x	A_θ	A_m	A_v	A_p'	B_x	B_θ	B_m	B_v	B_p'
0.0	2.435	−1.623	0.000	1.000	0.000	1.623	−1.750	1.000	0.000	0.000
0.1	2.273	−1.618	0.100	0.989	−0.227	1.453	−1.650	1.000	−0.007	−0.145
0.2	2.112	−1.603	0.198	0.956	−0.422	1.293	−1.550	0.999	−0.028	−0.259
0.3	1.952	−1.578	0.291	0.906	−0.586	1.143	−1.450	0.994	−0.058	−0.343
0.4	1.796	−1.545	0.379	0.840	−0.718	1.003	−1.351	0.987	−0.095	−0.401
0.5	1.644	−1.503	0.459	0.764	−0.822	0.873	−1.253	0.976	−0.137	−0.436
0.6	1.496	−1.454	0.532	0.677	−0.897	0.752	−1.156	0.960	−0.181	−0.451
0.7	1.353	−1.397	0.595	0.585	−0.947	0.642	−1.061	0.939	−0.226	−0.449
0.8	1.216	−1.335	0.649	0.489	−0.973	0.540	−0.968	0.914	−0.270	−0.432
0.9	1.086	−1.268	0.693	0.392	−0.977	0.448	−0.878	0.885	−0.312	−0.403
1.0	0.962	−1.197	0.727	0.295	−0.962	0.364	−0.792	0.852	−0.350	−0.364
1.2	0.738	−1.047	0.767	0.109	−0.885	0.223	−0.629	0.775	−0.414	−0.268
1.4	0.544	−0.893	0.772	−0.056	−0.761	0.112	−0.482	0.688	−0.456	−0.157
1.6	0.381	−0.741	0.746	−0.193	−0.609	0.029	−0.354	0.594	−0.477	−0.047
1.8	0.247	−0.596	0.696	−0.298	−0.445	−0.030	−0.245	0.498	−0.476	0.054
2.0	0.142	−0.464	0.628	−0.371	−0.283	−0.070	−0.155	0.404	−0.456	0.140
3.0	−0.075	−0.040	0.225	−0.349	0.226	−0.089	0.057	0.059	−0.213	0.268
4.0	−0.050	0.052	0.000	−0.106	0.201	−0.028	0.049	−0.042	0.017	0.112
5.0	−0.009	0.025	−0.033	0.015	0.046	0.000	−0.011	−0.026	0.029	−0.002

From *Drilled Pier Foundations*, by R. J. Woodwood, W. S. Gardner, and D. M. Greer. Copyright 1972 by McGraw-Hill. Used with the permission of McGraw-Hill Book Company.

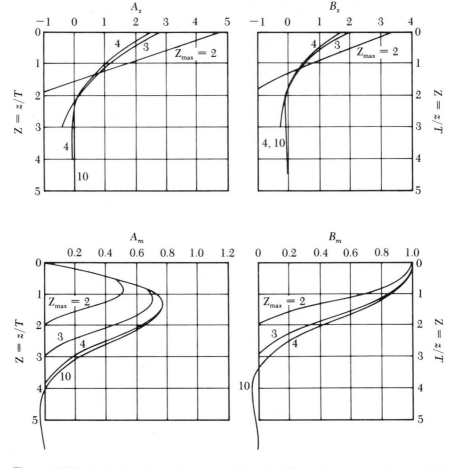

Figure 11.34 Variation of A_x, B_x, A_m, and B_m with Z (after Matlock and Reese, 1960)

Calculating the characteristic length T for the pile requires assuming a proper value of n_h. Table 11.13 gives some representative values.

Elastic solutions similar to those given in Eqs. 11.85–11.89 for piles embedded in *cohesive soil* were developed by Davisson and Gill (1963). Their equations are

and

$$x_z(z) = A_x' \frac{Q_g R^3}{E_p I_p} + B_x' \frac{M_g R^2}{E_p I_p} \tag{11.92}$$

and

$$M_z(z) = A_m' Q_g R + B_m' M_g \tag{11.93}$$

where A_x', B_x, A_m', and B_m' are coefficients

Table 11.13 Representative Values of n_h

Soil	n_k	
	kN/m³	lb/in³
Dry or moist sand		
Loose	1800–2200	6.5–8.0
Medium	5500–7000	20–25
Dense	15,000–18,000	55–65
Submerged sand		
Loose	1000–1400	3.5–5.0
Medium	3500–4500	12–18
Dense	9000–12,000	32–45

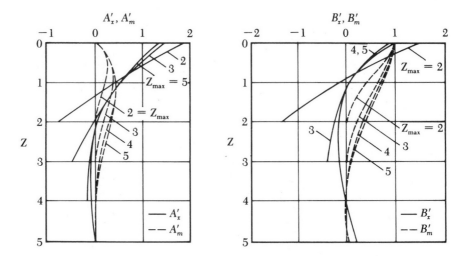

Figure 11.35 Variation of A_x, B'_x, A'_m, and B'_m with Z (after Davisson and Gill, 1963)

and

$$R = \sqrt[4]{\frac{E_p I_p}{k}} \qquad (11.94)$$

The values of the coefficients A' and B' are given in Figure 11.35. Note that

$$Z = \frac{z}{R} \qquad (11.95)$$

and

$$Z_{\max} = \frac{L}{R} \qquad (11.96)$$

The use of Eqs. (11.92) and (11.93) requires knowing the magnitude of the characteristic length, R. This can be calculated from Eq. (11.94), provided that the coefficient of the subgrade reaction is known. For sands, the coefficient of the subgrade reaction was given by Eq. (11.81), which showed a linear variation with depth. However, in cohesive soils, the subgrade reaction may be assumed to be approximately constant with depth. Vesic (1961) proposed the following equation to estimate the value of k:

$$k = 0.65 \sqrt[12]{\frac{E_s D^4}{E_p I_p}} \frac{E_s}{1 - \mu_s^2}$$ (11.97)

Here, E_s = modulus of elasticity of soil
D = pile width (or diameter)
μ_s = Poisson's ratio for the soil

Ultimate Load Analysis: Broms's Method

For laterally loaded piles, Broms (1965) developed a simplified solution based on the assumptions of (a) shear failure in soil, which is the case for short piles, and (b) bending of the pile, governed by the plastic yield resistance of the pile section, which is applicable to long piles. Broms's solution for calculating the ultimate load resistance, $Q_{u(g)}$, for *short piles* is given in Figure 11.36a. A similar solution

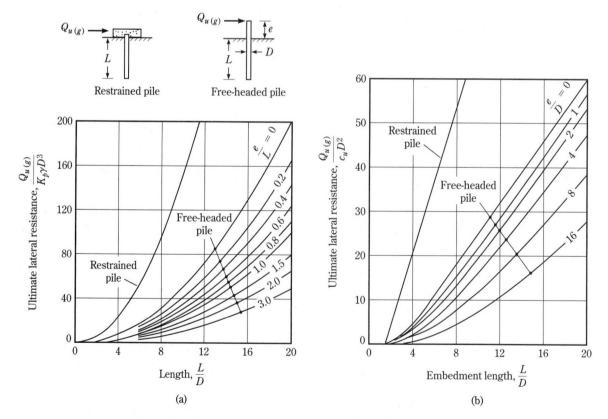

Figure 11.36 Broms's solution for ultimate lateral resistance of short piles (a) in sand and (b) in clay

for piles embedded in cohesive soil is shown in Figure 11.36b. In Figure 11.36a, note that

$$K_p = \text{Rankine passive earth pressure coefficient} = \tan^2\left(45 + \frac{\phi'}{2}\right) \quad (11.98)$$

Similarly, in Figure 11.36b,

$$c_u = \text{undrained cohesion} \approx \frac{0.75q_u}{\text{FS}} = \frac{0.75q_u}{2} = 0.375q_u \quad (11.99)$$

where   FS = factor of safety (=2)
   q_u = unconfined compression strength

Figure 11.37 shows Broms's analysis of long piles. In the figure, the yield moment for the pile is

$$M_y = SF_Y \quad (11.100)$$

where S = section modulus of the pile section
   F_Y = yield stress of the pile material

In solving a given problem, both cases (i.e., Figure 11.36 and Figure 11.37) should be checked.

The deflection of the pile head, $x_z(z = 0)$, under working load conditions can be estimated from Figure 11.38. In Figure 11.38a, the term η can be expressed as

$$\eta = \sqrt[5]{\frac{n_h}{E_p I_p}} \quad (11.101)$$

The range of n_h for granular soil is given in Table 11.13. Similarly, in Figure 11.38b, which is for clay, the term K is the horizontal soil modulus and can be defined as

$$K = \frac{\text{pressure (kN/m}^2 \text{ or lb/in}^2)}{\text{displacement (m or in.)}} \quad (11.102)$$

Also, the term β can be defined as

$$\beta = \sqrt[4]{\frac{KD}{4E_p I_p}} \quad (11.103)$$

Note that, in Figure 11.38, Q_g is the working load.

Ultimate Load Analysis: Meyerhof's Method

In 1995, Meyerhof offered solutions for laterally loaded rigid and flexible piles. (See Figure 11.39a.) According to Meyerhof's method, a pile can be defined as flexible if

$$K_r = \text{relative stiffness of pile} = \frac{E_p I_p}{E_s L^4} < 0.01 \quad (11.104)$$

where E_s = average horizontal soil modulus of elasticity

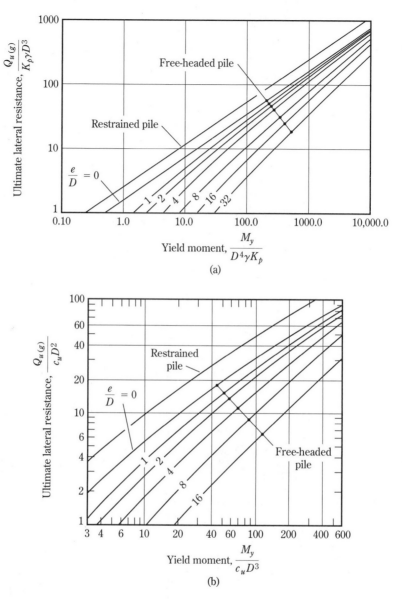

Figure 11.37 Broms's solution for ultimate lateral resistance of long piles (a) in sand and (b) in clay

Piles in Sand   For short (rigid) piles in *sand,* the ultimate load resistance can be given as

$$Q_{u(g)} = 0.12\gamma DL^2 K_{\mathrm{br}} \leqslant 0.4 p_l DL \tag{11.105}$$

where γ = unit weight of soil

K_{br} = resultant net soil pressure coefficient (Figure 11.39b)

p_l = limit pressure obtained from pressuremeter tests (see Chapter 2)

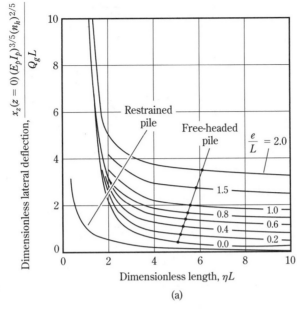

(a)

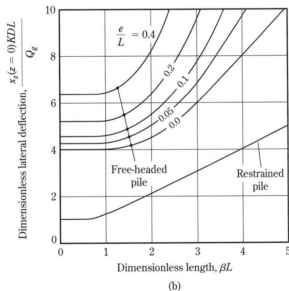

(b)

Figure 11.38 Broms's solution for estimating deflection of pile head (a) in sand and (b) in clay

The limit pressure can be given as

$$p_l = 0.4 p_a N_q \tan \phi' \text{ (for a Menard pressuremeter)} \qquad (11.106)$$

and

$$p_l = 0.6 p_a N_q \tan \phi' \qquad (11.107)$$

(for self-boring and full-displacement pressuremeters)

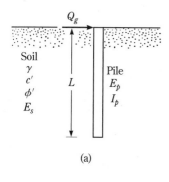

(a)

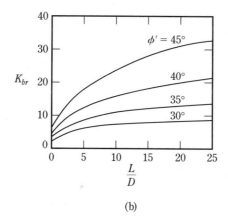

$\dfrac{L}{D}$

(b)

Figure 11.39 (a) Pile with lateral loading at ground level; (b) Variation of resultant net soil pressure coefficient K_{br}

where N_q = bearing capacity factor (see Table 3.4)

p_a = atmospheric pressure (≈ 100 kN/m^2 or 2000 lb/ft^2)

The maximum moment in the pile due to the lateral load, $Q_{u(g)}$, is

$$M_{\max} = 0.35Q_{u(g)}L \leq M_y$$
$$\uparrow$$
$$\text{Eq. (11.100)}$$
(11.108)

For long (flexible) piles in sand, the ultimate lateral load, $Q_{u(g)}$, can be estimated from Eq. (11.105) by substituting an effective length L_e for L, where

$$\frac{L_e}{L} = 1.65K_r^{0.12} \leq 1$$
(11.109)

The maximum moment in a flexible pile due to a working lateral load Q_g applied at the ground surface is

$$M_{\max} = 0.3K_r^{0.2}Q_gL \leq 0.3Q_gL \tag{11.110}$$

Piles in Clay The ultimate lateral load applied at the ground surface for short (rigid) piles embedded in clay can be given as

$$Q_{u(g)} = 0.4c_uK_{cr}DL \leq 0.4p_lDL \tag{11.111}$$

where p_l = limit pressure from pressuremeter test

K_{cr} = net soil pressure coefficient (see Figure 11.40)

The limit pressure in clay is

$$p_l \approx 6c_u \qquad \text{(for a Menard pressuremeter)} \tag{11.112}$$

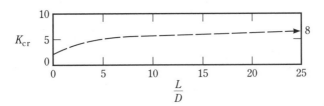

Figure 11.40 Variation of K_{cr}

and

$$p_l \approx 8c_u \quad \text{(for self-boring and full-displacement pressuremeters)} \quad (11.113)$$

The maximum bending moment in the pile due to $Q_{u(g)}$ is

$$M_{max} = 0.22Q_{u(g)}L \leqslant M_y \atop \underset{\text{Eq. (11.100)}}{\uparrow} \tag{11.114}$$

For long (flexible) piles, Eq. (11.111) can be used to estimate $Q_{u(g)}$ by substituting the effective length L_e in place of L, where

$$\frac{L_e}{L} = 1.5K_r^{0.12} \leqslant 1 \tag{11.115}$$

The maximum moment in a flexible pile due to a working lateral load Q_g applied at the ground surface is

$$M_{max} = 0.3K_r^{0.2}Q_gL \leqslant 0.15Q_gL \tag{11.116}$$

Example 11.8

Consider a steel H-pile (HP 250 × 85) 25 m long, embedded fully in a granular soil. Assume that $n_h = 12{,}000 \text{ kN/m}^3$. The allowable displacement at the top of the pile is 8 mm. Determine the allowable lateral load, Q_g. Let $M_g = 0$. Use the elastic solution.

Solution

From Table 11.1a, for an HP 250 × 85 pile,

$$I_p = 123 \times 10^{-6} \text{ m}^4 \quad \text{(about the strong axis)}$$

and let

$$E_p = 207 \times 10^6 \, \text{kN/m}^2$$

From Eq. (11.90),

$$T = \sqrt[5]{\frac{E_p I_p}{n_h}} = \sqrt[5]{\frac{(207 \times 10^6)(123 \times 10^{-6})}{12,000}} = 1.16 \, \text{m}$$

Here, $L/T = 25/1.16 = 21.55 > 5$, so the pile is a long one. Because $M_g = 0$, Eq. (11.85) takes the form

$$x_z(z) = A_x \frac{Q_g T^3}{E_p I_p}$$

and it follows that

$$Q_g = \frac{x_z(z) E_p I_p}{A_x T^3}$$

At $z = 0$, $x_z = 8 \, \text{mm} = 0.008 \, \text{m}$ and $A_x = 2.435$ (see Table 11.12), so

$$Q_g = \frac{(0.008)(207 \times 10^6)(123 \times 10^{-6})}{(2.435)(1.16^3)} = 53.59 \, \text{kN}$$

This magnitude of Q_g is based on the *limiting displacement condition only*. However, the magnitude of Q_g based on the *moment capacity* of the pile also needs to be determined. For $M_g = 0$, Eq. (11.87) becomes

$$M_z(z) = A_m Q_g T$$

According to Table 11.12, the maximum value of A_m at any depth is 0.772. The maximum allowable moment that the pile can carry is

$$M_{z(\text{max})} = F_Y \frac{I_p}{\dfrac{d_1}{2}}$$

Let $F_Y = 248,000 \, \text{kN/m}^2$. From Table 11.1a, $I_p = 123 \times 10^{-6} \, \text{m}^4$ and $d_1 = 0.254 \, \text{m}$, so

$$\frac{I_p}{\left(\dfrac{d_1}{2}\right)} = \frac{123 \times 10^{-6}}{\left(\dfrac{0.254}{2}\right)} = 968.5 \times 10^{-6} \, \text{m}^3$$

Now,

$$Q_g = \frac{M_{z(\text{max})}}{A_m T} = \frac{(968.5 \times 10^{-6})(248.000)}{(0.772)(1.16)} = 268.2 \, \text{kN}$$

Because $Q_g = 268.2 \, \text{kN} > 53.59 \, \text{kN}$, the deflection criteria apply. Hence, $Q_g =$ **53.59 kN.**

Example 11.9

Solve Example 11.8 by Broms's method. Assume that the pile is flexible and is free headed. Let the yield stress of the pile material, $F_y = 248$ MN/m^2; the unit weight of soil, $\gamma = 18$ kN/m^3; and the soil friction angle $\phi' = 35°$.

Solution
We check for bending failure. From Eq. (11.100),

$$M_y = SF_y$$

From Table 11.1a,

$$S = \frac{I_p}{\dfrac{d_1}{2}} = \frac{123 \times 10^{-6}}{\dfrac{0.254}{2}}$$

Also,

$$M_y = \left[\frac{123 \times 10^{-6}}{\dfrac{0.254}{2}} \right] (248 \times 10^3) = 240.2 \text{ kN-m}$$

and

$$\frac{M_y}{D^4 \gamma K_p} = \frac{M_y}{D^4 \gamma \tan^2\left(45 + \dfrac{\phi'}{2}\right)} = \frac{240.2}{(0.254)^4 (18) \tan^2\left(45 + \dfrac{35}{2}\right)} = 868.8$$

From Figure 11.37a, for $M_y/D^4 \gamma K_p = 868.8$, the magnitude of $Q_{u(g)}/K_p D^3 \gamma$ (for a free-headed pile with $e/D = 0$) is about 140, so

$$Q_{u(g)} = 140 K_p D^3 \gamma = 140 \tan^2\left(45 + \frac{35}{2}\right)(0.254)^3(18) = 152.4 \text{ kN}$$

Next, we check for pile head deflection. From Eq. (11.101),

$$\eta = \sqrt[5]{\frac{n_h}{E_p I_p}} = \sqrt[5]{\frac{12,000}{(207 \times 10^6)(123 \times 10^{-6})}} = 0.86 \text{ m}^{-1}$$

so

$$\eta L = (0.86)(25) = 21.5$$

From Figure 11.38a, for $\eta L = 21.5$, $e/L = 0$ (free-headed pile): thus,

$$\frac{x_o (E_p I_p)^{3/5} (n_h)^{2/5}}{Q_g L} \approx 0.15 \qquad \text{(by interpolation)}$$

and

$$Q_g = \frac{x_o(E_pI_p)^{3/5}(n_h)^{2/5}}{0.15L}$$

$$= \frac{(0.008)[(207 \times 10^6)(123 \times 10^{-6})]^{3/5}(12,000)^{2/5}}{(0.15)(25)} = 40.2 \text{ kN}$$

Hence, $Q_g =$ **40.2 kN (<152.4 kN).** ∎

11.20 *Pile-Driving Formulas*

To develop the desired load-carrying capacity, a point bearing pile must penetrate the dense soil layer sufficiently or have sufficient contact with a layer of rock. This requirement cannot always be satisfied by driving a pile to a predetermined depth, because soil profiles vary. For that reason, several equations have been developed to calculate the ultimate capacity of a pile during driving. These dynamic equations are widely used in the field to determine whether a pile has reached a satisfactory bearing value at the predetermined depth. One of the earliest such equations—commonly referred to as the *Engineering News (EN) Record formula*—is derived from the work—energy theory. That is,

Energy imparted by the hammer per blow =
(pile resistance)(penetration per hammer blow)

According to the EN formula, the pile resistance is the ultimate load Q_u, expressed as

$$Q_u = \frac{W_Rh}{S + C} \tag{11.117}$$

where $W_R =$ weight of the ram
$h =$ height of fall of the ram
$S =$ penetration of pile per hammer blow
$C =$ a constant

The pile penetration, S, is usually based on the average value obtained from the last few driving blows. In the equation's original form, the following values of C were recommended:

For drop hammers,

$$C = \begin{cases} 25.4 \text{ mm if } S \text{ and } h \text{ are in mm} \\ 1 \text{ in. if } S \text{ and } h \text{ are in inches} \end{cases}$$

For steam hammers,

$$C = \begin{cases} 2.54 \text{ mm if } S \text{ and } h \text{ are in mm} \\ 0.1 \text{ in. if } S \text{ and } h \text{ are in inches} \end{cases}$$

Also, a factor of safety FS = 6 was recommended for estimating the allowable pile capacity. Note that, for single- and double-acting hammers, the term $W_R h$ can be replaced by EH_E, where E is the efficiency of the hammer and H_E is the rated energy of the hammer. Thus,

$$Q_u = \frac{EH_E}{S + C} \qquad (11.118)$$

The EN formula has been revised several times over the years, and other pile-driving formulas also have been suggested. Some of them are tabulated in Table 11.14.

Table 11.14 Pile-Driving Formulas

Name	Formula
Modified EN formula	$Q_u = \dfrac{EW_R h}{S + C} \dfrac{W_R + n^2 W_p}{W_R + W_p}$ where $\quad E$ = efficiency of hammer $\qquad C$ = 2.54 mm if the units of S and h are in mm $\qquad C$ = 0.1 in. if the units of S and h are in in. $\qquad W_p$ = weight of the pile $\qquad n$ = coefficient of restitution between the ram and the pile cap

Typical values for E

Single- and double-acting hammers	0.7–0.85
Diesel hammers	0.8–0.9
Drop hammers	0.7–0.9

Typical values for n

Cast-iron hammer and concrete piles (without cap)	0.4–0.5
Wood cushion on steel piles	0.3–0.4
Wooden piles	0.25–0.3

Name	Formula
Michigan State Highway Commission formula (1965)	$Q_u = \dfrac{1.25 EH_E}{S + C} \dfrac{W_R + n^2 W_p}{W_R + W_p}$ where $\quad H_E$ = manufacturer's maximum rated hammer energy (lb-in.) $\qquad E$ = efficiency of hammer $\qquad C$ = 0.1 in. A factor of safety of 6 is recommended.
Danish formula (Olson and Flaate, 1967)	$Q_u = \dfrac{EH_E}{S + \sqrt{\dfrac{EH_E L}{2 A_p E_p}}}$

Table 11.14   (Continued)

Name	Formula
	where E = efficiency of hammer
	H_E = rated hammer energy
	E_p = modulud of elasticity of the pile material
	L = length of the pile
	A_p = cross-sectional area of the pile

Pacific Coast Uniform Building Code formula (International Conference of Building Officials, 1982)

$$Q_u = \frac{(EH_E)\left(\dfrac{W_R + nW_p}{W_R + W_p}\right)}{S + \dfrac{Q_uL}{AE_p}}$$

The value of n should be 0.25 for steel piles and 0.1 for all other piles. A factor of safety of 4 is generally recommended.

Janbu's formula (Janbu, 1953)

$$Q_u = \frac{EH_E}{K_u'S}$$

where

$$K_u' = C_d\left(1 + \sqrt{1 + \frac{\lambda'}{C_d}}\right)$$

$$C_d = 0.75 + 0.14\left(\frac{W_p}{W_R}\right)$$

$$\lambda' = \left(\frac{EH_EL}{A_pE_pS^2}\right)$$

Gates's formula (Gates, 1957)

$Q_u = a\sqrt{EH_E}(b - \log S)$
If Q_u is in kips, then S is in in., $a = 27$, $b = 1$, and H_E is in kip-ft.
If Q_u is in kN, then S is in mm, $a = 104.5$, $b = 2.4$, and H_E is in kN-m.
$E = 0.75$ for drop hammer; $E = 0.85$ for all other hammers
Use a factor of safety of 3.

Navy-McKay formula

$$Q_u = \frac{EH_E}{S\left(1 + 0.3\dfrac{W_P}{W_R}\right)}$$

Use a factor of safety of 6.

Example 11.10

A precast concrete pile 12 in. × 12 in. in cross section is driven by a hammer. The maximum rated hammer energy = 26 kip-ft, the weight of the ram = 8 kip, the total length of the pile = 65 ft, the hammer efficiency = 0.8, the coefficient of restitution = 0.45, the weight of the pile cap = 0.72 kip, and the number of blows for the last 1 in. of penetration = 5. Estimate the allowable pile capacity by using

a. the EN formula [Eq. (11.118)] with FS = 6.
b. the Modified EN formula from Table 11.14 with FS = 4.
c. the Danish formula from Table 11.14 with FS = 3.

Solution
Part a
From Eq. (11.118),

$$Q_u = \frac{EH_E}{S + C}$$

For

$$E = 0.8, H_E = 26 \text{ kip-ft}$$

and

$$S = \frac{1}{5} = 0.2 \text{ in.}$$

we obtain

$$Q_u = \frac{(0.8)\,\overbrace{(26)\,(12)}^{\text{kip-in.}}}{0.2 + 0.1} = 832 \text{ kip}$$

Hence,

$$Q_{\text{all}} = \frac{Q_u}{\text{FS}} = \frac{832}{6} = \textbf{138.7 kip}$$

Part b
According to the modified EN formula,

$$Q_u = \frac{EW_R h}{S + C} \frac{W_R + n^2 W_p}{W_R + W_p}$$

Now,

$$\text{Weight of piles} = W A_p \gamma_c = (65 \text{ ft})\,(1 \text{ ft} \times 1 \text{ ft})\,(150 \text{ lb/ft}^3)$$
$$= 9750 \text{ lb} = 9.75 \text{ kip}$$

Also,

$$W_p = \text{weight of pile} + \text{weight of cap}$$
$$= 9.75 + 0.72 = 10.47 \text{ kip}$$

So

$$Q_u = \left[\frac{(0.8)(26)(12)}{0.2 + 0.1} \right] \left[\frac{8 + (0.45)^2(10.47)}{8 + 10.47} \right]$$

$$= (832)(0.548) = 455.9 \text{ kip}$$

and

$$Q_{all} = \frac{Q_u}{FS} = \frac{455.9}{4} = \textbf{114 kip}$$

Part c

The Danish formula is

$$Q_u = \frac{EH_E}{S + \sqrt{\dfrac{EH_E L}{2A_p E_p}}}$$

Here, $E_p = 3 \times 10^6 \text{ lb/in}^2$, so

$$\sqrt{\frac{EH_E L}{2A_p E_p}} = \sqrt{\frac{(0.8)(26 \times 12)(65 \times 12)}{(2)(12 \times 12)\left(\dfrac{3 \times 10^6}{1000}\right)}} = 0.475 \text{ in.}$$

$$\underset{\text{kip/in}^2}{\uparrow}$$

Hence,

$$Q_u = \frac{(0.8)(26)(12)}{0.2 + 0.475} = 369.8 \text{ kip}$$

$$Q_{all} = \frac{Q_u}{FS} = \frac{369.8}{3} = \textbf{123.3 kip} \qquad \blacksquare$$

11.21   *Stress on Piles During Driving*

The maximum stress developed on a pile during the driving operation can be estimated from the pile-driving formulas presented in Table 11.14. To illustrate, we use the modified EN formula:

$$Q_u = \frac{EW_R h}{S + C} \frac{W_R + n^2 W_p}{W_R + W_p}$$

In this equation, S is the average penetration per hammer blow, which can also be expressed as

$$S = \frac{1}{N} \tag{11.119}$$

where S is in inches

   N = number of hammer blows per inch of penetration

Thus,

$$Q_u = \frac{EW_Rh}{(1/N) + 0.1} \frac{W_R + n^2W_p}{W_R + W_p} \tag{11.120}$$

   Different values of N may be assumed for a given hammer and pile, and Q_u may be calculated. The driving stress Q_u/A_p can then be calculated for each value of N. This procedure can be demonstrated with a set of numerical values. Suppose that a prestressed concrete pile 80 ft in length has to be driven by a hammer. The pile sides measure 10 in. From Table 11.3b, for this pile,

$$A_p = 100 \text{ in}^2$$

The weight of the pile is

$$A_pL\gamma_c = \left(\frac{100 \text{ in}^2}{144}\right)(80 \text{ ft})(150 \text{ lb/ft}^3) = 8.33 \text{ kip}$$

If the weight of the cap is 0.67 kip, then

$$W_p = 8.33 + 0.67 = 9 \text{ kip}$$

For the hammer, let

$$\text{Rated energy} = 19.2 \text{ kip-ft} = H_E = W_Rh$$

$$\text{Weight of ram} = 5 \text{ kip}$$

Assume that the hammer efficiency is 0.85 and that $n = 0.35$. Substituting these values into Eq. (11.120) yields

$$Q_u = \left[\frac{(0.85)(19.2 \times 12)}{\dfrac{1}{N} + 0.1}\right]\left[\frac{5 + (0.35)^2(9)}{5 + 9}\right] = \frac{85.37}{\dfrac{1}{N} + 0.1} \text{ kip}$$

Now the following table can be prepared:

N	Q_u (kip)	A_p (in²)	Q_u/A_p (kip/in²)
0	0	100	0
2	142.3	100	1.42
4	243.9	100	2.44
6	320.1	100	3.20
8	379.4	100	3.79
10	426.9	100	4.27
12	465.7	100	4.66
20	569.1	100	5.69

   Both the number of hammer blows per inch and the stress can be plotted in a graph, as shown in Figure 11.41. If such a curve is prepared, the number of blows per

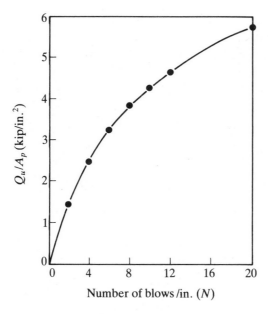

Figure 11.41 Plot of stress vs. blows/in.

inch of pile penetration corresponding to the allowable pile-driving stress can easily be determined.

Actual driving stresses in wooden piles are limited to about $0.7f_u$. Similarly, for concrete and steel piles, driving stresses are limited to about $0.6f'_c$ and $0.85f_y$, respectively.

In most cases, wooden piles are driven with a hammer energy of less than 60 kN-m ($\approx$45 kip-ft). Driving resistances are limited mostly to 4–5 blows per inch of pile penetration. For concrete and steel piles, the usual values of N are 6–8 and 12–14, respectively.

11.22 Pile Capacity For Vibration-Driven Piles

The principles of vibratory pile drivers (Figure 11.7e) were discussed briefly in Section 11.4. As mentioned there, the driver essentially consists of two counterrotating weights. The amplitude of the centrifugal driving force generated by a vibratory hammer can be given as

$$F_c = me\omega^2 \tag{11.121}$$

where m = total eccentric rotating mass

e = distance between the center of each rotating mass and the center of rotation

ω = operating circular frequency

Vibratory hammers typically include an isolated *bias weight* that can range from 4 to 40 kN. The bias weight is isolated from oscillation by springs, so it acts as a net downward load helping the driving efficiency by increasing the penetration rate of the pile.

The use of vibratory pile drivers began in the early 1930s. Installing piles with vibratory drivers produces less noise and damage to the pile, compared with impact driving. However, because of a limited understanding of the relationships between the load, the rate of penetration, and the bearing capacity of piles, this method has not gained popularity in the United States.

Vibratory pile drivers are patented. Some examples are the Bodine Resonant Driver (BRD), the Vibro Driver of the McKiernan-Terry Corporation, and the Vibro Driver of the L. B. Foster Company. Davisson (1970) provided a relationship for estimating the ultimate pile capacity in granular soil:

In SI units,

$$Q_u(\text{kN}) = \frac{0.746(H_p) + 98(v_p \, \text{m/s})}{(v_p \, \text{m/s}) + (S_L \, \text{m/cycle})(f \, \text{Hz})} \tag{11.122}$$

In English units,

$$Q_u(\text{lb}) = \frac{550(H_p) + 22{,}000(v_p \, \text{ft/s})}{(v_p \, \text{ft/s}) + (S_L \, \text{ft/cycle})(f \, \text{Hz})} \tag{11.123}$$

where H_p = horsepower delivered to the pile
v_p = final rate of pile penetration
S_L = loss factor
f = frequency, in Hz

The loss factor S_L for various types of granular soils is as follows (Bowles, 1996):

Closed-End Pipe Piles

- Loose sand: 0.244×10^{-3} m/cycle (0.0008 ft/cycle)
- Medium dense sand: 0.762×10^{-3} m/cycle (0.0025 ft/cycle)
- Dense sand: 2.438×10^{-3} m/cycle (0.008 ft/cycle)

H-Piles

- Loose sand: -0.213×10^{-3} m/cycle (-0.0007 ft/cycle)
- Medium dense sand: 0.762×10^{-3} m/cycle (0.0025 ft/cycle)
- Dense sand: 2.134×10^{-3} m/cycle (0.007 ft/cycle)

In 2000, Feng and Deschamps provided the following relationship for the ultimate capacity of vibrodriven piles in granular soil:

$$Q_u = \frac{3.6(F_c + 11W_B)}{1 + 1.8 \times 10^{10} \dfrac{v_p}{c} \sqrt{\text{OCR}}} \frac{L_E}{L} \tag{11.124}$$

Here, F_c = centrifugal force
W_B = bias weight
v_p = final rate of pile penetration
c = speed of light $[1.8 \times 10^{10}$ m/min $(5.91 \times 10^{10}$ ft/min$)]$
OCR = overconsolidation ratio
L_E = embedded length of pile
L = pile length

Example 11.11

Consider a 20-m-long steel pile driven by a Bodine Resonant Driver (Section HP 310×125) in a medium dense sand. If $H_p = 350$ horsepower, $v_p = 0.0016$ m/s, and $f = 115$ Hz, calculate the ultimate pile capacity, Q_u.

Solution
From Eq. (11.122),

$$Q_u = \frac{0.746 H_p + 98 v_p}{v_p + S_L f}$$

For an HP pile in medium dense sand, $S_L \approx 0.762 \times 10^{-3}$ m/cycle. So
$$Q_u = \frac{(0.746)(350) + (98)(0.0016)}{0.0016 + (0.762 \times 10^{-3})(115)} = \textbf{2928 kN}$$ ∎

11.23 *Negative Skin Friction*

Negative skin friction is a downward drag force exerted on a pile by the soil surrounding it. Such a force can exist under the following conditions, among others:

1. If a fill of clay soil is placed over a granular soil layer into which a pile is driven, the fill will gradually consolidate. The consolidation process will exert a downward drag force on the pile (see Figure 11.42a) during the period of consolidation.
2. If a fill of granular soil is placed over a layer of soft clay, as shown in Figure 11.42b, it will induce the process of consolidation in the clay layer and thus exert a downward drag on the pile.

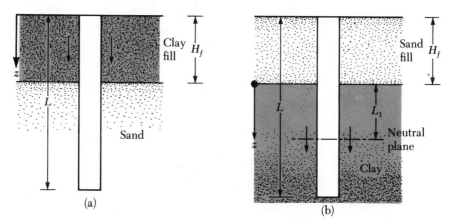

Figure 11.42 Negative skin friction

3. Lowering of the water table will increase the vertical effective stress on the soil at any depth, which will induce consolidation settlement in clay. If a pile is located in the clay layer, it will be subjected to a downward drag force.

In some cases, the downward drag force may be excessive and cause foundation failure. This section outlines two tentative methods for the calculation of negative skin friction.

Clay Fill over Granular Soil (Figure 11.42a)

Similar to the β method presented in Section 11.13, the negative (downward) skin stress on the pile is

$$f_n = K'\sigma_o' \tan \delta \tag{11.125}$$

where K' = earth pressure coefficient = $K_o = 1 - \sin \phi'$
   σ_o' = vertical effective stress at any depth $z = \gamma_f' z$
   γ_f' = effective unit weight of fill
   δ = soil–pile friction angle ≈ 0.5–$0.7\phi'$

Hence, the total downward drag force on a pile is

$$Q_n = \int_0^{H_f} (pK'\gamma_f' \tan \delta) z \, dz = \frac{pK'\gamma_f'H_f^2 \tan \delta}{2} \tag{11.126}$$

where H_f = height of the fill

If the fill is above the water table, the effective unit weight, γ_f', should be replaced by the moist unit weight.

Granular Soil Fill over Clay (Figure 11.42b)

In this case, the evidence indicates that the negative skin stress on the pile may exist from $z = 0$ to $z = L_1$, which is referred to as the *neutral depth*. (See Vesic, 1977, pp. 25–26.) The neutral depth may be given as (Bowles, 1982)

$$L_1 = \frac{(L - H_f)}{L_1} \left[\frac{L - H_f}{2} + \frac{\gamma_f'H_f}{\gamma'} \right] - \frac{2\gamma_f'H_f}{\gamma'} \tag{11.127}$$

where γ_f' and γ' = effective unit weights of the fill and the underlying clay layer, respectively

For end-bearing piles, the neutral depth may be assumed to be located at the pile tip (i.e., $L_1 = L - H_f$).

Once the value of L_1 is determined, the downward drag force is obtained in the following manner: The unit negative skin friction at any depth from $z = 0$ to $z = L_1$ is

$$f_n = K'\sigma_o' \tan \delta \tag{11.128}$$

where $K' = K_o = 1 - \sin \phi'$
$\sigma'_o = \gamma'_f H_f + \gamma' z$
$\delta = 0.5\text{–}0.7\phi'$

$$Q_n = \int_0^{L_1} pf_n \, dz = \int_0^{L_1} pK'(\gamma'_f H_f + \gamma' z) \tan \delta \, dz$$
$$= (pK'\gamma'_f H_f \tan \delta)L_1 + \frac{L_1^2 pK'\gamma' \tan \delta}{2} \tag{11.129}$$

If the soil and the fill are above the water table, the effective unit weights should be replaced by moist unit weights. In some cases, the piles can be coated with bitumen in the downdrag zone to avoid this problem.

A limited number of case studies of negative skin friction is available in the literature. Bjerrum et al. (1969) reported monitoring the downdrag force on a test pile at Sorenga in the harbor of Oslo, Norway (noted as pile G in the original paper). The study of Bjerrum et al. (1969) was also discussed by Wong and Teh (1995) in terms of the pile being driven to bedrock at 40 m. Figure 11.43a shows the soil profile and the pile. Wong and Teh estimated the following quantities:

– *Fill:* Moist unit weight, $\gamma_f = 16 \text{ kN/m}^3$
    Saturated unit weight, $\gamma_{\text{sat}(f)} = 18.5 \text{ kN/m}^3$
So

$$\gamma'_f = 18.5 - 9.81 = 8.69 \text{ kN/m}^3$$

and

$$H_f = 13 \text{ m}$$

– *Clay:* $K' \tan \delta \approx 0.22$
    Saturated effective unit weight, $\gamma' = 19 - 9.81 = 9.19 \text{ kN/m}^3$

– *Pile:* $L = 40 \text{ m}$
    Diameter, $D = 500 \text{ m}$

Thus, the maximum downdrag force on the pile can be estimated from Eq. (11.129). Since in this case the pile is a point bearing pile, the magnitude of $L_1 = 27$ m, and

$$Q_n = (p)(K' \tan \delta)[\gamma_f \times 2 + (13 - 2)\gamma'_f](L_1) + \frac{L_1^2 p\gamma'(K' \tan \delta)}{2}$$

or

$$Q_n = (\pi \times 0.5)(0.22)[(16 \times 2) + (8.69 \times 11)](27) + \frac{(27)^2(\pi \times 0.5)(9.19)(0.22)}{2}$$

$$= 2348 \text{ kN}$$

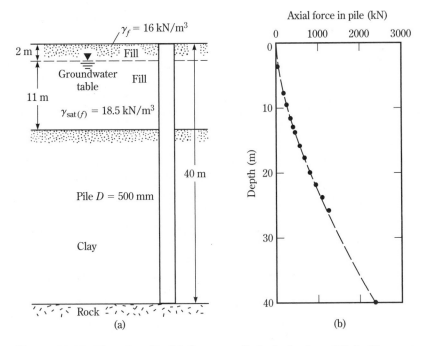

Figure 11.43 Negative skin friction on a pile in the harbor of Oslo, Norway [based on Bjerrum et al. (1969) and Wong and Teh (1995)]

The measured value of the maximum Q_n was about 2500 kN (Figure 11.43b), which is in good agreement with the calculated value.

Example 11.12

In Figure 11.42a, let $H_f = 2$ m. The pile is circular in cross section with a diameter of 0.305 m. For the fill that is above the water table, $\gamma_f = 16 \text{ kN/m}^3$ and $\phi' = 32°$. Determine the total drag force. Use $\delta = 0.6 \, \phi'$.

Solution
From Eq. (11.126),

$$Q_n = \frac{pK'\gamma_f H_f^2 \tan \delta}{2}$$

with

$$p = \pi(0.305) = 0.958 \text{ m}$$
$$K' = 1 - \sin \phi' = 1 - \sin 32 = 0.47$$

and

$$\delta = (0.6)(32) = 19.2°$$

Thus,

$$Q_n = \frac{(0.958)(0.47)(16)(2)^2 \tan 19.2}{2} = \textbf{5.02 kN}$$ ∎

Example 11.13

In Figure 11.42b, let $H_f = 2$ m, pile diameter $= 0.305$ m, $\gamma_f = 16.5$ kN/m^3, $\phi'_{clay} = 34°$, $\gamma_{sat(clay)} = 17.2$ kN/m^3, and $L = 20$ m. The water table coincides with the top of the clay layer. Determine the downward drag force. Assume that $\delta = 0.6\phi'_{clay}$.

Solution
The depth of the neutral plane is given in Eq. (11.127) as

$$L_1 = \frac{L - H_f}{L_1}\left(\frac{L - H_f}{2} + \frac{\gamma_f H_f}{\gamma'}\right) - \frac{2\gamma_f H_f}{\gamma'}$$

Note that γ'_f in Eq. (11.127) has been replaced by γ_f because the fill is above the water table, so

$$L_1 = \frac{(20 - 2)}{L_1}\left[\frac{(20 - 2)}{2} + \frac{(16.5)(2)}{(17.2 - 9.81)}\right] - \frac{(2)(16.5)(2)}{(17.2 - 9.81)}$$

or

$$L_1 = \frac{242.4}{L_1} - 8.93;\ L_1 = 11.75 \text{ m}$$

Now, from Eq. (11.129), we have

$$Q_n = (pK'\gamma_f H_f \tan \delta)L_1 + \frac{L_1^2 pK'\gamma' \tan \delta}{2}$$

with

$$p = \pi(0.305) = 0.958 \text{ m}$$

and

$$K' = 1 - \sin 34° = 0.44$$

Hence,

$$Q_n = (0.958)(0.44)(16.5)(2)[\tan(0.6 \times 34)](11.75)$$
$$+ \frac{(11.75)^2(0.958)(0.44)(17.2 - 9.81)[\tan(0.6 \times 34)]}{2}$$
$$= 60.78 + 79.97 = \textbf{140.75 kN}$$ ∎

Group Piles

11.24   *Group Efficiency*

In most cases, piles are used in groups, as shown in Figure 11.44, to transmit the structural load to the soil. A *pile cap* is constructed over *group piles*. The cap can be in contact with the ground, as in most cases (see Figure 11.44a), or well above the ground, as in the case of offshore platforms (see Figure 11.44b).

Determining the load-bearing capacity of group piles is extremely complicated and has not yet been fully resolved. When the piles are placed close to each other, a reasonable assumption is that the stresses transmitted by the piles to the soil will overlap (see Figure 11.44c), reducing the load-bearing capacity of the piles. Ideally, the piles in a group should be spaced so that the load-bearing capacity of the group is not less than the sum of the bearing capacity of the individual piles. In practice, the minimum center-to-center pile spacing, d, is $2.5D$ and, in ordinary situations, is actually about 3–$3.5D$.

The efficiency of the load-bearing capacity of a group pile may be defined as

$$\eta = \frac{Q_{g(u)}}{\Sigma\,Q_u} \tag{11.130}$$

where η = group efficiency
$Q_{g(u)}$ = ultimate load-bearing capacity of the group pile
Q_u = ultimate load-bearing capacity of each pile without the group effect

Many structural engineers use a simplified analysis to obtain the group efficiency for *friction piles*, particularly in sand. This type of analysis can be explained with the aid of Figure 11.44a. Depending on their spacing within the group, the piles may act in one of two ways: (1) as a *block*, with dimensions $L_g \times B_g \times L$, or (2) as *individual piles*. If the piles act as a block, the frictional capacity is $f_{av}p_gL \approx Q_{g(u)}$. [*Note:* p_g = perimeter of the cross section of block = $2(n_1 + n_2 - 2)d + 4D$, and f_{av} = average unit frictional resistance.] Similarly, for each pile acting individually, $Q_u \approx pLf_{av}$. (*Note: p* = perimeter of the cross section of each pile.) Thus,

$$\eta = \frac{Q_{g(u)}}{\Sigma\,Q_u} = \frac{f_{av}[2(n_1 + n_2 - 2)d + 4D]L}{n_1 n_2 pLf_{av}}$$

$$= \frac{2(n_1 + n_2 - 2)d + 4D}{pn_1 n_2} \tag{11.131}$$

Hence,

$$Q_{g(u)} = \left[\frac{2(n_1 + n_2 - 2)d + 4D}{pn_1 n_2}\right]\Sigma\,Q_u \tag{11.132}$$

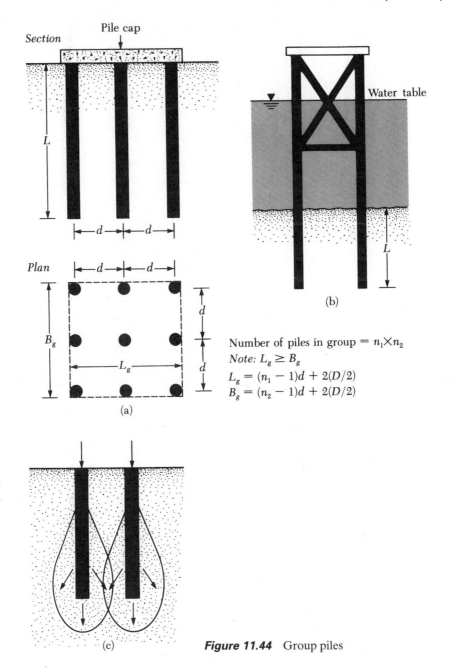

Figure 11.44 Group piles

From Eq. (11.132), if the center-to-center spacing d is large enough, $\eta > 1$. In that case, the piles will behave as individual piles. Thus, in practice, if $\eta < 1$, then

$$Q_{g(u)} = \eta \Sigma \, Q_u$$

and if $\eta \geq 1$, then

$$Q_{g(u)} = \Sigma \, Q_u$$

Table 11.15   Equations for Group Efficiency of Friction Piles

Name	Equation
Converse–Labarre equation	$\eta = 1 - \left[\dfrac{(n_1 - 1)n_2 + (n_2 - 1)n_1}{90 n_1 n_2} \right] \theta$
	where $\theta(\text{deg}) = \tan^{-1}(D/d)$
Los Angeles Group Action equation	$\eta = 1 - \dfrac{D}{\pi d n_1 n_2} [n_1(n_2 - 1)$
	$+ \; n_2(n_1 - 1)] + \sqrt{2}(n_1 - 1)(n_2 - 1)]$
Seiler–Keeney equation (Seiler and Keeney, 1944)	$\eta = \left\{ 1 - \left[\dfrac{11d}{7(d^2 - 1)} \right]\left[\dfrac{n_1 + n_2 - 2}{n_1 + n_2 - 1} \right] \right\} + \dfrac{0.3}{n_1 + n_2}$
	where d is in ft

There are several other equations like Eq. (11.132) for calculating the group efficiency of friction piles. Some of these are given in Table 11.15.

Feld (1943) suggested a method by which the load capacity of individual piles (when only frictional resistance is considered) in a group embedded in sand could be assigned. According to this method, the ultimate capacity of a pile is reduced by one-sixteenth by each adjacent diagonal or row pile. The technique can be explained if one examines Figure 11.45, which shows the plan of a group pile. For pile type *A*, there are eight adjacent piles, for pile type *B*, there are five, and for pile type *C*, there are three. With this in mind, the following table can be prepared:

Pile type	No. of Piles	No. of adjacent piles/pile	Reduction factor for each pile	Ultimate capacity[a]
A	1	8	$1 - \dfrac{8}{16}$	$0.5Q_u$
B	4	5	$1 - \dfrac{5}{16}$	$2.75Q_u$
C	4	3	$1 - \dfrac{3}{16}$	$3.25Q_u$
				$\Sigma\, 6.5Q_u = Q_{g(u)}$

[a](No of piles) (Q_u) (reduction factor)
Q_u = ultimate capacity for an isolated pile

Hence,

$$\eta = \frac{Q_{g(u)}}{\Sigma\, Q_u} = \frac{6.5Q_u}{9Q_u} = 72\%$$

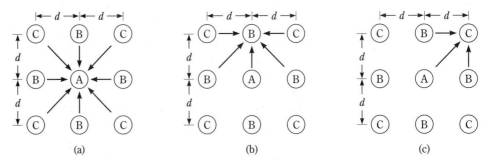

Figure 11.45 Feld's method for estimating the group capacity of friction piles

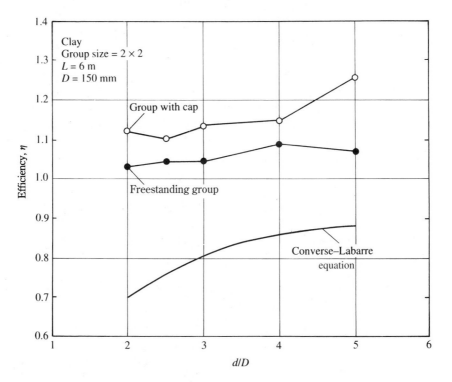

Figure 11.46 Variation of group efficiency with d/D (after Brand et al., 1972)

Figure 11.46 shows a comparison of field-test results in clay with the theoretical group efficiency calculated from the Converse–Labarre equation. (See Table 11.15.) Reported by Brand et al. (1972), the tests were conducted in soil, the details of which are given in Figure 3.7. Other parameters include the following:

– Length of piles = 6 m
– Diameter of piles = 150 mm
– Pile groups tested = 2 × 2
– Location of pile head = 1.5 m below the ground surface

Pile tests were conducted with and without a cap. (In the latter state, the group is a freestanding group.) Note that for $d/D \geq 2$, the magnitude of η was greater than 1.0. Also, for similar values of d/D, the group efficiency was greater with the pile cap than without the cap. Figure 11.47 shows the pile group settlement at various stages of the load test.

Figure 11.48 shows the variation of the group efficiency η for a 3 × 3 group pile in sand (Kishida and Meyerhof, 1965). It can be seen that, for loose and medium sands, the magnitude of the group efficiency is larger than unity. This is due primarily to the densification of sand surrounding the pile.

Liu et al. (1985) reported the results of field tests on 58 pile groups and 23 single piles embedded in granular soil. Included in the report were the following details:

- Pile length, $L = 8D$–$23D$
- Pile diameter, $D = 125$ mm–330 mm
- Type of pile installation = bored
- Spacing of piles in group, $d = 2D$–$6D$

Figure 11.49 shows the behavior of 3 × 3 pile groups with low-set and high-set pile caps in terms of the average skin friction, f_{av}. Figure 11.50 shows the variation of the average skin friction, based on the location of a pile in the group.

Based on the experimental observations of the behavior of group piles in sand to date, the following general conclusions may be drawn:

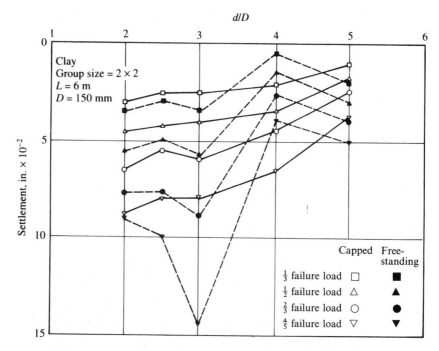

Figure 11.47 Variation of group pile settlement at various stages of load (after Brand et al., 1972)

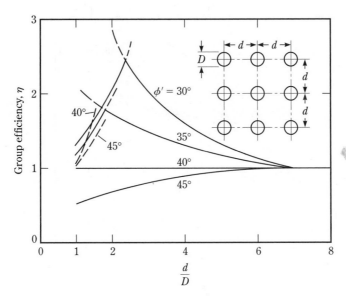

Figure 11.48 Variation of efficiency of pile groups in sand (based on Kishida and Meyerhof, 1965)

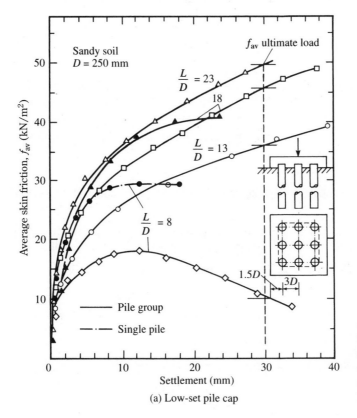

(a) Low-set pile cap

Figure 11.49 Behavior of low-set and high-set pile groups in terms of average skin friction (based on Liu et al., 1985)

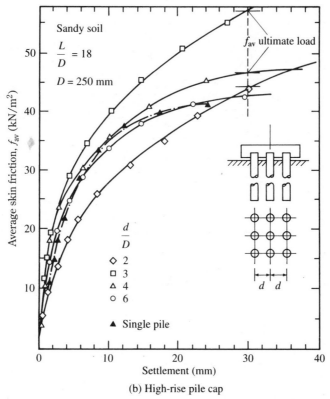

Figure 11.49
(Continued)

(b) High-rise pile cap

1. For *driven* group piles in *sand* with $d \geq 3D$, $Q_{g(u)}$ may be taken to be ΣQ_u, which includes the frictional and the point bearing capacities of individual piles.
2. For *bored* group piles in *sand* at conventional spacings ($d \approx 3D$), $Q_{g(u)}$ may be taken to be $\frac{2}{3}$ to $\frac{3}{4}$ times ΣQ_u (frictional and point bearing capacities of individual piles).

11.25   *Ultimate Capacity of Group Piles in Saturated Clay*

Figure 11.51 shows a group pile in saturated clay. Using the figure, one can estimate the ultimate load-bearing capacity of group piles in the following manner:

1. Determine $\Sigma Q_u = n_1 n_2 (Q_p + Q_s)$. From Eq. (11.19),

$$Q_p = A_p[9c_{u(p)}]$$

where $c_{u(p)}$ = undrained cohesion of the clay at the pile tip

Also, from Eq. (11.55),

$$Q_s = \Sigma \alpha p c_u \Delta L$$

So

$$\Sigma Q_u = n_1 n_2[9A_p c_{u(p)} + \Sigma \alpha p c_u \Delta L] \tag{11.133}$$

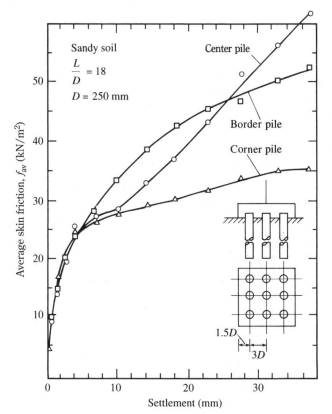

Figure 11.50 Average skin friction, based on pile location (after Liu et al., 1985)

2. Determine the ultimate capacity by assuming that the piles in the group act as a block with dimensions $L_g \times B_g \times L$. The skin resistance of the block is

$$\Sigma \, p_g c_u \Delta L = \Sigma \, 2(L_g + B_g)c_g \Delta L$$

Calculate the point bearing capacity:

$$A_p q_p = A_p c_{u(p)} N_c^* = (L_g B_g) c_{u(p)} N_c^*$$

Obtain the value of the bearing capacity factor N_c^* from Figure 11.52. Thus, the ultimate load is

$$\Sigma \, Q_u = L_g B_g c_{u(p)} N_c^* + \Sigma \, 2(L_g + B_g)c_u \Delta L \tag{11.134}$$

3. Compare the values obtained from Eqs. (11.133) and (11.134). The *lower* of the two values is $Q_{g(u)}$.

11.26 Piles in Rock

For point bearing piles resting on rock, most building codes specify that $Q_{g(u)} = \Sigma \, Q_u$, provided that the minimum center-to-center spacing of the piles is $D + 300$ mm. For H-piles and piles with square cross sections, the magnitude of D is equal to the diagonal dimension of the cross section of the pile.

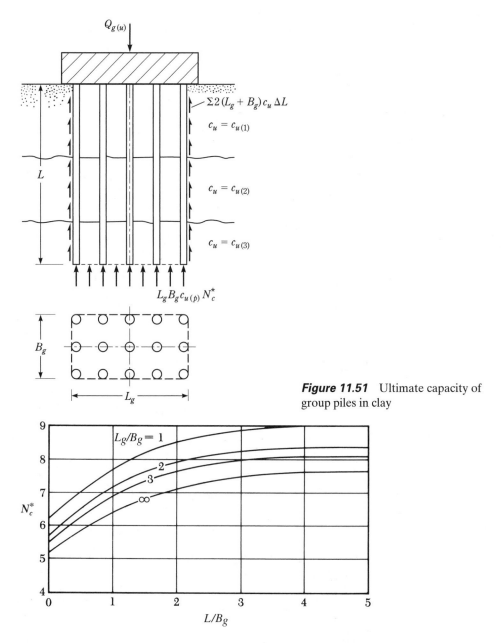

Figure 11.51 Ultimate capacity of group piles in clay

Figure 11.52 Variation of N_c^* with L_g/B_g and L/B_g

Example 11.14

In Figure 11.44a, let $n_1 = 4$, $n_2 = 3$, $D = 305$ mm, $d = 1220$ mm, and $L = 15$ m. Suppose the piles are square in cross section and are embedded in a homogeneous saturated clay with $c_u = 70$ kN/m^2. Using a factor of safety equal to 4, determine the allowable load-bearing capacity of the group pile. The unit weight of clay, $\gamma = 18.8 \text{ kN/m}^3$, and the groundwater table is located at a depth 18 m below the ground surface.

Solution
From Eq. (11.133),

$$\Sigma Q_u = n_1 n_2 [9A_p c_{u(p)} + \Sigma \alpha p c_u \Delta L]$$

with

$$A_p = (0.305)(0.305) = 0.093 \text{ m}^2$$

and

$$p = (4)(0.305) = 1.22 \text{ m}$$

The average value of the effective overburden pressure is

$$\overline{\sigma}_o' = \left(\frac{15}{2}\right)(18.8) = 141 \text{ kN/m}^2$$

Also, with

$$c_u = 70 \text{ kN/m}^2$$

it follows that

$$\frac{c_u}{\overline{\sigma}_o'} = \frac{70}{141} = 0.496$$

From Figure 11.24, for $\dfrac{c_u}{\overline{\sigma}_o'} = 0.496$, the magnitude of α is about 0.7. So

$$\Sigma Q_u = (4)(3)[(9)(0.093)(70) + (0.7)(1.22)(70)(15)]$$
$$= 12(58.59 + 896.7) \approx 11{,}463 \text{ kN}$$

Again, from Eq. (11.134), the ultimate block capacity is

$$L_g B_g c_{u(p)} N_c^* + \Sigma 2(L_g + B_g) c_u \Delta L$$

Now,

$$L_g = (n_1 - 1)d + 2\left(\frac{D}{2}\right) = (4 - 1)(1.22) + 0.305 = 3.965 \text{ m}$$

and

$$B_g = (n_2 - 1)d + 2\left(\frac{D}{2}\right) = (3 - 1)(1.22) + 0.305 = 2.745 \text{ m}$$

so we have

$$\frac{L}{B_g} = \frac{15}{2.745} = 5.46$$

and

$$\frac{L_g}{B_g} = \frac{3.965}{2.745} = 1.44$$

From Figure 11.52, $N_c^* \approx 8.6$. So

block capacity $= (3.965)(2.745)(70)(8.6) + 2(3.965 + 2.745)(70)(15)$

$$= 6552 + 14{,}091 = 20{,}643 \text{ kN}$$

Hence,

$$Q_{g(u)} = 11{,}463 \text{ kN} < 20{,}643 \text{ kN}$$

and

$$Q_{g(\text{all})} = \frac{Q_{g(u)}}{\text{FS}} = \frac{11{,}463}{4} \approx \textbf{2866 kN} \qquad \blacksquare$$

11.27 *Elastic Settlement of Group Piles*

In general, the settlement of a group pile under a similar working load per pile increases with the width of the group (B_g) and the center-to-center spacing of the piles (d). This fact is demonstrated in Figure 11.53, obtained from the experimental results of Meyerhof (1961) for group piles in sand. In the figure, $s_{g(e)}$ is the settlement of the group pile and s_e is the settlement of isolated piles under a similar working load.

Several investigations relating to the settlement of group piles have been reported in the literature, with widely varying results. The simplest relation for the settlement of group piles was given by Vesic (1969), namely,

$$s_{g(e)} = \sqrt{\frac{B_g}{D}}\, s_e \qquad (11.135)$$

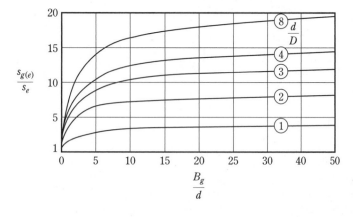

Figure 11.53 Settlement of group piles in sand (after Meyerhof, 1961)

where $s_{g(e)}$ = elastic settlement of group piles

B_g = width of group pile section

D = width or diameter of each pile in the group

s_e = elastic settlement of each pile at comparable working load (see Section 11.18)

For group piles in sand and gravel, for elastic settlement, Meyerhof (1976) suggested the empirical relation

$$s_{g(e)}(\text{in.}) = \frac{2q\sqrt{B_g}I}{(N_1)_{60}}$$ (11.136)

where $q = Q_g/(L_gB_g)$ (in U.S. ton/ft^2) (11.137)

L_g and B_g = length and width of the group pile section, respectively (ft)

$(N_1)_{60}$ = average corrected standard penetration number within seat of settlement ($\approx B_g$ deep below the tip of the piles)

I = influence factor = $1 - L/8B_g \geqslant 0.5$ (11.138)

L = length of embedment of piles

In SI units,

$$S_{g(e)}(\text{mm}) = \frac{0.96q\sqrt{B_g}I}{(N_1)_{60}}$$ (11.139)

where q is in kN/m^2 and B_g and L_g are in m, and

$$I = 1 - \frac{L(\text{m})}{8B_g(\text{m})}$$

Similarly, the group pile settlement is related to the cone penetration resistance by the formula

$$S_{g(e)} = \frac{qB_gI}{2q_c}$$ (11.140)

where q_c = average cone penetration resistance within the seat of settlement. (Note that, in Eq. (11.140), all quantities are expressed in consistent units.)

11.28 Consolidation Settlement of Group Piles

The consolidation settlement of a group pile in clay can be estimated by using the 2:1 stress distribution method. The calculation involves the following steps (see Figure 11.54):

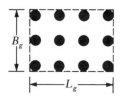

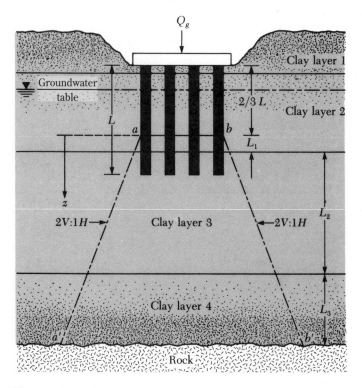

Figure 11.54   Consolidation settlement of group piles

1. Let the depth of embedment of the piles be L. The group is subjected to a total load of Q_g. If the pile cap is below the original ground surface, Q_g equals the total load of the superstructure on the piles, minus the effective weight of soil above the group piles removed by excavation.
2. Assume that the load Q_g is transmitted to the soil beginning at a depth of $2L/3$ from the top of the pile, as shown in the figure. The load Q_g spreads out along two vertical lines to one horizontal line from this depth. Lines aa' and bb' are the two 2:1 lines.
3. Calculate the increase in effective stress caused at the middle of each soil layer by the load Q_g. The formula is

$$\Delta\sigma_i' = \frac{Q_g}{(B_g + z_i)(L_g + z_i)} \tag{11.141}$$

where $\Delta\sigma'_i$ = increase in effective stress at the middle of layer i
L_g, B_g = length and width, respectively of the planned group piles
z_i = distance from $z = 0$ to the middle of the clay layer i

For example, in Figure 11.54, for layer 2, $z_i = L_1/2$; for layer 3, $z_i = L_1 + L_2/2$; and for layer 4, $z_i = L_1 + L_2 + L_3/2$. Note, however, that there will be no increase in stress in clay layer 1, because it is above the horizontal plane ($z = 0$) from which the stress distribution to the soil starts.

4. Calculate the consolidation settlement of each layer caused by the increased stress. The formula is

$$\Delta s_{c(i)} = \left[\frac{\Delta e_{(i)}}{1 + e_{o(i)}} \right] H_i \qquad (11.142)$$

where $\Delta s_{c(i)}$ = consolidation settlement of layer i
$\Delta e_{(i)}$ = change of void ratio caused by the increase in stress in layer i
e_o = initial void ratio of layer i (before construction)
H_i = thickness of layer i (*Note:* In Figure 11.54, for layer 2, $H_i = L_1$; for layer 3, $H_i = L_2$; and for layer 4, $H_i = L_3$.)

Relationships involving $\Delta e_{(i)}$ are given in Chapter 1.
5. The total consolidation settlement of the group piles is then

$$\Delta s_{c(g)} = \Sigma \Delta s_{c(i)} \qquad (11.143)$$

Note that the consolidation settlement of piles may be initiated by fills placed nearby, adjacent floor loads, or the lowering of water tables.

Example 11.15

A group pile in clay is shown in Figure 11.55. Determine the consolidation settlement of the piles. All clays are normally consolidated.

Solution
Because the lengths of the piles are 15 m each, the stress distribution starts at a depth of 10 m below the top of the pile. We are given that $Q_g = 2000$ kN.

Calculation of Settlement of Clay Layer 1
For normally consolidated clays,

$$\Delta s_{c(1)} = \left[\frac{C_{c(1)} H_1}{1 + e_{o(1)}} \right] \log \left[\frac{\sigma'_{o(1)} + \Delta\sigma'_{(1)}}{\sigma'_{o(1)}} \right]$$

$$\Delta\sigma'_{(1)} = \frac{Q_g}{(L_g + z_1)(B_g + z_1)} = \frac{2000}{(3.3 + 3.5)(2.2 + 3.5)} = 51.6 \text{ kN/m}^2$$

and

$$\sigma'_{o(1)} = 2(16.2) + 12.5(18.0 - 9.81) = 134.8 \text{ kN/m}^2$$

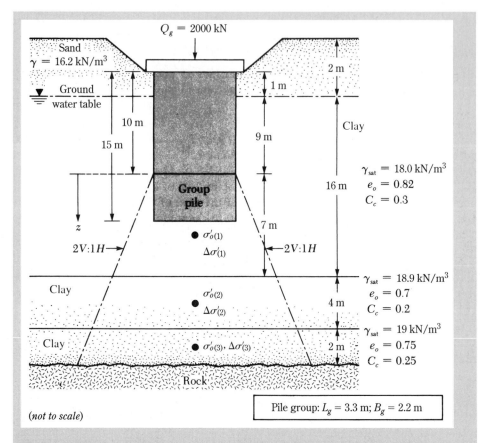

Figure 11.55 Consolidation settlement of a pile group

So

$$\Delta s_{c(1)} = \frac{(0.3)(7)}{1 + 0.82} \log \left[\frac{134.8 + 51.6}{134.8} \right] = 0.1624 \text{ m} = \mathbf{162.4 \text{ mm}}$$

Settlement of Layer 2
As with layer 1,

$$\Delta s_{c(2)} = \frac{C_{c(2)} H_2}{1 + e_{o(2)}} \log \left[\frac{\sigma'_{o(2)} + \Delta \sigma'_{(2)}}{\sigma'_{o(2)}} \right]$$

$$\sigma'_{o(2)} = 2(16.2) + 16(18.0 - 9.81) + 2(18.9 - 9.81) = 181.62 \text{ kN/m}^2$$

and

$$\Delta \sigma'_{(2)} = \frac{2000}{(3.3 + 9)(2.2 + 9)} = 14.52 \text{ kN/m}^2$$

Hence,

$$\Delta s_{c(2)} = \frac{(0.2)(4)}{1 + 0.7} \log \left[\frac{181.62 + 14.52}{181.62} \right] = 0.0157 \text{ m} = \mathbf{15.7mm}$$

Settlement of Layer 3
Continuing analogously, we have

$$\sigma'_{o(3)} = 181.62 + 2(18.9 - 9.81) + 1(19 - 9.81) = 208.99 \text{ kN/m}^2$$

$$\Delta \sigma'_{(3)} = \frac{2000}{(3.3 + 12)(2.2 + 12)} = 9.2 \text{ kN/m}^2$$

$$\Delta s_{c(3)} = \frac{(0.25)(2)}{1 + 0.75} \log \left[\frac{208.99 + 9.2}{208.99} \right] = 0.0054 \text{ m} = \mathbf{5.4 \text{ mm}}$$

Hence, the total settlement is

$$\Delta s_{c(g)} = 162.4 + 15.7 + 5.4 = \mathbf{183.5 \text{ mm}}$$ ■

Problems

11.1 A concrete pile is 25 m long and 305 mm $\times$ 305 mm in cross section. The pile is fully embedded in sand, for which $\gamma = 17.5$ kN/m^3 and $\phi' = 35°$. Calculate
  a. The ultimate point load, Q_p, by Meyerhof's method.
  b. The total frictional resistance [Eqs. (11.14), (11.37), (11.38), and (11.39)] for $K = 1.3$ and $\delta = 0.8\phi'$.
11.2 Solve Problem 11.1, Part (a), using Coyle and Castello's method.
11.3 Solve Problem 11.1, Part (a), using Vesic's method [Eq. (11.20)]. Take $I_r = I_{rr} = 50$.
11.4 Solve Problem 11.1, Part (a), using Janbu's method [Eq. (11.31)]. Take $\eta' = 90°$.
11.5 Use the results of Problems 11.1–11.4 to estimate an allowable value for the point load. Take FS = 4.
11.6 Redo Problem 11.1 for $\gamma = 18.5$ kN/m^3 and $\phi' = 40°$.
11.7 Solve Problem 11.6, Part (a), by Coyle and Castello's method.
11.8 A driven closed-ended pile, circular in cross section, is shown in Figure P11.8. Calculate
  a. The ultimate point load, using Meyerhof's procedure.
  b. The ultimate point load, using Vesic's procedure. Take $I_r = I_{rr} = 50$.
  c. An approximate ultimate point load, on the basis of Parts (a) and (b).
  d. The ultimate frictional resistance Q_s. [Use Eqs. (11.14), (11.37), (11.38), and (11.39), and take $K = 1.4$ and $\delta = 0.6\phi'$.]
  e. The allowable load of the pile. (Use FS = 4)
11.9 A concrete pile 20 m long having a cross section of 381 mm $\times$ 381 mm is fully embedded in a saturated clay layer for which $\gamma_{sat} = 18.5$ kN/m^3, $\phi = 0$,

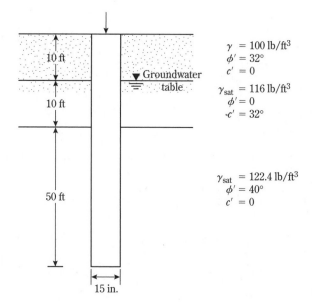

$\gamma = 100 \text{ lb/ft}^3$
$\phi' = 32°$
$c' = 0$

$\gamma_{sat} = 116 \text{ lb/ft}^3$
$\phi' = 0$
$c' = 32°$

$\gamma_{sat} = 122.4 \text{ lb/ft}^3$
$\phi' = 40°$
$c' = 0$

Figure P11.8

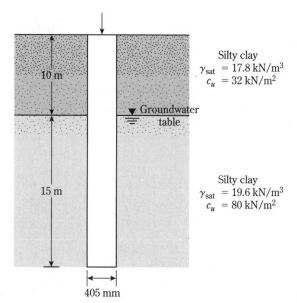

Silty clay
$\gamma_{sat} = 17.8 \text{ kN/m}^3$
$c_u = 32 \text{ kN/m}^2$

Silty clay
$\gamma_{sat} = 19.6 \text{ kN/m}^3$
$c_u = 80 \text{ kN/m}^2$

Figure P11.11

and $c_u = 70 \text{ kN/m}^2$. Assume that the water table lies below the tip of the pile. Determine the allowable load that the pile can carry. (Let FS = 3.) Use the α method to estimate the skin friction.

11.10 Redo Problem 11.9, using the λ method for estimating the skin friction.

11.11 A concrete pile 405 mm × 405 mm in cross section is shown in Figure P11.11. Calculate the ultimate skin resistance by using
a. the α method

b. the λ method

c. the β method

Let $\phi'_R = 25°$ for all clays, which are normally consolidated.

11.12 A steel pile (H section, HP 12 × 84, see Table 11.1b) is driven to a layer of sandstone. The length of the pile is 70 ft. The unconfined compression strength of the sandstone is $q_{u(\text{lab})} = 15,000$ lb/in.², and the angle of friction is 34°. Use a factor of safety of 3, and estimate the allowable point load that can be carried by the pile. Let $q_{u(\text{design})} = q_{u(\text{lab})}/6$.

11.13 The working load on a prestressed concrete pile 21 m long that has been driven into sand is 502 kN. The pile is octagonal in shape with $D = 356$ mm. (See Table 11.3a.) Skin resistance carries 350 kN of the allowable load, and point bearing carries the rest. Use $E_p = 21 × 10^6 \text{kN/m}^2$, $E_s = 25 × 10^3 \text{kN/m}^2$, $\mu_s = 0.35$, and $\xi = 0.62$ [Eq. (11.73)]. Determine the elastic settlement of the pile.

11.14 A concrete pile is 50 ft long and has a cross section of 16 in. × 16 in. The pile is embedded in a sand with $\gamma = 117$ lb/ft³ and $\phi' = 37°$. The working load is 180 kip. If 110 kip is contributed by the frictional resistance and 70 kip is from the point load, determine the elastic settlement of the pile. Take $E_p = 3 × 10^6 \text{lb/in.}^2$, $E_s = 5 × 10^3 \text{lb/in.}^2$, $\mu_s = 0.38$, and $\xi = 0.57$.

11.15 A 30-m-long concrete pile is 305 mm × 305 mm in cross section and is fully embedded in a sand deposit. If $n_h = 9200 \text{ kN/m}^2$, the moment at ground level, $M_g = 0$, the allowable displacement of pile head = 12 mm, $E_p = 21 × 10^6 \text{kN/m}^2$, and $F_{Y(\text{pile})} = 21,000 \text{ kN/m}^2$, calculate the allowable lateral load, Q_g, at the ground level. Use the elastic solution method.

11.16 Solve Problem 11.15 by Broms's method. Assume that the pile is flexible and free headed. Let the soil unit weight, $\gamma = 16 \text{ kN/m}^3$, the soil friction angle, $\phi' = 30°$, and the yield stress of the pile material, $F_Y = 21 \text{ MN/m}^2$.

11.17 A steel H pile (Section HP 13 × 100) is driven by a hammer. The maximum rated hammer energy is 36 kip-ft, the weight of the ram is 14 kip, and the length of the pile is 80 ft. Also, we have

$$\text{Coefficient of restitution} = 0.35$$
$$\text{Weight of the pile cap} = 1.2 \text{ kip}$$
$$\text{Hammer efficiency} = 0.85$$
$$\text{Number of blows for the last inch of penetration} = 10$$
$$E_p = 30 × 10^6 \text{ lb/in.}^2$$

Estimate the pile capacity, using Eq. (11.118). Take FS = 6.

11.18 Solve Problem 11.17, using the modified EN formula. (See Table 11.14.) Take FS = 4.

11.19 Solve Problem 11.17, using the Danish formula. (See Table 11.14.) Take FS = 3.

11.20 Figure 11.42a shows a pile. Let $L = 50$ ft, $D = 18$ in., $H_f = 11.5$ ft, $\gamma_f = 112$ lb/ft³, and $\phi'_{\text{fill}} = 28°$. Determine the total downward drag force on the pile. Assume that the fill is located above the water table and that $\delta = 0.6\phi'_{\text{fill}}$.

11.21 Redo Problem 11.20, assuming that the water table coincides with the top of the fill and $\gamma_{\text{sat(fill)}} = 124.5$ lb/ft³. If the other quantities remain the same, what would be the downward drag force on the pile? Assume that $\delta = 0.6\phi'_{fill}$.

11.22 In Figure 11.42b, let $L = 19$ m, $\gamma_{fill} = 15.2$ kN/m^3, $\gamma_{sat(clay)} = 19.5$ kN/m^3, $\phi'_{clay} = 30°$, $H_f = 3.2$ m, and $D = 0.46$ m. The water table coincides with the top of the clay layer. Determine the total downward drag on the pile. Assume that $\delta = 0.5\phi'_{clay}$.

11.23 Consider a group pile. (See Figure 11.44a.) If $n_1 = 4$, $n_2 = 3$, the pile diameter, $D = 400$ mm, and the pile spacing, $d = 900$ m, determine the efficiency of the group pile. Use Eq. (11.131).

11.24 Solve Problem 11.23, using the Converse–Labarre equation. (See Table 11.15.)

11.25 In Problem 11.23, if the center-to-center pile spacings are increased to 1200 mm, what will be the group efficiency?

11.26 Consider the group pile described in Problem 11.23. Assume that the piles are embedded in a saturated homogeneous clay having $c_u = 102$ kN/m^2. The length of the piles is 20 m. Find the allowable load-carrying capacity of the group pile. Use FS = 3. The ground water table is at a depth of 25 m below the ground surface. Take $\gamma_{sat} = 19$ kN/m^3.

11.27 A section of a 3×4 group pile in a layered saturated clay is shown in Figure P11.27. The piles are square in cross section (14 in. $\times$ 14 in.). The center-to-center spacing d of the piles is 35 in. Determine the allowable load-bearing capacity of the group pile. Use FS = 4.

11.28 Figure P11.28 shows a group pile in clay. Determine the consolidation settlement of the group. Use the 2:1 method to estimate the average effective stress in the clay layers.

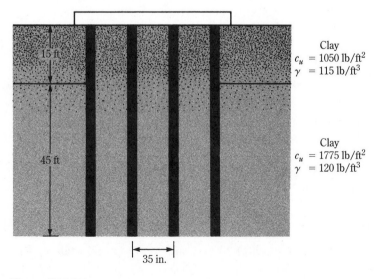

Figure P11.27

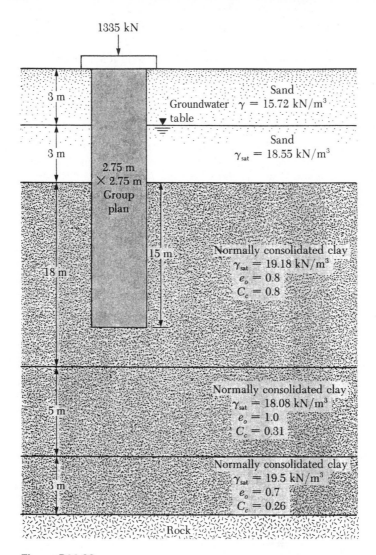

1335 kN

3 m — Sand, Groundwater $\gamma = 15.72$ kN/m³ table

3 m — Sand $\gamma_{sat} = 18.55$ kN/m³

2.75 m × 2.75 m Group plan

15 m — 18 m

Normally consolidated clay
$\gamma_{sat} = 19.18$ kN/m³
$e_o = 0.8$
$C_c = 0.8$

5 m — Normally consolidated clay
$\gamma_{sat} = 18.08$ kN/m³
$e_o = 1.0$
$C_c = 0.31$

3 m — Normally consolidated clay
$\gamma_{sat} = 19.5$ kN/m³
$e_o = 0.7$
$C_c = 0.26$

Rock

Figure P11.28

References

American Petroleum Institute (API). *Recommended Practice of Planning, Designing, and Construction of Fixed Offshore Platforms,* Report No. API-RF-2A, Dallas, 115 pp.

American Society of Civil Engineers (1959). "Timber Piles and Construction Timbers," *Manual of Practice,* No. 17, American Society of Civil Engineers, New York.

American Society of Civil Engineers (1993). *Design of Pile Foundations (Technical Engineering and Design Guides as Adapted from the U.S. Army Corps of Engineers,* No. 1), American Society of Civil Engineers, New York.

Baldi, G. et al. (1981). "Cone Resistance in Dry NC and OC Sands," *Proceedings, Cone Penetration Testing and Experience,* ASCE, pp. 145–177.

Bhusan, K. (1982). "Discussion: New Design Correlations for Piles in Sands," *Journal of the Geotechnical Engineering Division,* American Society of Civil Engineers, Vol. 108, No. GT11, pp. 1508–1510.

Bjerrum, L., Johannessen, I. J., and Eide, O. (1969). "Reduction of Skin Friction on Steel Piles to Rock," *Proceedings, Seventh International Conference on Soil Mechanics and Foundation Engineering,* Mexico City, Vol. 2, pp. 27–34.

Bowles, J. E. (1982). *Foundation Design and Analysis,* McGraw-Hill, New York.

Bowles, J. E. (1996). *Foundation Analysis and Design,* McGraw-Hill, New York.

Brand, E. W., Muktabhant, C., and Taechathummarak, A. (1972). "Load Test on Small Foundations in Soft Clay." *Proceedings, Performance of Earth and Earth-Supported Structures,* American Society of Civil Engineers, Vol. 1, Part 2, pp. 903–928.

Briaud, J. L., and Miran, J. (1991). *The Cone Penetration Test,* Report No. FHWA-TA-91-004, Federal Highway Administration, McLean, VA.

Briaud, J. L., Moore, B. H., and Mitchell, G. B. (1989). "Analysis of Pile Load Test at Lock and Dam 26," *Proceedings, Foundation Engineering: Current Principles and Practices,* American Society of Civil Engineers, Vol. 2, pp. 925–942.

Briaud, J. L., and Tucker, L. M. (1984). "Coefficient of Variation of *In-Situ* Tests in Sand," *Proceedings, Symposium on Probabilistic Characterization of Soil Properties,* Atlanta, pp. 119–139.

Briaud, J. L., Tucker, L., Lytton, R. L., and Coyle, H. M. (1985). *Behavior of Piles and Pile Groups,* Report No. FHWA/RD-83/038, Federal Highway Administration, Washington, DC.

Broms, B. B. (1965). "Design of Laterally Loaded Piles," *Journal of the Soil Mechanics and Foundations Division,* American Society of Civil Engineers, Vol. 91, No. SM3, pp. 79–99.

Bustamante, M., and Gianeselli, L. (1982). "Pile Bearing Capacity Prediction by Means of Static Penetrometer CPT," *Second European Symposium on Penetration Testing,* Vol. 2, pp. 493–500, Balkema, Amsterdam.

Coyle, H. M., and Castello, R. R. (1981). "New Design Correlations for Piles in Sand," *Journal of the Geotechnical Engineering Division,* American Society of Civil Engineers, Vol. 107, No. GT7, pp. 965–986.

Davisson, M. T. (1970). "BRD Vibratory Driving Formula," *Foundation Facts,* Vol. VI, No. 1, pp. 9–11.

Davisson, M. T., and Gill, H. L. (1963). "Laterally Loaded Piles in a Layered Soil System," *Journal of the Soil Mechanics and Foundations Division,* American Society of Civil Engineers, Vol. 89, No. SM3, pp. 63–94.

DeRuiter, J., and Beringen, F. L. (1979). "Pile Foundations for Large North Sea Structures," *Marine Geotechnology,* Vol. 3, No. 3, pp. 267–314.

Feld, J. (1943). "Friction Pile Foundations," *Discussion, Transactions,* American Society of Civil Engineers, Vol. 108.

Feng, Z., and Deschamps, R. J. (2000). "A Study of the Factors Influencing the Penetration and Capacity of Vibratory Driven Piles," *Soils and Foundations,* Vol. 40, No. 3, pp. 43–54.

Gates, M. (1957). "Empirical Formula for Predicting Pile Bearing Capacity." *Civil Engineering,* American Society of Civil Engineers, Vol. 27, No. 3, pp. 65–66.

Goodman, R. E. (1980). *Introduction to Rock Mechanics,* Wiley, New York.

Guang-Yu, Z. (1988). "Wave Equation Applications for Piles in Soft Ground," *Proceedings, Third International Conference on the Application of Stress-Wave Theory to Piles* (B. H. Fellenius, ed.), Ottawa, Ontario, Canada, pp. 831–836.

International Conference of Building Officials (1982). "Uniform Building Code," Whittier, CA.

Janbu, N. (1953). *An Energy Analysis of Pile Driving with the Use of Dimensionless Parameters,* Norwegian Geotechnical Institute, Oslo, Publication No. 3.

Janbu, N. (1976). "Static Bearing Capacity of Friction Piles," *Proceedings, Sixth European Conference on Soil Mechanics and Foundation Engineering,* Vol. 1.2, pp. 479–482.

Kishida, H., and Meyerhof, G. G. (1965). "Bearing Capacity of Pile Groups under Eccentric Loads in Sand," *Proceedings, Sixth International Conference on Soil Mechanics and Foundation Engineering,* Montreal, Vol. 2, pp. 270–274.

Liu, J. L., Yuan, Z. L., and Zhang, K. P. (1985). "Cap–Pile–Soil Interaction of Bored Pile Groups," *Proceedings, Eleventh International Conference on Soil Mechanics and Foundation Engineering,* San Francisco, Vol. 3, pp. 1433–1436.

Matlock, H., and Reese, L. C. (1960). "Generalized Solution for Laterally Loaded Piles," *Journal of the Soil Mechanics and Foundations Division,* American Society of Civil Engineers, Vol. 86, No. SM5, Part I, pp. 63–91.

McClelland, B. (1974). "Design of Deep Penetration Piles for Ocean Structures," *Journal of the Geotechnical Engineering Division,* American Society of Civil Engineers, Vol. 100, No. GT7, pp. 709–747.

Meyerhof, G. G. (1961). "Compaction of Sands and Bearing Capacity of Piles," *Transactions,* American Society of Civil Engineers, Vol. 126, Part 1, pp. 1292–1323.

Meyerhof, G. G. (1976). "Bearing Capacity and Settlement of Pile Foundations," *Journal of the Geotechnical Engineering Division,* American Society of Civil Engineers, Vol. 102, No. GT3, pp. 197–228.

Meyerhof, G. G. (1995). "Behavior of Pile Foundations under Special Loading Conditions: 1994 R. M. Hardy Keynote Address," *Canadian Geotechnical Journal,* Vol. 32, No. 2, pp. 204–222.

Michigan State Highway Commission (1965). *A Performance Investigation of Pile Driving Hammers and Piles,* Lansing, MI, 338 pp.

Nottingham, L. C., and Schmertmann, J. H. (1975). *An Investigation of Pile Capacity Design Procedures,* Research Report No. D629, Department of Civil Engineering, University of Florida, Gainesville, FL.

Olson, R. E., and Flaate, K. S. (1967). "Pile Driving Formulas for Friction Piles in Sand," *Journal of the Soil Mechanics and Foundations Division,* American Society of Civil Engineers, Vol. 93, No. SM6, pp. 279–296.

Randolph, M. F., and Murphy, B. S. (1985). "Shaft Capacity of Driven Piles in Clay," *Proceedings, Offshore Technology Conference,* Vol. 1, Houston, pp. 371–378.

Schmertmann, J. H. (1978). *Guidelines for Cone Penetration Test: Performance and Design,* Report FHWA-TS-78-209, Federal Highway Administration, Washington, DC.

Seiler, J. F., and Keeney, W. D. (1944). "The Efficiency of Piles in Groups," *Wood Preserving News,* Vol. 22, No. 11 (November).

Shioi, Y., and Fukui, J. (1982). "Application of *N*-Value to Design of Foundations in Japan," *Proceedings, Second European Symposium on Penetration Testing,* Vol. 1, pp. 159–164, Balkema, Amsterdam.

Skov, R., and Denver, H. (1988). "Time Dependence of Bearing Capacity of Piles," *Proceedings, Third International Conference on Application of Stress Wave Theory to Piles,* Ottawa, Canada, pp. 879–889.

Svinkin, M. (1996). Discussion on "Setup and Relaxation in Glacial Sand," *Journal of Geotechnical Engineering,* ASCE, Vol. 22, pp. 319–321.

Terzaghi, K., and Peck, R. B. (1967). *Soil Mechanics in Engineering Practice,* 2d ed., Wiley, New York.

Vesic, A. S. (1969). *Experiments with Instrumented Pile Groups in Sand,* American Society for Testing and Materials; Special Technical Publication No. 444, pp. 177–222.

Vesic, A. S. (1977). *Design of Pile Foundations,* National Cooperative Highway Research Program Synthesis of Practice No. 42, Transportation Research Board, Washington, DC.

Vijayvergiya, V. N., and Focht, J. A., Jr. (1972). *A New Way to Predict Capacity of Piles in Clay,* Offshore Technology Conference Paper 1718, Fourth Offshore Technology Conference, Houston.

Wong, K. S., and Teh, C. I. (1995). "Negative Skin Friction on Piles in Layered Soil Deposit," *Journal of Geotechnical and Geoenvironmental Engineering,* American Society of Civil Engineers, Vol. 121, No. 6, pp. 457–465.

Woodward, R. J., Gardner, W. S., and Greer, D. M. (1972). *Drilled Pier Foundations,* McGraw-Hill, New York.

12

Drilled-Shaft Foundations

12.1   Introduction

The terms *caisson, pier, drilled shaft,* and *drilled pier* are often used interchangeably in foundation engineering; all refer to a *cast-in-place pile generally having a diameter of about 750 mm* ($\approx$2.5 ft) or more, with or without steel reinforcement and with or without an enlarged bottom. Sometimes the diameter can be as small as 305 mm ($\approx$1 ft).

To avoid confusion, we use the term *drilled shaft* for a hole drilled or excavated to the bottom of a structure's foundation and then filled with concrete. Depending on the soil conditions, casings may be used to prevent the soil around the hole from caving in during construction. The diameter of the shaft is usually large enough for a person to enter for inspection.

The use of drilled-shaft foundations has several advantages:

1. A single drilled shaft may be used instead of a group of piles and the pile cap.
2. Constructing drilled shafts in deposits of dense sand and gravel is easier than driving piles.
3. Drilled shafts may be constructed before grading operations are completed.
4. When piles are driven by a hammer, the ground vibration may cause damage to nearby structures. The use of drilled shafts avoids this problem.
5. Piles driven into clay soils may produce ground heaving and cause previously driven piles to move laterally. This does not occur during the construction of drilled shafts.
6. There is no hammer noise during the construction of drilled shafts; there is during pile driving.
7. Because the base of a drilled shaft can be enlarged, it provides great resistance to the uplifting load.
8. The surface over which the base of the drilled shaft is constructed can be visually inspected.
9. The construction of drilled shafts generally utilizes mobile equipment, which, under proper soil conditions, may prove to be more economical than methods of constructing pile foundations.
10. Drilled shafts have high resistance to lateral loads.

There are also a couple of drawbacks to the use of drilled-shaft construction. For one thing, the concreting operation may be delayed by bad weather and always needs close supervision. For another, as in the case of braced cuts, deep excavations for drilled shafts may induce substantial ground loss and damage to nearby structures.

12.2 Types of Drilled Shafts

Drilled shafts are classified according to the ways in which they are designed to transfer the structural load to the substratum. Figure 12.1a shows a drilled *straight shaft*. It extends through the upper layer(s) of poor soil, and its tip rests on a strong load-bearing soil layer or rock. The shaft can be cased with steel shell or pipe when required (as it is with cased, cast-in-place concrete piles; see Figure 11.4). For such shafts, the resistance to the applied load may develop from end bearing and also from side friction at the shaft perimeter and soil interface.

A *belled shaft* (see Figures 12.1b and c) consists of a straight shaft with a bell at the bottom, which rests on good bearing soil. The bell can be constructed in the shape of a dome (see Figure 12.1b), or it can be angled (see Figure 12.1c). For angled bells, the underreaming tools that are commercially available can make 30° to 45° angles with the vertical. For the majority of drilled shafts constructed in the United States, the entire load-carrying capacity is assigned to the end bearing only. However, under certain circumstances, the end-bearing capacity and the side friction are taken into account. In Europe, both the side frictional resistance and the end-bearing capacity are always taken into account.

Straight shafts can also be extended into an underlying rock layer. (See Figure 12.1d.) In the calculation of the load-bearing capacity of such shafts, the end bearing and the shear stress developed along the shaft perimeter and rock interface can be taken into account.

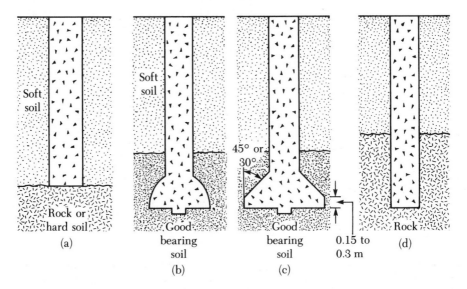

Figure 12.1 Types of drilled shaft: (a) straight shaft; (b) and (c) belled shaft; (d) straight-shaft socketed into rock

12.3 Construction Procedures

The most common construction procedure used in the United States involves rotary drilling. There are three major types of construction methods: the dry method, the casing method, and the wet method.

Dry Method of Construction

This method is employed in soils and rocks that are above the water table and that will not cave in when the hole is drilled to its full depth. The sequence of construction, shown in Figure 12.2, is as follows:

1. The excavation is completed (and belled if desired), using proper drilling tools, and the spoils from the hole are deposited nearby. (See Figure 12.2a.)
2. Concrete is then poured into the cylindrical hole. (See Figure 12.2b.)
3. If desired, a rebar cage is placed in the upper portion of the shaft. (See Figure 12.2c.)
4. Concreting is then completed, and the drilled shaft will be as shown in Figure 12.2d.

Casing Method of Construction

This method is used in soils or rocks in which caving or excessive deformation is likely to occur when the borehole is excavated. The sequence of construction is shown in Figure 12.3 and may be explained as follows:

1. The excavation procedure is initiated as in the case of the dry method of construction. (See Figure 12.3a.)
2. When the caving soil is encountered, bentonite slurry is introduced into the borehole. (See Figure 12.3b.) Drilling is continued until the excavation goes past the caving soil and a layer of impermeable soil or rock is encountered.
3. A casing is then introduced into the hole. (See Figure 12.3c.)
4. The slurry is bailed out of the casing with a submersible pump. (See Figure 12.3d.)
5. A smaller drill that can pass through the casing is introduced into the hole, and excavation continues. (See Figure 12.3e.)
6. If needed, the base of the excavated hole can then be enlarged, using an underreamer. (See Figure 12.3f.)
7. If reinforcing steel is needed, the rebar cage needs to extend the full length of the excavation. Concrete is then poured into the excavation and the casing is gradually pulled out. (See Figure 12.3g.)
8. Figure 12.3h shows the completed drilled shaft.

Wet Method of Construction

This method is sometimes referred to as the *slurry displacement method.* Slurry is used to keep the borehole open during the entire depth of excavation. (See Figure 12.4.) Following are the steps involved in the wet method of construction:

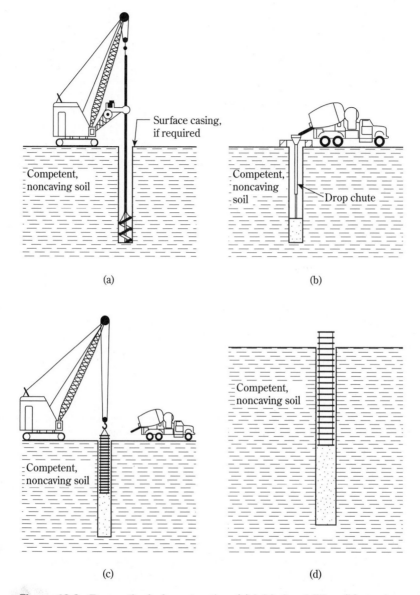

Figure 12.2   Dry method of construction: (a) initiating drilling; (b) starting concrete pour; (c) placing rebar cage; (d) completed shaft (after O'Neill and Reese, 1999)

1. Excavation continues to full depth with slurry. (See Figure 12.4a.)
2. If reinforcement is required, the rebar cage is placed in the slurry. (See Figure 12.4b.)
3. Concrete that will displace the volume of slurry is then placed in the drill hole. (See Figure 12.4c.)
4. Figure 12.4d shows the completed drilled shaft.

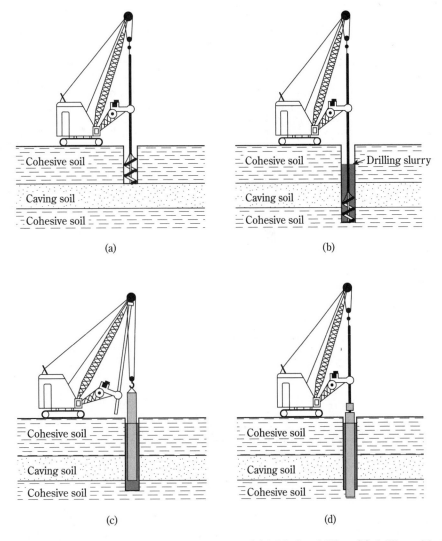

(a)

(b)

(c)

(d)

Figure 12.3 Casing method of construction: (a) initiating drilling; (b) drilling with slurry; (c) introducing casing; (d) casing is sealed and slurry is being removed from interior of casing; (e) drilling below casing; (f) underreaming; (g) removing casing; (h) completed shaft (after O'Neill and Reese, 1999)

12.4 *Other Design Considerations*

For the design of ordinary drilled shafts without casings, a minimum amount of vertical steel reinforcement is always desirable. Minimum reinforcement is 1% of the gross cross-sectional area of the shaft. In California, a reinforcing cage having a length of about 3.65 m (12 ft) is used in the top part of the shaft, and no reinforcement is provided at the bottom. This procedure helps in the construction process, because the cage is placed after the concreting is almost complete.

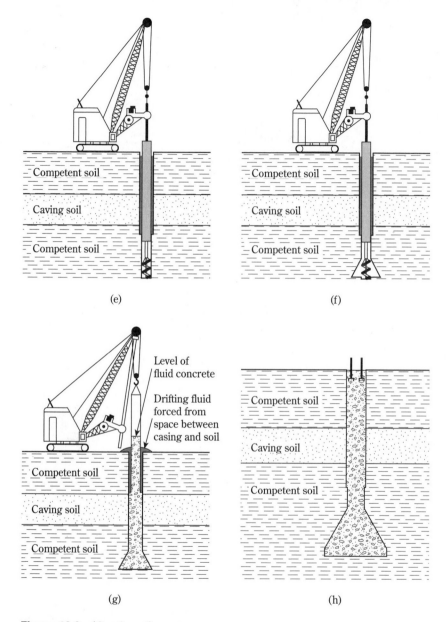

Figure 12.3 (Continued)

For drilled shafts with nominal reinforcement, most building codes suggest using a design concrete strength, f_c, on the order of $f'_c/4$. Thus, the minimum shaft diameter becomes

$$f_c = 0.25f'_c = \frac{Q_w}{A_{gs}} = \frac{Q_w}{\dfrac{\pi}{4}D_s^2}$$

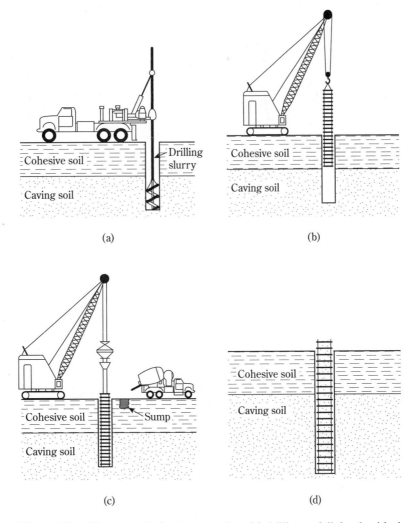

Figure 12.4   Slurry method of construction: (a) drilling to full depth with slurry; (b) placing rebar cage; (c) placing concrete; (d) completed shaft (after O'Neill and Reese, 1999)

or

$$D_s = \sqrt{\frac{Q_w}{\left(\dfrac{\pi}{4}\right)(0.25)f'_c}} = 2.257\sqrt{\frac{Q_u}{f'_c}} \qquad (12.1)$$

where D_s = diameter of the shaft
          f'_c = 28-day concrete strength
          Q_w = working load of the drilled shaft

A_{gs} = gross cross-sectional area of the shaft

If drilled shafts are likely to be subjected to tensile loads, reinforcement should be continued for the entire length of the shaft.

Concrete Mix Design

The concrete mix design for drilled shafts is not much different from that for any other concrete structure. When a reinforcing cage is used, consideration should be given to the ability of the concrete to flow through the reinforcement. In most cases, a concrete slump of about 15.0 mm (6 in.) is considered satisfactory. Also, the maximum size of the aggregate should be limited to about 20 mm (0.75 in.).

12.5 Load Transfer Mechanism

The load transfer mechanism from drilled shafts to soil is similar to that of piles, as described in Section 11.5. Figure 12.5 shows the load test results of a drilled shaft in a clay soil in Houston, Texas (Reese, Touma, and O'Neill, 1976). This drilled shaft had a diameter of 0.76 m (2.5 ft) and a depth of penetration of 7.04 m (23.1 ft). The soil profile at the site is shown in Figure 12.5a. Figure 12.5b shows the load–settlement curves. It can be seen that the total load carried by the drilled shaft was 1246 kN (140 tons). The load carried by side resistance was about 800 kN (90 tons), and the rest was carried by point bearing. It is interesting to note that, with a downward movement of about 6.35 mm (0.25 in.), full side resistance was mobilized. However, about 25 mm ($\approx$1 in.) of downward movement was required for mobilization of full point resistance. This situation is similar to that observed in the case of piles. Figure 12.5c shows the load-distribution curves for different stages of the loading.

12.6 Estimation of Load-Bearing Capacity

The ultimate load-bearing capacity of a drilled shaft (see Figure 12.6) is

$$Q_u = Q_p + Q_s \tag{12.2}$$

where Q_u = ultimate load
  Q_p = ultimate load-carrying capacity at the base
  Q_s = frictional (skin) resistance

The ultimate base load Q_p can be expressed in a manner similar to the way it is expressed in the case of shallow foundations [Eq. (3.21)], or

$$Q_p = A_p\left(c'N_cF_{cs}F_{cd}F_{cc} + q'N_qF_{qs}F_{qd}F_{qc} + \frac{1}{2}\gamma'N_\gamma F_{\gamma s}F_{\gamma d}F_{\gamma c}\right) \tag{12.3}$$

where c' = cohesion
  N_c, N_q, N_γ = bearing capacity factors (Table 3.4)
  $F_{cs}, F_{qs}, F_{\gamma s}$ = shape factors
  $F_{cd}, F_{qd}, F_{\gamma d}$ = depth factors

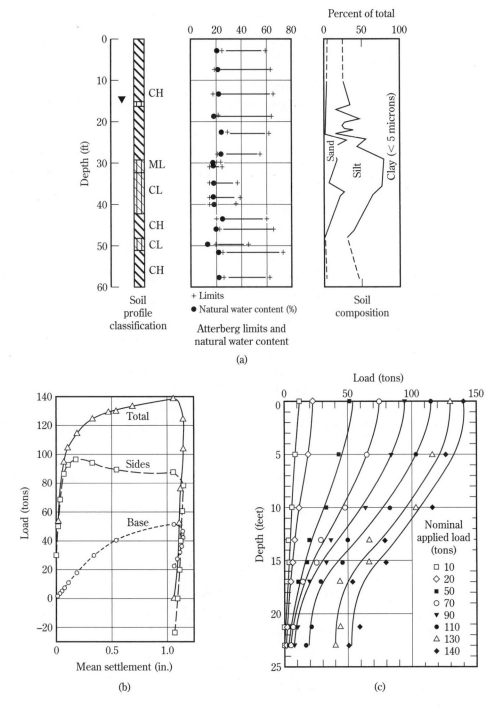

Figure 12.5 Load test results for a drilled shaft in Houston, Texas: (a) soil profile; (b) load–displacement curves; (c) load-distribution curves at various stages of loading (after Reese, Touma, and O'Neill, 1976)

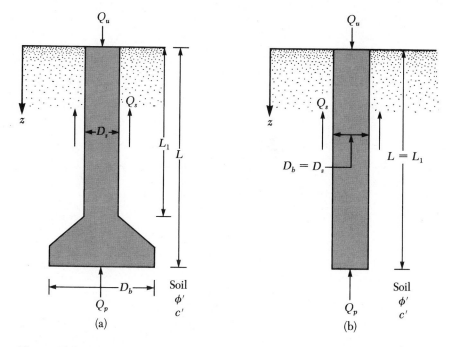

Figure 12.6 Ultimate bearing capacity of drilled shafts: (a) with bell; (b) straight shaft

$F_{cc}, F_{qc}, F_{\gamma c}$ = compressibility factors

γ' = effective unit weight of soil at the base of the shaft

q' = effective vertical stress at the base of the shaft

A_p = area of the base $= \dfrac{\pi}{4} D_b^2$

In most instances, the last term (the one containing N_γ) is neglected, except in the case of a relatively short drilled shaft. With this assumption, the net load-carrying capacity at the base (i.e., the gross load minus the weight of the drilled shaft) may be approximated as

$$Q_{p(\text{net})} = A_p [c' N_c F_{cs} F_{cd} F_{cc} + q'(N_q - 1) F_{qs} F_{qd} F_{qc}] \qquad (12.4)$$

Using Eqs. (3.25), (3.26), and (3.32), we obtain

$$F_{cs} = 1 + \frac{N_q}{N_c} \qquad (12.5)$$

$$F_{qs} = 1 + \tan \phi' \qquad (12.6)$$

$$F_{cd} = F_{qd} - \frac{1 - F_{qd}}{N_c \tan \phi'} \qquad (12.7)$$

and

$$F_{qd} = 1 + 2 \tan \phi'(1 - \sin \phi')^2 \underbrace{\tan^{-1}\left(\frac{L}{D_b}\right)}_{\text{radians}} \tag{12.8}$$

According to Chen and Kulhawy (1994), F_{cc} and F_{qc} can be calculated in the following manner:

1. Calculate the critical rigidity index as

$$I_{rc} = 0.5\exp[2.85\cot(45 - \frac{\phi'}{2})] \tag{12.9}$$

2. Calculate the reduced rigidity index as

$$I_{rr} = \frac{I_r}{1 + I_r\Delta} \tag{12.10}$$

where I_r = soil rigidity index

$$= \frac{E_s}{2(1 + \mu_s)q' \tan \phi'} \tag{12.11}$$

in which E_s = drained modulus of elasticity of soil
μ_s = drained Poisson's ratio of soil
Δ = volumetric strain within the plastic zone during loading

3. If $I_{rr} \geq I_{rc}$ then

$$F_{cc} = F_{qc} = 1 \tag{12.12}$$

However, if $I_{rr} < I_{rc}$, then

$$F_{cc} = F_{qc} - \frac{1 - F_{qc}}{N_c \tan \phi'} \tag{12.13}$$

and

$$F_{qc} = \exp\left\{(-3.8 \tan \phi') + \left[\frac{(3.07 \sin \phi')(\log_{10} 2I_{rr})}{1 + \sin \phi'}\right]\right\} \tag{12.14}$$

The expression for the frictional, or skin, resistance Q_s is similar to that for piles; that is,

$$Q_s = \int_0^{L_1} pf \, dz \tag{12.15}$$

where p = shaft perimeter = πD_s
f = unit frictional (or skin) resistance

The next two sections describe the procedures for obtaining the *ultimate* and *allowable* load-bearing capacities of drilled shafts in sand and saturated clay ($\phi = 0$).

| 12.7 | ***Drilled Shafts in Sand: Load-Bearing Capacity*** |

For drilled shafts in sand, $c' = 0$; hence, Eq. (12.4) simplifies to

$$Q_{p(\text{net})} = A_p[q'(N_q - 1)F_{qs}F_{qd}F_{qc}]$$ (12.16)

The value of N_q for a given soil friction angle ϕ' can be determined from Table 3.4. The shape factor F_{qs} and depth factor F_{qd} can be evaluated from Eqs. (12.6) and (12.8), respectively. To calculate the compressibility factor F_{qc}, Eqs. (12.9), (12.10), (12.11), (12.12), (12.13) and (12.14) will have to be used. The terms E_s, μ_s, and Δ can be estimated by the relationship (Chen and Kulhawy, 1994)

$$\frac{E_s}{p_a} = m$$ (12.17)

where p_a = atmospheric pressure ($\approx 100 \text{ kN/m}^2$ or 2000 lb/ft^2)

$$m = \begin{cases} 100 \text{ to } 200 \text{ (loose soil)} \\ 200 \text{ to } 500 \text{ (medium dense soil)} \\ 500 \text{ to } 1000 \text{ (dense soil)} \end{cases}$$

$$\mu_s = 0.1 + 0.3\left(\frac{\phi' - 25}{20}\right) \quad (\text{for } 25° \leq \phi' \leq 45°)$$ (12.18)

and the formula

$$\Delta = 0.005\left(1 - \frac{\phi' - 25}{20}\right)\left(\frac{q'}{p_a}\right)$$ (12.19)

The magnitude of $Q_{p(\text{net})}$ can also be reasonably estimated from a relationship based on the analysis of Berezantzev et al. (1961) that can be expressed as

$$Q_{p(\text{net})} = A_p q'(\omega N_q^* - 1)$$ (12.20)

where N_q^* = bearing capacity factor = $0.21e^{0.17\phi'}$ (12.21)
ω = correction factor = $f(L/D_b)$

In Eq. (12.21), ϕ' is in degrees. The variation of ω with L/D_b is given in Figure 12.7.

The frictional resistance at ultimate load, Q_s, developed in a drilled shaft may be calculated from the relation given in Eq. (12.15), in which

$$p = \text{shaft perimeter} = \pi D_s$$

$$f = \text{unit frictional (or skin) resistance} = K\sigma_o' \tan \delta$$ (12.22)

where K = earth pressure coefficient $\approx K_o = 1 - \sin \phi'$
σ_o' = effective vertical stress at any depth z

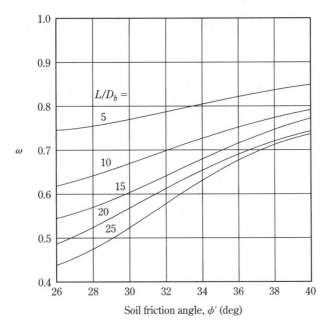

Figure 12.7 Variation of ω with ϕ' and L/D_b

Thus,

$$Q_s = \int_0^{L_1} pf\,dz = \pi D_s(1 - \sin\phi') \int_0^{L_1} \sigma_o' \tan\delta\,dz \qquad (12.23)$$

The value of σ_o' will increase to a depth of about $15D_s$ and will remain constant thereafter, as shown in Figure 11.18.

An appropriate factor of safety should be applied to the ultimate load to obtain the net allowable load, or

$$Q_{\text{all(net)}} = \frac{Q_{p(\text{net})} + Q_s}{\text{FS}} \qquad (12.24)$$

Load-Bearing Capacity Based on Settlement

On the basis of the performance of bored piles in sand with an average diameter of 750 mm (2.5 ft), Touma and Reese (1974) suggested a procedure for calculating the allowable load-carrying capacity. Their procedure, which is also applicable to drilled shafts in sand, is as follows:

For $L > 10D_b$ and a *base movement* of 25.4 mm (1 in.), the allowable net point load,

$$Q_{p\text{-all(net)}} = \frac{0.508A_p}{D_b}q_p \qquad (12.25)$$

where $Q_{p\text{-all(net)}}$ is in kN, A_p is in m^2, D_b is in m, and q_p is the unit point resistance in kN/m^2

In English units,

$$Q_{p\text{-all(net)}} = \frac{A_p}{0.6D_b}q_p \qquad (12.26)$$

where $Q_{p\text{-all(net)}}$ is in lb, A_p is in ft^2, D_b is in ft, and q_p is in lb/ft^2

The values of q_p, as recommended by Touma and Reese, are given in the following table:

Sand type	q_p (kN/m^2)	q_p (lb/ft^2)
Loose	0	0
Medium	1530	32,000
Very dense	3830	80,000

For sands of intermediate densities, linear interpolation can be used. The shaft friction resistance can be calculated as

$$Q_s = \int_0^{L_1} (0.7)\, p\sigma_o' \tan \phi'\, dz = 0.7(\pi D_s) \int_0^{L_1} \sigma_o' \tan \phi'\, dz$$
$$= 2.2D_s \int_0^{L_1} \sigma_o' \tan \phi'\, dz \qquad (12.27)$$

where ϕ' = effective soil friction angle
σ_o' = vertical effective stress at a depth z

For the definition of L_1, see Figure 12.6. Thus,

$$Q_{\text{all(net)}} = Q_{p\text{-all(net)}} + \frac{Q_s}{\text{FS}} \quad [\text{for a base movement of 25.4 mm (1 in.)}] \quad (12.28)$$

where FS = factor of safety (≈ 2)

On the basis of a database of 41 loading tests, Reese and O'Neill (1989) proposed a method for calculating the load-bearing capacity of drilled shafts that is based on settlement. The method is applicable to the following ranges:

1. Shaft diameter: D_s = 0.52 m to 1.2 m (1.7 ft to 3.93 ft)
2. Bell depth: L = 4.7 m to 30.5 m (15.4 ft to 100 ft)
3. Field standard penetration resistance: N_{60} = 5 to 60
4. Concrete slump = 100 mm to 225 mm (4 in. to 9 in.)

Reese and O'Neill's procedure (see Figure 12.8) gives

$$Q_{u(net)} = \sum_{i=1}^{N} f_i p \Delta L_i + q_p A_p \qquad (12.29)$$

where f_i = ultimate unit shearing resistance in layer i
  p = perimeter of the shaft = πD_s
  q_p = unit point resistance
  A_p = area of the base = $(\pi/4) D_b^2$

Following are the relationships for determining $Q_{u(net)}$ from Eq. (12.29) for granular soils. We have

$$f_i = \beta \sigma'_{ozi} \leq 4 \text{ kip/ft}^2 \qquad (12.30)$$

where σ'_{ozi} = vertical effective stress at the middle of layer i
  $\beta = 1.5 - 0.135 z_i^{0.5}$ $(0.25 \leq \beta \leq 1.2)$ (12.31)
  z_i = depth to the middle of layer i (ft)

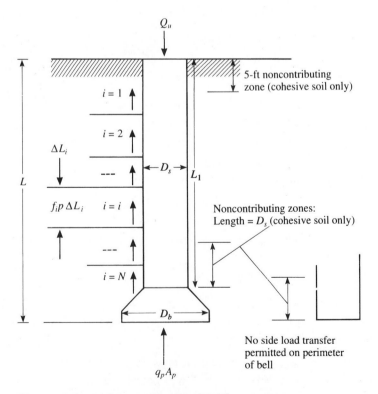

Figure 12.8 Development of Eq. (12.29)

The point bearing capacity is

$$q_p \ (\text{kip/ft}^2) = 1.2N_{60} \leqslant 90 \ \text{kip/ft}^2 \qquad (\text{for } D_b < 50 \ \text{in.}) \qquad (12.32)$$

where N_{60} = mean *uncorrected* standard penetration number within a distance of $2D_b$ below the base of the drilled shaft

If D_b is equal to or greater than 50 in., excessive settlement may occur. In that case, q_p may be replaced by

$$q_{\text{pr}} = \frac{50}{D_b(\text{in.})} \, q_p \qquad (\text{for } D_b \geqslant 50 \ \text{in.}) \qquad (12.33)$$

Figures 12.9 and 12.10 may now be used to calculate the allowable load $Q_{\text{all(net)}}$, based on the desired level of settlement.

In SI units, Eqs. (12.30) through (12.33) will be of the form

$$f_i = \beta\sigma'_{ozi} \leqslant \quad 192 \ \text{kN/m}^2 \qquad (12.34)$$

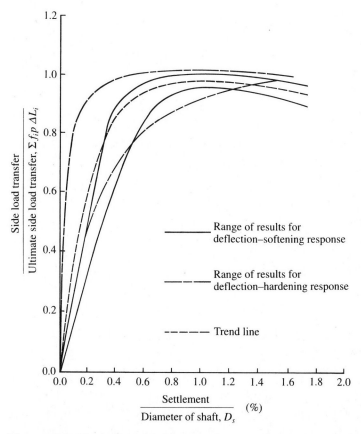

Figure 12.9 Normalized side load transfer vs. settlement for cohesionless soil (after Reese and O'Neill, 1989)

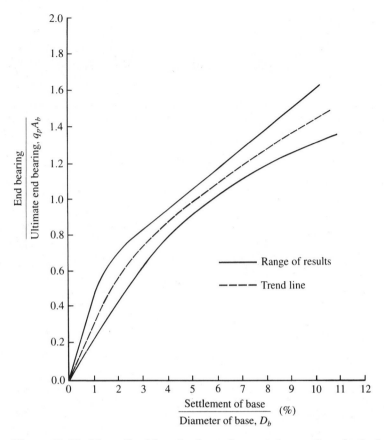

Figure 12.10 Normalized base load transfer vs. settlement for cohesionless soil (after Reese and O'Neill, 1989)

$$\beta = 1.5 - 0.244z_i^{0.5} \qquad (0.25 \le \beta \le 1.2) \tag{12.35}$$

(where z_i is in m)

$$q_p \ (\text{kN/m}^2) = 57.5N_{60} \le 4310 \ \text{kN/m}^2 \qquad (\text{for} D_b < 1.27 \ \text{m}) \tag{12.36}$$

and

$$q_{\text{pr}} = \frac{1.27}{D_b(\text{m})}q_p \tag{12.37}$$

Example 12.1

A soil profile is shown in Figure 12.11. A point bearing drilled shaft with a bell is placed in a layer of dense sand and gravel. Determine the allowable load the drilled shaft could carry. Use Eq. (12.16) and a factor of safety of 4. Take $D_s = 1$ m and $D_b = 1.75$ m. For the dense sand layer, $\phi' = 36°$; $E_s = 500p_a$. Ignore the frictional resistance of the shaft.

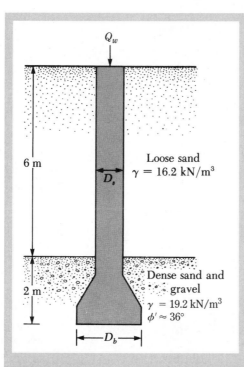

Figure 12.11 Allowable load of drilled shaft

Solution

We have

$$Q_{p(\text{net})} = A_p[q'(N_q - 1)F_{qs}F_{qd}F_{qc}]$$

and

$$q' = (6)(16.2) + (2)(19.2) = 135.6 \text{ kN/m}^2$$

For $\phi' = 36°$, from Table 3.4, $N_q = 37.75$. Also,

$$F_{qs} = 1 + \tan \phi' = 1 + \tan 36 = 1.727$$

and

$$F_{qd} = 1 + 2 \tan \phi'(1 - \sin \phi')^2 \tan^{-1}\left(\frac{L}{D_b}\right)$$

$$= 1 + 2 \tan 36(1 - \sin 36)^2 \tan^{-1}\left(\frac{8}{1.75}\right) = 1.335$$

From Eq. (12.9),

$$I_{rc} = 0.5 \exp\left[2.85 \cot\left(45 - \frac{\phi'}{2}\right)\right] = 134.3$$

From Eq. (12.17), $E_s = mp_a$. With $m = 500$, we have

$$E_s = (500)(100) = 50,000 \text{ kN/m}^2$$

From Eq. (12.18),

$$\mu_s = 0.1 + 0.3\left(\frac{\phi' - 25}{20}\right) = 0.1 + 0.3\left(\frac{36 - 25}{20}\right) = 0.265$$

So

$$I_r = \frac{E_s}{2(1 + \mu_s)(q')(\tan \phi')} = \frac{50,000}{2(1 + 0.265)(135.6)(\tan 36)} = 200.6$$

From Eq. (12.10),

$$I_{rr} = \frac{I_r}{1 + I_r\Delta}$$

with

$$\Delta = 0.005\left(1 - \frac{\phi' - 25}{20}\right)\frac{q'}{p_a} = 0.005\left(1 - \frac{36 - 25}{20}\right)\left(\frac{135.6}{100}\right) = 0.0031$$

it follows that

$$I_{rr} = \frac{200.6}{1 + (200.6)(0.0031)} = 123.7$$

I_{rr} is less than I_{rc}. So, from Eq. (12.14),

$$F_{qc} = \exp\left\{(-3.8 \tan \phi') + \left[\frac{(3.07 \sin \phi')(\log_{10} 2I_{rr})}{1 + \sin \phi'}\right]\right\}$$

$$= \exp\left\{(-3.8 \tan 36) + \left[\frac{(3.07 \sin 36) \log(2 \times 123.7)}{1 + \sin 36}\right]\right\} = 0.958$$

Hence,

$$Q_{p(net)} = \left[\left(\frac{\pi}{4}\right)(1.75)^2\right](135.6)(37.75 - 1)(1.727)(1.335)(0.958) = 26,474 \text{ kN}$$

and

$$Q_{p(all)} = \frac{Q_{p(net)}}{\text{FS}} = \frac{26,474}{4} \approx \mathbf{6619 \text{ kN}} \qquad \blacksquare$$

Example 12.2

Solve Example 12.1 using Eq. (12.20).

Solution

Equation (12.20) asserts that

$$Q_{p(net)} = A_p q'(\omega N_q^* - 1)$$

We have

$$N_q^* = 0.21 e^{0.17\phi'} = 0.21 e^{(0.17)(36)} = 95.52$$

and

$$\frac{L}{D_b} = \frac{8}{1.75} = 4.57$$

From Figure 12.7, for $\phi' = 36°$ and $L/D_b = 4.57$, the value of ω is about 0.83. So

$$Q_{p(net)} = \left[\left(\frac{\pi}{4} \right)(1.75)^2 \right](135.6)[(0.83)(95.52) - 1] = 25,532 \text{ kN}$$

and

$$Q_{p(all)} = \frac{25,532}{4} = \textbf{6383 kN} \quad \blacksquare$$

Example 12.3

A drilled shaft is shown in Figure 12.12. The uncorrected average standard penetration number (N_{60}) within a distance of $2D_b$ below the base of the shaft is about 30. Determine

a. The ultimate load-carrying capacity
b. The load-carrying capacity for a settlement of 0.5 in. Use Reese and O'Neill's method.

Solution

Part a
From Eqs. (12.30) and (12.31),

$$f_i = \beta \sigma'_{ozi}$$

and

$$\beta = 1.5 - 0.135 z_i^{0.5}$$

For this problem, $z_i = 20/2 = 10$ ft, so

$$\beta = 1.5 - (0.135)(10)^{0.5} = 1.07$$

and

$$\sigma'_{ozi} = \gamma z_i = (100)(10) = 1000 \text{ lb/ft}^2$$

Thus,

$$f_i = (1000)(1.07) = 1070 \text{ lb/ft}^2$$

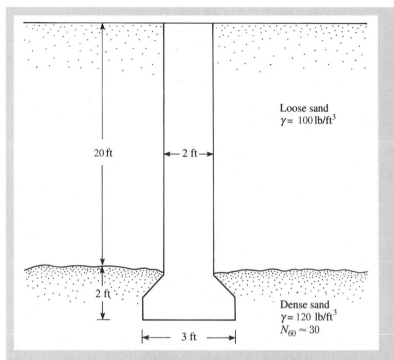

Figure 12.12 Drilled shaft supported by a dense layer of sand

and

$$\Sigma f_i p \, \Delta L_i = (1070)(\pi \times 2)(20) = 134{,}460 \text{ lb} = 134.46 \text{ kip}$$

From Eq. (12.32)

$$q_p = 1.2N_{60} = (1.2)(30) = 36 \text{ kip/ft}^2$$

so

$$q_p A_p = (36)\left[\frac{\pi}{4}(3)^2\right] = 254.47 \text{ kip}$$

Hence,

$$Q_{u(\text{net})} = q_p A_p + \Sigma f_i p \, \Delta L_i = 254.47 + 134.46 = \textbf{388.9 kip}$$

Part b
We have

$$\frac{\text{Allowable settlement}}{D_s} = \frac{0.5}{(2)(12)} = 0.021 = 2.1\%$$

The trend line shown in Figure 12.9 indicates that, for a normalized settlement of 2.1%, the normalized side load is about 0.9. Thus, the side load transfer is (0.9)(134.46) ≈ 121 kip. Similarly,

$$\frac{\text{Allowable settlement}}{D_b} = \frac{0.5}{(3)(12)} = 0.014 = 1.4\%$$

The trend line shown in Figure 12.10 indicates that, for a normalized settlement of 1.4%, the normalized base load is 0.312. So the base load is $(0.312)(254.47) = 79.4$ kip. Hence, the total load is

$$Q = 121 + 79.4 \approx \textbf{200 kip}$$

∎

12.8 Drilled Shafts in Clay: Load-Bearing Capacity

For saturated clays with $\phi = 0$, the bearing capacity factor N_q in Eq. (12.4) is equal to unity. Thus, for this case, Eq. (12.4) will be of the form

$$Q_{p(\text{net})} = A_p c_u N_c F_{cs} F_{cd} F_{cc} \tag{12.38}$$

where c_u = undrained cohesion

Assuming that $L \geqslant 3D_b$, we can rewrite Eq. (12.38) as

$$Q_{p(\text{net})} = A_p c_u N_c^* \tag{12.39}$$

where $\qquad N_c^* = N_c F_{cs} F_{cd} F_{cc} = 1.33[(\ln I_r) + 1] \tag{12.40}$

in which $\qquad I_r$ = soil rigidity index

The soil rigidity index was defined in Eq. (12.11). For $\phi = 0$,

$$I_r = \frac{E_s}{3c_u} \tag{12.41}$$

O'Neill and Reese (1999) provided an approximate relationship between c_u and $E_s/3c_u$. This relationship is shown in Figure 12.13. For all practical purposes, if c_u/p_a is equal to or greater than unity (p_a = atmospheric pressure $\approx 100 \text{ kN/m}^2$ or 2000 lb/ft^2), then the magnitude of N_c^* can be taken to be 9.

Experiments by Whitaker and Cooke (1966) showed that, for belled shafts, the full value of $N_c^* = 9$ is realized with a base movement of about 10%–15% of D_b. Similarly, for straight shafts ($D_b = D_s$), the full value of $N_c^* = 9$ is obtained with a base movement of about 20% of D_b.

The expression for the skin resistance of drilled shafts in clay is similar to Eq. (11.54), or

$$Q_s = \sum_{L=0}^{L=L_1} \alpha^* c_u p \, \Delta L \tag{12.42}$$

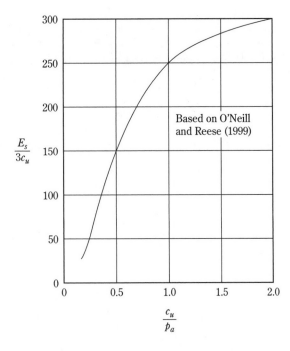

Figure 12.13   Approximate variation of $\dfrac{E_s}{3c_u}$ with c_u/p_a (*Note: p_a = atmospheric pressure*)

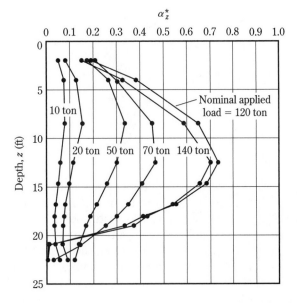

Figure 12.14   Variation of α_z^* with depth for the drilled shaft load test shown in Figure 12.5 (after Reese, Touma, and O'Neill, 1976)

where p = perimeter of the shaft cross section

The value of α^* that can be used in Eq. (12.42) has not yet been fully established. However, the field-test results available at this time indicate that α^* may vary between 1.0 and 0.3. Figure 12.14 shows the variation of α^* with depth (α_z^*) at various stages of loading for the case of the drilled shaft presented in Figure 12.5. The values of α_z^* were derived from Figure 12.5c. At ultimate load, the peak value of α_z^* is about 0.7, with an average of $\alpha^* \approx 0.5$.

Kulhawy and Jackson (1989) reported the field-test result of 106 straight drilled shafts—65 in uplift and 41 in compression. The magnitudes of α^* obtained from these tests are shown in Figure 12.15. The best correlation obtained from the results is

$$\alpha^* = 0.21 + 0.25\left(\frac{p_a}{c_u}\right) \leq 1 \tag{12.43}$$

where p_a = atmospheric pressure $\approx 1 \text{ ton/ft}^2$ ($\approx 100 \text{ kN/m}^2$)
So, conservatively, we may assume that

$$\alpha^* = 0.4 \tag{12.44}$$

Load-Bearing Capacity Based on Settlement

Reese and O'Neill (1989) suggested a procedure for estimating the ultimate and allowable (based on settlement) bearing capacities for drilled shafts in clay. According to this procedure, we can use Eq. (12.29) for the net ultimate load, or

$$Q_{u(\text{net})} = \sum_{i=1}^{n} f_i p\, \Delta L_i + q_p A_p$$

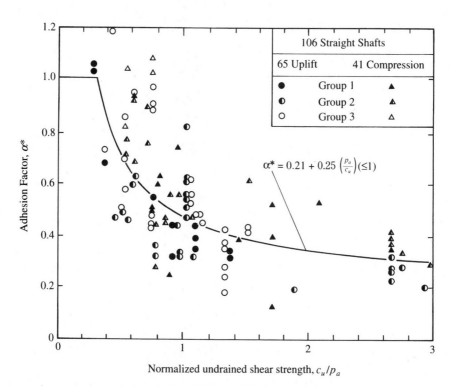

Figure 12.15   Variation of α^* with c_u/p_a (after Kulhawy and Jackson, 1989)

The unit skin friction resistance can be given as

$$f_i = \alpha_i^* c_{u(i)} \tag{12.45}$$

The following values are recommended for α_i^*:

$\alpha_i^* = 0$ for the top 1.5 m (5 ft) and bottom 1 diameter, D_s, of the drilled shaft. (*Note*: If $D_b > D_s$, then $\alpha^* = 0$ for 1 diameter above the top of the bell and for the peripheral area of the bell itself.)

$\alpha_i^* = 0.55$ elsewhere.

$$q_p = 6c_{ub}\left(1 + 0.2\frac{L}{D_b}\right) \leq 9c_{ub} \leq 80\,\text{kip/ft}^2 \tag{12.46}$$

where c_{ub} = average undrained cohesion within $2D_b$ below the base.

If D_b is large, excessive settlement will occur at the ultimate load per unit area, q_p, as given by Eq. (12.46). Thus, for $D_b > 1.91$ m (75 in.), q_p may be replaced by

$$q_{pr} = F_r q_p \tag{12.47}$$

where $F_r = \dfrac{2.5}{\psi_1 D_b\,(\text{in.}) + \psi_2} \leq 1$ $\tag{12.48}$

in which

$$\psi_1 = 0.0071 + 0.0021\left(\frac{L}{D_b}\right) \leq 0.015 \tag{12.49}$$

and

$$\psi_2 = \underset{\underset{\text{kip/ft}^2}{\uparrow}}{0.45(c_{ub})^{0.5}} \quad (0.5 \leq \psi_2 \leq 1.5) \tag{12.50}$$

In SI units, Eqs. (12.46), (12.48), (12.49), and (12.50) can be expressed as

$$q_p = 6c_{ub}\left(1 + 0.2\frac{L}{D_b}\right) \leq 9c_{ub} \leq 3830\,\text{kN/m}^2 \tag{12.51}$$

$$F_r = \frac{2.5}{\psi_1 D_b\,(\text{mm}) + \psi_2} \leq 1 \tag{12.52}$$

$$\psi_1 = 2.78 \times 10^{-4} + 8.26 \times 10^{-5}\left(\frac{L}{D_b}\right) \leq 5.9 \times 10^{-4} \tag{12.53}$$

and

$$\psi_2 = 0.065[c_{ub} \ (kN/m^2)]^{0.5} \tag{12.54}$$

Figures 12.16 and 12.17 may now be used to evaluate the allowable load-bearing capacity, based on settlement. (Note that the ultimate bearing capacity in Figure 12.17 is q_p, not q_{pr}.) To do so,

1. Select a value of settlement, s.
2. Calculate $\sum_{i=1}^{N} f_i p \ \Delta L_i$ and $q_p A_p$.
3. Using Figures 12.16 and 12.17 and the calculated values in Step 2, determine the *side load* and the *end bearing load*.
4. The sum of the side load and the end bearing load gives the total allowable load.

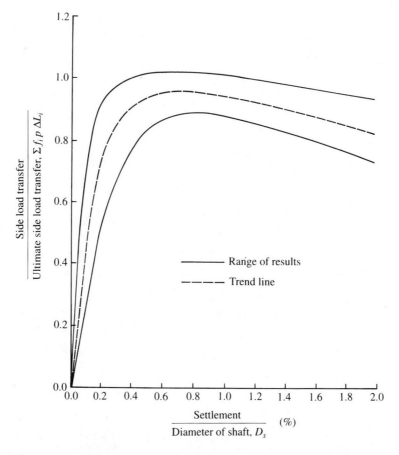

Figure 12.16 Normalized side load transfer vs. settlement for cohesive soil (after Reese and O'Neill, 1989)

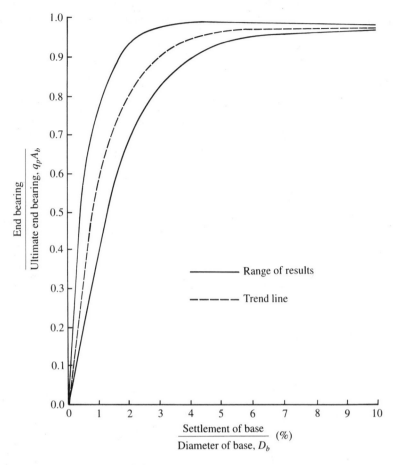

Figure 12.17 Normalized base load transfer vs. settlement for cohesive soil (after Reese and O'Neill, 1989)

Example 12.4

Figure 12.18 shows a drilled shaft without a bell. Here, $L_1 = 27$ ft, $L_2 = 8.5$ ft, $D_s = 3.3$ ft, $c_{u(1)} = 1000$ lb/ft², and $c_{u(2)} = 2175$ lb/ft². Determine

a. The net ultimate point bearing capacity
b. The ultimate skin resistance
c. The working load, Q_w (FS = 3)

Use Eqs. (12.39), (12.42), and (12.44).

Solution
Part a
From Eq. (12.39),

$$Q_{p(\text{net})} = A_p c_u N_c^* = A_p c_{u(2)} N_c^* = \left[\left(\frac{\pi}{4} \right) (3.3)^2 \right] (2175)(9)$$

$$= 167{,}425 \text{ lb} \approx \mathbf{167.4 \ kip}$$

(*Note:* Since $c_u / p_a > 1$, $N_c^* \approx 9$.)

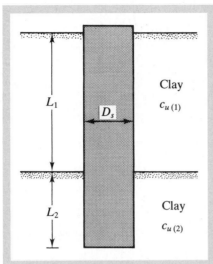

Figure 12.18 A drill shaft without a bell

Part b
From Eq. (12.42),

$$Q_s = \Sigma \alpha^* c_u p \Delta L$$

From Eq. (12.44),

$$\alpha^* = 0.4$$
$$p = \pi D_s = (3.14)(3.3) = 10.37 \text{ ft}$$

and

$$Q_s = (0.4)(10.37)[(1000 \times 27) + (2175 \times 8.5)]$$
$$= 188,682 \text{ lb} \approx \mathbf{188.7 \text{ kip}}$$

Part c

$$Q_w = \frac{Q_{p(net)} + Q_s}{\text{FS}} = \frac{167.4 + 188.7}{3} = \mathbf{118.7 \text{ kip}}$$ ∎

Example 12.5

A drilled shaft in a cohesive soil is shown in Figure 12.19. Use Reese and O'Neill's method to determine

a. The ultimate load-carrying capacity (Eqs. 12.45 through 12.50)
b. The load-carrying capacity for an allowable settlement of 0.5 in.

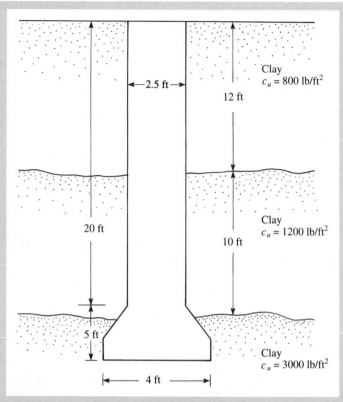

Figure 12.19   A drilled shaft in layered clay

Solution

Part a

From Eq. (12.45),

$$f_i = \alpha_i^* c_{u(i)}$$

From Figure 12.19

$$\Delta L_1 = 12 - 5 = 7 \text{ ft}$$

$$\Delta L_2 = (20 - 12) - D_s = (20 - 12) - 2.5 = 5.5 \text{ ft}$$

$$c_{u(1)} = 800 \text{ lb/ft}^2$$

and

$$c_{u(2)} = 1200 \text{ lb/ft}^2$$

Hence,

$$\Sigma f_i p \Delta L_i = \Sigma \alpha_i^* c_{u(i)} p \Delta L_i$$

$$= (0.55)(800)(\pi \times 2.5)(7) + (0.55)(1200)(\pi \times 2.5)(5.5)$$

$$= 52,700 \text{ lb} = 52.7 \text{ kip}$$

Again, from Eq. (12.46),

$$q_p = 6c_{ub}\left(1 + 0.2\frac{L}{D_b}\right) = (6)(3000)\left[1 + 0.2\left(\frac{20 + 5}{4}\right)\right] = 40,500 \text{ lb/ft}^2$$

$$= 40.5 \text{ kip/ft}^2$$

A check reveals that

$$q_p = 9c_{ub} = (9)(3000) = 27,000 \text{ lb/ft}^2 = 27 \text{ kip/ft}^2 < 40.5 \text{ kip/ft}^2$$

So we use $q_p = 27 \text{ kip/ft}^2$:

$$q_pA_p = q_p\left(\frac{\pi}{4}D_b^2\right) = (27)\left[\left(\frac{\pi}{4}\right)(4)^2\right] \approx 339.3 \text{ kip}$$

Hence,

$$Q_u = \Sigma\alpha_i^*c_{u(i)}p\Delta L_i + q_pA_p = 52.7 + 339.3 = \textbf{392 kip}$$

Part b
We have

$$\frac{\text{Allowable settlement}}{D_s} = \frac{0.5}{(2.5)(12)} = 0.167 = 1.67\%$$

The trend line shown in Figure 12.16 indicates that, for a normalized settlement of 1.67%, the normalized side load is about 0.89. Thus, the side load is

$$(0.89)(\Sigma f_ip\Delta L_i) = (0.89)(52.7) = 46.9 \text{ kip}$$

Again,

$$\frac{\text{Allowable settlement}}{D_b} = \frac{0.5}{(4)(12)} = 0.0104 = 1.04\%$$

The trend line shown in Figure 12.17 indicates that, for a normalized settlement of 1.04%, the normalized end bearing is about 0.57, so

$$\text{Base load} = (0.57)(q_pA_p) = (0.57)(339.3) = 193.4 \text{ kip}$$

Thus, the total load is

$$Q = 46.9 + 193.4 = \textbf{240.3 kip} \qquad \blacksquare$$

12.9 *Settlement of Drilled Shafts at Working Load*

The settlement of drilled shafts at working load is calculated in a manner similar to that outlined in Section 11.18. In many cases, the load carried by shaft resistance is small compared with the load carried at the base. In such cases, the contribution of

s_3 may be ignored. Note that in Eqs. (11.74) and (11.75) the term D should be replaced by D_b for drilled shafts.

Example 12.6

For the drilled shaft of Example 12.4, estimate the elastic settlement at working loads (i.e., $Q_w = 118.7$ kip). Use Eqs. (11.73), (11.75), and (11.76). Take $\xi = 0.65$, $E_p = 3 \times 10^6$ lb/in^2, $E_s = 2000$ lb/in^2, $\mu_s = 0.3$, and $Q_{wp} = 24.35$ kip.

Solution
From Eq. (11.73),

$$s_{e(1)} = \frac{(Q_{wp} + \xi Q_{ws})L}{A_p E_p}$$

Now,

$$Q_{ws} = 118.7 - 24.35 = 94.35 \text{ kip}$$

so

$$s_{e(1)} = \frac{[24.35 + (0.65 \times 94.35)](35.5)}{\left(\dfrac{\pi}{4} \times 3.3^2\right)\left(\dfrac{3 \times 10^6 \times 144}{1000}\right)} = 0.000823 \text{ ft} = 0.0099 \text{ in.}$$

From Eq. (11.75),

$$s_{e(2)} = \frac{Q_{wp}C_p}{D_b q_p}$$

From Table 11.11, for stiff clay, $C_p \approx 0.04$; also,

$$q_p = c_{u(b)}N_c^* = (2.175 \text{ kip/ft}^2)(9) = 19.575 \text{ kip/ft}^2$$

Hence,

$$s_{e(2)} = \frac{(24.35)(0.04)}{(3.3)(19.575)} = 0.015 \text{ ft} = 0.18 \text{ in.}$$

Again, from Eqs. (11.76) and (11.77),

$$s_{e(3)} = \left(\frac{Q_{ws}}{pL}\right)\left(\frac{D_s}{E_s}\right)(1 - \mu_s^2)I_{ws}$$

where $I_{ws} = 2 + 0.35\sqrt{\dfrac{L}{D_s}} = 2 + 0.35\sqrt{\dfrac{35}{3.3}} = 3.15$

So

$$s_{e(3)} = \left[\frac{94.35}{(\pi \times 3.3)(35.5)}\right]\left(\frac{3.3}{\dfrac{2000 \times 144}{1000}}\right)(1 - 0.3^2)(3.15) = 0.0084 \text{ ft} = 0.1 \text{ in.}$$

The total settlement is

$$s_e = s_{e(1)} + s_{e(2)} + s_{e(3)} = 0.0099 + 0.18 + 0.1 \approx \textbf{0.29 in.}$$ ■

12.10 *Lateral Load-Carrying Capacity*

Several methods for analyzing the lateral load-carrying capacity of piles, as well as the load-carrying capacity of drilled shafts, were presented in Section 11.9; therefore, they will not be repeated here. In 1994, Duncan et al. developed a *characteristic load method* for estimating the lateral load capacity for drilled shafts that is fairly simple to use. We describe this method next.

According to the characteristic load method, the *characteristic load* Q_c and *moment* M_c form the basis for the dimensionless relationship that can be given by the following correlations:

Characteristic Load

$$Q_c = 7.34 D_s^2 \, (E_p R_I) \left(\frac{c_u}{E_p R_I} \right)^{0.68} \qquad \text{(for clay)} \qquad (12.55)$$

$$Q_c = 1.57 D_s^2 \, (E_p R_I) \left(\frac{\gamma' D_s \phi' K_p}{E_p R_I} \right)^{0.57} \qquad \text{(for sand)} \qquad (12.56)$$

Characteristic Moment

$$M_c = 3.86 D_s^3 \, (E_p R_I) \left(\frac{c_u}{E_p R_I} \right)^{0.46} \qquad \text{(for clay)} \qquad (12.57)$$

$$M_c = 1.33 D_s^3 (E_p R_I) \left(\frac{\gamma' D_s \phi' K_p}{E_p R_I} \right)^{0.40} \qquad \text{(for sand)} \qquad (12.58)$$

In these equations, D_s = diameter of drilled shafts
E_p = modulus of elasticity of drilled shafts
R_I = ratio of moment of inertia of drilled shaft section to moment of inertia of a solid section (*Note: R_I = 1 for uncracked shaft without central void*)
γ' = effective unit weight of sand
ϕ' = effective soil friction angle (degrees)
K_p = Rankine passive pressure coefficient = $\tan^2(45 + \phi'/2)$

Deflection Due to Load Q_g Applied at the Ground Line

Table 12.1 gives the variation of x_o/D_s with Q_g/Q_c and Figure 12.20 gives the plot of Q_g/Q_c versus x_o/D_s for drilled shafts in sand and clay due to the load Q_g applied at the ground surface. Note that x_o is the ground line deflection. If the magnitudes of Q_g and

Table 12.1 Variation of x_o/D_s with Q_g/Q_c due to load Q_g applied at the ground surface

$\dfrac{x_o}{D_s}$	Clay		Sand	
	Free head $\dfrac{Q_g}{Q_c}$	Fixed head $\dfrac{Q_g}{Q_c}$	Free head $\dfrac{Q_g}{Q_c}$	Fixed head $\dfrac{Q_g}{Q_c}$
0.0000	0.0000	0.0000	0.0000	0.0000
0.0025	0.0040	0.0088	0.0008	0.0016
0.0050	0.0065	0.0133	0.0013	0.0028
0.0075	0.0078	0.0168	0.0017	0.0039
0.0100	0.0091	0.0197	0.0021	0.0049
0.0150	0.0113	0.0247	0.0027	0.0065
0.0200	0.0135	0.0289	0.0033	0.0079
0.0300	0.0171	0.0359	0.0043	0.0104
0.0400	0.0200	0.0419	0.0052	0.0125
0.0500	0.0226	0.0471	0.0060	0.0144
0.0600	0.0250	—	0.0068	—
0.0800	0.0292	—	0.0083	—
0.1000	0.0332	—	0.0097	—
0.1500	0.0412	—	0.0124	—

After Duncan et al. (1994)

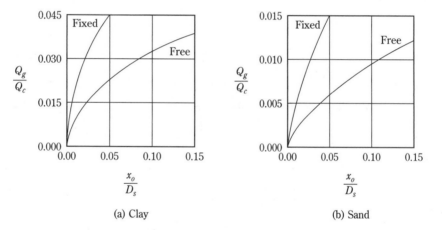

(a) Clay (b) Sand

Figure 12.20 Plot of Q_g/Q_c vs. x_o/D_s (after Duncan et al., 1994)

Q_c are known, the ratio Q_g/Q_c can be calculated. The table or the figure can then be used to estimate the corresponding value of x_o/D_s and, hence, x_o.

Deflection Due to Moment Applied at the Ground Line

Table 12.2 gives the variation of x_o/D_s with M_g/M_c and Figure 12.21 gives the variation plot of M_g/M_c with x_o/D_s for drilled shafts in sand and clay due to an applied moment M_g at the ground line. Again, x_o is the ground line deflection. If the magnitudes of M_g, M_c, and D_s are known, the value of x_o can be calculated with the use of the table or the figure.

Table 12.2   Moment–deflection coefficients

$\dfrac{x_o}{D_s}$	Applied Moment	
	Clay M_g/M_c	Sand M_g/M_c
0.00	0.0000	0.0000
0.01	0.0048	0.0019
0.02	0.0074	0.0032
0.03	0.0097	0.0044
0.04	0.0119	0.0055
0.05	0.0139	0.0065
0.06	0.0158	0.0075
0.08	0.0193	0.0094
0.10	0.0226	0.0113
0.15	0.0303	0.0150

After Duncan et al. (1994)

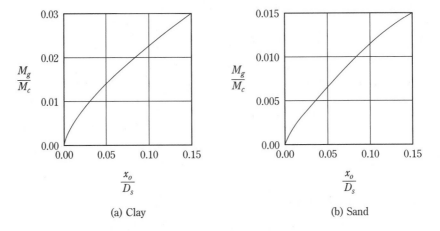

(a) Clay   (b) Sand

Figure 12.21   Plot of M_g/M_c vs. x_o/D_s (after Duncan et al., 1994)

Deflection Due to Load Applied Above the Ground Line

When a load Q is applied above the ground line, it induces both a load $Q_g = Q$ and a moment $M_g = Qe$ at the ground line, as shown in Figure 12.22. A superposition solution can now be used to obtain the ground line deflection. The step-by-step procedure is as follows:

1. Calculate Q_g and M_g.
2. Calculate the deflection x_{oQ} (see Figure 12.22a) that would be caused by the load Q_g acting alone. Use Table 12.1 or Figure 12.20.
3. Calculate the deflection x_{oM} (see Figure 12.22b) that would be caused by the moment acting alone. Use Table 12.2 or Figure 12.21.
4. Determine the value of a load Q_{gM} that would cause the same deflection as the moment (i.e., x_{oM}). Use Table 12.1 or Figure 12.20. This is schematically shown in Figure 12.22c.

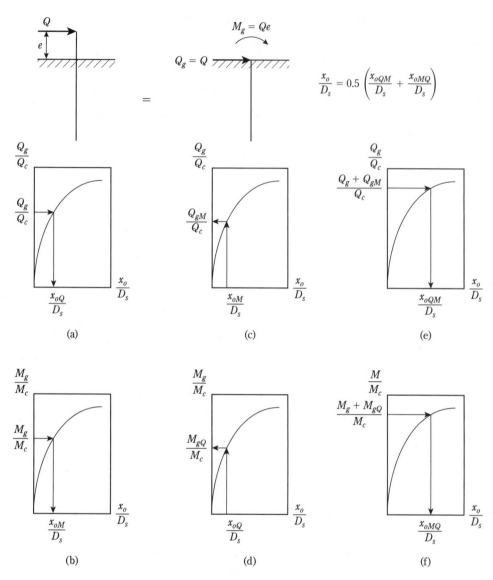

Figure 12.22   Nonlinear superposition of deflections due to load and moment (after Duncan et al., 1994)

5. Determine the value of a moment M_{gQ} that would cause the same deflection as the load (i.e, x_{oQ}), as shown in Figure 12.22d. Use Table 12.2 or Figure 12.21.

6. Calculate $(Q_g + Q_{gM})/Q_c$. Use Table 12.1 or Figure 12.20 to determine x_{oQM}/D_s. (See Figure 12.22e.)

7. Calculate $(M_g + M_{gQ})/M_c$. Use Table 12.2 or Figure 12.21 to determine x_{oMQ}/D_s. (See Figure 12.22f.)

8. Calculate the combined deflection: $x_{o(\text{combined})} = 0.5(x_{oQM} + x_{oMQ})$ (12.59)

Maximum Moment in Drilled Shaft Due to Ground Line Load Only

Figure 12.23 shows the plot of Q_g/Q_c with M_{max}/M_c and Table 12.3 shows the variation of M_{max}/M_c with Q_g/Q_c for fixed- and free-headed drilled shafts due only to the application of a ground line load Q_g. For fixed-headed shafts, the maximum moment in the shaft, M_{max}, occurs at the ground line. For this condition, if Q_c, M_c, and Q_g are known, the magnitude of M_{max} can be easily calculated.

Maximum Moment Due to Load and Moment at Ground Line

If a load Q_g and a moment M_g are applied at the ground line, the maximum moment in the drilled shaft can be determined in the following manner:

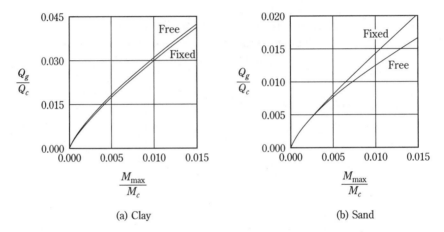

Figure 12.23 Variation of Q_g/Q_c with M_{max}/M_c (after Duncan et al., 1994)

Table 12.3 Load–moment coefficients

	Clay		Sand	
M_{max}/M_c	Free head $\dfrac{Q_g}{Q_c}$	Fixed head $\dfrac{Q_g}{Q_c}$	Free head $\dfrac{Q_g}{Q_c}$	Fixed head $\dfrac{Q_g}{Q_c}$
0.00	0.0000	0.0000	0.0000	0.0000
0.001	0.0050	0.0041	0.0021	0.0019
0.002	0.0090	0.0078	0.0038	0.0037
0.003	0.0125	0.0112	0.0052	0.0052
0.004	0.0157	0.0144	0.0065	0.0067
0.005	0.0185	0.0175	0.0076	0.0080
0.006	0.0212	0.0204	0.0087	0.0093
0.008	0.0264	0.0258	0.0107	0.0117
0.010	0.0319	0.0308	0.0126	0.0138
0.015	0.0432	0.0419	0.0168	0.0186

After Duncan et al. (1994)

1. Using the procedure described before, calculate $x_{o(\text{combined})}$ from Eq. (12.59).
2. To solve for the characteristic length T, use the following equation:

$$x_{o(\text{combined})} = \frac{2.43Q_g}{E_pI_p}T^3 + \frac{1.62M_g}{E_pI_p}T^2 \tag{12.60}$$

3. The moment in the shaft at a depth z below the ground surface can be calculated as

$$M_z = A_mQ_gT + B_mM_g \tag{12.61}$$

where A_m, B_m = dimensionless moment coefficients (Matlock and Reese, 1961); see Figure 12.24

The value of the maximum moment M_{max} can be obtained by calculating M_z at various depths in the upper part of the drilled shaft.

The characteristic load method just described is valid only if L/D_s has a certain minimum value. If the actual L/D_s is less than $(L/D_s)_{\text{min}}$, then the ground line deflections will be underestimated and the moments will be overestimated. The values of $(L/D_s)_{\text{min}}$ for drilled shafts in sand and clay are given in the following table:

Clay	$\dfrac{E_pR_I}{c_u}$	$(L/D_s)_{\text{min}}$
	1×10^5	6
	3×10^5	10
	1×10^6	14
	3×10^6	18

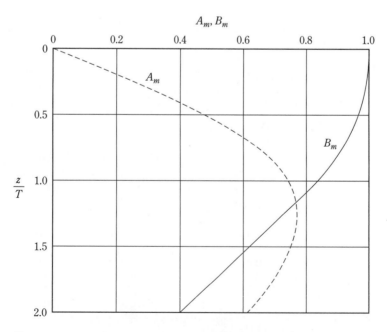

Figure 12.24   Variation of A_m and B_m with z/T

Sand	$\dfrac{E_p R_I}{\gamma' D_s \phi' K_p}$	$(L/D_s)_{\min}$
	1×10^4	8
	4×10^4	11
	2×10^5	14

Example 12.7

Figure 12.25 shows a drilled shaft in sand. If $L = 6$ m, $D_s = 800$ mm, average horizontal soil modulus $E_s = 35 \times 10^3$ kN/m², and $E_p = 20.7 \times 10^6$ kN/m², estimate the ultimate lateral load, $Q_{u(g)}$, applied at the ground surface. Use Meyerhof's method given in Section 11.19, and check your results with Eq. (11.106).

Solution

From Eq. (11.104), the relative stiffness of the shaft is

$$K_r = \frac{E_p I_p}{E_s L^4}$$

with

$$I_p = \frac{\pi}{64} D_s^4 = \frac{\pi}{64} \left(\frac{800}{1000} \right)^4 = 0.02 \text{ m}^4$$

it follows that

$$K_r = \frac{(20.7 \times 10^6)(0.02)}{(35 \times 10^3)(6)^4} = 0.009$$

Since K_r is less than 0.01, this is a flexible drilled shaft. From Eq. (11.109),

$$\frac{L_e}{L} = 1.65 \, K_r^{0.12}$$

so

$$L_e = (1.65)(0.009)^{0.12}(6) = 5.63 \text{ m}$$

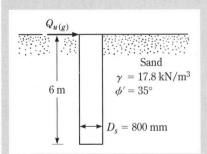

Figure 12.25 Ultimate lateral load of a drilled shaft

Thus, from Eq. (11.105)

$$Q_{u(g)} = 0.12\gamma \, D_s L_e^2 K_{br} \leq 0.4 p_l D L_e$$

Hence,

$$\frac{L_e}{D_s} = \frac{5.63}{0.8} = 7.04$$

From Figure 11.39, for $L_e/D_s = 7.04$ and $\phi' = 35°$, the value of $K_{br} \approx 10$, so

$$Q_{u(g)} = (0.12)(17.8)(0.8)(5.63)^2(10) = 541.6 \text{ kN}$$

As a check, we have

$$Q_{u(g)} = 0.4 p_l D_s L_e = (0.4)(40 N_q \tan \phi') D_s L_e$$
$$\uparrow$$
$$\text{Eq. 11.106}$$

For $\phi' = 35°$, $N_q = 33.3$ (see Table 3.4); thus,

$$Q_{u(g)} = (0.4)(40)(33.3)(\tan 35)(0.8)(5.63) = 1680.3 \text{ kN}$$

Consequently,

$$Q_{u(g)} = \mathbf{541.6 \text{ kN}} \qquad \blacksquare$$

Example 12.8

A free-headed drilled shaft in clay is shown in Figure 12.26. Let $E_p = 22 \times 10^6 \text{ kN/m}^2$. Determine

a. the ground line deflection, $x_{o(\text{combined})}$
b. the maximum bending moment in the drilled shaft
c. the maximum tensile stress in the shaft
d. the minimum penetration of the shaft needed for this analysis

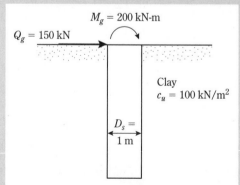

$M_g = 200 \text{ kN-m}$
$Q_g = 150 \text{ kN}$
Clay
$c_u = 100 \text{ kN/m}^2$
$D_s = 1 \text{ m}$

Figure 12.26 Free-headed drilled shaft

Solution
We are given

$$D_s = 1 \text{ m}$$

$$c_u = 100 \text{ kN/m}^2$$

$$R_I = 1$$

$$E_p = 22 \times 10^6 \text{ kN/m}^2$$

and

$$I_p = \frac{\pi D_s^4}{64} = \frac{(\pi)(1)^4}{64} = 0.049 \text{ m}^4$$

Part a
From Eq. (12.55),

$$Q_c = 7.34 D_s^2 \left(E_p R_I\right) \left(\frac{c_u}{E_p R_I}\right)^{0.68}$$

$$= (7.34)(1)^2 [(22 \times 10^6)(1)] \left[\frac{100}{(22 \times 10^6)(1)}\right]^{0.68}$$

$$= 37{,}607 \text{ kN}$$

From Eq. (12.57),

$$M_c = 3.86 D_s^3 \left(E_p R_I\right) \left(\frac{c_u}{E_p R_I}\right)^{0.46}$$

$$= (3.86)(1)^3 [(22 \times 10^6)(1)] \left[\frac{100}{(22 \times 10^6)(1)}\right]^{0.46}$$

$$= 296{,}139 \text{ kN-m}$$

Thus,

$$\frac{Q_g}{Q_c} = \frac{150}{37{,}607} = 0.004$$

From Table 12.1, $x_{oQ} \approx (0.0025) D_s = 0.0025 \text{ m} = 2.5 \text{ mm}$. Also,

$$\frac{M_g}{M_c} = \frac{200}{296{,}139} = 0.000675$$

From Table 12.2, $x_{oM} \approx (0.0014) D_s = 0.0014 \text{ m} = 1.4 \text{ mm}$, so

$$\frac{x_{oM}}{D_s} = \frac{0.0014}{1} = 0.0014$$

From Table 12.1, for $x_{oM}/D_s = 0.0014$, the value of $Q_{gM}/Q_c \approx 0.002$. Hence,

$$\frac{x_{oQ}}{D_s} = \frac{0.0025}{1} = 0.0025$$

From Table 12.2, for $x_{oQ}/D_s = 0.0025$, the value of $M_{gQ}/M_c \approx 0.0013$, so

$$\frac{Q_g}{Q_c} + \frac{Q_{gM}}{Q_c} = 0.004 + 0.002 = 0.006$$

From Table 12.1, for $(Q_g + Q_{gM})/Q_c = 0.006$, the value of $x_{oQM}/D_s \approx 0.0046$. Hence,

$$x_{oQM} = (0.0046)(1) = 0.0046 \text{ m} = 4.6 \text{ mm}$$

Thus, we have

$$\frac{M_g}{M_c} + \frac{M_{gQ}}{M_c} = 0.000675 + 0.0013 \approx 0.00198$$

From Table 12.2, for $(M_g + M_{gQ})/M_c = 0.00198$, the value of $x_{oMQ}/D_s \approx 0.0041$. Hence,

$$x_{oMQ} = (0.0041)(1) = 0.0041 \text{ m} = 4.1 \text{ mm}$$

Consequently,

$$x_{o \text{ (combined)}} = 0.5(x_{oQM} + x_{oMQ}) = (0.5)(4.6 + 4.1) = \textbf{4.35 mm}$$

Part b

From Eq. (12.60),

$$x_{o \text{ (combined)}} = \frac{2.43Q_g}{E_p I_p} T^3 + \frac{1.62 M_g}{E_p I_p} T^2$$

so

$$0.00435 \text{ m} = \frac{(2.43)(150)}{(22 \times 10^6)(0.049)} T^3 + \frac{(1.62)(200)}{(22 \times 10^6)(0.049)} T^2$$

or

$$0.00435 \text{ m} = 338 \times 10^{-6} T^3 + 300.6 \times 10^{-6} T^2$$

and it follows that

$$T \approx 2.05 \text{ m}$$

From Eq. (12.61),

$$M_z = A_m Q_g T + B_m M_g = A_m(150)(2.05) + B_m(200) = 307.5 A_m + 200 B_m$$

$$(a)$$

Now the following table can be prepared:

$\dfrac{z}{T}$	A_m (Figure 12.24)	B_m (Figure 12.24)	M_z (kN-m) [Eq. (a)]
0	0	1.0	200
0.4	0.36	0.98	306.7
0.6	0.52	0.95	349.9
0.8	0.63	0.9	373.7
1.0	0.75	0.845	399.6
1.1	0.765	0.8	395.2
1.25	0.75	0.73	376.6

So the maximum moment is 399.4 kN-m $\approx$ 400 kN-m and occurs at $z/T \approx 1$. Hence,

$$z = (1)(T) = (1)(2.05 \text{ m}) = \textbf{2.05 m}$$

Part c
The maximum tensile stress is

$$\sigma_{tensile} = \frac{M_{max}\left(\dfrac{D_s}{2}\right)}{I_p} = \frac{(400)\left(\dfrac{1}{2}\right)}{0.049} = \textbf{4081.6 kN/m}^2$$

Part d
We have

$$\frac{E_p R_I}{c_u} = \frac{(22 \times 10^6)(1)}{100} = 2.2 \times 10^5$$

By interpolation, for $(E_p R_I)/c_u = 2.2 \times 10^5$, the value of $(L/D_s)_{min} \approx 8.5$. So

$$L \approx (8.5)(1) = \textbf{8.5 m} \qquad \blacksquare$$

12.11 Drilled Shafts Extending into Rock

In Section 12.1, we noted that drilled shafts can be extended into rock. In the current section, we describe the principles of analysis of the load-bearing capacity of such drilled shafts, based on the procedure developed by Reese and O'Neill (1988, 1989). Figure 12.27 shows a drilled shaft whose depth of embedment in rock is equal to L. In the design process to be recommended, it is assumed that *there is either side resistance between the shaft and the rock or point resistance at the bottom, but not both.* Following is a step-by-step procedure for estimating the ultimate bearing capacity:

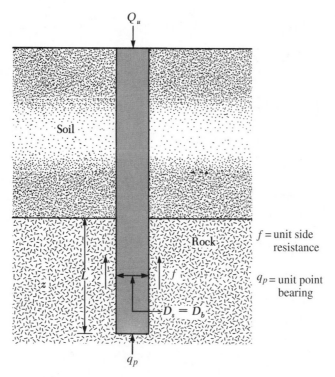

Figure 12.27 Drilled shaft socketed into rock

1. Calculate the ultimate unit side resistance as

$$f \text{ (lb/in}^2) = 2.5q_u^{0.5} \leq 0.15q_u \tag{12.62}$$

where q_u = unconfined compression strength of a rock core of NW size or larger, or of the drilled shaft concrete, whichever is smaller (in lb/in²)

In SI units, Eq (12.62) can be expressed as

$$f \text{ (kN/m}^2) = 6.564q_u^{0.5} \text{ (kN/m}^2) \leq 0.15q_u \text{ (kN/m}^2) \tag{12.63}$$

2. Calculate the ultimate capacity based on side resistance only, or

$$Q_u = \pi D_s L f \tag{12.64}$$

3. Calculate the settlement s_e of the shaft at the top of the rock socket, or

$$s_e = s_{e(s)} + s_{e(b)} \tag{12.65}$$

where $s_{e(s)}$ = elastic compression of the drilled shaft within the socket, assuming no side resistance

$s_{e(b)}$ = settlement of the base

However,

$$s_{e(s)} = \frac{Q_u L}{A_c E_c} \tag{12.66}$$

and

$$s_{e(b)} = \frac{Q_u I_f}{D_s E_{\text{mass}}} \tag{12.67}$$

where Q_u = ultimate load obtained from Eq. (12.62) or Eq. (12.63) (this assumes that the contribution of the overburden to the side shear is negligible)

A_c = cross-sectional area of the drilled shaft in the socket   (12.68)
$$= \frac{\pi}{4} D_s^2$$

E_c = Young's modulus of the concrete and reinforcing steel in the shaft

E_{mass} = Young's modulus of the rock mass into which the socket is drilled

I_f = elastic influence coefficient (see Figure 12.28)

The magnitude of E_{mass} can be determined from the average plot shown in Figure 12.29. In this figure, E_{core} is the Young's modulus of intact specimens of rock cores of NW size or larger. However, unless the socket is very long (O'Neill, 1997),

$$s_e \approx s_{e(b)} = \frac{Q_u I_f}{D_s E_{\text{mass}}} \tag{12.69}$$

4. If s_e is less than 10 mm ($\approx$0.4 in.), then the ultimate load-carrying capacity is that calculated by Eq. (12.64). If $s_e \geq 10$ mm. (0.4 in.), then go to Step 5.

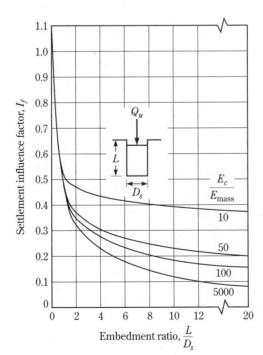

Figure 12.28   Variation of I_f (after Reese and O'Neill, 1989)

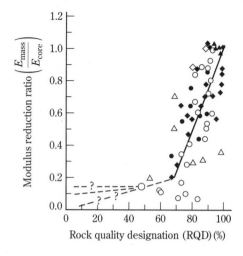

Figure 12.29 Plot of E_{mass}/E_{core} vs. RQD (after Reese and O'Neill, 1989)

5. If $s_e \geq 10$ mm (0.4 in.), there may be rapid, progressive side shear failure in the rock socket, resulting in a complete loss of side resistance. In that case, the ultimate capacity is equal to the point resistance, or

$$Q_u = 3A_p \left[\frac{3 + \dfrac{c_s}{D_s}}{10\left(1 + 300\dfrac{\delta}{c_s}\right)^{0.5}} \right] q_u \tag{12.70}$$

where c_s = spacing of discontinuities (same unit as D_s)
δ = thickness of individual discontinuity (same unit as D_s)
q_u = unconfined compression strength of the rock beneath the base of the socket, or the drilled shaft concrete, whichever is smaller

Note that Eq. (12.70) applies for horizontally stratified discontinuities with $c_s > 305$ mm (12 in.) and $\delta < 5$ mm (0.2 in.).

Example 12.9

Consider the case of a drilled shaft extending into rock, as shown in Figure 12.30. Let $L = 15$ ft, $D_s = 3$ ft, q_u (rock) = 10,500 lb/in^2, q_u (concrete) = 3000 lb/in^2, $E_c = 3 \times 10^6$ lb/in^2, RQD (rock) = 80%, E_{core} (rock) = 0.36×10^6 lb/in^2, $c_s = 18$ in., and $\delta = 0.15$ in. Estimate the allowable load-bearing capacity of the drilled shaft. Use a factor of safety (FS) = 3.

Solution

Step 1. From Eq. (12.62),

$$f \, (\text{lb/in}^2) = 2.5 \, q_u^{0.5} \leq 0.15 q_u$$

Since q_u (concrete) $<$ q_u (rock), use q_u (concrete) in Eq. (12.62). Hence,

$$f = 2.5(3000)^{0.5} = 136.9 \text{ lb/in}^2$$

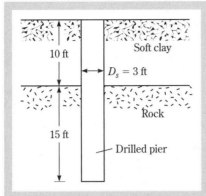

Figure 12.30 Drilled shaft extending into rock

As a check, we have

$$f = 0.15q_u = (0.15)(3000) = 450 \text{ lb/in}^2 > 136.9 \text{ lb/in}^2$$

So use $f = 136.9 \text{ lb/in}^2$.

Step 2. From Eq. (12.64),

$$Q_u = \pi D_s L f = [(\pi)(3 \times 12)(15 \times 12)(136.9)]\frac{1}{1000} = 2787 \text{ kip}$$

Step 3. From Eqs. (12.65), (12.66), and (12.67),

$$s_e = \frac{Q_u L}{A_c E_c} + \frac{Q_u I_f}{D_s E_{mass}}$$

For RQD $\approx 80\%$, from Figure 12.29, the value of $E_{mass}/E_{core} \approx 0.5$; thus,

$$E_{mass} = 0.5 E_{core} = (0.5)(0.36 \times 10^6) = 0.18 \times 10^6 \text{ lb/in}^2$$

so

$$\frac{E_c}{E_{mass}} = \frac{3 \times 10^6}{0.18 \times 10^6} \approx 16.7$$

and

$$\frac{L}{D_s} = \frac{15}{3} = 5$$

From Figure 12.28, for $E_c/E_{mass} = 16.7$ and $L/D_s = 5$, the magnitude of I_f is about 0.35. Hence,

$$s_e = \frac{(2787 \times 10^3 \text{ lb})(15 \times 12 \text{ in.})}{\frac{\pi}{4}(3 \times 12 \text{ in.})^2(3 \times 10^6 \text{ lb/in}^2)} + \frac{(2787 \times 10^3 \text{ lb})(0.35)}{(3 \times 12 \text{ in.})(0.18 \times 10^6 \text{ lb/in}^2)}$$

$$= 0.315 \text{ in.} < 0.4 \text{ in.}$$

Therefore,

$$Q_u = 2787 \text{ kip}$$

and

$$Q_{\text{all}} = \frac{Q_u}{\text{FS}} = \frac{2787}{3} = \textbf{929 kip} \qquad \blacksquare$$

Problems

12.1 A drilled shaft is shown in Figure P12.1. Use Eq. (12.20) and determine the net allowable point bearing capacity. Assume the following values:

$$
\begin{array}{ll}
D_b = 2 \text{ m} & \gamma_c = 15.6 \text{ kN/m}^3 \\
D_s = 1.2 \text{ m} & \gamma_s = 17.6 \text{ kN/m}^3 \\
L_1 = 6 \text{ m} & \phi' = 35° \\
L_2 = 3 \text{ m} & c_u = 35 \text{ kN/m}^3 \\
\end{array}
$$

Factor of safety = 4

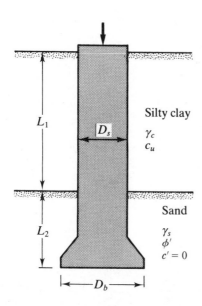

Silty clay
γ_c
c_u

Sand
γ_s
ϕ'
$c' = 0$

L_1 D_s

L_2

D_b

Figure P12.1

12.2 Redo Problem 12.1, this time using Eq. (12.16). Let $E_s = 600 p_a$.

12.3 Redo Problem 12.1 with the following data:

$$
\begin{array}{ll}
D_b = 1.75 \text{ m} & \gamma_c = 17.8 \text{ kN/m}^3 \\
D_s = 1 \text{ m} & \gamma_s = 18.2 \text{ kN/m}^3 \\
L_1 = 6.25 \text{ m} & \phi' = 32° \\
L_2 = 2.5 \text{ m} & c_u = 32 \text{ kN/m}^3 \\
\end{array}
$$

Factor of safety = 4

12.4 Solve Problem 12.3 using Eq. (12.16). Let $E_s = 400 p_a$.

12.5 For the drilled shaft described in Problem 12.1, what skin resistance would develop in the top 6 m, which is in clay?
   a. Use Eqs. (12.42) and (12.44).
   b. Use Eq. (12.45).

12.6 For the drilled shaft described in Problem 12.3, what skin resistance would develop in the top 6.25 m?
   a. Use Eqs. (12.42) and (12.44).
   b. Use Eq. (12.45).

12.7 Figure P12.7 shows a drilled shaft without a bell. Assume the following values:

$$L_1 = 6 \text{ m} \quad c_{u(1)} = 45 \text{ kN/m}^2$$
$$L_2 = 5 \text{ m} \quad c_{u(2)} = 74 \text{ kN/m}^2$$
$$D_s = 1.5 \text{ m}$$

Determine
   a. The net ultimate point bearing capacity [use Eqs. (12.39) and (12.40)]
   b. The ultimate skin friction [use Eqs. (12.42) and (12.44)]
   c. The working load Q_w (factor of safety = 3)

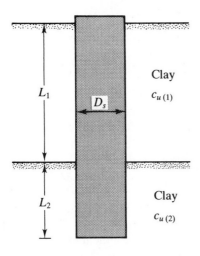

Figure P12.7

12.8 Repeat Problem 12.7 with the following data:

$$L_1 = 25 \text{ ft} \quad c_{u(1)} = 1200 \text{ lb/ft}^2$$
$$L_2 = 10 \text{ ft} \quad c_{u(2)} = 2000 \text{ lb/ft}^2$$
$$D_s = 3.5 \text{ ft}$$

Use Eqs. (12.45) and (12.46).

12.9 A drilled shaft in a medium sand is shown in Figure P12.9. Using the method proposed by Reese and O'Neill, determine the following:
   a. The net allowable point resistance for a base movement of 25 mm
   b. The shaft frictional resistance for a base movement of 25 mm
   c. The total load that can be carried by the drilled shaft for a total base movement of 25 mm

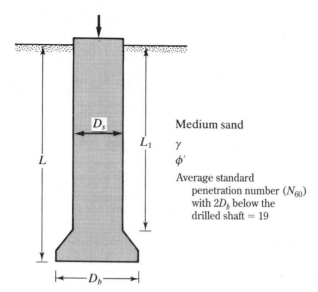

Medium sand

γ

ϕ'

Average standard
penetration number (N_{60})
with $2D_b$ below the
drilled shaft = 19

Figure P12.9

Assume the following values:

$$L = 12 \text{ m} \quad \gamma = 18 \text{ kN/m}^3$$
$$L_1 = 11 \text{ m} \quad \phi' = 38°$$
$$D_s = 1 \text{ m} \quad D_r = 65\% \text{ (medium sand)}$$
$$D_b = 2 \text{ m}$$

12.10 In Figure P12.9, let $L = 25$ ft, $L_1 = 20$ ft, $D_s = 3.5$ ft, $D_b = 5$ ft, $\gamma = 110$ lb/ft³, and $\phi' = 35°$. The average uncorrected standard penetration number (N_{60}) within $2D_b$ below the drilled shaft is 29. Determine
  a. The ultimate load-carrying capacity
  b. The load-carrying capacity for a settlement of 1 in.
  Use Reese and O'Neill's method.

12.11 For the drilled shaft described in Problem 12.7, determine
  a. The ultimate load-carrying capacity
  b. The load carrying capacity for a settlement of 12 mm
  Use the procedure outlined by Reese and O'Neill. (See Figures 12.16 and 12.17.)

12.12 For the drilled shaft described in Problem 12.7, estimate the total elastic settlement at working load. Use Eqs. (11.73), (11.75), and (11.76). Assume that $E_p = 20 \times 10^6$ kN/m², $C_p = 0.03$, $\xi = 0.65$, $\mu_s = 0.3$, $E_s = 12,000$ kN/m², and $Q_{ws} = 0.78 Q_w$. Use the value of Q_w from Part (c) of Problem 12.7.

12.13 For the drilled shaft described in Problem 12.8, estimate the total elastic settlement at working load. Use Eqs. (11.73), (11.75), and (11.76). Assume that $E_p = 3 \times 10^6$ lb/in², $C_p = 0.03$, $\xi = 0.65$, $\mu_s = 0.3$, $E_s = 2000$ lb/in², and $Q_{ws} = 0.83 Q_w$. Use the value of Q_w from Part (c) of Problem 12.8.

12.14 Suppose the drilled shaft shown in Figure P12.14 is a point bearing shaft with a working load of 650 kip. Calculate the drilled shaft settlement from Eqs. (11.73) and (11.74) for $E_p = 3 \times 10^6$ lb/in², $\mu_s = 0.35$, and $E_s = 6000$ lb/ft².

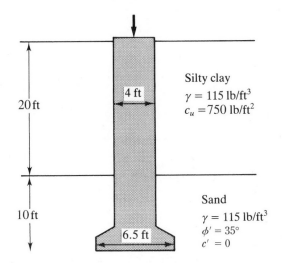

Silty clay
$\gamma = 115\ \text{lb/ft}^3$
$c_u = 750\ \text{lb/ft}^2$

20 ft

4 ft

10 ft

Sand
$\gamma = 115\ \text{lb/ft}^3$
$\phi' = 35°$
$c' = 0$

6.5 ft

Figure P12.14

12.15 Figure P12.15 shows a drilled shaft extending to rock. Assume the following values:

$q_{u \text{ (concrete)}} = 24{,}000\ \text{kN/m}^2 \quad E_{\text{(concrete)}} = 22\ \text{GN/m}^2$
$q_{u \text{ (rock)}} = 52{,}100\ \text{kN/m}^2 \quad E_{\text{core(rock)}} = 12.1\ \text{GN/m}^2$
$\text{RQD}_{\text{(rock)}} = 75\%$
Spacing of discontinuity in rock $= 550$ mm
Thickness of individual discontinuity in rock $= 2.5$ mm
Estimate the allowable load-bearing capacity of the drilled shaft. Use FS = 4.

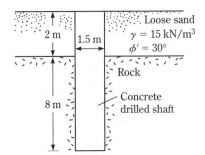

Loose sand
$\gamma = 15\ \text{kN/m}^3$
$\phi' = 30°$

2 m

1.5 m

Rock

8 m

Concrete
drilled shaft

Figure P12.15

12.16 A free-headed drill shaft is shown in Figure P12.16. Let $Q_g = 260$ kN, $M_g = 0$, $\gamma = 17.5\ \text{kN/m}^3$, $\phi' = 35°$, $c' = 0$, and $E_p = 22 \times 10^6\ \text{kN/m}^2$. Determine
a. the ground line deflection, x_o
b. the maximum bending moment in the drilled shaft
c. the maximum tensile stress in the shaft
d. the minimum penetration of the shaft needed for this analysis

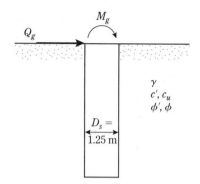

Figure P12.16

References

Berezantzev, V. G., Khristoforov, V. S., and Golubkov, V. N. (1961). "Load Bearing Capacity and Deformation of Piled Foundations," *Proceedings, Fifth International Conference on Soil Mechanics and Foundation Engineering,* Paris, Vol. 2, pp. 11–15.

Chen, Y.-J., and Kulhawy, F. H. (1994). "Case History Evaluation of the Behavior of Drilled Shafts under Axial and Lateral Loading," *Final Report, Project 1493-04, EPRI TR-104601,* Geotechnical Group, Cornell University, Ithaca, NY, December.

Duncan, J. M., Evans, L. T., Jr., and Ooi, P. S. K. (1994). "Lateral Load Analysis of Single Piles and Drilled Shafts," *Journal of Geotechnical Engineering,* ASCE, Vol. 120, No. 6, pp. 1018–1033.

Kulhawy, F. H., and Jackson, C. S. (1989). "Some Observations on Undrained Side Resistance of Drilled Shafts," *Proceedings, Foundation Engineering: Current Principles and Practices,* American Society of Civil Engineers, Vol. 2, pp. 1011–1025.

Matlock, H., and Reese, L. C. (1961). "Foundation Analysis of Offshore Pile-Supported Structures," in *Proceedings, Fifth International Conference on Soil Mechanics and Foundation Engineering,* Vol. 2, Paris, pp. 91–97.

O'Neill, M. W. (1997). Personal communication.

O'Neill, M. W., and Reese, L. C. (1999). *Drilled Shafts: Construction Procedure and Design Methods,* FHWA Report No. IF-99-025.

Reese, L. C., and O'Neill, M. W. (1988). *Drilled Shafts: Construction and Design,* FHWA, Publication No. HI-88-042.

Reese, L. C., and O'Neill, M. W. (1989). "New Design Method for Drilled Shafts from Common Soil and Rock Tests," *Proceedings, Foundation Engineering: Current Principles and Practices,* American Society of Civil Engineers, Vol. 2, pp. 1026–1039.

Reese, L. C., Touma, F. T., and O'Neill, M. W. (1976). "Behavior of Drilled Piers under Axial Loading," *Journal of Geotechnical Engineering Division,* American Society of Civil Engineers, Vol. 102, No. GT5, pp. 493–510.

Touma, F. T., and Reese, L. C. (1974). "Behavior of Bored Piles in Sand," *Journal of the Geotechnical Engineering Division,* American Society of Civil Engineers, Vol. 100, No. GT7, pp. 749–761.

Whitaker, T., and Cooke, R. W. (1966). "An Investigation of the Shaft and Base Resistance of Large Bored Piles in London Clay," *Proceedings, Conference on Large Bored Piles,* Institute of Civil Engineers, London, pp. 7–49.

13

Foundations on Difficult Soils

13.1 Introduction

In many areas of the United States and other parts of the world, certain soils make the construction of foundations extremely difficult. For example, expansive or collapsible soils may cause high differential movements in structures through excessive heave or settlement. Similar problems can also arise when foundations are constructed over sanitary landfills. Foundation engineers must be able to identify difficult soils when they are encountered in the field. Although not all the problems caused by all soils can be solved, preventive measures can be taken to reduce the possibility of damage to structures built on them. This chapter outlines the fundamental properties of three major soil conditions—collapsible soils, expansive soils, and sanitary landfills—and methods of careful construction of foundations.

Collapsible Soil

13.2 Definition and Types of Collapsible Soil

Collapsible soils, which are sometimes referred to as *metastable soils,* are unsaturated soils that undergo a large change in volume upon saturation. The change may or may not be the result of the application of additional load. The behavior of collapsing soils under load is best explained by the typical void ratio effective pressure plot (*e* against log σ') for a collapsing soil, as shown in Figure 13.1. Branch *ab* is determined from the consolidation test on a specimen at its natural moisture content. At an effective pressure level of σ'_w, the equilibrium void ratio is e_1. However, if water is introduced into the specimen for saturation, the soil structure will collapse. After saturation, the equilibrium void ratio at the same effective pressure level σ'_w is e_2; *cd* is the branch of the *e*–log σ' curve under additional load after saturation. Foundations that are constructed on such soils may undergo large and sudden settlement if the soil under them becomes saturated with an unanticipated supply of moisture. The moisture may come from any of several sources, such as (a) broken water pipelines, (b) leaky sewers, (c) drainage from reservoirs and swimming pools,

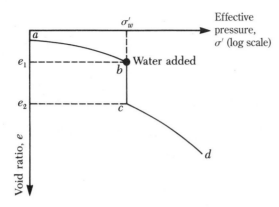

Figure 13.1 Nature of variation of void ratio with pressure for a collapsing soil

(d) a slow increase in groundwater, and so on. This type of settlement generally causes considerable structural damage. Hence, identifying collapsing soils during field exploration is crucial.

The majority of naturally occurring collapsing soils are *aeolian*—that is, wind-deposited sands or silts, such as loess, aeolic beaches, and volcanic dust deposits. The deposits have high void ratios and low unit weights and are cohesionless or only slightly cohesive. *Loess* deposits have silt-sized particles. The cohesion in loess may be the result of clay coatings surrounding the silt-size particles. The coatings hold the particles in a rather stable condition in an unsaturated state. The cohesion may also be caused by the presence of chemical precipitates leached by rainwater. When the soil becomes saturated, the clay binders lose their strength and undergo a structural collapse. In the United States, large parts of the Midwest and the arid West have such types of deposit. Loess deposits are also found over 15%–20% of Europe and over large parts of China. Figure 13.2 shows the major loessial areas in the United States and the places where the collapse of other types of soil have been reported (Dudley, 1970).

Many collapsing soils may be residual soils that are products of the weathering of parent rocks. Weathering produces soils with a large range of particle-size distribution. Soluble and colloidal materials are leached out by weathering, resulting in large void ratios and thus unstable structures. Many parts of South Africa and Zimbabwe have residual soils that are decomposed granites. Sometimes collapsing soil deposits may be left by flash floods and mudflows. These deposits dry out and are poorly consolidated. An excellent review of collapsing soils is that of Clemence and Finbarr (1981).

13.3   *Physical Parameters for Identification*

Several investigators have proposed various methods for evaluating the physical parameters of collapsing soils for identification. Some of these methods are discussed briefly in Table 13.1.

Jennings and Knight (1975) suggested a procedure for describing the *collapse potential* of a soil: An undisturbed soil specimen is taken at its natural moisture content in a consolidation ring. Step loads are applied to the specimen up to a pressure level σ'_w of 200 kN/m² ($\approx$29 lb/ft²). (In Figure 13.1, this is σ'_w.) At that pressure, the specimen is flooded for saturation and left for 24 hours. This test provides the void

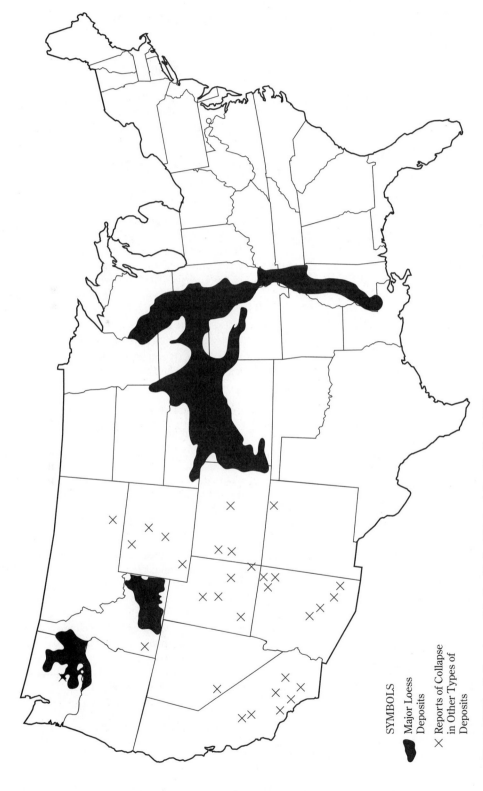

SYMBOLS

● Major Loess
Deposits

✕ Reports of Collapse
in Other Types of
Deposits

Figure 13.2 Collapsing soil in the United States (after Dudley, 1970)

ratios e_1 and e_2 before and after flooding, respectively. The collapse potential may now be calculated as

$$C_p = \Delta\varepsilon = \frac{e_1 - e_2}{1 + e_o} \tag{13.1}$$

Table 13.1 Reported Criteria for Identification of Collapsing Soil[a]

Investigator	Year	Criteria
Denisov	1951	Coefficient of subsidence: $$K = \frac{\text{void ratio at liquid limit}}{\text{natural void ratio}}$$ $K = 0.5$–0.75: highly collapsible $K = 1.0$: noncollapsible loam $K = 1.5$–2.0: noncollapsible soils
Clevenger	1958	If dry unit weight is less than $12.6\ \text{kN/m}^3\,(\approx 80\ \text{lb/ft}^3)$, settlement will be large; if dry unit weight is greater than $14\ \text{kN/m}^3\,(\approx 90\ \text{lb/ft}^3)$ settlement will be small.
Priklonski	1952	$$K_D = \frac{\text{natural moisture content} - \text{plastic limit}}{\text{plasticity index}}$$ $K_D < 0$: highly collapsible soils $K_D > 0.5$: noncollapsible soils $K_D > 1.0$: swelling soils
Gibbs	1961	Collapse ratio, $R = \dfrac{\text{saturation moisture content}}{\text{liquid limit}}$ This was put into graph form.
Soviet Building Code	1962	$$L = \frac{e_o - e_L}{1 + e_o}$$ where e_o = natural void ratio and e_L = void ratio at liquid limit. For natural degree of saturation less than 60%, if $L > -0.1$, the soil is a collapsing soil.
Feda	1964	$$K_L = \frac{w_o}{S_r} - \frac{\text{PL}}{\text{PI}}$$ where w_o = natural water content, S_r = natural degree of saturation, PL = plastic limit, and PI = plasticity index. For $S_r < 100\%$, if $K_L > 0.85$, the soil is a subsident soil.
Benites	1968	A dispersion test in which 2 g of soil are dropped into 12 ml of distilled water and specimen is timed until dispersed; dispersion times of 20 to 30 s were obtained for collapsing Arizona soils.
Handy	1973	Iowa loess with clay (<0.002 mm) contents: $<16\%$: high probability of collapse 16–24%: probability of collapse 24–32%: less than 50% probability of collapse $>32\%$: usually safe from collapse

[a]Modified after Lutenegger and Saber (1988)

where e_o = natural void ratio of the soil

$\Delta \varepsilon$ = vertical strain

The severity of foundation problems associated with a collapsible soil have been correlated with the collapse potential C_p by Jennings and Knight (1975). They were summarized by Clemence and Finbarr (1981) and are given in Table 13.2.

Holtz and Hilf (1961) suggested that a loessial soil that has a void ratio large enough to allow its moisture content to exceed its liquid limit upon saturation is susceptible to collapse. So, for collapse,

$$w_{(saturated)} \geqslant LL \tag{13.2}$$

where LL = liquid limit

However, for saturated soils,

$$e_o = wG_s \tag{13.3}$$

where G_s = specific gravity of soil solids

Combining Eqs. (13.2) and (13.3) for collapsing soils yields

$$e_o \geqslant (LL)(G_s) \tag{13.4}$$

The natural dry unit weight of the soil required for its collapse is

$$\gamma_d \leqslant \frac{G_s \gamma_w}{1 + e_o} = \frac{G_s \gamma_w}{1 + (LL)(G_s)} \tag{13.5}$$

For an average value of $G_s = 2.65$, the limiting values of γ_d for various liquid limits may now be calculated from Eq. (13.5).

Figure 13.3 shows a plot of the preceding limiting values of dry unit weights against the corresponding liquid limits. For any soil, if the natural dry unit weight falls below the limiting line, the soil is likely to collapse.

Care should be taken to obtain undisturbed samples for determining the collapse potentials and dry unit weights—preferably block samples cut by hand. The reason is that samples obtained by thin-walled tubes may undergo some compression during the sampling process. However, if cut block samples are used, the boreholes should be made *without water*.

Table 13.2 Relation of Collapse Potential to the Severity of Foundation Problems[a]

$C_p(\%)$	Severity of problem
0–1	No problem
1–5	Moderate trouble
5–10	Trouble
10–20	Severe trouble
20	Very severe trouble

[a]After Clemence and Finbarr (1981)

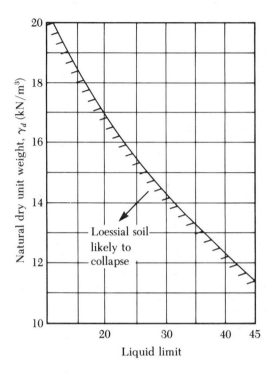

Figure 13.3 Loessial soil likely to collapse

Procedure for Calculating Collapse Settlement

Jennings and Knight (1975) proposed the following laboratory procedure for determining the collapse settlement of structures upon saturation of soil:

1. Obtain *two* undisturbed soil specimens for tests in a standard consolidation test apparatus (oedometer).
2. Place the two specimens under $1 \text{ kN/m}^2 (0.15 \text{ lb/in}^2)$ pressure for 24 hours.
3. After 24 hours, saturate one specimen by flooding. Keep the other specimen at its natural moisture content.
4. After 24 hours of flooding, resume the consolidation test on both specimens by doubling the load (the same as in the standard consolidation test) to the desired pressure level.
5. Plot the e–$\log \sigma'$ graphs for both specimens (Figures 13.4a and b).
6. Calculate the *in situ* effective pressure, σ'_o. Draw a vertical line corresponding to the pressure σ'_o.
7. From the e–$\log \sigma'_o$ curve of the soaked specimen, determine the preconsolidation pressure, σ'_c. If $\sigma'_c/\sigma'_o = 0.8$–$1.5$, the soil is normally consolidated; however, if $\sigma'_c/\sigma'_o > 1.5$, the soil is preconsolidated.
8. Determine e'_o, corresponding to σ'_o from the e–$\log \sigma'_o$ curve of the soaked specimen. (This procedure for normally consolidated and overconsolidated soils is shown in Figures 13.4a and b, respectively.)
9. Through point (σ'_o, e'_o) draw a curve that is similar to the e–$\log \sigma'_o$ curve obtained from the specimen tested at its natural moisture content.

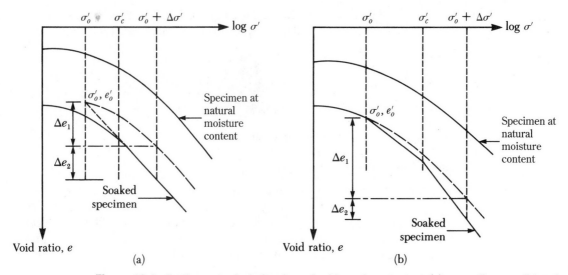

Figure 13.4 Settlement calculation from double oedometer test: (a) normally consolidated soil; (b) overconsolidated soil

10. Determine the incremental pressure, $\Delta\sigma'$, on the soil caused by the construction of the foundation. Draw a vertical line corresponding to the pressure of $\sigma'_o + \Delta\sigma'$ in the e–log σ'_o curve.

11. Now, determine Δe_1 and Δe_2. The settlement of soil without change in the natural moisture content is

$$S_{c(1)} = \frac{\Delta e_1}{1 + e'_o} (H)$$ (13.6)

where H = thickness of soil susceptible to collapse

Also, the settlement caused by collapse in the soil structure is

$$S_{c(2)} = \frac{\Delta e_2}{1 + e'_o} (H)$$ (13.7)

13.5   *Foundation Design in Soils Not Susceptible to Wetting*

For actual foundation design purposes, some standard field load tests may also be conducted. Figure 13.5 shows the relation of the nature of load per unit arca versus settlement in a field load test in a loessial deposit. Note that the load–settlement relationship is essentially linear up to a certain critical pressure, σ'_{cr}, at which there is a breakdown of the soil structure and hence a large settlement. Sudden breakdown of soil structure is more common with soils having a high natural moisture content than with normally dry soils.

If enough precautions are taken in the field to prevent moisture from increasing under structures, spread foundations and raft foundations may be built on collapsible soils. However, the foundations must be proportioned so that the critical stresses (see Figure 13.5) in the field are never exceeded. A factor of safety of about 2.5 to 3 should be used to calculate the allowable soil pressure, or

$$\sigma'_{all} = \frac{\sigma'_{cr}}{FS} \tag{13.8}$$

where σ'_{all} = allowable soil pressure
        FS = factor   of safety

The differential and total settlements of these foundations should be similar to those of foundations designed for sandy soils.

   Continuous foundations may be safer than isolated foundations over collapsible soils in that they can effectively minimize differential settlement. Figure 13.6 shows a typical procedure for constructing continuous foundations. This procedure uses footing beams and longitudinal load-bearing beams.

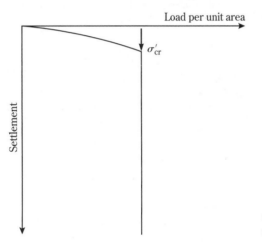

Figure 13.5   Field load test in loessial soil: load per unit area vs. settlement

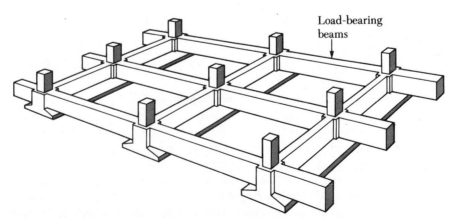

Figure 13.6   Continuous foundation with load-bearing beams (after Clemence and Finbarr, 1981)

In the construction of heavy structures, such as grain elevators, over collapsible soils, settlements up to about 0.3 m ($\approx$1 ft) are sometimes allowed (Peck, Hanson, and Thornburn, 1974). In this case, tilting of the foundation is not likely to occur, because there is no eccentric loading. The total expected settlement for such structures can be estimated from standard consolidation tests on specimens of field moisture content. Without eccentric loading, the foundations will exhibit uniform settlement over loessial deposits; however, if the soil is of residual or colluvial nature, settlement may not be uniform. The reason is the nonuniformity generally encountered in residual soils.

Extreme caution must be used in building heavy structures over collapsible soils. If large settlements are expected, drilled-shaft and pile foundations should be considered. These types of foundation can transfer the load to a stronger load-bearing stratum.

13.6 *Foundation Design in Soils Susceptible to Wetting*

If the upper layer of soil is likely to get wet and collapse sometime after construction of the foundation, several design techniques to avoid failure of the foundation may be considered:

1. If the expected depth of wetting is about 1.5 to 2 m ($\approx$5 to 6.5 ft) from the ground surface, the soil may be moistened and recompacted by heavy rollers. Spread footings and rafts may be constructed over the compacted soil. An alternative to recompaction by heavy rollers is *heavy tamping,* which is sometimes referred to as *dynamic compaction.* (See Chapter 14.) Heavy tamping consists primarily of dropping a heavy weight repeatedly on the ground. The height of the drop can vary from 8 to 30 m ($\approx$25 to 100 ft). The stress waves generated by the dropping weight help in the densification of the soil.
2. If conditions are favorable, foundation trenches can be flooded with solutions of sodium silicate and calcium chloride to stabilize the soil chemically. The soil will then behave like a soft sandstone and resist collapse upon saturation. This method is successful only if the solutions can penetrate to the desired depth; thus, it is most applicable to fine sand deposits. Silicates are rather costly and are not generally used. However, in some parts of Denver, silicates have been used successfully.

   The injection of a sodium silicate solution to stabilize collapsible soil deposits has been used extensively in the former Soviet Union and Bulgaria (Houston and Houston, 1989). This process, which is used for dry collapsible soils and for wet collapsible soils that are likely to compress under the added weight of the structure to be built, consists of three steps:

   Step 1. Injection of carbon dioxide to remove any water that is present and for preliminary activation of the soil
   Step 2. Injection of sodium silicate grout
   Step 3. Injection of carbon dioxide to neutralize alkalies.

3. When the soil layer is susceptible to wetting to a depth of about 10 m ($\approx$30 ft), several techniques may be used to cause collapse of the soil *before* construction of the foundation is begun. Two of these techniques are *vibroflotation* and *ponding* (also called *flooding*). Vibroflotation is used successfully in free-draining soil. (See Chapter 14.) Ponding—by constructing low dikes—is utilized at sites that have no impervious layers. However, even after saturation and collapse of

the soil by ponding, some additional settlement of the soil may occur after construction of the foundation is begun. Additional settlement may also be caused by incomplete saturation of the soil at the time of construction. Ponding may be used successfully in the construction of earth dams.

4. If precollapsing the soil is not practical, foundations may be extended beyond the zone of possible wetting, although the technique may require drilled shafts and piles. The design of drilled shafts and piles must take into consideration the effect of negative skin friction resulting from the collapse of the soil structure and the associated settlement of the zone of subsequent wetting.

In some cases, a *rock-column type of foundation* (*vibroreplacement*) may be considered. Rock columns are built with large boulders that penetrate the collapsible soil layer. They act as piles in transferring the load to a more stable soil layer.

13.7   *Case Histories of Stabilization of Collapsible Soil*

Use of Dynamic Compaction

Lutenegger (1986) reported the use of dynamic compaction to stabilize a thick layer of friable loess before the construction of a foundation in Russe, Bulgaria. During field exploration, the water table was not encountered to a depth of 10 m ($\approx$30 ft), and the natural moisture content was below the plastic limit. Initial density measurements made on undisturbed soil specimens indicated that the moisture content at saturation would exceed the liquid limit, a property usually encountered in collapsible loess.

For dynamic compaction of the soil, the upper 1.7 m (5.6 ft) of crust material was excavated. A circular concrete weight of 133 kN ($\approx$30,000 lb) was used as a hammer. At each grid point, compaction was achieved by dropping the hammer 7 to 12 times through a vertical distance of 2.5 m (8.2 ft).

Figure 13.7a shows the dry density of the soil before and after compaction. Figure 13.7b shows the increase in field standard penetration resistance before and after compaction. The increase in both dry density and standard penetration resistance of the soil shows that dynamic compaction can be used effectively to stabilize collapsible soil.

Chemical Stabilization

Semkin et al. (1986) reported on the chemical stabilization of a loessial soil deposit with a carbon dioxide–sodium silicate–carbon dioxide injection scheme. (See also Houston and Houston, 1989.) The site is that of the Interregional Center in Tashkent, Uzbekistan, which consists of two buildings—one three stories high with strip foundations and the other one story high with spread foundations. The loessial soil deposit at the site was about 35 m (115 ft) thick, and the water table was at a depth of about 18 m (60 ft). The natural moisture content and porosity of the soil above the water table were 10%–25% and 0.48, respectively.

The Interregional Center was constructed in 1973. Unanticipated leakage from conduits in the center caused differential settlement in 1974. Without shutting down the center, chemical stabilization was used effectively, and the differential settlement was stopped.

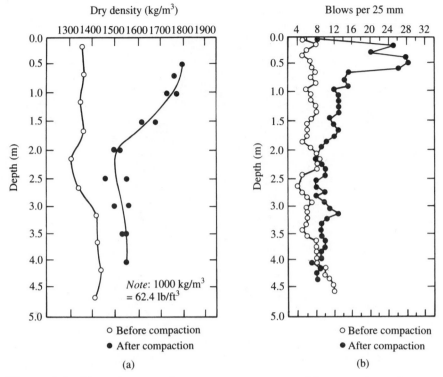

Figure 13.7 (a) Dry density before and after compaction; (b) penetration resistance before and after compaction (after Lutenegger, 1986)

Expansive Soils

13.8 *General Nature of Expansive Soils*

Many plastic clays swell considerably when water is added to them and then shrink with the loss of water. Foundations constructed on such clays are subjected to large uplifting forces caused by the swelling. These forces induce heaving, cracking, and the breakup of both building foundations and slab-on-grade members. Expansive clays cover large parts of the United States, South America, Africa, Australia, and India. In the United States, these clays are predominant in Texas, Oklahoma, and the upper Missouri Valley. In general, expansive clays have liquid limits and plasticity indices greater than about 40 and 15, respectively.

As noted, an increase in moisture content causes clay to swell. The depth in a soil to which periodic changes of moisture occur is usually referred to as the *active zone* (see Figure 13.8). The depth of the active zone varies, depending on the location of the site. Some typical active-zone depths in American cities are given in Table 13.3. In some clays and clay shales in the western United States, the depth of the active zone can be as much as 15 m (≈50 ft). The active-zone depth can easily be determined by plotting the liquidity index against the depth of the soil profile

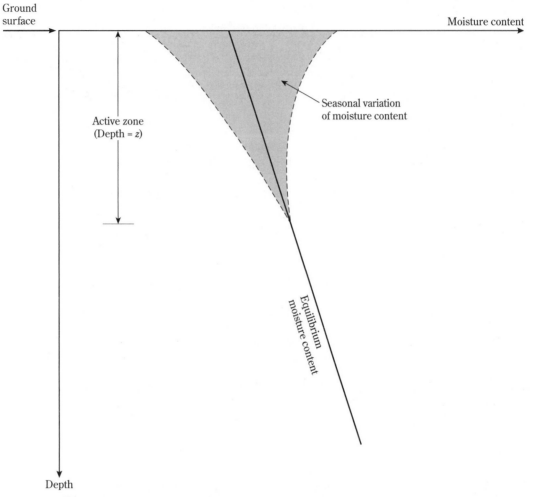

Figure 13.8   Definition of active zone

Table 13.3   Typical Active-Zone Depths in Some
U.S. Cities[a]

City	Depth of active zone	
	(m)	**(ft)**
Houston	1.5 to 3	5 to 10
Dallas	2.1 to 4.6	7 to 15
San Antonio	3 to 9	10 to 30
Denver	3 to 4.6	10 to 15

[a] After O'Neill and Poormoayed (1980)

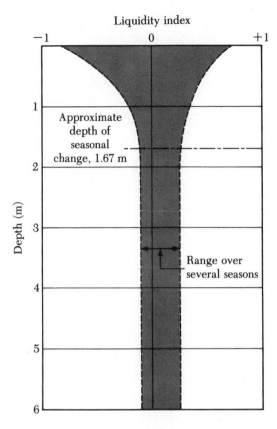

Figure 13.9 Active zone in Houston area, Beaumont formation (after O'Neill and Poormoayed, 1980)

over several seasons. Figure 13.9 shows such a plot for the Beaumont formation in the Houston area.

13.9 *Laboratory Measurement of Swell*

To study the magnitude of possible swell in a clay, simple laboratory oedometer tests can be conducted on undisturbed specimens. Two common tests are the unrestrained swell test and the swelling pressure test.

Unrestrained Swell Test

In the *unrestrained swell test,* the specimen is placed in an oedometer under a small surcharge of about 6.9 kN/m^2 (1 lb/in^2). Water is then added to the specimen, and the expansion of the volume of the specimen (i.e., its height; the area of cross section is constant) is measured until equilibrium is reached. The percentage of free swell may than be expressed as the ratio

$$s_{w(\text{free})}(\%) = \frac{\Delta H}{H}(100) \tag{13.9}$$

where $s_{w(\text{free})}$ = free swell, as a percentage

  ΔH = height of swell due to saturation

  H = original height of the specimen

Vijayvergiya and Ghazzaly (1973) analyzed various soil test results obtained in this manner and prepared a correlation chart of the free swell, liquid limit, and natural moisture content, as shown in Figure 13.10. O'Neill and Poormoayed (1980) developed a relationship for calculating the free surface swell from this chart:

$$\Delta S_F = 0.0033 Z s_{w(\text{free})} \tag{13.10}$$

where ΔS_F = free surface swell

  Z = depth of active zone

  $s_{w(\text{free})}$ = free swell, as a percentage (see Figure 13.10)

Swelling Pressure Test

The *swelling pressure test* can be conducted by taking a specimen in a consolidation ring and applying a pressure equal to the effective overburden pressure, σ'_o, plus the approximate anticipated surcharge caused by the foundation, σ'_s. Water is then added

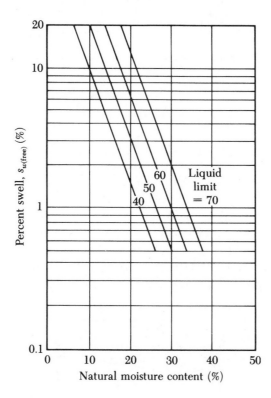

Figure 13.10 Relation between percentage of free swell, liquid limit, and natural moisture content (after Vijayvergiya and Ghazzaly, 1973)

to the specimen. As the specimen starts to swell, pressure is applied in small increments to prevent swelling. Pressure is maintained until full swelling pressure is developed on the specimen, at which time the total pressure is

$$\sigma'_T = \sigma'_o + \sigma'_s + \sigma'_1 \tag{13.11}$$

where σ'_T = total pressure applied to prevent swelling, or zero swell pressure
σ'_1 = further pressure applied to prevent swelling after addition of water

Figure 13.11 shows the variation of the percentage of swell with effective pressure during a swelling pressure test. (For more information on this type of test, see Sridharan et al., 1986.)

A σ'_T of about 20–30 kN/m^2 (400–650 lb/ft^2) is considered to be low, and a σ'_T of 1500–2000 kN/m^2 (30,000–40,000 lb/ft^2) is considered to be high. After zero swell pressure is attained, the soil specimen can be unloaded in steps to the level of the effective overburden pressure, σ'_o. The unloading process will cause the specimen to swell. The equilibrium swell for each pressure level is also recorded. The variation of the swell, s_w in percent, and the applied pressure on the specimen will be like that shown in Figure 13.11.

The swelling pressure test can be used to determine the surface heave, ΔS, for a foundation (O'Neill and Poormoayed, 1980) as given by the formula

$$\Delta S = \sum_{i=1}^{n} [s_{w(1)} \, (\%)](H_i)(0.01) \tag{13.12}$$

where $s_{w(1)}(\%)$ = swell, in percent, for layer i under a pressure of $\sigma'_o + \sigma'_s$ (see Figure 13.11)
ΔH_i = thickness of layer i

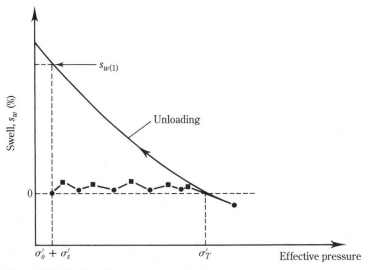

Figure 13.11 Swelling pressure test

Modified Free Swell Index Test

In 1987, Sivapullaiah et al. suggested a new test for obtaining a *modified free-swell index* for clays that appears to give a better indication of the swelling potential of clayey soils. This test begins with an oven-dried soil with a mass of about 10 g. The soil mass is well pulverized and transferred into a 100-ml graduated jar containing distilled water. After 24 hours, the swollen sediment volume is measured. The *modified free-swell index* is then calculated as

$$\text{Modified free swell index} = \frac{V - V_s}{V_s} \tag{13.13}$$

where V = soil volume after swelling

V_s = volume of soil solid = $\dfrac{M_s}{G_s \rho_w}$

in which M_s = mass of oven-dried soil

G_s = specific gravity of soil solids

ρ_w = density of water

Based on the modified free swell index, the swelling potential of a soil may be qualitatively classified as follows:

Modified free swell index	Swelling potential
<2.5	Negligible
2.5 to 10	Moderate
10 to 20	High
>20	Very high

Example 13.1

A soil profile has an active zone of expansive soil of 6 ft. The liquid limit and the average natural moisture content during the construction season are 50% and 20%, respectively. Determine the free surface swell.

Solution

From Figure 13.10 for LL = 50% and w = 20%, $s_{w(\text{free})}$ = 3%. From Eq. (13.10),

$$\Delta S_F = 0.0033 Z s_{w(\text{free})}$$

Hence,

$$\Delta S_F = 0.0033(6)(3)(12) = \textbf{0.71 in.} \qquad \blacksquare$$

Example 13.2

A soil profile's active-zone depth is 3.5 m. If a foundation is to be placed 0.5 m below the ground surface, what would be the estimated total swell? The following data were obtained from laboratory tests:

Depth (m)	Swell under effective overburden and estimated foundation surcharge pressure, $s_{w(1)}$(%)
0.5	2
1	1.5
2	0.75
3	0.25

Solution

The values of $s_{w(1)}(\%)$ are plotted with depth in Figure 13.12a. The area of this diagram will be the total swell. The trapezoidal rule yields

$$\Delta S = \frac{1}{100}\left[\frac{1}{2}(1)(0 + 0.5) + \frac{1}{2}(1)(0.5 + 1.1) + \frac{1}{2}(1)(1.1 + 2)\right]$$

$$= 0.026 \text{ m} = \textbf{26 mm.} \qquad \blacksquare$$

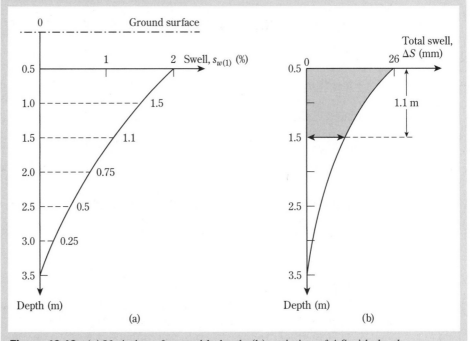

Figure 13.12 (a) Variation of $s_{w(1)}$ with depth; (b) variation of ΔS with depth

Example 13.3

In Example 13.2, if the allowable total swell is 10 mm, what would be the undercut necessary to reduce the total swell?

Solution

Using the procedure outlined in Example 13.2, we calculate the total swell at various depths below the foundation from Figure 13.12a as follows:

Depth (m)	Total swell, ΔS(mm)
3.5	0
3	$0 + \left[\frac{1}{2}(0.5)(0.25)\right]\frac{1}{100} = 0.000625$ m $= 0.625$ mm
2.5	$0.000625 + \frac{1}{100}\left[\frac{1}{2}(0.5)(0.25 + 0.5)\right] = 0.0025$ m $= 2.5$ mm
1.5	$0.0025 + \frac{1}{100}\left[\frac{1}{2}(1)(0.5 + 1.1)\right] = 0.0105$ m $= 10.5$ mm
0.5	26 mm

Plotted in Figure 13.12b, these total settlements show that a total swell of 10 mm corresponds to a depth of 1.6 m below the ground surface.

Hence, the undercut below the foundation is **1.6 − 0.5 = 1.1 m**. This soil should be excavated, replaced by nonswelling soil, and recompacted. ∎

13.10 Classification of Expansive Soil on the Basis of Index Tests

Classification systems for expansive soils are based on the problems they create in the construction of foundations (potential swell). Most of the classifications contained in the literature are summarized in Figure 13.13 and Table 13.4. However, the classification system developed by the U.S. Army Waterways Experiment Station (Snethen et al., 1977) is the one most widely used in the United States. It has also been summarized by O'Neill and Poormoayed (1980); see Table 13.5.

13.11 Foundation Considerations for Expansive Soils

If a soil has a low swell potential, standard construction practices may be followed. However, if the soil possesses a marginal or high swell potential, precautions need to be taken, which may entail

1. Replacing the expansive soil under the foundation
2. Changing the nature of the expansive soil by compaction control, prewetting, installation of moisture barriers, or chemical stabilization
3. Strengthening the structures to withstand heave, constructing structures that are flexible enough to withstand the differential soil heave without failure, or constructing isolated deep foundations below the depth of the active zone

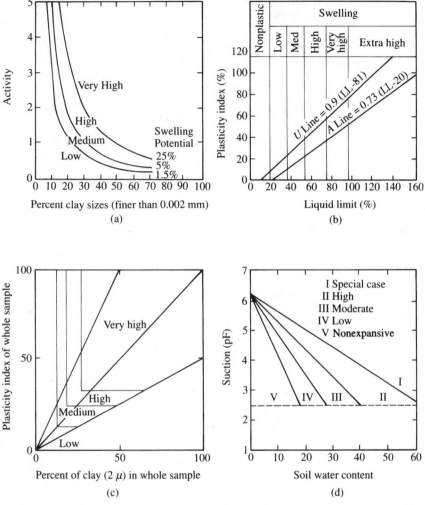

Figure 13.13 Commonly used criteria for determining swell potential (after Abduljauwad and Al-Sulaimani, 1993)

One particular method may not be sufficient in all situations. Combining several techniques may be necessary, and local construction experience should always be considered. Following are some details regarding the commonly used techniques for dealing with expansive soils.

Replacement of Expansive Soil

When shallow, moderately expansive soils are present at the surface, they can be removed and replaced by less expansive soils and then compacted properly.

Changing the Nature of Expansive Soil

1. *Compaction:* The heave of expansive soils decreases substantially when the soil is compacted to a lower unit weight on the high side of the optimum moisture

Table 13.4 Summary of Some Criteria for Identifying Swell Potential (after Abduljauwad and Al-Sulaimani, 1993)

Reference	Criteria	Remarks
Holtz (1959)	CC > 28, PI > 35, and SL < 11 (very high) $20 \leq$ CC $\leq 31, 25 \leq$ PI ≤ 41, and $7 \leq$ SL ≤ 12 (high) $13 \leq$ CC $\leq 23, 15 \leq$ PI ≤ 28, and $10 \leq$ SL ≤ 16 (medium) CC ≤ 15, PI ≤ 18, and SL ≥ 15 (low)	Based on CC, PI, and SL
Seed et al. (1962)	See Figure 13.13a	Based on oedometer test using compacted specimen, percentage of clay <2 μm, and activity
Altmeyer (1955)	LS < 5, SI > 12, and PS < 0.5 (noncritical) $5 \leq$ LS $\leq 8, 10 \leq$ SL ≤ 12, and $0.5 \leq$ PS ≤ 1.5 (marginal) LS > 8, SL < 10, and PS > 1.5 (critical)	Based on LS, SL, and PS Remolded sample ($\rho_{d(\max)}$ and w_{opt}) soaked under 6.9 kPa surcharge
Dakshanamanthy and Raman (1973)	See Figure 13.13b	Based on plasticity chart
Raman (1967)	PI > 32 and SI > 40 (very high) $23 \leq$ PI ≤ 32 and $30 \leq$ SI ≤ 40 (high) $12 \leq$ PI ≤ 23 and $15 \leq$ SI ≤ 30 (medium) PI < 12 and SI < 15 (low)	Based on PI and SI
Sowers and Sowers (1970)	SL < 10 and PI > 30 (high) $10 \leq$ SL ≤ 12 and $15 \leq$ PI ≤ 30 (moderate) SL > 12 and PI < 15 (low)	Little swell will occur when w_o results in LI of 0.25
Van Der Merwe (1964)	See Figure 13.13c	Based on PI, percentage of clay <2 μm, and activity
Uniform Building Code, 1968	EI > 130 (very high) and $91 \leq$ EI ≤ 130 (high) $51 \leq$ EI ≤ 90 (medium) and $21 \leq$ EI ≤ 50 (low) $0 \leq$ EI ≤ 20 (very low)	Based on oedometer test on compacted specimen with degree of saturation close to 50% and surcharge of 6.9 kPa
Snethen (1984)	LL > 60, PI > 35, $\tau_{nat} > 4$, and SP > 1.5 (high) $30 \leq$ LL $\leq 60, 25 \leq$ PI $\leq 35, 1.5 \leq \tau_{nat} \leq 4$, and $0.5 \leq$ SP ≤ 1.5 (medium) LL < 30, PI < 25, $\tau_{nat} < 1.5$, and SP < 0.5 (low)	PS is representative for field condition and can be used without τ_{nat}, but accuracy will be reduced
Chen (1988)	PI ≥ 35 (very high) and $20 \leq$ PI ≤ 55 (high) $10 \leq$ PI ≤ 35 (medium) and PI ≤ 15 (low)	Based on PI
McKeen (1992)	Figure 13.13d	Based on measurements of soil water content, suction, and change in volume on drying
Vijayvergiya and Ghazzaly (1973)	$\log$ SP $= (1/12)(0.44$ LL $- w_o + 5.5)$	Empirical equations
Nayak and Christensen (1974)	SP $= (0.00229$ PI$)(1.45C)/w_o + 6.38$	Empirical equations
Weston (1980)	SP $= 0.00411(\text{LL}_w)^{4.17} q^{-3.86} w_o^{-2.33}$	Empirical equations

Table 13.4 (Continued)

Note: C = clay, %
CC = colloidal content, %
EI = Expansion index = 100 × percent swell × fraction passing no. 4 sieve
LI = liquidity index, %
LL = liquid limit, %
LL_w = weighted liquid limit, %
LS = linear shrinkage, %
PI = plasticity index, %
PS = probable swell, %

q = surcharge
SI = shrinkage index = LL − SL, %
SL = shrinkage limit, %
SP = swell potential, %
w_o = natural soil moisture
w_{opt} = optimum moisture content, %
τ_{nat} = natural soil suction in tsf
$\rho_{d(max)}$ = max dry density

Table 13.5 Expansive Soil Classification System[a]

Liquid limit	Plasticity index	Potential swell (%)	Potential swell classification
<50	<25	<0.5	Low
50–60	25–35	0.5–1.5	Marginal
>60	>35	>1.5	High
Potential swell = vertical swell under a pressure equal to overburden pressure			

[a] Compiled from O'Neill and Poormoayed (1980)

content (possibly 3–4% above the optimum moisture content). Even under such conditions, a slab-on-ground type of construction should not be considered when the total probable heave is expected to be about 38 mm (1.5 in.) or more. Figure 13.14a shows the recommended limits of soil compaction in the field for reduction of heave. Note that the recommended dry unit weights are based on climatic ratings. According to U.S. Weather Bureau data, a climatic rating of 15 represents an extremely unfavorable climatic condition; a rating of 45 represents a favorable climatic condition. The isobars of climatic rating for the continental United States are shown in Figure 13.14b.

2. *Prewetting:* One technique for increasing the moisture content of the soil is ponding and hence achieving most of the heave before construction. However, this technique may be time consuming because the seepage of water through highly plastic clays is slow. After ponding, 4–5% of hydrated lime may be added to the top layer of the soil to make it less plastic and more workable (Gromko, 1974).

3. *Installation of moisture barriers:* The long-term effect of the differential heave can be reduced by controlling the moisture variation in the soil. This is achieved by providing vertical moisture barriers about 1.5 m (≈5 ft) deep around the perimeter of slabs for the slab-on-grade type of construction. These moisture barriers may be constructed in trenches filled with gravel, lean concrete, or impervious membranes.

4. *Stabilization of soil:* Chemical stabilization with the aid of lime and cement has often proved useful. A mix containing about 5% lime is sufficient in most cases. Lime or cement and water are mixed with the top layer of soil and compacted.

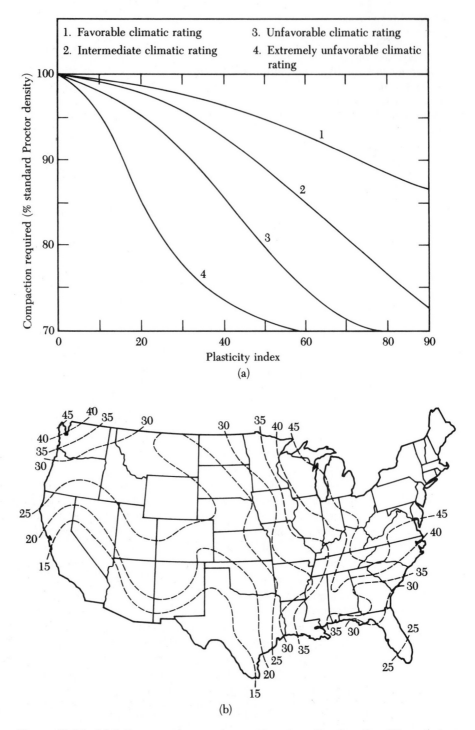

Figure 13.14 (a) Soil compaction requirement based on climatic rating; (b) equivalent climatic rating of the United States (after Gromko, 1974)

The addition of lime or cement will decrease the liquid limit, the plasticity index, and the swell characteristics of the soil. This type of stabilization work can be done to a depth of 1 to 1.5 m ($\approx$3 to 5 ft). Hydrated high-calcium lime and dolomite lime are generally used for lime stabilization.

Another method of stabilization of expansive soil is the *pressure injection* of lime slurry or lime–fly ash slurry into the soil, usually to a depth of 4 to 5 m or (12 to 16 ft) and occasionally deeper to cover the active zone. Further details of the pressure injection technique are presented in Chapter 14. Depending on the soil conditions at a site, single or multiple injections can be planned, as shown in Figure 13.15. Figure 13.16 shows the slurry pressure injection work for a building pad.

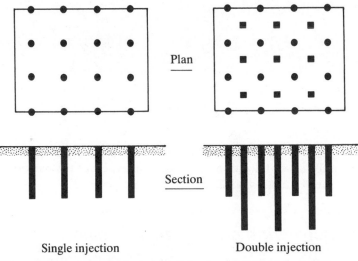

Figure 13.15 Multiple lime slurry injection planning for a building pad

Figure 13.16 Pressure injection of lime slurry for a building pad (Courtesy of GKN Hayward Baker, Inc., Woodbine Division, Ft. Worth, Texas)

Figure 13.17 Slope stabilization of a canal bank by pressure injection of lime–fly ash slurry (Courtesy of GKN Hayward Baker, Inc., Woodbine Division, Ft. Worth, Texas)

The stakes that are marked are the planned injection points. Figure 13.17 shows lime–fly ash stabilization by pressure injection of the bank of a canal that had experienced sloughs and slides.

13.12   *Construction on Expansive Soils*

Care must be exercised in choosing the type of foundation to be used on expansive soils. Table 13.6 shows some recommended construction procedures based on the total predicted heave, ΔS, and the length-to-height ratio of the wall panels. For example, the table proposes the use of waffle slabs as an alternative in designing rigid buildings that are capable of tolerating movement. Figure 13.18 shows a schematic diagram of a waffle slab. In this type of construction, the ribs hold the structural load. The waffle voids allow the expansion of soil.

Table 13.6 also suggests the use of a drilled shaft foundation with a suspended floor slab when structures are constructed independently of movement of the soil. Figure 13.19a shows a schematic diagram of such an arrangement. The bottom of the shafts should be placed below the active zone of the expansive soil. For the design of the shafts, the uplifting force, U, may be estimated (see Figure 13.19b) from the equation

$$U = \pi D_s Z \sigma_T' \tan \phi_{ps}' \tag{13.14}$$

where D_s = diameter of the shaft
Z = depth of the active zone
ϕ_{ps}' = effective angle of plinth–soil friction
σ_T' = pressure for zero horizontal swell (see Figure 13.11;
$\sigma_T' = \sigma_o' + \sigma_s' + \sigma_1'$)

Table 13.6   Construction Procedures for Expansive Clay Soils[a]

Total predicted heave (mm)		Recommended construction	Method	Remarks
L/H = 1.25	L/H = 2.5			
0 to 6.35	12.7	No precaution		
6.35 to 12.7	12.7 to 50.8	Rigid building tolerating movement (steel reinforcement as necessary)	*Foundations:* Pads Strip footings Raft (waffle)	Footings should be small and deep, consistent with the soil-bearing capacity. Rafts should resist bending.
			Floor slabs: Waffle Tile	Slabs should be designed to resist bending and should be independent of grade beams.
			Walls:	Walls on a raft should be as flexible as the raft. There should be no vertical rigid connections. Brickwork should be strengthened with tie bars or bands.
12.7 to 50.8	50.8 to 101.6	Building damping movement	*Joints:* Clear Flexible	Contacts between structural units should be avoided, or flexible, waterproof material may be inserted in the joints.
			Walls: Flexible Unit construction Steel frame	Walls or rectangular building units should heave as a unit.
			Foundations: Three point Cellular Jacks	Cellular foundations allow slight soil expansion to reduce swelling pressure. Adjustable jacks can be inconvenient to owners. Three-point loading allows motion without duress.
>50.8	>101.6	Building independent of movement	*Foundation drilled shaft:* Straight shaft Bell bottom	Smallest-diameter and widely spaced piers compatible with the load should be placed. Clearance should be allowed under grade beams.
			Suspended floor:	Floor should be suspended on grade beams 305 to 460 mm above the soil.

[a] After Gromko, 1974

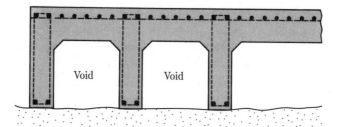

Figure 13.18   Waffle slab

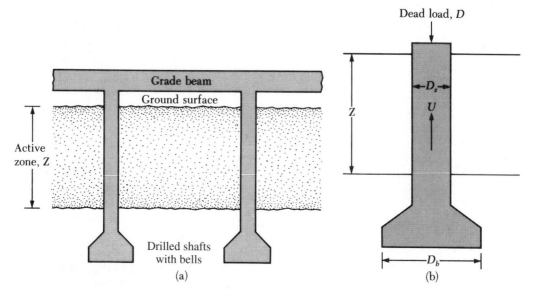

Figure 13.19   (a) Construction of drilled shafts with bells and grade beam; (b) definition of parameters in Eq. (13.14)

In most cases, the value of ϕ'_{ps} varies between $10°$ and $20°$. An average value of the zero horizontal swell pressure must be determined in the laboratory. In the absence of laboratory results, $\sigma'_T \tan \phi'_{ps}$ may be considered equal to the undrained shear strength of clay, c_u, in the active zone.

The belled portion of the drilled shaft will act as an anchor to resist the uplifting force. Ignoring the weight of the drilled shaft, we have

$$Q_{net} = U - D \tag{13.15}$$

where Q_{net} = net uplift load
$\quad\quad\quad D$ = dead load

Now,

$$Q_{net} \approx \frac{c_u N_c}{FS}\left(\frac{\pi}{4}\right)(D_b^2 - D_s^2) \tag{13.16}$$

where c_u = undrained cohesion of the clay in which the bell is located
  N_c = bearing capacity factor
  FS = factor of safety
  D_b = diameter of the bell of the drilled shaft

Combining Eqs. (13.15) and (13.16) gives

$$U - D = \frac{c_u N_c}{FS}\left(\frac{\pi}{4}\right)(D_b^2 - D_s^2) \qquad (13.17)$$

Conservatively, from Table 3.4 and Eq. (3.25),

$$N_c \approx N_{c(\text{strip})}F_{cs} = N_{c(\text{strip})}\left(1 + \frac{N_q B}{N_c L}\right) \approx 5.14\left(1 + \frac{1}{5.14}\right) = 6.14$$

A drilled-shaft design is examined in Example 13.4.

Example 13.4

Figure 13.20 shows a drilled shaft. The depth of the active zone is 15 ft. The zero swell pressure of the swelling clay (σ_T') is 70 lb/in². For the shaft, the dead load is 135 kip and the live load is 68 kip. Assume $\phi_{ps}' = 12°$

a. Determine the diameter of the bell, D_b.
b. Determine the reinforcement required for the pier shaft.
c. Check the bearing capacity of the shaft, assuming a zero uplift force.

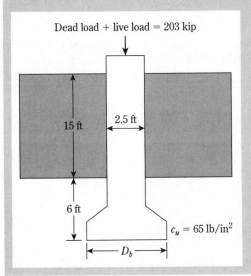

Dead load + live load = 203 kip

15 ft   2.5 ft

6 ft

$c_u = 65$ lb/in²

D_b

Figure 13.20 Drill shaft in a swelling clay

Solution

Part a: Determining the Bell Diameter, D_b
The uplift force is given by Eq. 13.14:

$$U = \pi D_s Z \sigma'_T \tan \phi'_{ps}$$

Assume that $\phi'_{ps} \approx 12°$, $Z = 15$ ft, and $\sigma'_T = 70$ lb/in^2. Then

$$U = \pi(2.5 \times 12)(15 \times 12)(70) \tan 12° = 252{,}416 \text{ lb} \approx 252.4 \text{ kip}$$

Assume the dead load and live load to be zero and FS in Eq. (13.17) to be 1.25. Then, from Eq. (13.17),

$$U = \frac{c_u N_c}{\text{FS}}\left(\frac{\pi}{4}\right)(D_b^2 - D_s^2)$$

or

$$252.4 = \frac{\left(\dfrac{65}{1000}\right)(6.14)}{1.25}\left(\frac{\pi}{4}\right)(D_b^2 - 30^2)$$

Thus,

$$D_b = 43.66 \text{ in.; use } \textbf{45 in.}$$

The factor of safety against uplift with the dead load also should be checked. A factor of safety of at least 3 is desirable. So, from Eq. (13.17),

$$\text{FS} = \frac{c_u N_c\left(\dfrac{\pi}{4}\right)(D_b^2 - D_s^2)}{U - D} = \frac{\left(\dfrac{65}{1000}\right)(6.14)\left(\dfrac{\pi}{4}\right)(45^2 - 30^2)}{252.4 - 135} = \textbf{3.01} > \textbf{3, OK}$$

Part b: Reinforcement
Reinforcement should be provided for the total uplift load, or 252.4 kip (assuming that the dead load and live load are both zero). So the area of steel required is

$$A_s = \frac{U}{\left(\dfrac{\text{yield stress of steel}}{\text{factor of safety}}\right)}$$

An adequate FS is 1.25, and the yield stress of steel $\approx 40{,}000$ lb/in^2. Hence,

$$A_s = \frac{(252.4)(1.25)}{40} = 7.9 \text{ in}^2$$

Part c: Checking for Bearing Capacity
Assume that $U = 0$. Then

$$\text{Dead load + live load} = 135 + 68 = 203 \text{ kip}$$

and

$$\text{Downward load per unit area} = \frac{203}{(\pi/4)(D_b^2)} = \frac{203}{(\pi/4)(45^2)}$$
$$= 0.128 \text{ kip/in}^2 = 128 \text{ lb/in}^2$$

Now,

$$\text{Net bearing capacity of the soil under the bell} = q_u = c_u N_c = (65)(6.14)$$
$$= 399 \text{ lb/in}^2$$

Thus, the factor of safety against bearing capacity failure is
$399/128 = \mathbf{3.12 > 3, OK}$ ∎

Sanitary Landfills

13.13 General Nature of Sanitary Landfills

Sanitary landfills provide a way to dispose of refuse on land without endangering public health. Sanitary landfills are used in almost all countries, to varying degrees of success. The refuse disposed of in sanitary landfills may contain organic, wood, paper, and fibrous wastes, or demolition wastes such as bricks and stones. The refuse is dumped and compacted at frequent intervals and is then covered with a layer of soil, as shown in Figure 13.21. In the compacted state, the average unit weight of the refuse may vary between 5 and 10 kN/m³ (32–64 lb/ft³). A typical city in the United States, with a population of 1 million, generates about $3.8 \times 10^6 \text{ m}^3 (\approx 135 \times 10^6 \text{ ft}^3)$ of compacted landfill material per year.

As property values continue to increase in densely populated areas, constructing structures over sanitary landfills becomes more and more tempting. In some instances, a visual site inspection may not be enough to detect an old sanitary landfill. However, construction of foundations over sanitary landfills is generally

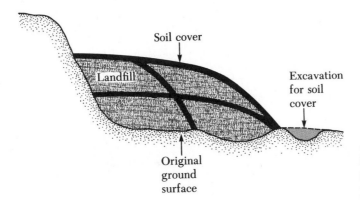

Figure 13.21
Schematic diagram of a sanitary landfill in progress

problematic because of poisonous gases (e.g., methane), excessive settlement, and low inherent bearing capacity.

13.14   *Settlement of Sanitary Landfills*

Sanitary landfills undergo large continuous settlements over a long time. Yen and Scanlon (1975) documented the settlement of several landfill sites in California. After completion of the landfill, the settlement rate (Figure 13.22) may be expressed as

$$m = \frac{\Delta H_f}{\Delta t} \tag{13.18}$$

where m = settlement rate

H_f = maximum height of the sanitary landfill

On the basis of several field observations, Yen and Scanlon determined the following empirical correlations for the settlement rate:

$m = 0.0268 - 0.0116 \log t_1$ (for fill heights ranging from 12 to 24 m)   (13.19)

$m = 0.038 - 0.0155 \log t_1$ (for fill heights ranging from 24 to 30 m)   (13.20)

$m = 0.0433 - 0.0183 \log t_1$ (for fill heights larger than 30 m)   (13.21)

In these equations, m is in m/mo.

t_1 is the median fill age, in months

The median fill age may be defined from Figure 13.22 as

$$t_1 = t - \frac{t_c}{2} \tag{13.22}$$

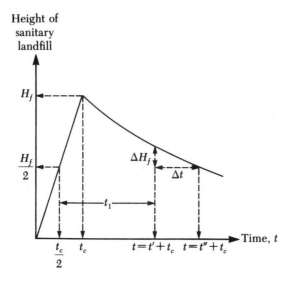

Figure 13.22   Settlement of sanitary landfills

where t = time from the beginning of the landfill
   t_c = time for completion of the landfill

Equations (13.19), (13.20), and (13.21) are based on field data from landfills for which t_c varied from 70 to 82 months. To get an idea of the approximate length of time required for a sanitary landfill to undergo complete settlement, consider Eq. (13.19). For a fill 12 m high and for $t_c = 72$ months,

$$m = 0.0268 - 0.0116 \log t_1$$

so

$$\log t_1 = \frac{0.0268 - m}{0.0116}$$

If $m = 0$ (zero settlement rate), $\log t_1 = 2.31$, or $t_1 \approx 200$ months. Thus, settlement will continue for $t_1 - t_c/2 = 200 - 36 = 164$ months (≈ 14 years) after completion of the fill—a fairly long time. This calculation emphasizes the need to pay close attention to the settlement of foundations constructed on sanitary landfills.

A comparison of Eqs. (13.19) through (13.21) for rates of settlement shows that the value of m increases with the height of the fill. However, for fill heights greater than about 30 m, the rate of settlement should not be much different from that obtained from Eq. (13.21). The reason is that the decomposition of organic matter close to the surface is mainly the result of an anaerobic environment. For deeper fills, the decomposition is slower. Hence, for fill heights greater than about 30 m, the rate of settlement does not exceed that for fills that are about 30 m in height.

In English units, Eqs. (13.19)–(13.21) can be expressed in the following form:

$$m = 0.088 - 0.038 \log t_1 \qquad \text{(for fill heights 40–80 ft)} \qquad (13.23)$$

$$m = 0.125 - 0.051 \log t_1 \qquad \text{(for fill heights 80–100 ft)} \qquad (13.24)$$

$$m = 0.142 - 0.06 \log t_1 \qquad \text{(for fill heights greater than 100 ft)} \qquad (13.25)$$

Sowers (1973) also proposed a formula for calculating the settlement of a sanitary landfill, namely,

$$\Delta H_f = \frac{\alpha H_f}{1 + e} \log \left(\frac{t''}{t'} \right) \qquad (13.26)$$

where H_f = height of the fill
   e = void ratio
   α = a coefficient for settlement
   t'', t' = times (see Figure 13.22)
   ΔH_f = settlement between times t' and t''

The coefficients α fall between

$$\alpha = 0.09e \quad \text{(for conditions favorable to decomposition)} \qquad (13.27)$$

and

$$\alpha = 0.03e \quad \text{(for conditions unfavorable to decomposition)} \quad (13.28)$$

Equation (13.26) is similar to the equation for secondary consolidation settlement [Equation (5.72)].

Problems

13.1   Use Figure 13.3, which is based on Eq. (13.5) and $G_s = 2.65$, to draw a similar curve plotting γ_d (in kN/m³) against the liquid limit with $G_s = 2.69$. Show the zone in which the loessial soils are likely to collapse on saturation.

13.2   A natural loessial soil deposit has a dry unit weight of 13.2 kN/m³. The liquid limit of the soil is 31 and $G_s = 2.64$. Is collapse likely to occur in this soil?

13.3   A collapsible soil layer is 10 ft thick. The average effective overburden pressure on the soil layer is 1200 lb/ft². An undisturbed specimen of the soil was subjected to a double oedometer test. (See Figure 13.4.) The preconsolidation pressure of the specimen, as determined from the soaked specimen, was 1700 lb/ft². Is this soil normally consolidated or preconsolidated?

13.4   An expansive soil has an active-zone thickness of 8 m. The natural moisture content of the soil is 20%, and its liquid limit is 50. Calculate the free surface swell of the expansive soil upon saturation.

13.5   Repeat Problem 13.4 for a liquid limit of the soil of 55. All other quantities are the same.

13.6   Following are the results of a modified free swell index test:

> Mass of dry soil = 10 grams
> $G_s = 2.74$
> Volume of swollen sediment after 24 hours in water = 28.6 cm³

a. Determine the modified free swell index.
b. Describe the swell potential.

13.7   An expansive soil profile has an active-zone thickness of 5.2 m. A shallow foundation is to be constructed 1.2 m below the ground surface. A swelling pressure test provided the following data:

Depth below ground surface (m)	Swell under overburden and estimated foundation surcharge pressure, $s_{w(1)}$ (%)
1.2	3.0
2.2	2.0
3.2	1.2
4.2	0.55
5.2	0.0

Estimate the total possible swell under the foundation.

13.8   Repeat Problem 13.7 for an active-zone thickness of 25 ft, a shallow foundation depth of 2.5 ft, and the following data:

Depth below ground surface (ft)	Swell under overburden and estimated foundation surcharge pressure, $s_{w(1)}$(%)
2.5	4.7
5.0	3.85
10.0	2.45
15.0	1.3
20.0	0.4
25.0	0.0

13.9 In Problem 13.7, if the allowable total swell is 15 mm, what would be the necessary undercut?

13.10 In Problem 13.8, if the allowable total swell is 1.0 in., what would be the necessary undercut?

13.11 For the drilled shaft with bell shown in Figure 13.19(b), let

Thickness of active zone, $Z = 30$ ft
Dead load $= 300$ kip
Live load $= 60$ kip
Diameter of the shaft, $D_s = 3.5$ ft
Zero swell pressure for the clay in the active zone $= 6$ ton/ft^2
Average angle of plinth-soil friction, $\phi'_{ps} = 15°$

and

Average undrained cohesion of the clay around the bell $= 3020$ lb/ft^2

Determine the diameter of the bell, D_b. A factor of safety of 2 against uplift is required, together with the assumption that the dead load plus live load is equal to zero.

13.12 In Problem 13.11, if an additional requirement is that the factor of safety against uplift is at least 3 with the dead load on (live load $= 0$), what should be the diameter of the bell?

References

Abduljauwad, S. N., and Al-Sulaimani, G. J. (1993). "Determination of Swell Potential of Al-Qatif Clay," *Geotechnical Testing Journal,* American Society for Testing and Materials, Vol. 16, No. 4, pp. 469–484.

Altmeyer, W. T. (1955). "Discussion of Engineering Properties of Expansive Clays," *Journal of the Soil Mechanics and Foundations Division,* American Society of Civil Engineers, Vol. 81, No. SM2, pp. 17–19.

Benites, L. A. (1968). "Geotechnical Properties of the Soils Affected by Piping near the Benson Area, Cochise County, Arizona," M. S. Thesis, University of Arizona, Tucson.

Chen, F. H. (1988). *Foundations on Expansive Soils,* Elsevier, Amsterdam.

Clemence, S. P., and Finbarr, A. O. (1981). "Design Considerations for Collapsible Soils," *Journal of the Geotechnical Engineering Division,* American Society of Civil Engineers, Vol. 107, No. GT3, pp. 305–317.

Clevenger, W. (1958). "Experience with Loess as Foundation Material," *Transactions,* American Society of Civil Engineers, Vol. 123, pp. 151–170.

Dakshanamanthy, V., and Raman, V. (1973). "A Simple Method of Identifying an Expansive Soil," *Soils and Foundations,* Vol. 13, No. 1, pp. 97–104.

Denisov, N. Y. (1951). *The Engineering Properties of Loess and Loess Loams,* Gosstroiizdat, Moscow.

Dudley, J. H. (1970). "Review of Collapsing Soil," *Journal of the Soil Mechanics and Foundation Engineering Division,* American Society of Civil Engineers, Vol. 96, No. SM3, pp. 925–947.

Feda, J. (1964). "Colloidal Activity, Shrinking and Swelling of Some Clays," *Proceedings, Soil Mechanics Seminar,* Loda, Illinois, pp. 531–546.

Gibbs, H. J. (1961). *Properties Which Divide Loose and Dense Uncemented Soils,* Earth Laboratory Report EM-658, Bureau of Reclamation, U.S. Department of the Interior, Washington, DC.

Gromko, G. J. (1974). "Review of Expansive Soils," *Journal of the Geotechnical Engineering Division,* American Society of Civil Engineers, Vol. 100, No. GT6, pp. 667–687.

Handy, R. L. (1973). "Collapsible Loess in Iowa," *Proceedings, Soil Science Society of America,* Vol. 37, pp. 281–284.

Holtz, W. G. (1959). "Expansive Clays—Properties and Problems," *Journal of the Colorado School of Mines,* Vol. 54, No. 4, pp. 89–125.

Holtz, W. G., and Hilf, J. W. (1961). "Settlement of Soil Foundations Due to Saturation," *Proceedings, Fifth International Conference on Soil Mechanics and Foundation Engineering,* Paris, Vol. 1, 1961, pp. 673–679.

Houston, W. N., and Houston, S. L. (1989). "State-of-the-Practice Mitigation Measures for Collapsible Soil Sites," *Proceedings, Foundation Engineering: Current Principles and Practices,* American Society of Civil Engineers, Vol. 1, pp. 161–175.

Jennings, J. E., and Knight, K. (1975). "A Guide to Construction on or with Materials Exhibiting Additional Settlements Due to 'Collapse' of Grain Structure," *Proceedings, Sixth Regional Conference for Africa on Soil Mechanics and Foundation Engineering,* Johannesburg, pp. 99–105.

Lutenegger, A. J. (1986). "Dynamic Compaction in Friable Loess," *Journal of Geotechnical Engineering,* American Society of Civil Engineers, Vol. 112, No. GT6, pp. 663–667.

Lutenegger, A. J., and Saber, R. T. (1988). "Determination of Collapse Potential of Soils," *Geotechnical Testing Journal,* American Society for Testing and Materials, Vol. 11. No. 3, pp. 173–178.

McKeen, R. G. (1992). "A Model for Predicting Expansive Soil Behavior," *Proceedings, Seventh International Conference on Expansive Soils,* Dallas, Vol. 1, pp. 1–6.

Nayak, N. V., and Christensen, R. W. (1974). "Swell Characteristics of Compacted Expansive Soils," *Clay and Clay Minerals,* Vol. 19, pp. 251–261.

O'Neill, M. W., and Poormoayed, N. (1980). "Methodology for Foundations on Expansive Clays," *Journal of the Geotechnical Engineering Division,* American Society of Civil Engineers, Vol. 106, No. GT12, p. 1345–1367.

Peck, R. B., Hanson, W. E., and Thornburn, T. B. (1974). *Foundation Engineering,* Wiley, New York.

Priklonskii, V. A. (1952). *Gruntovedenic, Vtoraya Chast',* Gosgeolzdat, Moscow.

Raman, V. (1967). "Identification of Expansive Soils from the Plasticity Index and the Shrinkage Index Data," *The Indian Engineer,* Vol. 11, No. 1, pp. 17–22.

Seed, H. B., Woodward, R. J., Jr., and Lundgren, R. (1962). "Prediction of Swelling Potential for Compacted Clays," *Journal of the Soil Mechanics and Foundations Division,* American Society of Civil Engineers, Vol. 88, No. SM3, pp. 53–87.

Semkin, V. V., Ermoshin, V. M., and Okishev, N. D. (1986). "Chemical Stabilization of Loess Soils in Uzbekistan," *Soil Mechanics and Foundation Engineering* (trans. from Russian), Vol. 23, No. 5, pp. 196–199.

Sivapullaiah, P. V., Sitharam, T. G., and Rao, K. S. S. (1987). "Modified Free Swell Index for Clay," *Geotechnical Testing Journal,* American Society for Testing and Materials, Vol. 11, No. 2, pp. 80–85.

Snethen, D. R. (1984). "Evaluation of Expedient Methods for Identification and Classification of Potentially Expansive Soils," *Proceedings, Fifth International Conference on Expansive Soils,* Adelaide, pp. 22–26.

Snethen, D. R., Johnson, L. D., and Patrick, D. M. (1977). *An Evaluation of Expedient Methodology for Identification of Potentially Expansive Soils,* Report No. FHWA-RD-77-94, U.S. Army Engineers Waterways Experiment Station, Vicksburg, MS.

Sowers, G. F. (1973). "Settlement of Waste Disposal Fills," *Proceedings, Eighth International Conference on Soil Mechanics and Foundation Engineering,* Moscow, pp. 207–210.

Sowers, G. B., and Sowers, G. F. (1970). *Introductory Soil Mechanics and Foundations,* 3d ed. Macmillan, New York.

Sridharan, A., Rao, A. S., and Sivapullaiah, P. V. (1986), "Swelling Pressure of Clays," *Geotechnical Testing Journal,* American Society for Testing and Materials, Vol. 9, No. 1, pp. 24–33.

Uniform Building Code (1968). *UBC Standard No. 29-2.*

Van Der Merwe, D. H. (1964), "The Prediction of Heave from the Plasticity Index and Percentage Clay Fraction of Soils," *Civil Engineer in South Africa,* Vol. 6, No. 6, pp. 103–106.

Vijayvergiya, V. N., and Ghazzaly, O. I. (1973). "Prediction of Swelling Potential of Natural Clays," *Proceedings, Third International Research and Engineering Conference on Expansive Clays,* pp. 227–234.

Weston, D. J. (1980). "Expansive Roadbed Treatment for Southern Africa," *Proceedings, Fourth International Conference on Expansive Soils,* Vol. 1, pp. 339–360.

Yen, B. C., and Scanlon, B. (1975). "Sanitary Landfill Settlement Rates," *Journal of the Geotechnical Engineering Division,* American Society of Civil Engineers, Vol. 101, No. GT5, pp. 475–487.

14

Soil Improvement and Ground Modification

14.1 Introduction

The soil at a construction site may not always be totally suitable for supporting structures such as buildings, bridges, highways, and dams. For example, in granular soil deposits, the *in situ* soil may be very loose and indicate a large elastic settlement. In such a case, the soil needs to be densified to increase its unit weight and thus its shear strength.

Sometimes the top layers of soil are undesirable and must be removed and replaced with better soil on which the structural foundation can be built. The soil used as fill should be well compacted to sustain the desired structural load. Compacted fills may also be required in low-lying areas to raise the ground elevation for construction of the foundation.

Soft saturated clay layers are often encountered at shallow depths below foundations. Depending on the structural load and the depth of the layers, unusually large consolidation settlement may occur. Special soil-improvement techniques are required to minimize settlement.

In Chapter 13, we mentioned that the properties of expansive soils could be altered substantially by adding stabilizing agents such as lime. Improving *in situ* soils by using additives is usually referred to as *stabilization*.

Various techniques are used to

1. Reduce the settlement of structures
2. Improve the shear strength of soil and thus increase the bearing capacity of shallow foundations
3. Increase the factor of safety against possible slope failure of embankments and earth dams
4. Reduce the shrinkage and swelling of soils

This chapter discusses some of the general principles of soil improvement, such as compaction, vibroflotation, precompression, sand drains, wick drains, and stabilization by admixtures, as well as the use of stone columns and sand compaction piles in weak clay to construct foundations.

14.2   *General Principles of Compaction*

If a small amount of water is added to a soil that is then compacted, the soil will have a certain unit weight. If the moisture content of the same soil is gradually increased and the energy of compaction is the same, the dry unit weight of the soil will gradually increase. The reason is that water acts as a lubricant between the soil particles, and under compaction it helps rearrange the solid particles into a denser state. The increase in dry unit weight with increase of moisture content for a soil will reach a limiting value beyond which the further addition of water to the soil will result in a *reduction* in dry unit weight. The moisture content at which the *maximum dry unit weight* is obtained is referred to as the *optimum moisture content.*

The standard laboratory tests used to evaluate maximum dry unit weights and optimum moisture contents for various soils are

a. the Standard Proctor test (ASTM designation D-698) and
b. the Modified Proctor test (ASTM designation D-1557)

The soil is compacted in a mold in several layers by a hammer. The moisture content of the soil, w, is changed, and the dry unit weight, γ_d, of compaction for each test is determined. The maximum dry unit weight of compaction and the corresponding optimum moisture content are determined by plotting a graph of γ_d against w (%). The standard specifications for the two types of Proctor test are given in Tables 14.1 and 14.2.

Table 14.1   Specifications for Standard Proctor Test (Based on ASTM Designation 698)

Item	Method A	Method B	Method C
Diameter of mold	101.6 mm (4 in.)	101.6 mm (4 in.)	152.4 mm (6 in.)
Volume of mold	944 cm^3($\frac{1}{30}$ ft^3)	944 cm^3($\frac{1}{30}$ ft^3)	2124 cm^3(0.075 ft^3)
Weight of hammer	2.5 kg (5.5 lb)	2.5 kg (5.5 lb)	2.5 kg (5.5 lb)
Height of hammer drop	304.8 mm (12 in.)	304.8 mm (12 in.)	304.8 mm (12 in.)
Number of hammer blows per layer of soil	25	25	56
Number of layers of compaction	3	3	3
Energy of compaction	600 kN·m/m^3 (12,400 ft·lb/ft^3)	600 kN·m/m^3 (12,400 ft·lb/ft^3)	600 kN·m/m^3 (12,400 ft·lb/ft^3)
Soil to be used	Portion passing no. 4 (4.57-mm) sieve. May be used if 20% *or less* by weight of material is retained on no. 4 sieve.	Portion passing $\frac{3}{8}$-in. (9.5-mm) sieve. May be used if soil retained on no. 4 sieve *is more* than 20% and 20% *or less* by weight is retained on 9.5-mm ($\frac{3}{8}$-in.) sieve.	Portion passing $\frac{3}{4}$-in. (19.0-mm) sieve. May be used if *more than* 20% by weight of material is retained on 9.5 mm ($\frac{3}{8}$ in.) sieve and *less than* 30% by weight is retained on 19.00-mm ($\frac{3}{4}$-in.) sieve.

Table 14.2 Specifications for Modified Proctor Test (Based on ASTM Designation 1557)

Item	Method A	Method B	Method C
Diameter of mold	101.6 mm (4 in.)	101.6 mm (4 in.)	152.4 mm (6 in.)
Volume of mold	944 cm³ ($\frac{1}{30}$ ft³)	944 cm³ ($\frac{1}{30}$ ft³)	2124 cm³ (0.075 ft³)
Weight of hammer	4.54 kg (10 lb)	4.54 kg (10 lb)	4.54 kg (10 lb)
Height of hammer drop	457.2 mm (18 in.)	457.2 mm (18 in.)	457.2 mm (18 in.)
Number of hammer blows per layer of soil	25	25	56
Number of layers of compaction	5	5	5
Energy of compaction	2700 kN·m/m³ (56,000 ft·lb/ft³)	2700 kN·m/m³ (56,000 ft·lb/ft³)	2700 kN·m/m³ (56,000 ft·lb/ft³)
Soil to be used	Portion passing no. 4 (4.57-mm) sieve. May be used if 20% *or less* by weight of material is retained on no. 4 sieve.	Portion passing 9.5-mm ($\frac{3}{8}$-in.) sieve. May be used if soil retained on no. 4 sieve *is more* than 20% and 20% *or less* by weight is retained on 9.5-mm ($\frac{3}{8}$-in.) sieve.	Portion passing 19.0-mm ($\frac{3}{4}$-in.) sieve. May be used if *more than* 20% by weight of material is retained on 9.5-mm ($\frac{3}{8}$-in.) sieve and *less than* 30% by weight is retained on 19-mm ($\frac{3}{4}$-in.) sieve.

Figure 14.1 shows a plot of γ_d against w (%) for a clayey silt obtained from standard and modified Proctor tests (method A). The following conclusions may be drawn:

1. The maximum dry unit weight and the optimum moisture content depend on the degree of compaction.
2. The higher the energy of compaction, the higher is the maximum dry unit weight.
3. The higher the energy of compaction, the lower is the optimum moisture content.
4. No portion of the compaction curve can lie to the right of the zero-air-void line. The zero-air-void dry unit weight, γ_{zav}, at a given moisture content is the theoretical maximum value of γ_d, which means that all the void spaces of the compacted soil are filled with water, or

$$\gamma_{zav} = \frac{\gamma_w}{\dfrac{1}{G_s} + w} \tag{14.1}$$

where γ_w = unit weight of water
 G_s = specific gravity of the soil solids
 w = moisture content of the soil

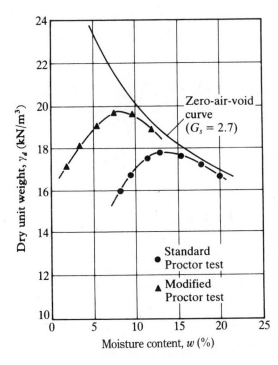

Figure 14.1 Standard and modified Proctor compaction curves for a clayey silt (method A)

5. The maximum dry unit weight of compaction and the corresponding optimum moisture content will vary from soil to soil.

Using the results of laboratory compaction (γ_d versus w), specifications may be written for the compaction of a given soil in the field. In most cases, the contractor is required to achieve a relative compaction of 90% or more on the basis of a specific laboratory test (either the standard or the modified Proctor compaction test). The relative compaction is defined as

$$RC = \frac{\gamma_{d(\text{field})}}{\gamma_{d(\text{max})}} \tag{14.2}$$

Chapter 1 introduced the concept of relative density (for the compaction of granular soils), defined as

$$D_r = \left[\frac{\gamma_d - \gamma_{d(\text{min})}}{\gamma_{d(\text{max})} - \gamma_{d(\text{min})}} \right] \frac{\gamma_{d(\text{max})}}{\gamma_d}$$

where γ_d = dry unit weight of compaction in the field

$\gamma_{d(\text{max})}$ = maximum dry unit weight of compaction as determined in the laboratory

$\gamma_{d(\text{min})}$ = minimum dry unit weight of compaction as determined in the laboratory

For granular soils in the field, the degree of compaction obtained is often measured in terms of relative density. Comparing the expressions for relative density and relative compaction reveals that

$$RC = \frac{A}{1 - D_r(1 - A)} \tag{14.3}$$

where $A = \dfrac{\gamma_{d(min)}}{\gamma_{d(max)}}$

Lee and Singh (1971) reviewed 47 different soils and, on the basis of their review, presented the following correlation:

$$D_r(\%) = \frac{(RC - 80)}{0.2} \tag{14.4}$$

14.3 *Correction for Compaction of Soils with Oversized Particles*

As shown in Tables 14.1 and 14.2, depending on the method used, certain oversized particles (such as material retained on a no. 4 sieve or a $\frac{3}{4}$-in. sieve) may need to be removed from the soil to conduct laboratory compaction tests. ASTM test designation D-4718 provides a method for making corrections for maximum dry unit weight and optimum moisture content in the presence of oversized particles. This is helpful in writing specifications for field compaction. According to the method, the corrected maximum dry unit weight may be calculated as

$$\gamma_{d(max)\text{-}C} = \frac{100\gamma_w}{\dfrac{P_C}{G_m} + \dfrac{\gamma_w(100 - P_C)}{\gamma_{d(max)\text{-}F}}} \tag{14.5}$$

where γ_w = unit weight of water
P_C = percentage of oversized particles by weight
G_m = bulk specific gravity of the oversized particles
$\gamma_{d(max)\text{-}F}$ = maximum dry unit weight of the finer fraction used in the laboratory compaction

The corrected optimum moisture content can be expressed as

$$w_{\text{optimum-}C}(\%) = w_{\text{optimum-}F}(100 - P_C) + w_C P_C \tag{14.6}$$

where $w_{\text{optimum-}F}$ = optimum moisture content of the finer fraction determined in the laboratory

w_C = saturated surface dry moisture content of the oversized particles (fraction)

The preceding equations for corrected maximum dry unit weight and optimum moisture content are valid when the oversized particles are about 30% or less (by weight) of the total soil sample.

14.4 *Field Compaction*

Ordinary compaction in the field is done by rollers. Of the several types of roller used, the most common are

1. Smooth-wheel rollers (or smooth drum rollers)
2. Pneumatic rubber-tired rollers
3. Sheepsfoot rollers
4. Vibratory rollers

Figure 14.2 shows a *smooth-wheel roller* that can also create vertical vibration during compaction. Smooth-wheel rollers are suitable for proof-rolling subgrades and for finishing the construction of fills with sandy or clayey soils. They provide 100% coverage under the wheels, and the contact pressure can be as high as 300–400 kN/m^2 ($\approx$45–60 lb/in^2). However, they do not produce a uniform unit weight of compaction when used on thick layers.

Figure 14.2 Vibratory smooth-wheel rollers (Courtesy of Tampo Manufacturing Co., Inc., San Antonio, Texas)

Figure 14.3 Pneumatic rubber-tired roller (Courtesy of Tampo Manufacturing Co., Inc., San Antonio, Texas)

Pneumatic rubber-tired rollers (Figure 14.3) are better in many respects than smooth-wheel rollers. Pneumatic rollers, which may weigh as much as 2000 kN (450 kip), consist of a heavily loaded wagon with several rows of tires. The tires are closely spaced—four to six in a row. The contact pressure under the tires may range up to 600–700 kN/m^2 ($\approx$85–100 lb/in^2), and they give about 70–80% coverage. Pneumatic rollers, which can be used for sandy and clayey soil compaction, produce a combination of pressure and kneading action.

Sheepsfoot rollers (Figure 14.4) consist basically of drums with large numbers of projections. The area of each of the projections may be 25–90 cm^2 (4–14 in^2). These rollers are *most effective in compacting cohesive soils*. The contact pressure under the projections may range from 1500–7500 kN/m^2 ($\approx$215–1100 lb/in^2). During compaction in the field, the initial passes compact the lower portion of a lift. Later, the middle and top of the lift are compacted.

Vibratory rollers are efficient in compacting granular soils. Vibrators can be attached to smooth-wheel, pneumatic rubber-tired or sheepsfoot rollers to send vibrations into the soil being compacted. Figures 14.2 and 14.4 show vibratory smooth-wheel rollers and a vibratory sheepsfoot roller, respectively.

In general, compaction in the field depends on several factors, such as the type of compactor, type of soil, moisture content, lift thickness, towing speed of the compactor, and number of passes the roller makes.

Figure 14.5 shows the variation of the unit weight of compaction with depth for a poorly graded dune sand compacted by a vibratory drum roller. Vibration was produced by mounting an eccentric weight on a single rotating shaft within the drum cylinder. The weight of the roller used for this compaction was 55.7 kN

Figure 14.4   Vibratory sheepsfoot roller (Courtesy of Tampo Manufacturing Co., Inc., San Antonio, Texas)

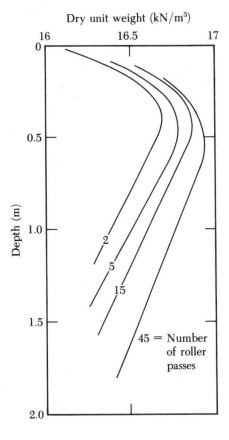

Figure 14.5   Vibratory compaction of a sand: Variation of dry unit weight with depth and number of roller passes; lift thickness = 2.44 m (after D'Appolonia et al., 1969)

(12.5 kip), and the drum diameter was 1.19 m (3.9 ft). The lifts were kept at 2.44 m (8 ft). Note that, at any depth, the dry unit weight of compaction increases with the number of passes the roller makes. However, the rate of increase in unit weight gradually decreases after about 15 passes. Note also the variation of dry unit weight with depth by the number of roller passes. The dry unit weight and hence the relative density, D_r, reach maximum values at a depth of about 0.5 m ($\approx$1.6 ft) and then gradually decrease as the depth decreases. The reason is the lack of confining pressure toward the surface. Once the relation between depth and relative density (or dry unit weight) for a soil for a given number of passes is determined, for satisfactory compaction based on a given specification, the approximate thickness of each lift can be easily estimated.

14.5 *Compaction Control for Clay Hydraulic Barriers*

Compacted clays are commonly used as hydraulic barriers in cores of earth dams, liners and covers of landfills, and liners of surface impoundments. Since the primary purpose of a barrier is to minimize flow, the hydraulic conductivity, k, is the controlling factor. In many cases, it is desired that the hydraulic conductivity be less than 10^{-7} cm/s. This can be achieved by controlling the minimum degree of saturation during compaction, a relation that can be explained by referring to the compaction characteristics of three soils described in Table 14.3 (Othman and Luettich, 1994).

Figures 14.6, 14.7, and 14.8 show the standard and modified Proctor test results and the hydraulic conductivities of compacted specimens. Note that the solid symbols represent specimens with hydraulic conductivities of 10^{-7} cm/s or less. As can be seen from these figures, the data points plot generally parallel to the line of full saturation. Figure 14.9 shows the effect of the degree of saturation during compaction on the hydraulic conductivity of the three soils. It is evident from the figure that, if it is desired that the maximum hydraulic conductivity be 10^{-7} cm/s, then all soils should be compacted at a minimum degree of saturation of 88%.

In field compaction at a given site, soils of various composition may be encountered. Small changes in the content of fines will change the magnitude of hydraulic conductivity. Hence, considering the various soils likely to be encountered at a given site the procedure just described aids in developing a minimum-degree-of-saturation criterion for compaction to construct hydraulic barriers.

Table 14.3 Characteristics of Soils Reported in Figures 14.6, 14.7, and 14.8

Soil	Classification	Liquid limit	Plasticity index	Percent finer than no. 200 sieve (0.075 mm)
Wisconsin A	CL	34	16	85
Wisconsin B	CL	42	19	99
Wisconsin C	CH	84	60	71

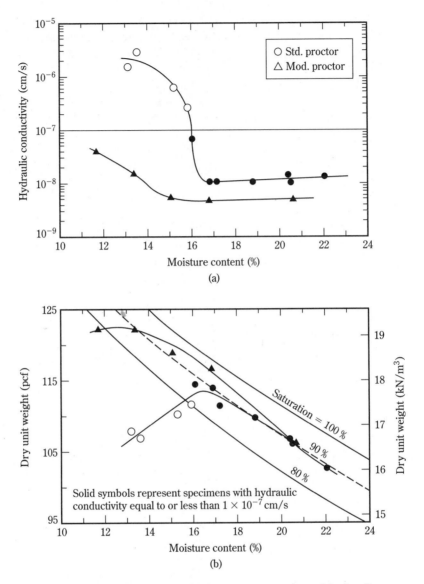

Figure 14.6 Standard and modified Proctor test results and hydraulic conductivity of Wisconsin A soil (after Othman and Luettich, 1994)

14.6 *Vibroflotation*

Vibroflotation is a technique developed in Germany in the 1930s for *in situ* densifica-tion of thick layers of loose granular soil deposits. Vibroflotation was first used in the United States about 10 years later. The process involves the use of a *vibroflot* (called the *vibrating unit*), as shown in Figure 14.10. The device is about 2 m (6 ft) in length. This vibrating unit has an eccentric weight inside it and can develop a centrifugal

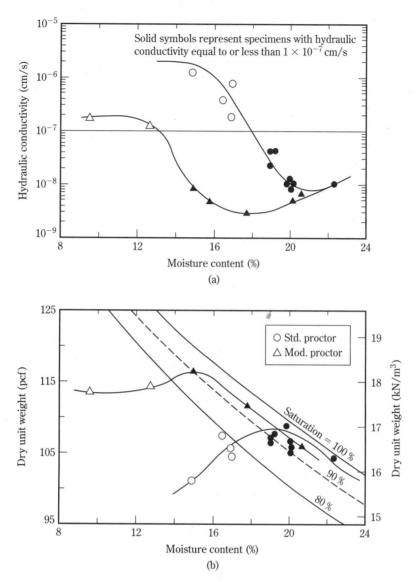

Figure 14.7 Standard and modified Proctor test results and hydraulic conductivity of Wisconsin B soil (after Othman and Luettich, 1994)

force. The weight enables the unit to vibrate horizontally. Openings at the bottom and top of the unit are for water jets. The vibrating unit is attached to a follow-up pipe. The figure shows the vibroflotation equipment necessary for compaction in the field.

The entire compaction process can be divided into four steps (see Figure 14.11):

Step 1. The jet at the bottom of the vibroflot is turned on, and the vibroflot is lowered into the ground.

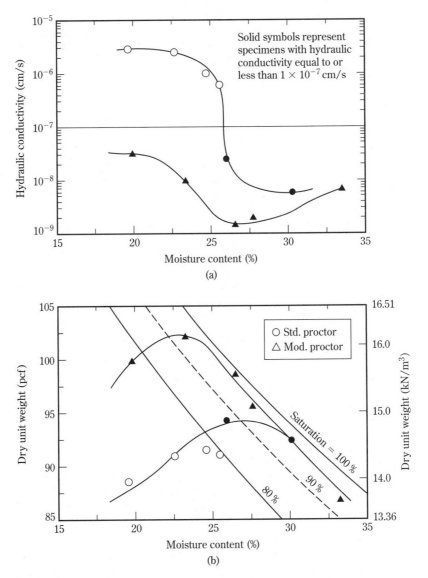

Figure 14.8 Standard and modified Proctor test results and hydraulic conductivity of Wisconsin C soil (after Othman and Luettich, 1994)

Step 2. The water jet creates a quick condition in the soil, which allows the vibrating unit to sink.

Step 3. Granular material is poured into the top of the hole. The water from the lower jet is transferred to the jet at the top of the vibrating unit. This water carries the granular material down the hole.

Step 4. The vibrating unit is gradually raised in about 0.3-m (1-ft) lifts and is held vibrating for about 30 seconds at a time. This process compacts the soil to the desired unit weight.

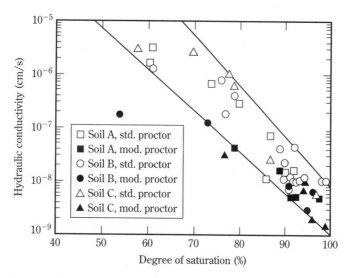

Figure 14.9 Effect of degree of saturation on hydraulic conductivity of Wisconsin A, B, and C soils (after Othman and Luettich, 1994)

Table 14.4 gives the details of various types of vibroflot unit used in the United States. The 30-HP electric units have been used since the latter part of the 1940s. The 100-HP units were introduced in the early 1970s. The zone of compaction around a single probe will vary according to the type of vibroflot used. The cylindrical zone of compaction will have a radius of about 2 m (6 ft) for a 30-HP unit and about 3 m (10 ft) for a 100-HP unit. Compaction by vibroflotation involves various probe spacings, depending on the zone of compaction. (See Figure 14.12.) Mitchell (1970) and Brown (1977) reported several successful cases of foundation design that used vibroflotation.

The success of densification of *in situ* soil depends on several factors, the most important of which are the grain-size distribution of the soil and the nature of the backfill used to fill the holes during the withdrawal period of the vibroflot. The range of the grain-size distribution of *in situ* soil marked Zone 1 in Figure 14.13 is most suitable for compaction by vibroflotation. Soils that contain excessive amounts of fine sand and silt-size particles are difficult to compact; for such soils, considerable effort is needed to reach the proper relative density of compaction. Zone 2 in Figure 14.13 is the approximate lower limit of grain-size distribution for compaction by vibroflotation. Soil deposits whose grain-size distribution falls into Zone 3 contain appreciable amounts of gravel. For these soils, the rate of probe penetration may be rather slow, so compaction by vibroflotation might prove to be uneconomical in the long run.

The grain-size distribution of the backfill material is one of the factors that control the rate of densification. Brown (1977) defined a quantity called *suitability number* for rating a backfill material. The suitability number is given by the formula

$$S_N = 1.7\sqrt{\frac{3}{(D_{50})^2} + \frac{1}{(D_{20})^2} + \frac{1}{(D_{10})^2}} \qquad (14.7)$$

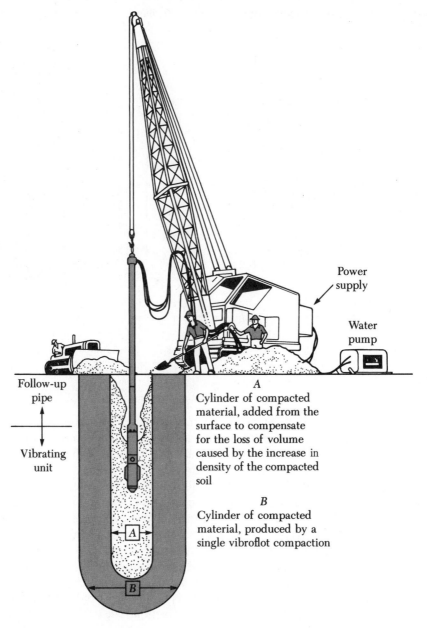

Figure 14.10 Vibroflotation unit (after Brown, 1977)

where D_{50}, D_{20}, and D_{10} are the diameters (in mm) through which 50%, 20%, and 10%, respectively, of the material is passing. The smaller the value of S_N, the more desirable is the backfill material. Following is a backfill rating system proposed by Brown (1977):

Range of S_N	Rating as backfill
0–10	Excellent
10–20	Good
20–30	Fair

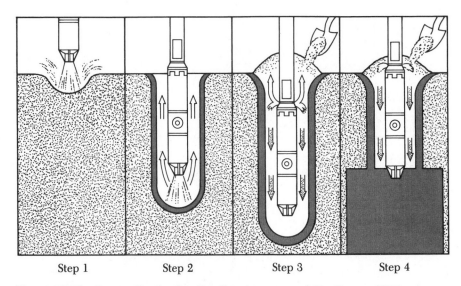

Step 1   Step 2   Step 3   Step 4

Figure 14.11 Compaction by the vibroflotation process (after Brown, 1977)

Range of S_N	Rating as backfill
30–50	Poor
>50	Unsuitable

An excellent case study that evaluated the benefits of vibroflotation was presented by Basore and Boitano (1969). Densification of granular subsoil was necessary for the construction of a three-story office building at the Treasure Island Naval Station in San Francisco, California. The top 9 m ($\approx$30 ft) of soil at the site was loose to medium-dense sand fill that had to be compacted. Figure 14.14 shows the layout of the vibroflotation compaction points and the location of the test borings. Sixteen compaction points were arranged in groups of four, with 1.22 m (4 ft), 1.52 m (5 ft), 1.83 m (6 ft), and 2.44 m (8 ft) spacing. Prior to compaction, standard penetration tests were conducted at the centers of groups of three compaction points. After the completion of compaction by vibroflotation, the variation of the standard penetration resistance with depth was determined at the same points.

Figure 14.15 shows the variation of the standard penetration resistance, N_{60}, with depth before and after compaction. The 1.22-m (4-ft) spacing produced the greatest increase in density of the sand, whereas the 2.44-m (8-ft) spacing had practically no effect. Figure 14.16 shows the relationship between the standard penetration resistance before and after compaction, the spacing between compaction points, and the depth below the ground surface. For any given spacing between compaction points, the standard penetration resistance decreased with an increase in depth. The increase in the standard penetration resistance at any depth indicates the increase in the relative density of compaction, D_r, of sand. Figure 14.17 shows the variation of D_r before and after compaction for depths up to 9.15 m (30 ft).

During the past 30 to 35 years, the vibroflotation technique has been used successfully on large projects to compact granular subsoils, thereby controlling structural settlement.

Table 14.4 Types of Vibrating Units[a]

	100-HP electric and hydraulic motors	30-HP electric motors
(a) Vibrating tip		
Length	2.1 m	1.86 m
	(7 ft)	(6.11 ft)
Diameter	406 mm	381 mm
	(16 in.)	(15 in.)
Weight	18 kN	18 kN
	(4000 lb)	(4000 lb)
Maximum movement when free	12.5 mm	7.6 mm
	(0.49 in.)	(0.3 in.)
Centrifugal force	160 kN	90 kN
	(18 ton)	(10 ton)
(b) Eccentric		
Weight	1.16 kN	0.76 kN
	(260 lb)	(170 lb)
Offset	38 mm	32 mm
	(1.5 in.)	(1.25 in.)
Length	610 mm	387 mm
	(24 in.)	(15.3 in.)
Speed	1800 rpm	1800 rpm
(c) Pump		
Operating flow rate	0–1.6 m³/min	0–6 m³/min
	(0–400 gal/min)	(0–150 gal/min)
Pressure	690–1035 kN/m²	690–1035 kN/m²
	(100–150 lb/in²)	(100–150 lb/in²)
(d) Lower follow-up pipe and extensions		
Diameter	305 mm	305 mm
	(12 in.)	(12 in.)
Weight	3.65 kN/m	3.65 kN/m
	(250 lb/ft)	(250 lb/ft)

[a] After Brown (1977)

14.7   *Precompression*

When highly compressible, normally consolidated clayey soil layers lie at a limited depth and large consolidation settlements are expected as the result of the construction of large buildings, highway embankments, or earth dams, precompression of soil may be used to minimize postconstruction settlement. The principles of precompression are best explained by reference to Figure 14.18. Here, the proposed structural load per unit area is $\Delta\sigma'_{(p)}$, and the thickness of the clay layer undergoing

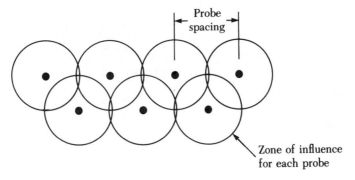

Figure 14.12 Nature of probe spacing for vibroflotation

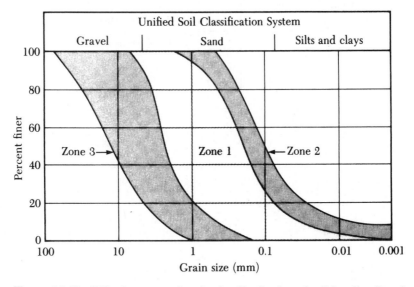

Figure 14.13 Effective range of grain-size distribution of soil for vibroflotation

consolidation is H_c. The maximum primary consolidation settlement caused by the structural load is then

$$S_{c(p)} = \frac{C_c H_c}{1 + e_o} \log \frac{\sigma'_o + \Delta\sigma'_{(p)}}{\sigma'_o} \tag{14.8}$$

The settlement–time relationship under the structural load will be like that shown in Figure 14.18b. However, if a surcharge of $\Delta\sigma'_{(p)} + \Delta\sigma'_{(f)}$ is placed on the ground, the primary consolidation settlement will be

$$S_{c(p+f)} = \frac{C_c H_c}{1 + e_o} \log \frac{\sigma'_o + [\Delta\sigma'_{(p)} + \Delta\sigma'_{(f)}]}{\sigma'_o} \tag{14.9}$$

The settlement–time relationship under a surcharge of $\Delta\sigma'_{(p)} + \Delta\sigma'_{(f)}$ is also shown in Figure 14.18b. Note that a total settlement of $S_{c(p)}$ would occur at time t_2, which

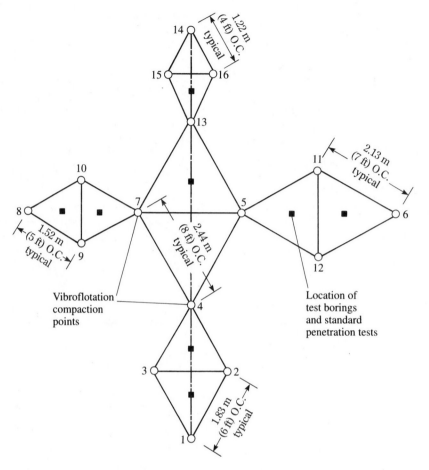

Figure 14.14 Layout of vibroflotation compaction points and test borings (after Basore and Boitano, 1969)

is much shorter than t_1. So, if a temporary total surcharge of $\Delta\sigma'_{(p)} + \Delta\sigma'_{(f)}$ is applied on the ground surface for time t_2, the settlement will equal $S_{c(p)}$. At that time, if the surcharge is removed and a structure with a permanent load per unit area of $\Delta\sigma'_{(p)}$ is built, no appreciable settlement will occur. The procedure just described is called *precompression*. The total surcharge $\Delta\sigma'_{(p)} + \Delta\sigma'_{(f)}$ can be applied by means of temporary fills.

Derivation of Equations for Obtaining $\Delta\sigma'_{(f)}$ and t_2

Figure 14.18b shows that, under a surcharge of $\Delta\sigma'_{(p)} + \Delta\sigma'_{(f)}$, the degree of consolidation at time t_2 after the application of the load is

$$U = \frac{S_{c(p)}}{S_{c(p+f)}} \tag{14.10}$$

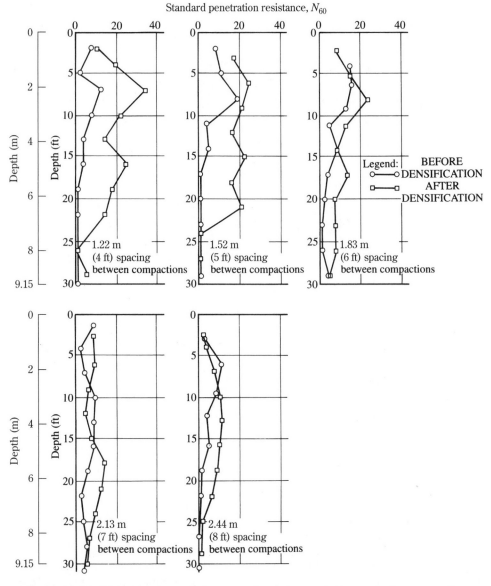

Figure 14.15 Variation of standard penetration resistance before and after compaction (after Basore and Boitano, 1969)

Substitution of Eqs. (14.8) and (14.9) into Eq. (14.10) yields

$$U = \frac{\log\left[\dfrac{\sigma_o' + \Delta\sigma_{(p)}'}{\sigma_o'}\right]}{\log\left[\dfrac{\sigma_o' + \Delta\sigma_{(p)}' + \Delta\sigma_{(f)}'}{\sigma_o'}\right]} = \frac{\log\left[1 + \dfrac{\Delta\sigma_{(p)}'}{\sigma_o'}\right]}{\log\left\{1 + \dfrac{\Delta\sigma_{(p)}'}{\sigma_o'}\left[1 + \dfrac{\Delta\sigma_{(f)}'}{\Delta\sigma_{(p)}'}\right]\right\}} \tag{14.11}$$

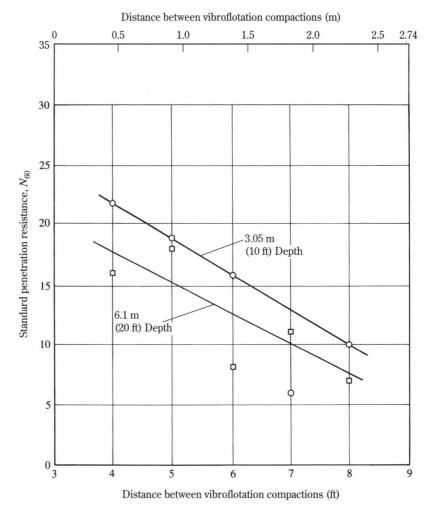

Figure 14.16 Variation of field standard penetration resistance after compaction with spacing and depth (after Basore and Boitano, 1969)

Figure 14.19 gives magnitudes of U for various combinations of $\Delta\sigma'_{(p)}/\sigma'_o$, and $\Delta\sigma'_{(f)}/\Delta\sigma'_{(p)}$. The degree of consolidation referred to in Eq. (14.11) is actually the average degree of consolidation at time t_2, as shown in Figure 14.18b. However, if the average degree of consolidation is used to determine t_2, some construction problems might occur. The reason is that, after the removal of the surcharge and placement of the structural load, the portion of clay close to the drainage surface will continue to swell, and the soil close to the midplane will continue to settle. (See Figure 14.20.) In some cases, net continuous settlement might result. A conservative approach may solve the problem; that is, assume that U in Eq. (14.11) is the midplane degree of consolidation (Johnson, 1970a). Now, from Eq. (1.65),

$$U = f(T_v) \tag{1.65}$$

where T_v = time factor = $C_v t_2 / H^2$

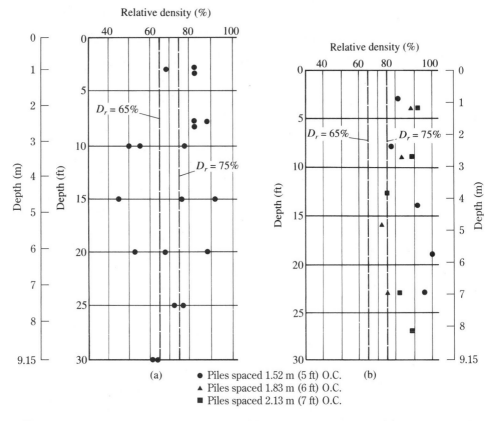

● Piles spaced 1.52 m (5 ft) O.C.
▲ Piles spaced 1.83 m (6 ft) O.C.
■ Piles spaced 2.13 m (7 ft) O.C.

Figure 14.17 Variation of relative density (a) before compaction and (b) after compaction (after Basore and Boitano, 1969)

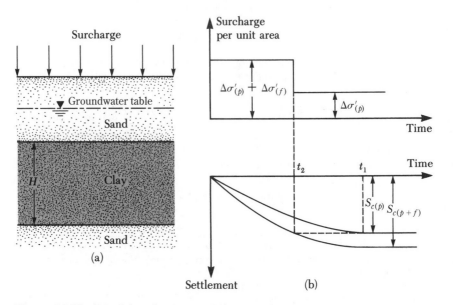

Figure 14.18 Principles of precompression

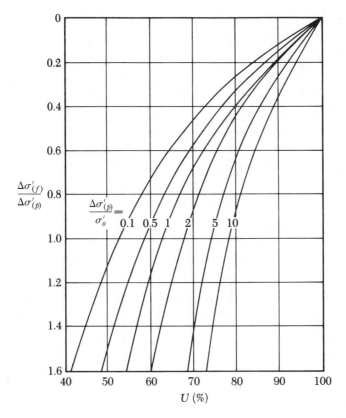

Figure 14.19 Plot of $\Delta\sigma'_{(f)}/\Delta\sigma'_{(p)}$ against U for various values of $\Delta\sigma'_{(p)}/\sigma'_o$—Eq. (14.11)

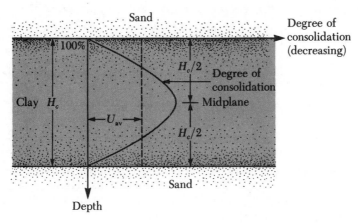

Figure 14.20

C_v = coefficient of consolidation
t_2 = time
H = maximum drainage path ($=H_c/2$ for two-way drainage and H_c for one-way drainage)

The variation of U (the midplane degree of consolidation) with T_v is given in Figure 14.21.

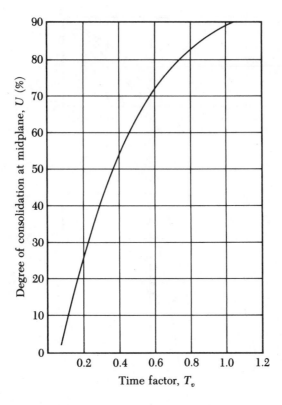

Figure 14.21 Plot of midplane degree of consolidation against T_v

Procedure for Obtaining Precompression Parameters

Two problems may be encountered by engineers during precompression work in the field:

1. The value of $\Delta\sigma'_{(f)}$ is known, but t_2 must be obtained. In such a case, obtain σ'_o, $\Delta\sigma_{(p)}$, and solve for U, using Eq. (14.11) or Figure 14.19. For this value of U, obtain T_v from Figure 14.21. Then

$$t_2 = \frac{T_v H^2}{C_v} \qquad (14.12)$$

2. For a specified value of t_2, $\Delta\sigma'_{(f)}$ must be obtained. In such a case, calculate T_v. Then use Figure 14.21 to obtain the midplane degree of consolidation, U. With the estimated value of U, go to Figure 14.19 to get the required value of $\Delta\sigma'_{(f)}/\Delta\sigma'_{(p)}$, and then calculate $\Delta\sigma'_{(f)}$.

Examples of Precompression and General Comments

Johnson (1970a) presented an excellent review of the use of precompression for improving foundation soils for several projects, including the Morganza Floodway Control Structure near Baton Rouge, Louisiana, the Old River Low-Sill Control Structure near Natchez, Mississippi, and the Old River Overbank Control Structure, Port Elizabeth Marine Terminal, New York. Figure 14.22 shows the subsoil condi-

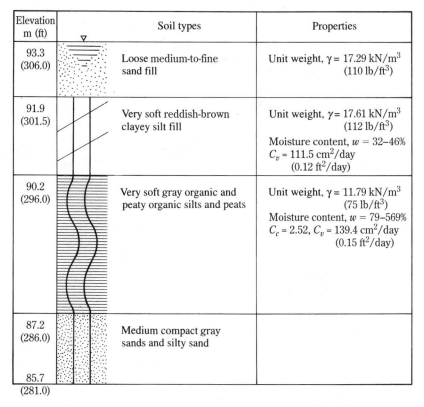

Elevation m (ft)		Soil types	Properties
93.3 (306.0)		Loose medium-to-fine sand fill	Unit weight, γ = 17.29 kN/m^3 (110 lb/ft^3)
91.9 (301.5)		Very soft reddish-brown clayey silt fill	Unit weight, γ = 17.61 kN/m^3 (112 lb/ft^3) Moisture content, w = 32–46% C_v = 111.5 cm^2/day (0.12 ft^2/day)
90.2 (296.0)		Very soft gray organic and peaty organic silts and peats	Unit weight, γ = 11.79 kN/m^3 (75 lb/ft^3) Moisture content, w = 79–569% C_c = 2.52, C_v = 139.4 cm^2/day (0.15 ft^2/day)
87.2 (286.0) 85.7 (281.0)		Medium compact gray sands and silty sand	

Figure 14.22 Subsoil conditions at Port Elizabeth Marine Terminal (after Johnson, 1970a)

tions encountered near the Port Elizabeth Marine Terminal before the construction of warehouse buildings No. 131 and 132 located in the Jersey Meadows just west of New York City. Details of the precompression of the subsoil before the construction of the buildings are given in Figure 14.23. Also, the theoretical variation of the settlement with time is shown.

In most cases, the predicted settlement exceeded the actual consolidation settlement. The reason is that several variables are involved in proper precompression design and performance. The information obtained from only a handful of borings is used in the calculation of both the surcharge load and the time necessary for removal of the surcharge. Hence, precise numbers for precompression design may be difficult to obtain. Settlement observations should be continued during the period that surcharge is applied, because they may dictate design changes.

Example 14.1

Examine Figure 14.18. During the construction of a highway bridge, the average permanent load on the clay layer is expected to increase by about 115 kN/m^2. The average effective overburden pressure at the middle of the clay layer is

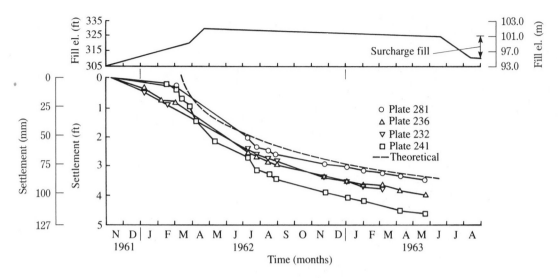

Figure 14.23 Precompression for support of warehouses No. 131 and 132, Port Elizabeth Marine Terminal, New York (after Johnson, 1970a)

210 kN/m². Here, $H_c = 6$ m, $C_c = 0.28$, $e_o = 0.9$, and $C_v = 0.36$ m²/mo. The clay is normally consolidated. Determine

a. The total primary consolidation settlement of the bridge without precompression
b. The surcharge, $\Delta\sigma'_{(f)}$, needed to eliminate the entire primary consolidation settlement in nine months by precompression.

Solution

Part a
The total primary consolidation settlement may be calculated from Eq. (14.8):

$$S_{c(p)} = \frac{C_c H_c}{1 + e_o} \log\left[\frac{\sigma'_o + \Delta\sigma'_{(p)}}{\sigma'_o}\right] = \frac{(0.28)(6)}{1 + 0.9} \log\left[\frac{210 + 115}{210}\right]$$

$$= 0.1677 \text{ m} = \textbf{167.7 mm}$$

Part b
We have

$$T_v = \frac{C_v t_2}{H^2}$$

$$C_v = 0.36 \text{ m}^2/\text{mo.}$$

$$H = 3 \text{ m (two-way drainage)}$$

$$t_2 = 9 \text{ mo.}$$

Hence,

$$T_v = \frac{(0.36)(9)}{3^2} = 0.36$$

According to Figure 14.21, for $T_v = 0.36$, the value of U is 47%. Now,

$$\Delta\sigma'_{(p)} = 115 \text{ kN/m}^2$$

and

$$\sigma'_o = 210 \text{ kN/m}^2$$

so

$$\frac{\Delta\sigma'_{(p)}}{\sigma'_o} = \frac{115}{210} = 0.548$$

According to Figure 14.19, for $U = 47\%$ and $\Delta\sigma'_{(p)}/\sigma'_o = 0.548$, $\Delta\sigma'_{(f)}/\Delta\sigma'_{(p)} \approx 1.8$; thus,

$$\Delta\sigma'_{(f)} = (1.8)(115) = \textbf{207 kN/m}^2 \qquad \blacksquare$$

14.8 *Sand Drains*

The use of sand drains is another way to accelerate the consolidation settlement of soft, normally consolidated clay layers and achieve precompression before the construction of a desired foundation. Sand drains are constructed by drilling holes through the clay layer(s) in the field at regular intervals. The holes are then backfilled with sand. This can be achieved by several means, such as (a) rotary drilling and then backfilling with sand; (b) drilling by continuous-flight auger with a hollow stem and backfilling with sand (through the hollow steam); and (c) driving hollow steel piles. The soil inside the pile is then jetted out, after which backfilling with sand is done. Figure 14.24 shows a schematic diagram of sand drains. After backfilling the drill holes with sand, a surcharge is applied at the ground surface. The surcharge will increase the pore water pressure in the clay. The excess pore water pressure in the clay will be dissipated by drainage—both vertically and radially to the sand drains—thereby accelerating settlement of the clay layer. In Figure 14.24a, note that the radius of the sand drains is r_w. Figure 14.24b shows the plan of the layout of the sand drains. The effective zone from which the radial drainage will be directed toward a given sand drain is approximately cylindrical, with a diameter of d_e.

To determine the surcharge that needs to be applied at the ground surface and the length of time that it has to be maintained, see Figure 14.18 and use the corresponding equation, Eq. (14.11):

$$U_{v,r} = \frac{\log\left[1 + \dfrac{\Delta\sigma'_{(p)}}{\sigma'_o}\right]}{\log\left\{1 + \dfrac{\Delta\sigma'_{(p)}}{\sigma'_o}\left[1 + \dfrac{\Delta\sigma'_{(f)}}{\Delta\sigma'_{(p)}}\right]\right\}} \qquad (14.13)$$

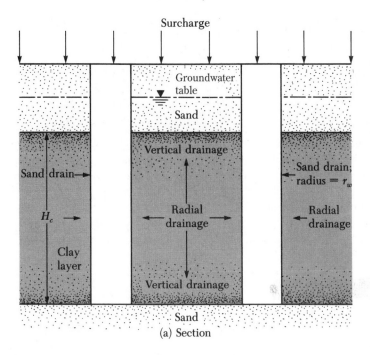

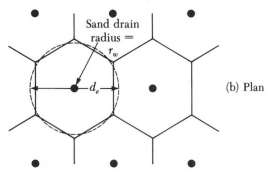

Figure 14.24 Sand drains

The notations $\Delta\sigma'_{(p)}$, σ'_o, and $\Delta\sigma'_{(f)}$ are the same as those in Eq. (14.11); however, the left-hand side of Eq. (14.13) is the *average degree* of consolidation instead of the degree of consolidation at midplane. Both *radial* and *vertical* drainage contribute to the average degree of consolidation. If $U_{v,r}$ can be determined for any time t_2 (see Figure 14.18b), the total surcharge $\Delta\sigma'_{(f)} + \Delta\sigma'_{(p)}$ may be obtained easily from Figure 14.19. The procedure for determining the average degree of consolidation ($U_{v,r}$) follows:

For a given surcharge and duration, t_2, the average degree of consolidation due to drainage in the vertical and radial directions is

$$U_{v,r} = 1 - (1 - U_r)(1 - U_v) \tag{14.14}$$

where U_r = average degree of consolidation with radial drainage only

  U_v = average degree of consolidation with vertical drainage only

The successful use of sand drains has been described in detail by Johnson (1970b). As with precompression, constant field settlement observations may be necessary during the period the surcharge is applied.

Average Degree of Consolidation Due to Radial Drainage Only

Figure 14.25 shows a schematic diagram of a sand drain. In the figure, r_w is the radius of the sand drain and $r_e = d_e/2$ is the radius of the effective zone of drainage. It is also important to realize that, during the installation of sand drains, a certain zone of clay surrounding them is smeared, thereby changing the hydraulic conductivity of the clay. In the figure, r_s is the radial distance from the center of the sand drain to the farthest point of the smeared zone. Now, for the average-degree-of-consolidation relationship, we will use the *theory of equal strain.* Two cases may arise that relate to the nature of the application of surcharge, and they are shown in Figure 14.26. (See the notations shown in Figure 14.18). Either (a) the entire surcharge is applied instantaneously (see Figure 14.26a), or (b) the surcharge is applied in the form of a ramp load (see Figure 14.26b). When the entire surcharge is applied instantaneously (Barron, 1948),

$$U_r = 1 - \exp\left(\frac{-8T_r}{m}\right) \qquad (14.15)$$

where

$$m = \frac{n^2}{n^2 - S^2} \ln\left(\frac{n}{S}\right) - \frac{3}{4} + \frac{S^2}{4n^2} + \frac{k_h}{k_s}\left(\frac{n^2 - S^2}{n^2}\right) \ln S \qquad (14.16)$$

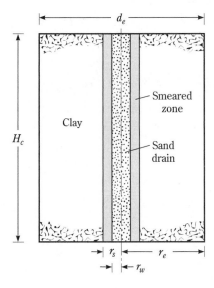

Figure 14.25 Schematic diagram of a sand drain

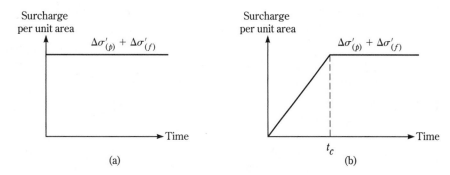

Figure 14.26   Nature of application of surcharge

in which

$$n = \frac{d_e}{2r_w} = \frac{r_e}{r_w} \tag{14.17}$$

$$S = \frac{r_s}{r_w} \tag{14.18}$$

k_h = hydraulic conductivity of clay in the horizontal
        direction in the unsmeared zone

k_s = horizontal hydraulic conductivity in the smeared zone

T_r = nondimensional time factor for radial drainage only = $\dfrac{C_{vr}t_2}{d_e^2}$ (14.19)

C_{vr} = coefficient of consolidation for radial drainage

$$= \frac{k_h}{\left[\dfrac{\Delta e}{\Delta\sigma'(1 + e_{av})}\right]\gamma_w} \tag{14.20}$$

For a *no-smear case*, $r_s = r_w$ and $k_h = k_s$, so $S = 1$ and Eq. (14.16) becomes

$$m = \left(\frac{n^2}{n^2 - 1}\right) \ln(n) - \frac{3n^2 - 1}{4n^2} \tag{14.21}$$

Table 14.5 gives the values of U_r for various values of T_r and n.

   If the surcharge is applied in the form of a *ramp* and *there is no smear*, then
(Olson, 1977)

$$U_r = \frac{T_r - \dfrac{1}{A}[1 - \exp(-AT_r)]}{T_{rc}} \qquad \text{(for } T_r \leq T_{rc}) \tag{14.22}$$

Table 14.5 Variation of U_r for Various Values of T_r and n, No-Smear Case [Eqs. (14.15) and (14.21)]

Degree of consolidation U_r (%)	Time factor T_r for value of n $(= r_e/r_w)$				
	5	10	15	20	25
0	0	0	0	0	0
1	0.0012	0.0020	0.0025	0.0028	0.0031
2	0.0024	0.0040	0.0050	0.0057	0.0063
3	0.0036	0.0060	0.0075	0.0086	0.0094
4	0.0048	0.0081	0.0101	0.0115	0.0126
5	0.0060	0.0101	0.0126	0.0145	0.0159
6	0.0072	0.0122	0.0153	0.0174	0.0191
7	0.0085	0.0143	0.0179	0.0205	0.0225
8	0.0098	0.0165	0.0206	0.0235	0.0258
9	0.0110	0.0186	0.0232	0.0266	0.0292
10	0.0123	0.0208	0.0260	0.0297	0.0326
11	0.0136	0.0230	0.0287	0.0328	0.0360
12	0.0150	0.0252	0.0315	0.0360	0.0395
13	0.0163	0.0275	0.0343	0.0392	0.0431
14	0.0177	0.0298	0.0372	0.0425	0.0467
15	0.0190	0.0321	0.0401	0.0458	0.0503
16	0.0204	0.0344	0.0430	0.0491	0.0539
17	0.0218	0.0368	0.0459	0.0525	0.0576
18	0.0232	0.0392	0.0489	0.0559	0.0614
19	0.0247	0.0416	0.0519	0.0594	0.0652
20	0.0261	0.0440	0.0550	0.0629	0.0690
21	0.0276	0.0465	0.0581	0.0664	0.0729
22	0.0291	0.0490	0.0612	0.0700	0.0769
23	0.0306	0.0516	0.0644	0.0736	0.0808
24	0.0321	0.0541	0.0676	0.0773	0.0849
25	0.0337	0.0568	0.0709	0.0811	0.0890
26	0.0353	0.0594	0.0742	0.0848	0.0931
27	0.0368	0.0621	0.0776	0.0887	0.0973
28	0.0385	0.0648	0.0810	0.0926	0.1016
29	0.0401	0.0676	0.0844	0.0965	0.1059
30	0.0418	0.0704	0.0879	0.1005	0.1103
31	0.0434	0.0732	0.0914	0.1045	0.1148
32	0.0452	0.0761	0.0950	0.1087	0.1193
33	0.0469	0.0790	0.0987	0.1128	0.1239
34	0.0486	0.0820	0.1024	0.1171	0.1285
35	0.0504	0.0850	0.1062	0.1214	0.1332
36	0.0522	0.0881	0.1100	0.1257	0.1380
37	0.0541	0.0912	0.1139	0.1302	0.1429
38	0.0560	0.0943	0.1178	0.1347	0.1479
39	0.0579	0.0975	0.1218	0.1393	0.1529
40	0.0598	0.1008	0.1259	0.1439	0.1580
41	0.0618	0.1041	0.1300	0.1487	0.1632
42	0.0638	0.1075	0.1342	0.1535	0.1685
43	0.0658	0.1109	0.1385	0.1584	0.1739

Table 14.5 (Continued)

Degree of consolidation U_r (%)	Time factor T_r for value of n ($= r_e/r_w$)				
	5	10	15	20	25
44	0.0679	0.1144	0.1429	0.1634	0.1793
45	0.0700	0.1180	0.1473	0.1684	0.1849
46	0.0721	0.1216	0.1518	0.1736	0.1906
47	0.0743	0.1253	0.1564	0.1789	0.1964
48	0.0766	0.1290	0.1611	0.1842	0.2023
49	0.0788	0.1329	0.1659	0.1897	0.2083
50	0.0811	0.1368	0.1708	0.1953	0.2144
51	0.0835	0.1407	0.1758	0.2020	0.2206
52	0.0859	0.1448	0.1809	0.2068	0.2270
53	0.0884	0.1490	0.1860	0.2127	0.2335
54	0.0909	0.1532	0.1913	0.2188	0.2402
55	0.0935	0.1575	0.1968	0.2250	0.2470
56	0.0961	0.1620	0.2023	0.2313	0.2539
57	0.0988	0.1665	0.2080	0.2378	0.2610
58	0.1016	0.1712	0.2138	0.2444	0.2683
59	0.1044	0.1759	0.2197	0.2512	0.2758
60	0.1073	0.1808	0.2258	0.2582	0.2834
61	0.1102	0.1858	0.2320	0.2653	0.2912
62	0.1133	0.1909	0.2384	0.2726	0.2993
63	0.1164	0.1962	0.2450	0.2801	0.3075
64	0.1196	0.2016	0.2517	0.2878	0.3160
65	0.1229	0.2071	0.2587	0.2958	0.3247
66	0.1263	0.2128	0.2658	0.3039	0.3337
67	0.1298	0.2187	0.2732	0.3124	0.3429
68	0.1334	0.2248	0.2808	0.3210	0.3524
69	0.1371	0.2311	0.2886	0.3300	0.3623
70	0.1409	0.2375	0.2967	0.3392	0.3724
71	0.1449	0.2442	0.3050	0.3488	0.3829
72	0.1490	0.2512	0.3134	0.3586	0.3937
73	0.1533	0.2583	0.3226	0.3689	0.4050
74	0.1577	0.2658	0.3319	0.3795	0.4167
75	0.1623	0.2735	0.3416	0.3906	0.4288
76	0.1671	0.2816	0.3517	0.4021	0.4414
77	0.1720	0.2900	0.3621	0.4141	0.4546
78	0.1773	0.2988	0.3731	0.4266	0.4683
79	0.1827	0.3079	0.3846	0.4397	0.4827
80	0.1884	0.3175	0.3966	0.4534	0.4978
81	0.1944	0.3277	0.4090	0.4679	0.5137
82	0.2007	0.3383	0.4225	0.4831	0.5304
83	0.2074	0.3496	0.4366	0.4992	0.5481
84	0.2146	0.3616	0.4516	0.5163	0.5668
85	0.2221	0.3743	0.4675	0.5345	0.5868
86	0.2302	0.3879	0.4845	0.5539	0.6081
87	0.2388	0.4025	0.5027	0.5748	0.6311
88	0.2482	0.4183	0.5225	0.5974	0.6558
89	0.2584	0.4355	0.5439	0.6219	0.6827
90	0.2696	0.4543	0.5674	0.6487	0.7122

Table 14.5 (Continued)

Degree of consolidation U_r (%)	Time factor T_r for value of n (= r_e/r_w)				
	5	10	15	20	25
91	0.2819	0.4751	0.5933	0.6784	0.7448
92	0.2957	0.4983	0.6224	0.7116	0.7812
93	0.3113	0.5247	0.6553	0.7492	0.8225
94	0.3293	0.5551	0.6932	0.7927	0.8702
95	0.3507	0.5910	0.7382	0.8440	0.9266
96	0.3768	0.6351	0.7932	0.9069	0.9956
97	0.4105	0.6918	0.8640	0.9879	1.0846
98	0.4580	0.7718	0.9640	1.1022	1.2100
99	0.5391	0.9086	1.1347	1.2974	1.4244

and

$$U_r = 1 - \frac{1}{AT_{rc}}[\exp(AT_{rc}) - 1]\exp(-AT_{rc}) \quad (\text{for } T_r \geq T_{rc}) \qquad (14.23)$$

where

$$T_{rc} = \frac{C_{vr}t_c}{d_e^2} \text{ (see Figure 14.26b for the definition of } t_c) \qquad (14.24)$$

and

$$A = \frac{2}{m} \qquad (14.25)$$

Average Degree of Consolidation Due to Vertical Drainage Only

Using Figure 14.26a, for instantaneous application of a surcharge, we may obtain the average degree of consolidation due to vertical drainage only from Eqs. (1.66) and (1.67). We have

$$T_v = \frac{\pi}{4}\left[\frac{U_v(\%)}{100}\right]^2 \quad (\text{for } U_v = 0\text{–}60\%) \qquad (1.66)$$

and

$$T_v = 1.781 - 0.933 \log (100 - U_v(\%)) \quad (\text{for } U_v > 60\%) \qquad (1.67)$$

where U_v = average degree of consolidation due to vertical drainage only

$$T_v = \frac{C_v t_2}{H^2} \qquad (1.61)$$

C_v = coefficient of consolidation for vertical drainage

For the case of ramp loading, as shown in Figure 14.26b, the variation of $U_v(\%)$ with T_v and T_c (Olson, 1977) is given in Figure 14.27. Note that

$$T_c = \frac{C_v t_c}{H^2} \qquad (14.26)$$

where H = length of maximum vertical drainage path

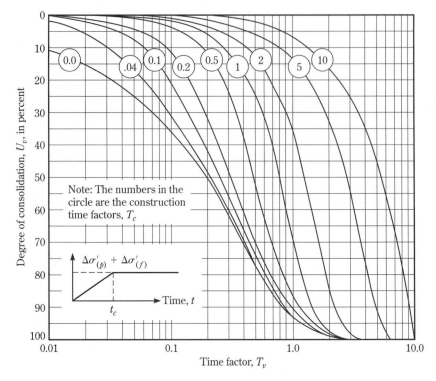

Figure 14.27 Variation of U_v with T_v and T_c (after Olson, 1977)

14.9 An Example of a Sand Drain Application

Aboshi and Monden (1963) provided details on the field performance of 2700 sand drains used to construct the Toya Quay Wall on reclaimed land in Japan in a study that was summarized by Johnson (1970b). The location of the site is shown in the inset of Figure 14.28. The soil at the site consisted of 30-m ($\approx$98-ft) thick soft, normally consolidated clayey silt. The following data are for the *in situ* soil and the sand drains:

In situ soil:	Liquid limit (LL) = 110
	Plastic limit (PL) = 48
	Natural moisture content, w = 74%–65%
Sand drains:	Total number used = 2700
	Length = 15 m
	d_e = 3.15 m
	r_w = 0.225 m

$$\left.\begin{array}{c} C_{vr} \\ \hline C_v \end{array}\right\} = \begin{array}{l} 2.7 \quad \text{(from triaxial test)} \\ 1.7 \quad \text{(from consolidometer)} \end{array}$$

The top portion of the figure shows the variation of the surcharge load application with time. The bottom portion shows the observed and theoretical variation of settlement due to *sand drains only*. The agreement appears to be excellent.

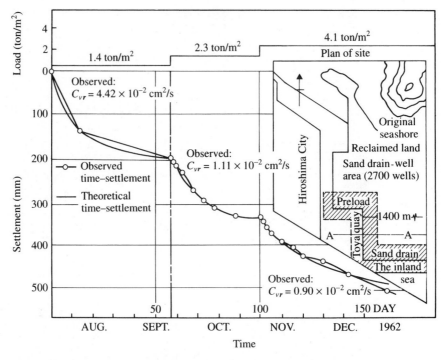

Figure 14.28   Comparison of observed and theoretical settlements due to sand drawn only for Toya Quay wall construction, Japan (after Johnson, 1970b)

Example 14.2

Redo Example 14.1, with the addition of some sand drains. Assume that $r_w = 0.1$ m, $d_e = 3$ m, $C_v = C_{vr}$, and the surcharge is applied instantaneously. (See Figure 14.26a). Also, assume that this is a no-smear case.

Solution

Part a
The total primary consolidation settlement will be 167.7 mm, as before.

Part b
From Example 14.1, $T_v = 0.36$. Using Eq. (1.66), we obtain

$$T_v = \frac{\pi}{4}\left[\frac{U_v(\%)}{100}\right]^2$$

or

$$U_v = \sqrt{\frac{4T_v}{\pi}} \times 100 = \sqrt{\frac{(4)(0.36)}{\pi}} \times 100 = 67.7\%$$

Also,

$$n = \frac{d_e}{2r_w} = \frac{3}{2 \times 0.1} = 15$$

Again,

$$T_r = \frac{C_{vr}t_2}{d_e^2} = \frac{(0.36)(9)}{(3)^2} = 0.36$$

From Table 14.5 for $n = 15$ and $T_r = 0.36$, the value of U_r is about 77%, Hence,

$$U_{v,r} = 1 - (1 - U_v)(1 - U_r) = 1 - (1 - 0.67)(1 - 0.77)$$

$$= 0.924 = 92.4\%$$

Now, from Figure 14.19, for $\Delta\sigma'_p/\sigma'_o = 0.548$ and $U_{v,r} = 92.4\%$, the value of $\Delta\sigma'_f/\Delta\sigma'_p \approx 0.12$. Hence,

$$\Delta\sigma'_{(f)} = (115)(0.12) = \mathbf{13.8 \ kN/m^2} \qquad \blacksquare$$

Example 14.3

Suppose that, for the sand drain project of Figure 14.24, the clay is normally consolidated. We are given the following data:

Clay: $H_c = 15$ ft (two-way drainage)

$C_c = 0.31$

$e_o = 1.1$

Effective overburden pressure at the middle of the clay layer

$= 1000 \ lb/ft^2$

$C_v = 0.115 \ ft^2/$ day

Sand drain: $r_w = 0.3$ ft

$d_e = 6$ ft

$C_v = C_{vr}$

A surcharge is applied as shown in Figure 14.29. Assume this to be a no-smear case. Calculate the degree of consolidation 30 days after the surcharge is first applied. Also, determine the consolidation settlement at that time due to the surcharge.

Solution
From Eq. (14.26),

$$T_c = \frac{C_v t_c}{H^2} = \frac{(0.115 \ ft^2/day)(60)}{\left(\dfrac{15}{2}\right)^2} = 0.123$$

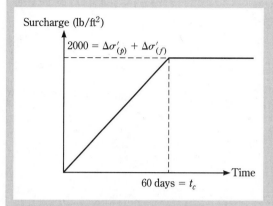

$2000 = \Delta\sigma'_{(p)} + \Delta\sigma'_{(f)}$

60 days = t_c

Figure 14.29 Ramp load for a sand drain project

and

$$T_v = \frac{C_v t_2}{H^2} = \frac{(0.115)(30)}{\left(\dfrac{15}{2}\right)^2} = 0.061$$

Using Figure 14.27 for $T_c = 0.123$ and $T_v = 0.061$, we have $U_v \approx 9\%$. For the sand drain,

$$n = \frac{d_e}{2r_w} = \frac{6}{(2)(0.3)} = 10$$

From Eq. (14.24),

$$T_{rc} = \frac{C_{vr} t_c}{d_e^2} = \frac{(0.115)(60)}{(6)^2} = 0.192$$

and

$$T_r = \frac{C_{vr} t_2}{d_e^2} = \frac{(0.115)(30)}{(6)^2} = 0.096$$

Again, from Eq. (14.22),

$$U_r = \frac{T_r - \dfrac{1}{A}[1 - \exp(-AT_r)]}{T_{rc}}$$

Also, for the no-smear case,

$$m = \frac{n^2}{n^2 - 1} \ln(n) - \frac{3n^2 - 1}{4n^2} = \frac{10^2}{10^2 - 1} \ln(10) - \frac{3(10)^2 - 1}{4(10)^2} = 1.578$$

and

$$A = \frac{2}{m} = \frac{2}{1.578} = 1.267$$

so

$$U_r = \frac{0.096 - \dfrac{1}{1.267}[1 - \exp(-1.267 \times 0.096)]}{0.192} = 0.03 = 3\%$$

From Eq. (14.14),

$$U_{v,r} = 1 - (1 - U_r)(1 - U_v) = 1 - (1 - 0.03)(1 - 0.09) = 0.117 = \mathbf{11.7\%}$$

The total primary settlement is thus

$$S_{c(p)} = \frac{C_c H_c}{1 + e_o} \log \left[\frac{\sigma_o' + \Delta\sigma_{(p)}' + \Delta_{\sigma f}'}{\sigma_o'} \right]$$

$$= \frac{(0.31)(15)}{1 + 1.1} \log \left(\frac{1000 + 2000}{1000} \right) = 1.056 \text{ ft}$$

and the settlement after 30 days is

$$S_{c(p)} U_{v,r} = (1.056)(0.117)(12) = \mathbf{1.48 \text{ in.}} \quad \blacksquare$$

14.10 *Prefabricated Vertical Drains*

Prefabricated vertical drains (PVDs), also referred to as *wick* or *strip drains,* were originally developed as a substitute for the commonly used sand drain. With the advent of materials science, these drains began to be manufactured from synthetic polymers such as polypropylene and high-density polyethylene. PVDs are normally manufactured with a corrugated or channeled synthetic core enclosed by a geotextile filter, as shown schematically in Figure 14.30. Installation rates reported in the literature are on the order of 0.1 to 0.3 m/s, excluding equipment mobilization and setup time. PVDs have been used extensively in the past for expedient consolidation of low-permeability soils under surface surcharge. The main advantage of PVDs over sand drains is that they do not require drilling; thus, installation is much faster. Figures 14.31a and b are photographs of the installation of PVDs in the field.

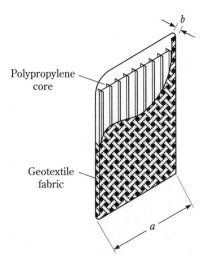

Polypropylene core

Geotextile fabric

Figure 14.30 Prefabricated vertical drain (PVD)

(a)

(b)

Figure 14.31 Installation of PVDs in the field. *Note:* (b) is a closeup view of (a). (Courtesy of E. C. Shin, Incheon, South Korea)

Design of PVDs

The relationships for the average degree of consolidation due to radial drainage into sand drains are given in Eqs. (14.15) through (14.20) for equal-strain cases. Yeung (1997) used these relationships to develop design curves for PVDs. The theoretical developments used by Yeung are given next.

Figure 14.32 shows the layout of a square-grid pattern of prefabricated vertical drains. See also Figure 14.30 for the definition of a and b). The equivalent diameter of a PVD can be given as

$$d_w = \frac{2(a + b)}{\pi} \tag{14.27}$$

Now, Eq. (14.15) can be rewritten as

$$U_r = 1 - \exp\left(-\frac{8C_{vr}t}{d_w^2} \frac{d_w^2}{d_e^2 m}\right) = 1 - \exp\left(-\frac{8T_r'}{\alpha'}\right) \tag{14.28}$$

where d_e = diameter of the effective zone of drainage = $2r_e$

Also,

$$T_r' = \frac{C_{vr}t}{d_w^2} \tag{14.29}$$

$$\alpha' = n^2 m = \frac{n^4}{n^2 - S^2} \ln\left(\frac{n}{S}\right) - \left(\frac{3n^2 - S^2}{4}\right) + \frac{k_h}{k_s}(n^2 - S^2) \ln S \tag{14.30}$$

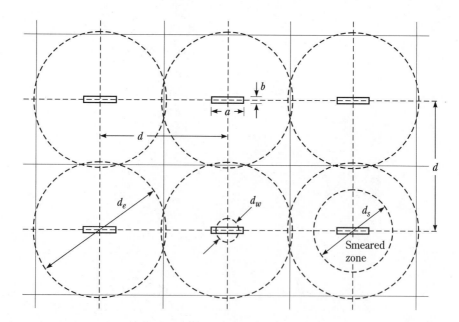

Figure 14.32 Layout of square-grid pattern of prefabricated vertical drains (after Yeung, 1997)

and

$$n = \frac{d_e}{d_w} \qquad (14.31)$$

From Eq. (14.28),

$$T_r' = -\frac{\alpha'}{8} \ln (1 - U_r)$$

or

$$(T_r')_1 = \frac{T_r'}{\alpha'} = -\frac{\ln (1 - U_r)}{8} \qquad (14.32)$$

Figure 14.33 shows a plot of U_r versus $(T_r')_1$. Also, Figure 14.34 shows a plot of n versus α' from Eq. (14.30).

Following is a step-by-step procedure for the design of prefabricated vertical drains:

1. Determine time t_2 available for the consolidation process and the $U_{v,r}$ required therefore [Eq. (14.13)]
2. Determine U_v at time t_2 due to vertical drainage. From Eq. (14.14)

$$U_r = 1 - \frac{1 - U_{v,r}}{1 - U_v} \qquad (14.33)$$

3. For the PVD that is to be used, calculate d_w from Eq. (14.27).
4. Determine $(T_r')_1$ from Eqs. (14.32) and (14.33).
5. Determine T_r' from Eq. (14.29).
6. Determine

$$\alpha' = \frac{T_r'}{(T_r')_1}.$$

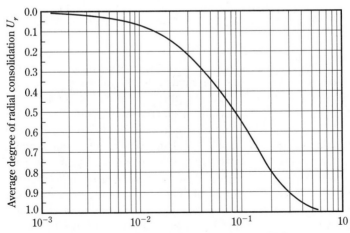

Figure 14.33 Plot of U_r vs. $(T_r')_1$ (after Yeung, 1977)

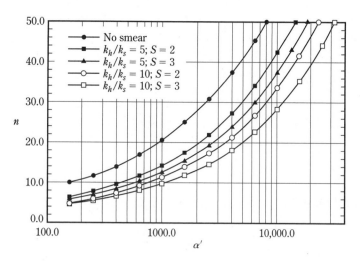

Figure 14.34 Relationship between α' and n (after Yeung, 1997)

7. Using Figure 14.34 and α' determined from Step 6, determine n.

8. From Eq. (14.31),

$$d_e = n \quad d_w$$
$$\quad\;\uparrow \quad\;\; \uparrow$$
$$\quad\text{Step 7} \quad \text{Step 3}$$

9. Choose the drain spacing:

$$d = \frac{d_e}{1.05} \quad \text{(for triangular pattern)}$$

$$d = \frac{d_e}{1.128} \quad \text{(for square pattern)}$$

14.11 Lime Stabilization

As mentioned in Section 14.1, admixtures are occasionally used to stabilize soils in the field—particularly fine-grained soils. The most common admixtures are lime, cement, and lime–fly ash. The main purposes of stabilizing the soil are to (a) modify the soil, (b) expedite construction, and (c) improve the strength and durability of the soil.

The types of *lime* commonly used to stabilize fine-grained soils are hydrated high-calcium lime $[\text{Ca(OH)}_2]$, calcitic quicklime (CaO), monohydrated dolomitic lime $[\text{Ca(OH)}_2 \cdot \text{MgO}]$, and dolomitic quicklime. The quantity of lime used to stabilize most soils usually is in the range from 5 to 10%. When lime is added to clayey soils, two *pozzolanic* chemical reactions occur: *cation exchange* and *flocculation–agglomeration*. In the cation exchange and flocculation–agglomeration reactions, the *monovalent* cations generally associated with clays are replaced by the *divalent* calcium ions. The cations can be arranged in a series based on their affinity for exchange:

$$\text{Al}^{3+} > \text{Ca}^{2+} > \text{Mg}^{2+} > \text{NH}_4^+ > \text{K}^+ > \text{Na}^+ > \text{Li}^+$$

Any cation can replace the ions to its right. For example, calcium ions can replace potassium and sodium ions from a clay. Flocculation–agglomeration produces a change in the texture of clay soils. The clay particles tend to clump together to form larger particles, thereby (a) decreasing the liquid limit, (b) increasing the plastic limit, (c) decreasing the plasticity index, (d) increasing the shrinkage limit, (e) increasing the workability, and (f) improving the strength and deformation properties of a soil.

Pozzolanic reaction between soil and lime involves a reaction between lime and the silica and alumina of the soil to form cementing material. One such reaction is

$$Ca(OH)_2 + SiO_2 \rightarrow CSH$$
$$\uparrow$$
$$\text{Clay silica}$$

where $C = CaO$
$S = SiO_2$
$H = H_2O$

The pozzolanic reaction may continue for a long time.

Figure 14.35 shows the variation of the liquid limit, the plasticity index, and the shrinkage limit of a clay with the percentage of lime admixture. The first 2–3% lime (on the dry-weight basis) substantially influences the workability and the property (such as plasticity) of the soil. The addition of lime to clayey soils also affects their compaction characteristics. Lime stabilization in the field can be done in three ways:

1. The *in situ* material or the borrowed material can be mixed with the proper amount of lime at the site and then compacted after the addition of moisture.
2. The soil can be mixed with the proper amount of lime and water at a plant and then hauled back to the site for compaction.
3. Lime slurry can be pressure injected into the soil to a depth of 4 to 5 m (12 to 16 ft). Figure 14.36 shows a vehicle used for pressure injection of lime slurry.

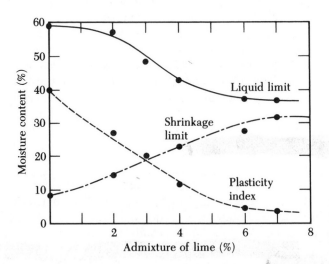

Figure 14.35 Variation of liquid limit, plasticity index, and shrinkage of a clay with lime additive

Figure 14.36 Equipment for pressure injection of lime slurry (Courtesy of GKN Hayward Baker, Inc., Woodbine Division, Ft. Worth, Texas)

The slurry-injection mechanical unit is mounted to the injection vehicle. A common injection unit is a hydraulic-lift mast with crossbeams that contain the injection rods. The rods are pushed into the ground by the action of the lift mast beams. The slurry is generally mixed in a batching tank about 3 m (10 ft) in diameter and 12 m (36 ft) long and is pumped at high pressure to the injection rods. Figure 14.37 is a photograph of the lime slurry pressure-injection process. The ratio typically specified for the preparation of lime slurry is 1.13 kg (2.5 lb) of dry lime to a gallon of water. (For more information on this technique, see Blacklock and Pengelly, 1988).

Because the addition of hydrated lime to soft clayey soils immediately increases the plastic limit, thus changing the soil from plastic to solid and making it appear to "dry up," limited amounts of the lime can be thrown on muddy and troublesome construction sites. This action improves trafficability and may save money and time. Quicklimes have also been successfully used in drill holes having diameters of 100 mm to 150 mm (4 in. to 6 in.) for stabilization of subgrades and slopes. For this type of work, holes are drilled in a grid pattern and then filled with quicklime.

14.12 Cement Stabilization

Cement is being increasingly used as a stabilizing material for soil, particularly in the construction of highways and earth dams. The first controlled soil–cement construction in the United States was carried out near Johnsonville, South Carolina, in 1935. Cement can be used to stabilize sandy and clayey soils. As in the case of lime, cement helps decrease the liquid limit and increase the plasticity index and workability of clayey soils. Cement stabilization is effective for clayey soils when the liquid

Figure 14.37 Pressure injection of lime slurry (Courtesy of GKN Hayward Baker, Inc., Woodbine Division, Ft. Worth, Texas)

limit is less than 45–50 and the plasticity index is less than about 25. The optimum requirements of cement by volume for effective stabilization of various types of soil are given in Table 14.6.

Like lime, cement helps increase the strength of soils, and strength increases with curing time. Table 14.7 presents some typical values of the unconfined compressive strength of various types of untreated soil and of soil–cement mixtures made with approximately 10% cement by weight.

Granular soils and clayey soils with low plasticity obviously are most suitable for cement stabilization. Calcium clays are more easily stabilized by the addition of cement, whereas sodium and hydrogen clays, which are expansive in nature, respond better to lime stabilization. For these reasons, proper care should be given in the selection of the stabilizing material.

For field compaction, the proper amount of cement can be mixed with soil either at the site or at a mixing plant. If the latter approach is adopted, the mixture can

Table 14.6 Cement Requirement by Volume for Effective Stabilization of Various Soils[a]

Soil type		Percent cement by volume
AASHTO classification	Unified classification	
A-2 and A-3	GP, SP, and SW	6–10
A-4 and A-5	CL, ML, and MH	8–12
A-6 and A-7	CL, CH	10–14

[a] After Mitchell and Freitag (1959)

Table 14.7 Typical Compressive Strengths of Soils and Soil–Cement Mixtures[a]

Material	Unconfined compressive strength range	
	kN/m²	lb/in²
Untreated soil:		
Clay, peat	Less than 350	Less than 50
Well-compacted sandy clay	70–280	10–40
Well-compacted gravel, sand, and clay mixtures	280–700	40–100
Soil–cement (10% cement by weight):		
Clay, organic soils	Less than 350	Less than 50
Silts, silty clays, very poorly graded sands, slightly organic soils	350–1050	50–150
Silty clays, sandy clays, very poorly graded sands, and gravels	700–1730	100–250
Silty sands, sandy clays, sands, and gravels	1730–3460	250–500
Well-graded sand–clay or gravel–sand–clay mixtures and sands and gravels	3460–10,350	500–1500

[a] After Mitchell and Freitag (1959)

then be carried to the site. The soil is compacted to the required unit weight with a predetermined amount of water.

Similar to lime injection, cement slurry made of portland cement and water (in a water–cement ratio of 0.5:5) can be used for pressure grouting of poor soils under foundations of buildings and other structures. Grouting decreases the hydraulic conductivity of soils and increases their strength and load-bearing capacity. For the design of low-frequency machine foundations subjected to vibrating forces, stiffening the foundation soil by grouting and thereby increasing the resonant frequency is sometimes necessary.

14.13   Fly-Ash Stabilization

Fly ash is a by-product of the pulverized coal combustion process usually associated with electric power-generating plants. It is a fine-grained dust and is composed primarily of silica, alumina, and various oxides and alkalies. Fly ash is pozzolanic in nature and can react with hydrated lime to produce cementitious products. For that reason, lime–fly-ash mixtures can be used to stabilize highway bases and subbases.

Effective mixes can be prepared with 10–35% fly ash and 2–10% lime. Soil–lime–fly-ash mixes are compacted under controlled conditions, with proper amounts of moisture to obtain stabilized soil layers.

A certain type of fly ash, referred to as "Type C" fly ash, is obtained from the burning of coal primarily from the western United States. This type of fly ash contains a fairly large proportion (up to about 25%) of free lime that, with the addition of water, will react with other fly-ash compounds to form cementitious products. Its use may eliminate the need to add manufactured lime.

14.14 Stone Columns

A method now being used to increase the load-bearing capacity of shallow foundations on soft clay layers is the construction of stone columns. This generally consists of water-jetting a vibroflot (see Section 14.6) into the soft clay layer to make a circular hole that extends through the clay to firmer soil. The hole is then filled with an imported gravel. The gravel in the hole is gradually compacted as the vibrator is withdrawn. The gravel used for the stone column has a size range of 6–40 mm (0.25–1.6 in.). Stone columns usually have diameters of 0.5–0.75 m (1.6–2.5 ft) and are spaced at about 1.5–3 m (5–10 ft) center to center. Figure 14.38 shows the construction of a stone column.

After stone columns are constructed, a fill material should always be placed over the ground surface and compacted before the foundation is constructed. The stone columns tend to reduce the settlement of foundations at allowable loads. Several case histories of construction projects using stone columns are presented in Hughes and Withers (1974), Hughes et al. (1975), Mitchell and Huber (1985), and other works.

Stone columns work more effectively when they are used to stabilize a large area where the undrained shear strength of the subsoil is in the range of 10 to 50 kN/m^2 (200–1000 lb/ft^2) than to improve the bearing capacity of structural foundations (Bachus and Barksdale, 1989). Subsoils weaker than that may not provide sufficient lateral support for the columns. For large-site improvement, stone columns are most effective to a depth of 6–10 m (20–30 ft). However, they have been constructed to a depth of 31 m (100 ft). Bachus and Barksdale provided the following general guidelines for the design of stone columns to stabilize large areas.

Figure 14.39a shows the plan view of several stone columns, and Figure 14.39b depicts the unit cell idealization of a stone column. The area replacement ratio for the stone columns may be expressed as

$$a_s = \frac{A_s}{A} \tag{14.34}$$

where A_s = area of the stone column
A = total area within the unit cell

For an *equilateral triangular pattern* of stone columns,

$$a_s = 0.907\left(\frac{D}{s}\right)^2 \tag{14.35}$$

where D = diameter of the stone column
s = spacing between the columns

Figure 14.38 Construction of a stone column (Courtesy of The Reinforced Earth Company, Vienna, Virginia and Menard Soil Treatment Inc., Orange, California)

When a uniform stress by means of a fill operation is applied to an area with stone columns to induce consolidation, a stress concentration occurs due to the change in the stiffness between the stone columns and the surrounding soil. (See Figure 14.39c.) The stress concentration factor is defined as

$$n' = \frac{\sigma_s'}{\sigma_c'}$$
(14.36)

where σ_s' = effective stress in the stone column
σ_c' = effective stress in the subgrade soil

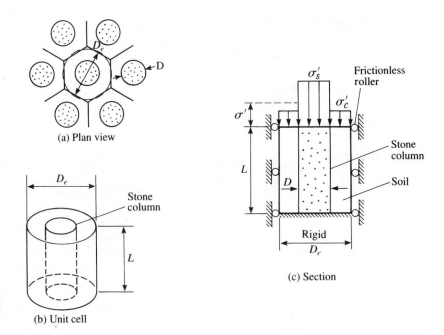

Figure 14.39 Plan view and unit cell idealization of stone column (after Bachus and Barksdale, 1989)

The relationships for σ_s' and σ_c' are

$$\sigma_s' = \sigma'\left[\frac{n'}{1 + (n' - 1)a_s}\right] = \mu_s\sigma' \qquad (14.37)$$

and

$$\sigma_c' = \sigma'\left[\frac{1}{1 + (n' - 1)a_s}\right] = \mu_c\sigma' \qquad (14.38)$$

where σ' = average effective vertical stress
     μ_s, μ_c = stress concentration coefficients

The variation of μ_c with a_s and n' is shown in Figure 14.40. The improvement in the soil owing to the stone columns may be expressed as

$$\frac{S_{e(t)}}{S_e} = \mu_c \qquad (14.39)$$

where $S_{e(t)}$ = settlement of the treated soil
     S_e = total settlement of the untreated soil

Load-Bearing Capacity of Stone Columns

If a foundation is constructed over a stone column as shown in Figure 14.41, failure will occur by bulging of the column at ultimate load. The bulging will occur within a length of 2.5*D* to 3*D,* measured from the top of the stone column, where *D* is the diameter of the column.

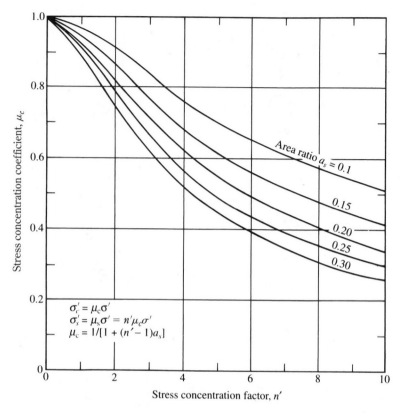

Figure 14.40   Variation of μ_c with a_s and n' (after Bachus and Barksdale, 1989)

Hughes et al. (1975) provided an approximate relationship for the ultimate bearing capacity of stone columns, which can be given as

$$q_u = \tan^2\left(45 + \frac{\phi'}{2}\right)(4c_u + \sigma_r') \tag{14.40}$$

where c_u = undrained shear strength of the clay
   σ_r' = effective radial stress as measured by a pressuremeter ($\approx 2c_u$)
   ϕ' = effective stress friction angle of the stone column material

Thus, assuming that the stone column carries the entire load of the foundation, the ultimate load can be given as

$$Q_u = \frac{\pi}{4}D^2 \tan^2\left(45 + \frac{\phi'}{2}\right)(4c_u + \sigma_r') \tag{14.41}$$

On the basis of large-scale-model tests, Christoulas et al. (2000) suggested that

$$Q_u = \pi D L c_u \tag{14.42}$$

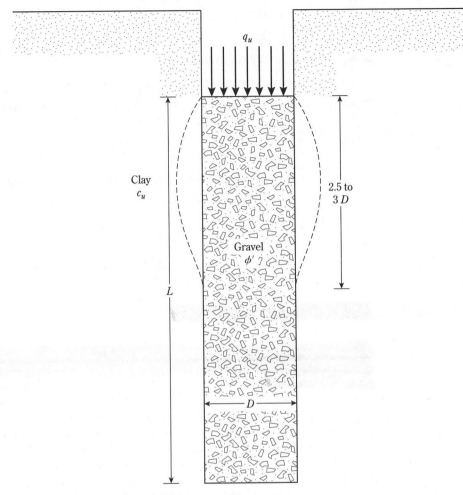

Figure 14.41 Bearing capacity of stone column

In my own opinion, the lower of the two values of Q_u obtained from Eqs. (14.41) and (14.42) should be used for actual design purposes. The allowable load can then be given as

$$Q_{\text{all}} = \frac{Q_u}{\text{FS}} \tag{14.43}$$

where FS = factor of safety ($\approx$1.5 to 2)

Christoulas et al. (2000) also suggested a relationship between the load Q and the elastic settlement S_e for stone columns that can be expressed as

$$S_e = \left(\frac{Q}{LE_{\text{clay}}}\right)I_d \qquad (\text{for } Q \leq Q_1) \tag{14.44}$$

and

$$S_e = \left(\frac{Q_1}{LE_{\text{clay}}}\right)I_d + \left(\frac{Q - Q_1}{4LE_{\text{clay}}}\right)I_d \qquad (\text{for } Q_1 \leq Q \leq Q_u) \tag{14.45}$$

where $\quad Q_1 = \dfrac{0.1 D L E_{clay}}{I_d}$ $\qquad\qquad\qquad$ (14.46)

in which $\quad E_{clay}$ = modulus of elasticity of clay

$\qquad\qquad I_d$ = influence factor (Mattes and Poulos, 1969)

The influence factor proposed by Mattes and Poulos is a function of three quantities:

a. $K = \dfrac{E_{col}}{E_{clay}}$

$\qquad$ where $\quad E_{col}$ = modulus of elasticity of the stone column material

b. $\dfrac{L}{D}$

$\qquad$ **c.** Poisson's ratio of clay, μ_{clay}. A value of $\mu_{clay} = 0.5$ will give a conservative result.

The variation of I_d (for $\mu_{clay} = 0.5$) with K is shown in Figure 14.42.

14.15 Sand Compaction Piles

Sand compaction piles are similar to stone columns, and they can be used in marginal sites to improve stability, control liquefaction, and reduce the settlement of various structures. Built in soft clay, these piles can significantly accelerate the pore water pressure-dissipation process and hence the time for consolidation.

Sand piles were first constructed in Japan between 1930 and 1950 (Ichimoto, 1981). Large-diameter compacted sand columns were constructed in 1955, using the Compozer technique (Aboshi et al., 1979). The Vibro-Compozer method of sand pile construction was developed by Murayama in Japan in 1958 (Murayama, 1962).

Sand compaction piles are constructed by driving a hollow mandrel with its bottom closed during driving. On partial withdrawal of the mandrel, the bottom doors open. Sand is poured from the top of the mandrel and is compacted in steps by applying air pressure as the mandrel is withdrawn. The piles are usually 0.46–0.76 m (1.5–2.5 ft) in diameter and are placed at about 1.5–3 m (5–10 ft) center to center. The pattern of layout of sand compaction piles is the same as for stone columns. Figure 14.43 shows the construction of sand compaction piles in the harbor of Yokohama, Japan.

The case history of a successful sand compaction pile project in Korea was reported by Shin, Shin, and Das (1992), and it is summarized next. Sand compaction piles were first used in Korea in 1984 for the construction of the Kwang-Yang Steel Mill complex. The project site, Kwang-Yang, is located on the South Sea about 300 km (187 miles) south of Seoul, where a delta is formed at the convergence of the Sum Jin and Su Oh Rivers. (See Figure 14.44a.) The site was selected for the steel mill complex because it is directly connected to the sea, which facilitates the import of raw material and export of finished products.

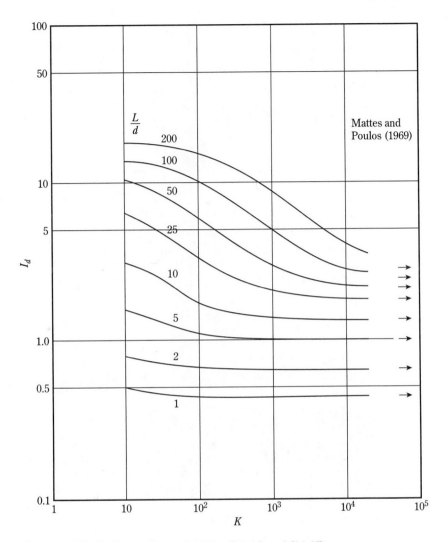

Figure 14.42 Influence factor I_d—Eqs. (14.44) and (14.45)

Figure 14.44b shows the general nature of the *in situ* soil, along with a backfilled sand of about 16.5 ft (5 m). Site improvement techniques used in this project included sand compaction piles, as well as some sand drains with preloading. The sand compaction piles and the sand drains had diameters of 0.7 m (2.3 ft) and 0.4 m (1.3 ft), respectively, with an average length of about 25 m (82 ft) each. The center-to-center spacing of the sand drains and sand compaction piles ranged from 1.75 m (5.8 ft) to 2.5 m (8.2 ft). The reclaimed area was about 1,450,000 m^2 (15,602,000 ft^2). The volume of sand used for the construction of sand piles and preloading, including the backfill sand layer, was about 11.6×10^6 m^3(15.2×10^6 yd^3). The improved site presently supports a stockyard of heavy material, a slab yard, oil tanks, embankments for roads and railways, and steel mills.

Figure 14.43 Construction of sand compaction pile in Yokohama, Japan, harbor (Courtesy of E.C. Shin, Korea)

Figure 14.45 shows the staged preloading, along with the variation with time of the pore water pressure and consolidation settlement at the stockyard. The strength of the subsoil was sufficiently improved, and after removal of the preload, all construction work progressed smoothly.

14.16 Dynamic Compaction

Dynamic compaction is a technique that is beginning to gain popularity in the United States for densification of granular soil deposits. The process primarily involves dropping a heavy weight repeatedly on the ground at regular intervals. The weight of the hammer used varies from 8 to 35 metric tons, and the height of the hammer drop varies between 7.5 and 30.5 m (≈ 25 and 100 ft). The stress waves gen-

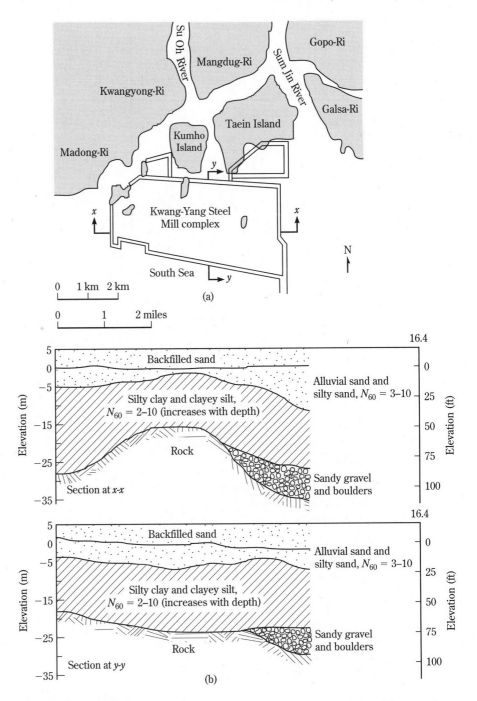

Figure 14.44 (a) Site location of the Kwang-Yang Steel Mill complex; (b) general nature of the soil profile. *Note:* N_{60} = field standard penetration number (after Shin, Shin, and Das, 1992)

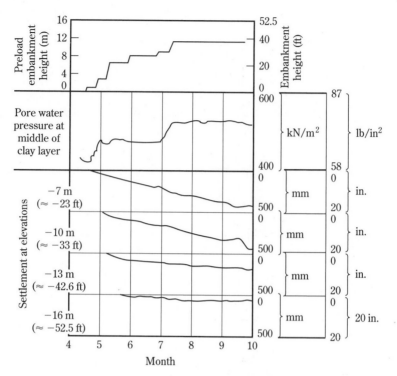

Figure 14.45   Pore water pressure and settlement measurement with staged preloading at the stockyard site (after Shin, Shin, and Das, 1992)

erated by the hammer drops help in the densification. The degree of compaction achieved depends on

   a. the weight of the hammer
   b. the height of the drop
   c. the spacing of the locations at which the hammer is dropped

Figure 14.46 shows a dynamic compaction in operation.

   Leonards et al. (1980) suggested that the significant depth of influence for compaction is approximately

$$DI \approx \tfrac{1}{2}\sqrt{W_H h} \qquad (14.47)$$

where   DI = significant depth of densification (m)
     W_H = dropping weight (metric ton)
     h = height of drop (m)
In English units, Eq. (14.47) becomes

$$DI = 0.61\sqrt{W_H h} \qquad (14.48)$$

where DI and h are in ft and W_H is in kip

   Partos et al. (1989) provided several case histories of site improvement that used dynamic compaction. Figure 14.47 shows the effect of dynamic compaction in improving

Figure 14.46 Dynamic compaction (Courtesy of The Reinforced Earth Company, Vienna, Virginia and Menard Soil Treatment Inc., Orange, California)

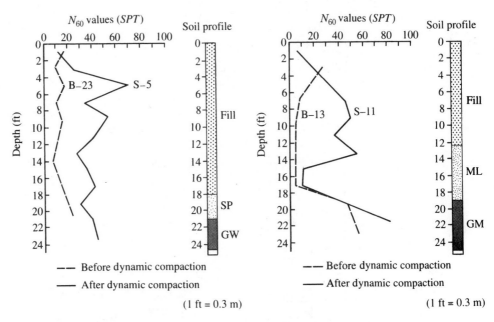

Figure 14.47 Standard penetration resistance before and after dynamic compaction, Riverview Executive Park, Trenton, New Jersey (after Partos et al., 1989)

the standard penetration resistance at the construction site of an office building in Riverview Executive Park in Trenton, New Jersey. For this dynamic compaction,

$$\text{Weight of the hammer, } W_H = 163 \text{ kN (18.5 ton)}$$

$$\text{Height of drop, } h = 26 \text{ m (85 ft)}$$

$$\text{Spacing between hammer drops} = 3.3 \text{ m (10.6 ft)}$$

$$\text{Number of drops at each location} = 7$$

In 1992, Poran and Rodriguez suggested a rational method for conducting dynamic compaction for granular soils in the field. According to their method, for a hammer of width D having a weight W_H and a drop h, the approximate shape of the densified area will be of the type shown in Figure 14.48 (i.e., a semiprolate spheroid). Note that in this figure $b = DI$. Figure 14.49 gives the design chart for a/D and b/D versus $NW_H h/Ab$ (D = width of the hammer if not circular in cross section; A = area of cross section of the hammer; N = number of required hammer drops). The method uses the following steps:

1. Determine the required significant depth of densification, $DI (=b)$.
2. Determine the hammer weight (W_H), height of drop (h), dimensions of the cross section, and thus the area A and the width D.

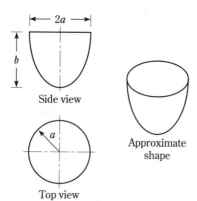

Figure 14.48 Approximate shape of the densified area due to dynamic compaction (after Poran and Rodriguez, 1992)

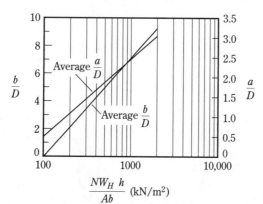

Figure 14.49 Plot of a/D and b/D vs. $NW_H h/Ab$ (after Poran and Rodriguez, 1992)

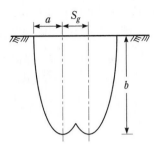

Figure 14.50 Approximate grid spacing for dynamic compaction

3. Determine $DI/D = b/D$.
4. Use Figure 14.49 and determine the magnitude of $NW_H h/Ab$ for the value of b/D obtained in Step 3.
5. Since the magnitudes of W_H, h, A, and b are known (or assumed) from Step 2, the number of hammer drops can be estimated from the value of $NW_H h/Ab$ obtained from Step 4.
6. With known values of $NW_H h/Ab$, determine a/D and thus a from Figure 14.49.
7. The grid spacing, S_g, for dynamic compaction may now be assumed to be equal to or somewhat less than a. (See Figure 14.50.)

Problems

14.1 Make the necessary calculations and prepare the zero-air-void unit weight curves (in lb/ft³) related to a Proctor compaction test for $G_s = 2.6, 2.65, 2.7,$ and 2.75.

14.2 The maximum dry unit weight of a soil was determined by a Proctor compaction test to be 17.8 kN/m³. If the same soil is used in the compaction of an embankment to a unit weight of 15 kN/m³, what would be the relative compaction?

14.3 A sandy soil has a maximum dry unit weight of 109.5 lb/ft³ and a dry unit weight of compaction in the field of 99 lb/ft³. Estimate
a. the relative compaction in the field
b. the relative density in the field
c. the minimum dry unit weight of the soil

14.4 The following are given for a natural soil deposit:

$$\text{Moist unit weight, } \gamma = 16 \text{ kN/m}^3$$
$$\text{Moisture content, } w = 14\%$$
$$G_s = 2.69$$

This soil is to be excavated and transported to a construction site for use in a compacted fill. If the specification calls for the soil to be compacted to a minimum dry unit weight of 17.2 kN/m³ at the same moisture content of 14%, how many cubic meters of soil from the excavation site are needed to produce 10,000 m³ of compacted fill?

14.5 A proposed embankment fill required 12,000 m³ of compacted soil. The void ratio of the compacted fill is specified as 0.65. The following table lists four available borrow pits, along with the void ratios of the soil and the cost per cubic meter for moving the soil to the proposed construction site:

Borrow pit	Void ratio	Cost ($/m³)
A	0.9	6
B	1.1	5
C	0.95	8
D	0.75	10

Make the calculations necessary to select the pit from which the soil should be brought to minimize the cost. Assume G_s to be the same for all borrow-pit soil.

14.6 For a vibroflotation work, the backfill to be used has $D_{50} = 1$ mm, $D_{20} = 0.5$ mm, and $D_{10} = 0.08$ mm. Determine the suitability number of the backfill. How would you rate the material?

14.7 Repeat Problem 14.6 for $D_{50} = 1.8$ mm, $D_{20} = 0.75$ mm, and $D_{10} = 0.12$ mm.

14.8 For the construction of an airport, a large fill operation is required. For the work, the average permanent load, $\Delta\sigma'_{(p)}$, on the clay layer will increase by about 70 kN/m³. The average effective overburden pressure on the clay layer before the fill operation is 95 kN/m². For the clay layer that is normally consolidated and drained at the top and bottom, $H_c = 5.5$ m, $C_c = 0.24$, $e_o = 0.81$, and $C_v = 0.5$ m²/month. Use Figure 14.18, and determine

    a. the primary consolidation settlement of the clay layer caused by the additional permanent load, $\Delta\sigma'_{(p)}$

    b. the time required for 90% of primary consolidation settlement under the additional permanent load only

    c. the temporary surcharge, $\Delta\sigma'_{(f)}$, that will be required to eliminate the entire primary consolidation settlement in six months by the precompression technique

14.9 Redo Part (c) of Problem 14.8 for a time of elimination of primary consolidation settlement of nine months.

14.10 Repeat Problem 14.8 with $\Delta\sigma'_{(p)} = 1200$ lb/ft², the average effective overburden pressure on clay layer $= 1000$ lb/ft², $H_c = 15$ ft, $C_c = 0.3$, $e_o = 1.0$, and $C_v = 1.5 \times 10^{-2}$ in.²/min.

14.11 For the sand drain project shown in Figures 14.24 and 14.25, let $r_w = 0.2$ m, $r_s = 0.3$ m, $d_e = 5$ m, $C_v = C_{vr} = 0.3$ m²/month, $k_h/k_s = 2$, and $H_c = 6$ m. Determine

    a. the degree of consolidation of the clay layer caused only by the sand drains after six months of application of the surcharge.

    b. the degree of consolidation of the clay layer caused by the combination of vertical drainage (drained at top and bottom) and radial drainage after six months of application of the surcharge. Assume that the surcharge is applied instantaneously.

14.12 A 10-ft-thick clay layer is drained at the top and bottom. Its characteristics are $C_{vr} = C_v$ (for vertical drainage) $= 0.042$ ft²/day, $r_w = 8$ in., and $d_e = 6$ ft. Estimate the degree of consolidation of the clay layer caused by the combination of vertical and radial drainage at t = 0.2, 0.4, 0.8, and 1 yr. Assume that the surcharge is applied instantaneously and there is no smear.

14.13 For a sand drain project (see Figure 14.24), the following are given:

Clay: Normally consolidated
  H_c = 20 ft (one-way drainage)
  H_c = 0.28
  e_o = 0.9
  C_v = 0.21 ft^2/day
  Effective overburden pressure at the middle of clay
  layer = 2000 lb/ft^2

Sand drain: r_w = 0.25 ft
  $r_w = r_s$
  d_e = 7.5 ft
  $C_v = C_{vr}$

A surcharge is applied as shown in Figure P14.13. Calculate the degree of consolidation and the consolidation settlement 40 days after the surcharge is first applied.

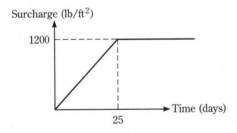

Figure P.14.13

References

Aboshi, H., Ichimoto, E., and Harada, K. (1979). "The Compozer—a Method to Improve Characteristics of Soft Clay by Inclusion of Large Diameter Sand Column," *Proceedings, International Conference on Soil Reinforcement, Reinforced Earth and Other Techniques,* Vol. 1, Paris, pp. 211–216.

Aboshi, H., and Monden, H. (1963). "Determination of the Horizontal Coefficient of Consolidation of an Alluvial Clay," *Proceedings, Fourth Australia–New Zealand Conference on Soil Mechanics and Foundation Engineering,* pp. 159–164.

American Society for Testing and Materials (1997). *Annual Book of Standards,* Vol. 04.08, West Conshohocken, PA.

Bachus, R. C., and Barksdale, R. D. (1989). "Design Methodology for Foundations on Stone Columns," *Proceedings, Foundation Engineering: Current Principles and Practices* American Society of Civil Engineers, Vol. 1, pp. 244–257.

Barron, R. A. (1948). "Consolidation of Fine-Grained Soils by Drain Wells," *Transactions,* American Society of Civil Engineers, Vol. 113, pp. 718–754.

Basore, C. E., and Boitano, J. D. (1969). "Sand Densification by Piles and Vibroflotation," *Journal of the Soil Mechanics and Foundations Division,* American Society of Civil Engineers, Vol. 95, No. SM6, pp. 1303–1323.

Blacklock, J. R., and Pengelly, A. D. (1988). "Soil Treatment for Foundations on Expansive Clay," *Special Topics in Foundations,* GSP No. 16 (ed. B. M. Das), American Society of Civil Engineers, pp. 73–92.

Brown, R. E. (1977). "Vibroflotation Compaction of Cohesionless Soils," *Journal of the Geotechnical Engineering Division,* American Society of Civil Engineers, Vol. 103, No. GT12, pp. 1437–1451.

Christoulas, S., Bouckovalas, G., and Giannaros, C. (2000). "An Experimental Study on Model Stone Columns," *Soils and Foundations,* Vol. 40, No. 6, pp. 11–22.

D'Appolonia, D. J., Whitman, R. V., and D'Appolonia, E. (1969). "Sand Compaction with Vibratory Rollers," *Journal of the Soil Mechanics and Foundations Division,* American Society of Civil Engineers, Vol. 95, No. SM1, pp. 263–284.

Hughes, J. M. O., and Withers, N. J. (1974). "Reinforcing of Soft Cohesive Soil with Stone Columns," *Ground Engineering,* Vol. 7, pp. 42–49.

Hughes, J. M. O., Withers, N. J., and Greenwood, D. A. (1975). "A Field Trial of Reinforcing Effects of Stone Columns in Soil," *Geotechnique,* Vol. 25, No. 1, pp. 31–34.

Ichimoto, A. (1981). "Construction and Design of Sand Compaction Piles," *Soil Improvement, General Civil Engineering Laboratory* (in Japanese), Vol. 5. pp. 37–45.

Johnson, S. J. (1970a). "Precompression for Improving Foundation Soils," *Journal of the Soil Mechanics and Foundations Division,* American Society of Civil Engineers. Vol. 96, No. SM1, pp. 114–144.

Johnson, S. J. (1970b). "Foundation Precompression with Vertical Sand Drains," *Journal of the Soil Mechanics and Foundations Division,* American Society of Civil Engineers. Vol. 96, No. SM1, pp. 145–175.

Lee, K. L., and Singh, A. (1971). "Relative Density and Relative Compaction," *Journal of the Soil Mechanics and Foundations Division,* American Society of Civil Engineers, Vol. 97, No. SM7, pp. 1049–1052.

Leonards, G. A., Cutter, W. A., and Holtz, R. D. (1980). "Dynamic Compaction of Granular Soils," *Journal of Geotechnical Engineering Division,* ASCE, Vol. 96, No. GT1, pp. 73–110.

Mattes, N. S., and Poulos, H. G. (1969). "Settlement of Single Compressible Pile," *Journal of the Soil Mechanics and Foundations Division,* ASCE, Vol. 95, No. SM1, pp. 189–208.

Mitchell, J. K. (1970). "In-Place Treatment of Foundation Soils," *Journal of the Soil Mechanics and Foundations Division,* American Society of Civil Engineers, Vol. 96, No. SM1, pp. 73–110.

Mitchell, J. K., and Freitag, D. R. (1959). "A Review and Evaluation of Soil–Cement Pavements," *Journal of the Soil Mechanics and Foundations Division,* American Society of Civil Engineers, Vol. 85, No. SM6, pp. 49–73.

Mitchell, J. K., and Huber, T. R. (1985). "Performance of a Stone Column Foundation," *Journal of Geotechnical Engineering,* American Society of Civil Engineers, Vol. 111, No. GT2, pp. 205–223.

Murayama, S. (1962). "An Analysis of Vibro-Compozer Method on Cohesive Soils," *Construction in Mechanization* (in Japanese), No. 150, pp. 10–15.

Olson, R. E. (1977). "Consolidation under Time-Dependent Loading," *Journal of Geotechnical Engineering Division,* ASCE, Vol. 102, No. GT1, pp. 55–60.

Othman, M. A., and Luettich, S. M. (1994). "Compaction Control Criteria for Clay Hydraulic Barriers," *Transportation Research Record,* No. 1462, National Research Council, Washington, DC, pp. 28–35.

Partos, A., Welsh, J. P., Kazaniwsky, P. W., and Sander, E. (1989). "Case Historics of Shallow Foundation on Improved Soil," *Proceedings, Foundation Engineering: Current Principles and Practices,* American Society of Civil Engineers, Vol. 1, pp. 313–327.

Poran, C. J., and Rodriguez, J. A. (1992). "Design of Dynamic Compaction," *Canadian Geotechnical Journal,* Vol. 2, No. 5, pp. 796–802.

Shin, E. C., Shin, B. W., and Das, B. M. (1992). "Site Improvement for a Steel Mill Complex," *Proceedings, Specialty Conference on Grouting, Soil Improvement and Geosynthetics,* ASCE, Vol. 2, pp. 816–828.

Yeung, A. T. (1997). "Design Curves for Prefabricated Vertical Drains," *Journal of Geotechnical and Geoenvironmental Engineering,* Vol. 123, No. 8, pp. 755–759.

Appendix A

Field Instruments

This appendix provides brief descriptions of some common instruments used in the field in geotechnical engineering projects: the piezometer, earth pressure cell, load cell, cone pressuremeter, dilatometer, and inclinometer.

Piezometer

The piezometer measures water levels and pore water pressure. The simplest piezometer is the Casagrande piezometer (shown schematically in Figure 2.16) which is installed in a borehole of 75–150 mm diameter. Figure A.1 shows a photograph of

Figure A.1 Components of Casagrande-type piezometer (Courtesy of N. Sivakugan, James Cook University, Australia)

the components of the Casagrande piezometer. The device consists of a plastic riser pipe joined to a filter tip that is placed in sand. A bentonite seal is placed above the sand to isolate the pore water pressure at the filter tip. The annular space between the riser pipe and the borehole is backfilled with a bentonite–cement grout to prevent vertical migration of water. Although they are simple, reliable, and inexpensive, the standpipe piezometers are slow to reflect the change in pore water pressures and require someone at the site to monitor their operation. The other types are pneumatic piezometers, vibrating wire piezometers, and hydraulic piezometers. Piezometers are used to monitor pore water pressures to determine safe rates of fill or excavation; to monitor pore pressure dissipation in ground improvement techniques such as PVDs, sand drains, or dynamic compaction; to monitor pore pressures to check the performance of embankments, landfills, and tailings dams; and to monitor water drawdown during pumping tests. Its nonmetallic construction makes the piezometer suitable in any adverse environment for long-term operations in most soils.

Earth Pressure Cell

An earth pressure cell (Figure A.2) measures the total earth pressure, which includes the effective stress and the pore water pressure. It is made by welding two circular stainless-steel diaphragm plates around their periphery. The space between the plates is filled with a de-aired fluid. The cell is installed with its sensitive surface in direct contact with the soil. Any change in earth pressure is transmitted to the fluid inside the cell, which is measured by a pressure transducer that is similar to a diaphragm-type piezometer. Earth pressure cells can be used to measure stresses within embankments and subgrades, foundation bearing pressures, and contact pressures on retaining walls, tunnel linings, railroad bases, piers, and bridge abutments.

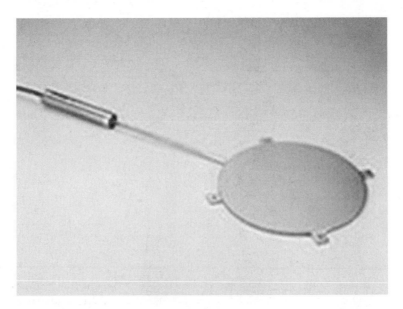

Figure A.2 Earth pressure cell (Courtesy of N. Sivakugan, James Cook University, Australia)

Figure A.3 Load cell (Courtesy of N. Sivakugan, James Cook University, Australia)

Load Cells

Load cells (Figure A.3) measure the loads acting on them. The load cell can be of the mechanical, hydraulic, vibrating wire, or electrical-resistant strain gauge type. Load cells measure strains or displacements under the applied loads that are translated into loads through calibration. Load cells are used in proof testing and performance monitoring of tiebacks, rock bolts, soil nails and other anchor systems, and pile load tests. They are also used to measure strut loads in braced excavations. In the laboratory, they are employed in triaxial devices and in load tests on model piles and footings.

Cone Pressuremeter

The cone gives a reliable classification of soils and good estimates of strength, but is poor in estimating soil stiffness. The pressuremeter, on the other hand, is well suited to measuring stiffness and strength parameters (Houlsby and Withers 1988). The cone pressuremeter (Figure A.4) exploits the features of a piezocone and a pressuremeter. It combines the continuous profiling capability of the cone with the ability of the pressuremeter to measure the load-deformation characteristics of soils—particularly the shear modulus. Approximately 2 m in length, the probe consists of a piezocone at the lower end of the penetrometer shaft and a pressuremeter

Figure A.5 Inclinometer system (Courtesy of
N. Sivakugan, James Cook University, Australia)

Figure A.4 Cone pressuremeter
(Courtesy of N. Sivakugan, James
Cook University, Australia)

of the same diameter above, as shown in the figure. The top of the probe can be connected by the cone penetration test rods to the ground, and the probe is generally
pushed into the ground by a cone truck or by jacking against a kentledge. The his-
torical developments of the cone pressuremeter are given in Dalton (1997).

Inclinometer

Inclinometers are quite popular for measuring lateral displacement in slopes, landslides, retaining walls, piles, and bridge abutments. An inclinometer system (Figure A.5)
is composed of a torpedo-shaped probe fitted with guide wheels and a tilt sensor

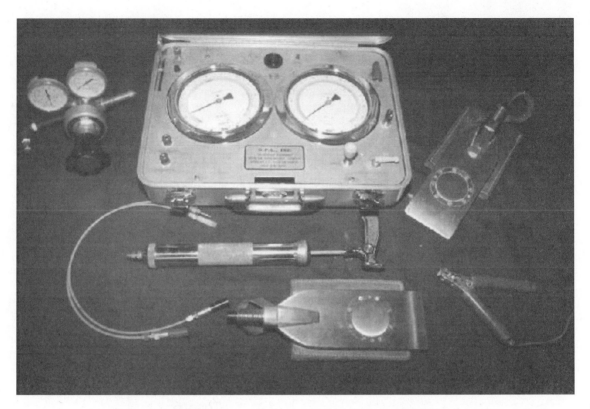

Figure A.6 Dilatometer and other equipment (Courtesy of N. Sivakugan, James Cook University, Australia)

and connected by a graduated cable to a readout unit. The probe is sent through the inclinometer casing installed vertically in a borehole, and the tilt sensor records the deviation between the probe axis and the vertical plane. By installing the inclinometer horizontally, the profiles of settlement or heave beneath embankments, storage tanks, and landfills can be obtained.

Dilatometer

The dilatometer test was discussed in detail in Section 2.6. Figure A.6 shows two flat dilatometers with other instruments for conducting the test.

References

Dalton, C. (1997). "Pressure for Change," *Ground Engineering,* Vol. 30, No. 9, pp. 22–23.
Houslby, G.T., and Withers, N.J. (1988). "Analysis of the Cone Penetrometer Tests in Clay," *Geotechnique,* Vol. 38, No. 4, pp. 575–587.

Answers to Selected Problems

Chapter 1

1.1   **a.** 0.41   **b.** 46.3%
     c. 15.58 kN/m^3

1.3   **a.** 0.74   **b.** 0.43
     c. 95.9 lb/ft^3 **d.** 110.3 lb/ft^3
     e. 54%

1.5 $\gamma_d = 102.2 \text{ lb/ft}^3$
     $\gamma = 112.4 \text{ lb/ft}^3$

1.7

Soil	Classification
A	A-1-a(0)
B	A-2-6(1)
C	A-7-5(19)
D	A-4(5)
E	A-2-7(1)
F	A-6(5)

1.9

Eq. No.	k (cm/s)
1.26	0.279
1.27	0.243
1.28	0.343

1.11

Point	σ (kN/m²)	u (kN/m²)	σ' (kN/m²)
A	0	0	0
B	34.66	0	34.66
C	74.52	19.62	54.9
D	129.84	49.05	80.79

1.13   52 mm

1.15   **a.** 66.65 kN/m^2 **b.** 0.299
     c. 106.8 mm

1.17   6.59 days

1.19   7.5 mm @ t = 30 days
     40.5 mm @ t = 120 days

1.21   30°

1.23 $\phi' = 28°$; $c' = 30 \text{ kN/m}^2$

Chapter 2

2.1   8.96%

2.3

Depth (m)	$(N_1)_{60}$
1.5	9
3	10
4.5	10
6	8
7.5	12
9	12

2.5   34°

2.7   31°

2.9 51.4 kN/m^2

2.11   **a.** 39.8°   **b.** 49%

2.13 4121.6 kN/m^2

2.15   40%

2.17 $v_1 = 678.4 \text{ ft/s}$; $Z_1 = 32.5 \text{ ft}$;
     $v_2 = 1748.25 \text{ ft/s}$

Chapter 3

3.1   **a.** 5278 lb/ft^2 **b.** 176.8 lb/ft^2

3.3 **a.** 5723 lb/ft^2 **b.** 184.7 kN/m^2
3.5 121.2 kip
3.7 5495 kN
3.9 **a.** 40° **b.** 1237 kip
3.11 537.4 kip
3.13 6.75 ft

Chapter 4

4.1 502.8 kip
4.3 13,399 lb/ft^2
4.5 495.5 kN
4.7 1282.5 kN
4.9 7425 lb/ft^2
4.11 3733 lb/ft^2

Chapter 5

5.1

z (m)	$\Delta\sigma$ (kN/m^2)
1.5	42.4
3	14.6

5.3

z (m)	$\Delta\sigma$ (kN/m^2)
2	63
4	22.5
6	11.25

5.5 **a.** 61.4 kN/m^2
      b. 48.71 kN/m^2

5.7

Location	$\Delta\sigma$ (kN/m^2)
A	160.5
B	153
C	14.45

5.9 34.5 mm
5.11 14.1 mm
5.13 19.3 mm
5.15 15.57 mm
5.17 700.5 kN/m^2
5.19 53.7 mm

5.21 **a.**

Depth (ft)	$(N_1)_{60}$
5	17
10	13
15	13
20	8
25	12

b. 6.49 kip/ft^2

Chapter 6

6.1 12.3 kip/ft^2
6.3 153.25 kN/m^2
6.5 9.23 m
6.7 236 mm
6.9 193 mm

6.11

Location	Pressure (kN/m^2)
A	36.81
B	31.86
C	26.91
D	25.19
E	30.14
F	35.09

6.13 21.36 lb/in.^{3}

Chapter 7

7.1 $P_o = 176.21$ kN/m; $\overline{z} = 1.84$ m
7.3 $P_o = 390.9$ kN/m; $\overline{z} = 2.11$ m
7.5 58.6 kN/m
7.7 12.6 kip/ft
7.9 5598 lb/ft
7.11 **a.** 2916 lb/ft **b.** 3559 lb/ft
7.13 **a.** $P_p = 51,376.5$ lb/ft
       b. $\overline{z} = 8.8$ ft
7.15 82.05 kip/ft
7.17 3289 kN/m
7.19 50.8 kip/ft

Chapter 8

8.1 Overturning: 3.41
      Sliding: 1.5
      Bearing: 5.5
8.3 Overturning: 2.81
      Sliding: 1.56
      Bearing: 3.14
8.5 Overturning: 6.2
      Sliding: 2.35

8.7   Overturning: 2.66
Sliding: 1.7
Bearing: 3.49

8.9

z (m)	σ'_a(kN/m^2)
2	24.15
4	25.54
6	30.79
8	38.43

8.11   **a.** 23.2   **b.** 4.37   **c.** 11.3
8.13   Overturning: 3.43
Sliding: 1.35
Bearing: 9.79

Chapter 9

9.1   **a.** 23.52 ft   **b.** 53.6 ft
c. 96.94 kip-ft/ft
9.3   **a.** 27.34 ft   **b.** 60.54 ft
c. 172.9 kip-ft/ft
9.5 $D_{\text{theory}} = 3.7$ m;
$M_{\text{max}} = 115.2$ kN-m/m
9.7   **a.** 8.25 m   **b.** 19.55 m
c. 587.44 kN-m/m
9.9 $D_{\text{theory}} = 1.1$ m;
$M_{\text{max}} = 32.18$ kN-m/m
9.11   **a.** 759 kN-m/m   **b.** PZ-35
9.13   PZ-27
9.15   **a.** 3.3 m   **b.** 102.6 kN/m
9.17   22.88 kip

Chapter 10

10.1

Location	Strut load (kip)
A	90.52
B	47.07
C	57.93

10.3

Location	Strut load (kip)
A	68.71
B	35.73
C	43.98

10.5   **a.** 14.98 kN/m^2
b. 42.5 kN/m^2; see Figure 10.6

10.7

Location	Strut load (kN)
A	306.5
B	405.55
C	413.45

10.9

Location	Strut load (kN)
A	306.5
B	439.35
C	218.9

10.11   3.57

Chapter 11

11.1   **a.** 423 kN   **b.** 1532 kN
11.3   1508 kN
11.5   129 kN
11.7   1291 kN
11.9   571 kN
11.11   **a.** 1737 kN   **b.** 1996 kN
c. 2033 kN
11.13   19.8 mm
11.15   54.7 kN
11.17   306 kip
11.19   223.8 kip
11.21   3096 lb
11.23   70.3%
11.25   90.2%
11.27   985 kip

Chapter 12

12.1   7437 kN
12.3   3509 kN
12.5   **a.** 316.7 kN   **b.** 326.6 kN
12.7   **a.** 1099.8 kN   **b.** 1206.4 kN
c. 768.7 kN
12.9   **a.** 893.5 kN   **b.** 3017 kN
c. 3910 kN
12.11   **a.** 2373 kN   **b.** 1795 kN
12.13   0.258 in.
12.15   27,865 kN

Chapter 13

13.1

LL	γ_d (kN/m³) below which collapse will occur
10	20.8
15	18.8
20	17.16
25	15.78
30	14.60
35	13.59
40	12.71

13.3 Normally consolidated
13.5 119 mm
13.7 52.7 mm
13.9 11.7 m
13.11 12.57 ft

Chapter 14

14.1

	γ_{zav} (lb/ft³)			
w (%)	$G_s = 2.6$	2.65	2.7	2.76
5	143.6	148.4	148.4	150.9
10	128.8	130.7	132.7	134.6
15	116.7	118.3	119.9	121.5
20	106.7	108.1	109.4	110.7

14.3 **a.** 90.4% **b.** 50%
    c. 89.68 lb/ft³
14.5 *B*
14.7 $S_n = 14.4$; Rating: Good
14.9 40.6 kN/m²
14.11 **a.** 23% **b.** 61.9%
14.13 $U_{v,r} = 18\%$; $S_c = 1.3$ in.

Index

Photo Credits

This page constitutes an extension of the copyright page. We have made every effort to trace the ownership of all copyrighted material and to secure permission from copyright holders. In the event of any question arising as to the use of any material, we will be pleased to make the necessary corrections in future printings. Thanks are due to the following authors, publishers, and agents for permission to use the material indicated.

Chapter 1. Page 52: Courtesy of Soiltest, Inc., Lake Bluff, Illinois

Chapter 2. Page 75: Courtesy of William B. Ellis, El Paso Engineering and Testing, Inc., El Paso, Texas **Page 76:** Courtesy of Danny R. Anderson, Danny R. Anderson Consultants, El Paso, Texas

Chapter 10. Page 441: Courtesy of Ralph B. Peck **Page 442:** Courtesy of Ralph B. Peck

Chapter 11. Page 486: Courtesy of Mike O'Neill **Page 487:** Courtesy of E. C. Shin, University of Incheon, Korea **Page 520:** Courtesy of E. C. Shin, University of Incheon, Korea

Chapter 13. Pages 651, 652: Courtesy of GKN Hayward Baker, Inc., Woodbine Division, Ft. Worth, Texas

Chapter 14. Pages 669–671: Courtesy of Tampo Manufacturing Co., Inc., San Antonio, Texas **Page 701:** Courtesy of E. C. Shin, University of Incheon, Korea **Pages 706, 707:** Courtesy of GKN Hayward Baker, Inc., Woodbine Division, Ft. Worth, Texas **Page 710:** Courtesy of The Reinforced Earth Company, Vienna, Virginia **Page 716:** Courtesy of E. C. Shin, University of Incheon, Korea **Page 719:** Courtesy of The Reinforced Earth Company, Vienna, Virginia

Appendix A. Pages 727–731: Courtesy of N. Sivakugan, James Cook University, Australia